LISS 2012

Zhenji Zhang • Runtong Zhang • Juliang Zhang
Editors

LISS 2012

Proceedings of 2nd International Conference on Logistics, Informatics and Service Science

Volume One

Beijing, China
July 12–15, 2012

Hosted by
School of Economics and Management, Beijing Jiaotong University

In Cooperation with
University of Reading, UK

Sponsored by
NSFC – National Natural Science Foundation of China
K.C. Wong Education Foundation, Hong Kong
Springer

Editors
Zhenji Zhang
School of Economics and Management
Department of Information Management
Beijing Jiaotong University
Beijing, People's Republic of China

Runtong Zhang
School of Economics and Management
Department of Information Management
Beijing Jiaotong University
Beijing, People's Republic of China

Juliang Zhang
School of Economics and Management
Department of Information Management
Beijing Jiaotong University
Beijing, People's Republic of China

ISBN 978-3-642-32053-8 ISBN 978-3-642-32054-5 (eBook)
DOI 10.1007/978-3-642-32054-5
Springer Berlin Heidelberg New York Dordrecht London

Library of Congress Control Number: 2012955057

Printed on acid-free paper

Springer is part of Springer Science+Business Media (www.springer.com)

Preface

This volume contains the proceedings of the 2012 International Conference on Logistics, Informatics and Services Sciences (LISS'2012), held in Beijing, China, hosted by Beijing Jiaotong University (BJTU) and sponsored by the National Natural Science Foundation of China (NSFC), K. C. Wong Education Foundation (Hong Kong), The University of Reading and Springer.

The conference was held in cooperation with Advances in Information Sciences and Service Sciences, Journal of Industrial Engineering and Management, Journal of Electronic Commerce in Organizations, Information Services and Use, International Journal of Physical Distribution & Logistics Management, Information Technology and Management.

This conference is a prime international forum for both researchers and industry practitioners to exchange latest fundamental advances in the state of the art and practice of logistics, informatics, service operations and service science, with four simultaneous tracks, covering different aspects, including: "Service Management", "Logistics Management", "Information Management" and "Engineering Management". It also had seven special session, Advanced Research on Logistics/Supply Chain Management, e-Infrastructures and Technology, Modeling and Its Application for Network Information Management, Modeling Methods for Logistics Systems & Supply Chain, Integrated Service Management, Theory of Industrial Security and Development, and Tourism Management, including 40 papers. Papers published in each track describe state-of-the-art research work that is often oriented towards real-world applications and highlight the benefits of related methods and techniques for the emerging field of service science, logistics and informatics development.

LISS 2012 received 361 paper submissions from 12 countries and regions; 163 papers were accepted and published after strict peer reviews. The total acceptance ratio was 45.1%. Additionally, a number of invited talks, presented by internationally recognized specialists in different areas, have positively contributed to reinforce the overall quality of the conference and to provide a deeper understanding of related areas.

The program for this conference required the dedicated effort of many people. Firstly, we must thank the authors, whose research and development efforts are recorded here. Secondly, we thank the members of the program committee and the additional reviewers for the valuable help with their expert reviewing of all submitted papers. Thirdly, we thank the invited speakers for their invaluable contribution and the time for preparing their talks. Fourthly, we thank the special session chairs whose collaboration with LISS was much appreciated. Finally many thanks are given to the colleagues from BJTU for their hard work in organizing this year event.

A final selection of papers, from those presented at LISS 2012 in Beijing, will be done based on the classifications and comments provided by the Program Committee and on the assessment provided by session chairs. Extended and revised versions of the selected papers will be published in special issues of the six international journals.

We wish you all enjoyed an exciting conference and an unforgettable stay in Beijing, China. We hope to meet you again next year for LISS 2013, and the details of which will soon be available at http://icir.bjtu.edu.cn/liss2013.

School of Economics and Management
Beijing Jiaotong University
Beijing, China

Zhenji Zhang
Runtong Zhang
Juliang Zhang

Organizing Committees

Organization Chairs

Jianqin Zhou	Beijing Jiaotong University, China
Jiayi Yao	Beijing Jiaotong University, China
Yisong Li	Beijing Jiaotong University, China

Financial Chairs

Bing Zhu	Beijing Jiaotong University, China

Special Session/Workshop Chairs

Guowei Hua	Beijing Jiaotong University, China
Xiaomin Zhu	Beijing Jiaotong University, China
Yacan Wang	Beijing Jiaotong University, China

Publicity Chairs

Xianliang Shi	Beijing Jiaotong University, China
Shifeng Liu	Beijing Jiaotong University, China
Jing Li	Beijing Jiaotong University, China
Mingcong Tang	The Chinese University of Hong Kong, Hong Kong, China

Web Master

Kun Zhang	Beijing Jiaotong University, China

International Steering Committee

International Program Committee

Name	Affiliation
Bernhard Bauer	University of Augsburg, Germany
Gabriele Bavota	University of Salerno, Italy
Lamia Hadrich Belguith	ANLP Research Group, MIRACL, University of Sfax, Tunisia
Noureddine Belkhatir	Grenoble University, France
Ida Bifulco	University of Salerno, Italy
Antonio Borghesi	Università Degli Studi Di Verona, Italy
Danielle Boulanger	IAE- Université Jean Moulin Lyon 3, France
Coral Calero	University of Castilla – La Mancha, Spain
Luis M. Camarinha-Matos	New University of Lisbon, Portugal
Angélica Caro	University of Bio-Bio, Chile
Nunzio Casalino	Università degli Studi Guglielmo Marconi, Italy
Maiga Chang	Athabasca University, Canada
Dan Chang	Beijing Jiaotong University, China
Sohail S. Chaudhry	Villanova University, U.S.A.
Zhixiong Chen	Mercy College, U.S.A.
Shiping Chen	Commonwealth Scientific and Industrial Research Organisation, Australia
Xuedong Chen	Beijing Jiaotong University, China
Jinjun Chen	Swinburne University of Technology, Australia
David Chen	Université Bordeaux 1, France
William Cheng-Chung Chu	Tunghai University, Taiwan, China
Daniela Barreiro Claro	Universidade Federal da Bahia (UFBA), Brazil
Francesco Colace	Università Degli Studi di Salerno, Italy
Rolland Colette	Universit paris1 Panthon Sorbonne, France
Cesar A. Collazos	Universidad del Cauca, Colombia

Jose Eduardo Corcoles	Castilla-La Mancha University, Spain
Bernard Coulette	University of Toulouse 2 – IRIT Laboratory, France
Sharon Cox	Birmingham City University, U.K.
Karl Cox	University of Brighton, U.K.
Wei Dai	Victoria University, Australia
Peter Dell	Curtin University, Australia
Kamil Dimililer	Near East University, Cyprus
José Javier Dolado	University of the Basque Country, Spain
Javier Dolado	University of the Basque Country, Spain
Ming Dong	Shanghai Jiao Tong University, China
Yingge Du	Beijing Jiaotong University, China
Wenji Fan	Research Institute of Highway, China
Edilson Ferneda Catholic	University of Brasília, Brazil
Maria João Silva Costa Ferreira	Universidade Portucalense, Portugal
Juan J. Flores	University of Michoacan, Mexico
Cipriano Forza	Università di Padova, Italy
Susan Foster	Monash University, Australia
Rita Francese	Università degli Studi di Salerno, Italy
Ana Fred	Technical University of Lisbon/IT, Portugal
Leonardo Garrido	Tecnológico de Monterrey, Campus Monterrey, Mexico
Alexander Gelbukh	National Polytechnic Institute, Mexico
Joseph Giampapa	Carnegie Mellon University, U.S.A.
Ergun Gide	Central Queensland University Sydney International Campus, Australia
Pascual Gonzalez	Universidad de Castilla-la Mancha, Spain
Gustavo Gonzalez-Sanchez	Mediapro Research, Spain
Juanqiong Gou	Beijing Jiaotong University, China
Feliz Gouveia	University Fernando Pessoa/Cerem, Portugal
Maria Carmen Penadés Gramaje	Universitat Politècnica de València, Spain
Chunfang Guo	Beijing Jiaotong University, China
Jatinder Gupta	University of Alabama in Huntsville, U.S.A.
Sami Habib	Kuwait University, Kuwait
Maki K. Habib	The American University in Cairo, Egypt
S. Hariharan	B.S. Abdur Rahman University, India
Sven Hartmann	Clausthal University of Technology, Germany
Paul Hawking	Victoria University, Australia

Peter Higgins	Swinburne University of Technology, Australia
Olli-Pekka Hilmola	Lappeenranta University of Technology, Finland
Wladyslaw Homenda	Warsaw University of Technology, Poland
Jun Hong	Queen's University Belfast, U.K.
Wei-Chiang Hong	Oriental Institute of Technology, Taiwan, China
Jiewu Hu	Beijing Jiaotong University, China
Xiaohua Tony Hu	Drexel University, U.S.A.
Guowei Hua	Beijing Jiaotong University, China
Kai-I Huang	Tunghai University, Taiwan, China
Alexander Ivannikov	State Research Institute For Information Technologies and Telecommunications, Russian Federation
Ivan Jelinek	Czech Technical University in Prague, Czech Republic
Michail Kalogiannakis	University of Crete, Greece
Dimitris Karagiannis	University of Vienna, Austria
Kap Hwan Kim	PU.S. An National University, Republic of Korea
Marite Kirikova	Riga Technical University, Latvia
Alexander Knapp	Universität Augsburg, Germany
Harold Krikke	Tilburg University, The Netherlands
Stan Kurkovsky	Central Connecticut State University, U.S.A.
Rob Kusters	Eindhoven University of Technology & Open University of the Netherlands, The Netherlands
Erwin van der Laan	RSM Erasmus University, The Netherlands
Hongjie Lan	Beijing Jiaotong University, China
Eng Wah Lee	Singapore Institute of Manufacturing Technology, Singapore
Der-Horng Lee	The National University of Singapore, Singapore
Kauko Leiviskä	University of Oulu, Finland
Daniel Lemire	UQAM – University of Quebec at Montreal, Canada
Hongchang Li	Beijing Jiaotong University, China
Xuemei Li	Beijing Jiaotong University, China
Dong Li	University of Liverpool, U.K.
Zhaojun Li	University of Washington, U.S.A.
Da-Yin Liao	National Chi-Nan University, Taiwan, China
Ching-Torng Lin	National Tsing Hua University, Taiwan, China
Cheng-Chang Lin	National Cheng Kung University, Taiwan, China
Shifeng Liu	Beijing Jiaotong University, China
Dehong Liu	Beijing Jiaotong University, China
Xiaochun Lu	Beijing Jiaotong University, China
Miguel R. Luaces	Universidade da Coruña, Spain
Xiyan Lv	Beijing Jiaotong University, China
Yongsheng Ma	University of Alberta, Canada

Cristiano Maciel	Universidade Federal de Mato Grosso, Brazil
K. L. Mak	University of Hong Kong, Hong Kong, China
Nuno Mamede	INESC-ID, Portugal
Yannis Manolopoulos	Aristotle University, Greece
Hamid Mcheick	University of Quebec at Chicoutimi, Canada
Subhas Misra	IIT Kanpur, India
Chiung Moon	Yonsei University, Korea
Paula Morais	Universidade Portucalense, Portugal
Fernando Moreira	Universidade Portucalense, Portugal
Haralambos Mouratidis	University of East London, U.K.
Pietro Murano	University of Salford, U.K.
Tomoharu Nakashima	Osaka Prefecture University, Japan
Paolo Napoletano	University of Salerno, Italy
Eric Ngai	China Polytechnic University, Hong Kong, China
Andreas Ninck	Berne University of Applied Sciences, Switzerland
David L. Olson	University of Nebraska, U.S.A.
Hichem Omrani	CEPS/INSTEAD, Luxembourg
Samia Oussena	Thames Valley University, U.K.
Tansel Ozyer	TOBB ETU, Turkey
Marcin Paprzycki	Polish Academy of Sciences, Poland
Rodrigo Paredes	Universidad de Talca, Chile
Namkyu Park	Wayne State University, U.S.A.
Gan Oon Peen	Southeastern Institute of Manufacturing and Technology, Singapore
Laurent Péridy	IMA-UCO, France
Dana Petcu	West University of Timisoara, Romania
Michael J. Piovoso	Pennsylvania State University, U.S.A.
Ángeles S. Places	University of A Coruña, Spain
Geert Poels	Ghent University, Belgium
Klaus Pohl	University of Duisburg-Essen, Germany
Qiuli Qin	Beijing Jiaotong University, China
Jolita Ralyte	University of Geneva, Switzerland
T. Ramayah	Universiti Sains Malaysia, Malaysia
Hajo A. Reijers	Eindhoven University of Technology, The Netherlands
Nuno Ribeiro	University Fernando Pessoa, Portugal
Nuno de Magalhães Ribeiro	Universidade Fernando Pessoa, Portugal
Michele Risi	University of Salerno, Italy
Daniel Rodriguez	University of Alcalá, Spain
Pilar Rodriguez	Universidad Autònoma de Madrid, Spain
Alfonso Rodriguez	University of Bio-Bio, Chile
Colette Rolland	Université Paris 1 Panthéon-Sorbonne, France

Hongji Yang	De Montfort University, U.K.
Jasmine Yeap	Universiti Sains Malaysia, Malaysia
Shang-Tae Yee	General Motors Research and Development Center, U.S.A.
David C. Yen	Miami University, U.S.A.
Ping Yu	University of Wollongong, Australia
Yugang Yu	Erasmus University, The Netherlands
Ruixue Zang	Beijing Jiaotong University, China
Daniel Zeng	University of Arizona Tucson, U.S.A.
Kai Zheng	Xinhua News Agency, China
Xiaomin Zhu	Beijing Jiaotong University, China
Bing Zhu	Beijing Jiaotong University, China
Eugenio Zimeo	University of Sannio, Italy
Li Zuo	Beijing Jiaotong University, China

Special Session Program Committee

Contents of Volume One

Contents of Volume Two

Part VI Theory of Industrial Security and Development

Part VII Tourism Management

Part X Modeling and Its Application for Network Information Management

Part XI Modeling Methods for Logistics Systems and Supply Chain

Part XII Integrated Service Management

Part I
Keynote Lectures

Constructing the Theoretical System of Comprehensive Material Flow (CMF) Engineering

Shoubo Xu

Abstract Based on the methodology of technological economics, a fundamental theoretical system of integration and optimization of CMF resources is constructed, with the objective law of the contradictory relationship between the materials and their flow as well as its development as the study object and logical starting point. On account of the integral properties of the material of commodities (i.e., the substance, value and information), a core theoretical system of goods MF is constructed about the commodity flow, capital flow and information flow, which change separately but also form a complete and indivisible system. In light of the space attributes of the flow of commodities, a sustainable applied theoretical system of CMF is constructed, to explore the operation mechanism of the MF in economic circle, social circle and natural circle which are in both competition and symbiosis, contradiction and adaptation, development and balance with one and another.

Keywords Technological economics • Material flow • Material flow engineering

---

S. Xu (✉)
Beijing Jiaotong University, Beijing, People's Republic of China
e-mail: shbxu@bjtu.edu.cn

Z. Zhang et al. (eds.), *LISS 2012: Proceedings of 2nd International Conference on Logistics, Informatics and Service Science*, DOI 10.1007/978-3-642-32054-5_1,

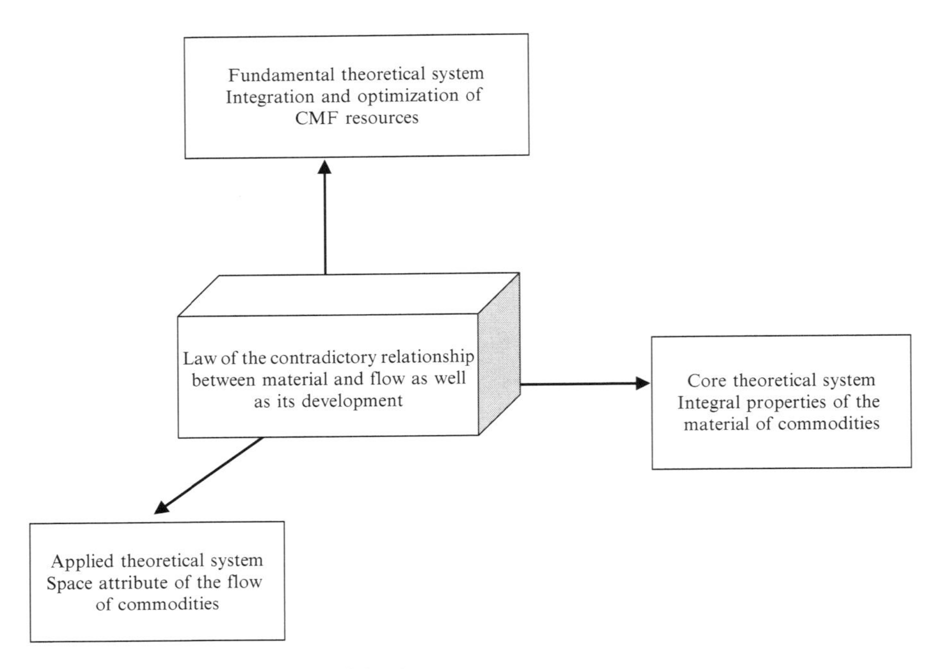

Illustration of theoretical system of CMF

Prof. Shoubo Xu was born in Shaoxing (a city which located at Zhengjiang province of China). He obtained a Bachelor's Degree in Power Engineering from Nanjing Institute of Technology in 1955. Then he graduated from the Energy Institute of the Academy of Science of USSR in 1960, with an Associate Doctorate Degree of Technological Science. Now he is honored as a professor, consultant and PHD supervisor of Economics and Management school at Beijing Jiao Tong University. And he also works as the Director of China Center of Technological Economics Research, the President of Comprehensive Energy Institute, Honorary Dean of the Material Flow School at Beijing Jiaotong University, and named as Chinese Director of the Sino-Austria Innovation Research Center. At the same time, Dr Xu is also regarded as the Chairman of Professors Association in the Economics and Management department. Besides that, he was the core initiator and co-founder of the Chinese Technological Economics and Comprehensive Energy Engineering, the pioneer of our nation's Comprehensive Material Flow Engineering and the science of Managing According to Reason MR.

For more than 50 years, academician Xu has made 422 achievements in theoretical and application aspects in the three new scientific fields of TE, ETE/CEE and MFTE/CMFE. More than 50 of his achievements have received awards, including the National Science Congress Award, the National Science and Technology Progress Award and various awards from the Chinese Academy of Science, Chinese Academy of Social Sciences, National Development and Reform Commission and City of Beijing, etc., he has received National Science and Technology Progress Award nine times (the first prize provincial, one time National Science and Technology Progress first prize and one time third prize, four times provincial second prize and three times third prize).

Decision and Process Intelligence for Services

Jan Vanthienen

Abstract To meet the needs of customers in services, intelligent decision procedures need to be installed and service processes need to be monitored closely. Research in these areas offers some very promising opportunities.

This keynote shows how decision services can be modeled in a straightforward way and flexibly implemented in the service process while maintaining its agility. Moreover, analyzing the processes using process intelligence enables to check and improve the processes and the decisions, such as to advance their effectiveness and efficiency for services.

Keywords Intelligent decision • Service processes • Process intelligence

---

Prof. Dr. Jan Vanthienen is professor of information management at KU Leuven, Department of Decision Sciences and Information Management, where he coordinates the Leuven Institute for Research in Information Systems (LIRIS).

His research interests include information and knowledge management, business rules, processes & decisions, business intelligence and intelligent systems. On these topics, he has published more than 100 full papers in reviewed international journals and conference proceedings.

He received the Belgian Francqui Chair 2009 at FUNDP and an IBM Faculty Award in 2011. He is chairholder of the PricewaterhouseCoopers Chair on E-Business at KU Leuven, co-chairholder of the Microsoft Research Chair on Intelligent Environments and co-founder and president-elect of the Benelux Association for Information Systems (BENAIS).

J. Vanthienen (✉)
Department of Decision Sciences and Information Management,
Faculty of Business and Economics, Katholieke Universiteit Leuven, Belgium
e-mail: jan.vanthienen@econ.kuleuven.ac.be

Z. Zhang et al. (eds.), *LISS 2012: Proceedings of 2nd International Conference on Logistics, Informatics and Service Science*, DOI 10.1007/978-3-642-32054-5_2,

Information Architecture in Pervasive Healthcare

Kecheng Liu

Abstract The concept of pervasive healthcare has emerged in the early twenty-first century. It aims to provide healthcare services to anyone, anytime and anywhere by integrating seamlessly the primary, secondary and tertiary healthcare service provisions. This ensures patients are covered with healthcare services when needed. In addition, clinicians can obtain the real time information by accessing the electronic patient record that supports decision making in providing health services. Pervasive healthcare involves real time systems integration, mobile devices and high speed wireless network connection. This can improve patient safety and healthcare quality and reduce the elevating operational costs where patients can be monitored remotely without needing to use the facilities in a hospital. Health informatics, according to the UK National Health Service (NHS) is defined as the knowledge, skills and tool which enable information to be collected, managed, used and shared to support the delivery of healthcare services. The importance of health informatics is to gain understanding of clinical, social, business and technical needs of hospital required to integrate systems, improve patient safety and care quality, and help the clinicians and patient working better together. The core element in health informatics is information itself. It is vital to understand the information flow and information field in a pervasive healthcare environment. Information flow is about knowing how information is transmitted across systems that support certain clinical pathways whereas information field contains a set of shared norms that portraits how clinicians and patients behave.

Information architecture (IA) is a high level map of information requirements of an organisation that also consists of the information flow and information field. The purpose of the IA is to provide right data at the right time, location and process to the right stakeholder with a right motivation. Organisational semiotics, a fundamental theory for information and communication, helps in understanding the

---

K. Liu (✉)
Henley Business School, University of Reading, UK
e-mail: k.liu@henley.ac.uk

Z. Zhang et al. (eds.), *LISS 2012: Proceedings of 2nd International Conference on Logistics, Informatics and Service Science*, DOI 10.1007/978-3-642-32054-5_3,

nature of information. It deals with information and information systems in a balanced way, taking account of both the physical space (when physical actions take place) and the information space (which are mainly characterised by information and communication using signs, symbols and data). Sharing information for multi-stakeholders for making decisions and taking appropriate actions is essential for pervasive healthcare. The information architecture can be reflected in information systems implementation such as electronic patient record (EPR) and other forms. This keynote speech aims to discuss the notion of information architecture for pervasive healthcare collaborating with multiple stakeholders (i.e. clinicians, patients) in decision support from semiotics perspective. An example will be shown towards the end of the presentation.

Keywords Pervasive healthcare • Health informatics • Information architecture

---

Dr. Kecheng Liu, Fellow of British Computer Society, is a full professor and holds a chair of Applied Informatics at University of Reading, UK. He is the Director of Informatics Research Centre, and Head of School of Business Informatics, Systems and Accounting, Henley Business School. He has published over 200 papers in conferences and journals. He is a world leading figure in organisational semiotics. His research interests span from requirements engineering, enterprise information systems management and engineering, business processing modelling, alignment of business and IT strategies, co-design of business and IT systems, pervasive informatics and intelligent spaces for working and living. He serves in several journal editorial boards, and guest edited many special issues.

He has been visiting Professor in a number of universities, including Southeast University of China, Fudan University, Beijing Jiaotong University, Dalian University of Technology, the Graduate School of Chinese Academy of Science, Shanghai University of Finance and Economics (current), Beijing Institute of Technology (current) and Renmin University of China (current). He is member of Senior Board of IBG (British Intelligent Buildings Group) and senior advisor on digital hospitals in a governmental healthcare organisation in China.

Part II
Logistics Management

The Discussion About the Development Strategy of Low-Carbon Logistic in Changzhutan Area

Fang Wei and Feng Ling

Abstract Nowadays, the concept of low-carbon gets more and more social attention, so Changzhutan area should combine energy conservation and environment protection with logistic operation. This article presents current situation of low-carbon logistics in Changzhutan, analyses the development strategy of low-carbon logistics.

Keywords Low-carbon logistics • Two-type-society • Sustainable development

Logistics, as the method of transport and value realization, plays an important role in social economy development, at the same time, logistic activities have important impact on environment. For example, transportation would cause exhaust gas, noise pollution, resource waste, which against the principle of sustainable development. In recent years, the economy of Changsha-zhuzhou-xiangtan(Changzhutan) urban agglomeration developed rapidly, so did the regional logistics, but low-carbon consciousness in the logistics industry haven't formed extensive understanding.

1 The Overview of Low-Carbon Logistics

In 2003, "Low-carbon economy" first appeared in the British energy white paper which pointed out that "our energy future: to create a low-carbon economy"; in July 2007, the United States senate proposed "a low-carbon economy act" [1]. In December 2009, after the United Nations climate change conference in Copenhagen, the low-carbon concept enters people's daily life.

F. Wei (✉) • F. Ling
Business School, Central South University of Forestry & Technology,
Changsha, People's Republic of China
e-mail: fangwei69@163.com; 1030245048@qq.com

Z. Zhang et al. (eds.), *LISS 2012: Proceedings of 2nd International Conference on Logistics, Informatics and Service Science*, DOI 10.1007/978-3-642-32054-5_4,

From an environmental perspective, the dramatic rise in carbon dioxide emissions from the burning of fossil fuels is raising the earth's temperature and threatening an unprecedented change in the chemistry of the planet and global climate, with potentially dramatic consequences for the future of human civilization and the ecosystems of the earth [2]. Low-carbon economy, generally means reducing the coal, oil or other high carbon energy consumption and greenhouse gas emissions by technological innovation, new energy development and such kinds of means, reconcile economic development and environmental protection in a win-win form. Low-carbon logistics is based on low-carbon economy theory, uses the advanced logistics technology to achieve the highest efficiency of resources using, minimum environmental impact [3].

2 The Necessity for Changzhutan Area to Develop Low-Carbon Logistics

2.1 The Requirement of Changzhutan Urban Agglomeration Economy Development

Developing low-carbon logistics, can not only save logistics resource and reduce transportation cost, but also can improve the consumers' satisfaction, and promote the development of social economy.

Changzhutan area is the main industrial production place in Hunan province, the raw materials, fuel and products all need the modern logistics to transit. Develop low-carbon logistics will be helpful for urban agglomeration harmonious and stable development. In 2009, the GDP in Changzhutan area reached 550.671 billion, with a growth rate of 14.5%, in 2010, the GDP grew to 671.591 billion, with the growth rate of 15.5%. GDP growth brings large freight volume, the higher the level of economy, the bigger the logistics demand, and the more need to develop modern logistics.

2.2 The Inevitable Requirement of Changzhutan "Two-Type-Society" Construction

In December 2007 Changzhutan urban agglomeration was approved as "national resource conservation and environment friendly society" construction synthetically reform testing district, it must change the past ways—"high investment, high consumption, high pollution", reduce the pollution which brings by economic construction, coordinate development between economy and environment.

At present, highway transportation is the main transport way in Changzhutan area, car emission is an important reason for haze, acid rain, air pollution, and other environmental problems [4]. Unreasonable logistics activities should be responsible for the waste of resources and environmental pollution. For the "two-type-society" construction, Changzhutan area should pay attention to the resource conservation and environmental protection in logistics operation. Low-carbon logistics emphasizes the logistics system efficiency and coordination, and the balance of enterprise interests and environment.

3 The Problems in Changzhutan Low-Carbon Logistics Development

In 2010, the GDP, the local fiscal revenue, the social goods retail sales, the social fixed assets investment of the provincial total in Changzhutan area are respective 42.2, 38.4, 43.2 and 45.9%, it becomes the most densely populated industry area in Hunan. The logistics planning and construction also developed well with the economy development, but there are still some problems.

3.1 Infrastructure Construction Is Not Perfect

By the end of 2010, the province road total mileage was 227,998 km, the highway mileage was 2,386 km, and there were 474,000 operating cars, and the gross of transport ships were 1.984 million tons. Changzhutan urban agglomeration formed half an hour high-speed commute circle. The highway, water transportation got a large degree development.

Although road traffic got extended, the existing traffic load still too heavy for the increasingly frequent trade, and the urban transportation congestion is serious, such as Zhuzhou clothing wholesale market and the Changsha Mawangdui market. And the Changzhutan modern logistics facilities are not perfect. On the one hand, Changzhutan logistic facilities are deficiency and aging, and freight vehicles and equipment are backward. On the storage aspect, most warehouses in enterprises are bungalow which were built before the 1970s and 1980s, few warehouses can do the mechanical work [5].

3.2 Changzhutan Transportation Energy Consumption Is Large

The materials flow along with the resources and energy consumption, and unreasonable logistics mode caused waste [6]. Unreasonable freight branches and distribution center layout would lead to the goods weave transport and increase the vehicle fuel consumption, intensify the exhaust pollution, noise pollution and urban traffic jams.

The energy consumption amount in Changzhutan area is large and the growth rate is high. The coal, oil, and other traditional energy, renewable energy are the major energy in Changzhutan area (Table 1).

Table 1 Hunan non-industrial main unit of energy in the first quarter energy consumption (2012)

Resources	Energy consumption	Increase rate %	
Electricity	0.615 billion/kw	3	↓
Natural gas	15491.8 thousands cubic meters	23.9	↑
Gasoline	40.7 thousand tons	16.5	↓
Diesel oil	190.5 thousand tons	5.2	↓
Transportation industry energy consumption	214.8 thousand tons/coal	3.4	↑
Total	509.4 thousand tons/coal	4.6	↓

The energy consumption of Changsha's traffic transportation, warehousing and the postal service rised from 1.614 t of sce/hundred million yuan in 2005 to 2.058 t of sce/hundred million yuan in 2010, energy consumption share rised from 6.7% in 2005 to 10.0% in 2010, the environmental problems caused by Changzhutan logistic development still serious.

3.3 The Reverse Logistics Construction Awareness Is Lack

In recent years, Changzhutan area pay more attention to the low-carbon economy's construction and circulation economy development. In Xiangtan, for example, Xiangtan HuaLing iron & steel Co., LTD recycled 2.8 million tons steel slag, the dust mud and wasteoil slag each year, and the solid waste resources comprehensive utilization was 86% or more, it realized comprehensive application value about 300 million yuan, the comprehensive profits about 200 million yuan.

For the whole area, some enterprises knew little about reverse logistics, and the consciousness of environmental protection is weak. Reverse logistics collection mainly includes waste products, by-products, expired products and no longer use product. Waste products and by-products are easy to collect. But expired products and no longer use products flew to the sales department or the users, it's difficult to collect [7]. And renewable resources industry usually gets very small profit or no profit, so few enterprises interest in the reverse logistics. Enterprises prefer larger scrap metal and waste paper recycling to waste batteries and waste plastics, etc.

3.4 Logistics Informatization Level Is Low

Logistics domain informationization includes the bar code, RFID, sensors, positioning and other data collection technology. Since the "eleventh five-year plan" published, large and medium size enterprises already used ERP, supply chain and information management system.

As the informatization strategy propulsion, the information construction in Changzhutan obtained many achievements, but there are still some problems, such as regional information network facilities utilization rate is low, the public and enterprise informatization level is not high. At present, logistic enterprise information system applications are gradually increased. But the public logistics information platform is not perfect, and the logistics information also can't be full sharing.

4 The Countermeasures to Develop Changzhutan Low-Carbon Logistics

4.1 The Government Should Strengthen the Low-Carbon Logistics Construction

At present, Hunan province government and Changzhutan governments haven't published unitary planning, promoting policies and any measures for the low-carbon construction of Changzhutan urban agglomeration. It should make unified incentive policy to promote low-carbon logistics construction.

Firstly, the governmental low-carbon logistics development planning should be formulated as early as possible. The administrative barriers should be broken and the government should worked out unified logistics development planning, build the infrastructure sharing mechanism, avoid repeated construction.

Secondly, encourage and promote the use of clean energy sources, construct circular economy system. New energy vehicle demonstration should be encouraged. At the same time, Changzhutan urban agglomeration should invest reverse logistics circular economy.

Thirdly, propagandize low-carbon idea, build low-carbon parks. Changzhutan area should adjust the industrial structure, cultivate the industry cluster and carry out low-carbon theme activities, strengthen the concept of low-carbon logistics propagandize, enhance the enterprises' low-carbon consciousness.

4.2 Develop the Third Party Logistics, Raise the Utilization Ratio of Resource

To develop low-carbon logistics, third party logistics and common distribution should be encourage. The Changzhutan inter-city logistics is a close transport and the car is the primary transportation, most distribution and warehousing are enterprise self-conducting logistics. It would cause traffic load, air pollution and traffic accidents. Therefore, we must develop socialized, specialized third-party logistics.

The logistics enterprise can consider more professional logistics rationalization, and simplify distribution link, to make the transport reasonable. It is helpful for carrying out the reasonable using and configuration in a broad range of logistics resources, and the TPL can avoid low transportation efficiency and alleviate the pressure of urban environment pollution.

4.3 Optimize Logistics Equipment, Promote the Logistics Information Construction

In Changzhutan area the highway transportation often be chosen as the transportation way which due to the seriously air pollution, noise pollution. Therefore, the government should strengthen the management of transportation vehicles and select rational transport vehicles, such as lower emissions trucks. Furthermore, it can make full use of the present resources, such as railway, waterway and low polluted transport way, develop railway transportation, or take the advantage of water transportation to reduce pollution.

4.4 Promote the Logistics Information Construction

Logistics information is the key to guarantee the timely transportation, timely supply and also the zero inventory. The area should strengthen the logistics software standardization construction, speed up the general standard system establishing, realize standard data transmission format and standard interface as quickly as possible. Using the network and information technology to connect users, manufacturers, suppliers and related units, and using information technology to track the logistics in order to realize the information sharing and effective control [8].

5 Conclusion

As a new idea, the low-carbon logistics has not formed systematic theory, and some experts and scholars have done much work on low-carbon logistics study, but it is not complete. This article sets forth some measures for the low-carbon logistics development in Changzhutan area on the aspect of government, industry, enterprises. To construct the "two-type-society", Changzhutan urban agglomeration should reduce the environment pollution while developing economy, and support the promotion of low-carbon logistics development.

References

1. BaiJing J (2010) The logistics development in low-carbon economy era. Logist Technol 2:48–50
2. Maria da Graca Carvalho, Matteo Bonifacio, Pierre Dechamps J (2011) Building a low-carbon society. Energy 4:1842–1847
3. LiDongHui J (2010) The China's low-carbon logistics problems, causes and counter measures analysis. Commer Cult 2:329–330
4. TaoJing J (2010) The discuss of low-carbon logistics under low-carbon economy. China Econ Trade Period 2:72–72
5. NieYuPeng, Tang Xiyu J (2010) The study about Changsha-zhuzhou-xiangtan logistic industry present development situation. Entrep World 2:117–118
6. QinWenZhan J (2010) The Changzhutan green logistics system construction under the "two type society" construction. Public SciTechnol 2:216–217
7. LvDongMei J (2008) The countermeasures of regional reverse logistics development. Railw Freight 5:9–10
8. TuLing, PengTongLi J (2009) The research of Changsha-zhuzhou-xiangtan urban agglomeration logistics resource sharing. Logist Procure Res 23:59–61

Trade Off Under Low-Carbon Economy: Ocean Shipping Vessel Speed, Carbon Emission or Shipowner's Profit

Gang Li, Huan Xu, and Wei Liu

Abstract This paper studies ships' optimum operating speed under low-carbon economy. It analyzes the relations between ship's carbon emission and the operating speed, gets the optimum speed under minimum carbon emission, establishes the relations between the shipowner's profit and the speed, and derives the maximum profit from the speed under which the entire fleet minimum carbon emission.

Keywords Shipping industry • Low carbon emission • Vessel speed • Shipowner's profit

1 Introduction

In these years, the rapid increase of greenhouse gases in the air leads to the global climate change, which directly influence human's life and development [1]. According to the report released by IMO, the entire seaborne industry released 1.04 billion tons of CO_2 in 2007, accounting for 3.3% of the total quantity for global CO_2 emission, and 870 million carbon released by international shipment, which accounts for 2.7% of the total [2]. Many scholars have studied the CO_2 emission in shipping industry, for example, Shuang et al. [3] mentioned that 'a 4% reduction in operating speed will lead to a 13% reduction in GHG emission for vessels'. Corbett et al. [4] analyze the cost-effective on slowing down operating speed in the emission reduction seaborne industry.

G. Li (✉)
College of Transport and Communications, Shanghai Maritime University, Shanghai 201306, People's Republic of China

Wuhan University of Technology, Wuhan, People's Republic of China
e-mail: lililiganganggang@gmail.com

H. Xu • W. Liu
College of Transport and Communications, Shanghai Maritime University, Shanghai 201306, People's Republic of China

Z. Zhang et al. (eds.), *LISS 2012: Proceedings of 2nd International Conference on Logistics, Informatics and Service Science*, DOI 10.1007/978-3-642-32054-5_5,

As known to all, operating speed is a sensitive problem between carrier and shipper, vessels should sail in a competitive speed to win a proper share of the market. Low-speed navigation will bring some chain reactions. In one hand, lower speed will lead to longer voyage time, and then ship owner has to put more vessels to carry the same quantity of cargo and same service frequency. In the other hand, in some conditions, shipper may choose another mode of transportation instead of shipping after comparing the cost and profit. Both two aspects make it difficult to reduce carbon emissions. Therefore, we need to search for a balance between low-carbon emission under low operating speed and high-carbon emission caused by more vessels [5–7].

2 Relation Among Operating Speed, Host Power and Oil Consumption, and CO_2 Emission

2.1 Relation Among Operating Speed, Host Power and Oil Consumption

Vessel host power is divided into indicated power and effective power. There're some relations among ship displacement, host power and operating speed.

$$N_e = \frac{D^{\frac{2}{3}} \cdot V^3}{C} \tag{1}$$

D means ship displacement (t), V means vessel operating speed (kn), N_e means vessel host power (kw), C is admiralty coefficient.

From this equation we can see, vessel host power is proportional to the cube of the speed, which means a tiny alteration on speed will have an obvious influence on host power. The operational states of the vessel decide the operating speed, as well as oil consumption.

As the relation between speed and power mentioned above, we can see there're some quantitative relations between operating speed and oil consumption for vessels.

Because a vessel has constant shape, size, and displacement no matter what speed it sails. We can achieve Eq. (2) from Eq. (1).

$$\frac{V_0^3}{N_{e0}^3} = \frac{V^3}{N_e^3} \tag{2}$$

V_0 means vessel's design speed, N_{e0} means the host power of a vessel under the design speed, V means vessel's operating speed, N_e means the host power under speed V.

According to Eq. (2), after changing speed, or say, under speed V, the host power is,

$$N_e = N_{e0}\left(\frac{V}{V_0}\right)^3 \tag{3}$$

G stands for the daily oil consumption of the vessel main engine, it is proportional to the host power of the vessel N_e,

$$G = g_e \times N_e \times 24 \times 10^3 = g_e \times N_{e0} \times \left(\frac{V}{V_0}\right)^3 \times 24 \times 10^3 = G_0 \times \left(\frac{V}{V_0}\right)^3 \tag{4}$$

G_0 means the daily oil consumption of the vessel's main engine under speed V_0 (kg), g_e means the oil consumption ratio of the vessel's main engine (g/kw·h).

In fact, it has a much more complicated relation between speed and oil consumption, we can come to a conclusion via reference theories, oil consumption is proportional to the square or cube of operating speed.

2.2 *Relation Between Vessel Operating Speed and CO_2 Emission*

As known to all, seaborne fuel is carbon compounds. When derv burns, nearly all carbon turn into CO_2. So, we can calculate the amount of CO_2 with fuel consumption.

The total fuel consumption per vessel per voyage equals to the sum of fuel consumed by both main engine and auxiliary engine, so the equation represents the fuel consumption per vessel per round voyage is as follows.

$$F_{ijk} = MF_k \cdot \left(\frac{V}{V_0}\right)^3 \cdot \frac{2d_{ij}}{24V} + \left(AF_{k1} \cdot \frac{2d_{ij}}{24V} + AF_{k2} \cdot t_s\right) \tag{5}$$

MF_k means the daily fuel consumption of vessel k's main engine under the design speed of V_0, AF_{k1} means the daily derv consumption of vessel k's auxiliary engine per voyage day, AF_{k2} means the daily derv consumption of vessel k's auxiliary engine per anchor day, i, j means port of departure and destination respectively. d_{ij} means the distance of this voyage, t_s means the layover time in voyage.

Once the oil consumption per vessel per voyage is worked out, we can multiply it with the ratio of carbon content, which is defined as 86.4%, and CO_2 conversion rate (44/12) to calculate CO_2 emission per vessel per voyage.

$$CO_2 = 0.8645 \cdot \frac{44}{12} \cdot \sum\nolimits_{i,j,k} F_{ijk} = 3.17 \cdot \sum\nolimits_{i,j,k} F_{ijk} \tag{6}$$

Combined with Eq. (5), we have Eq. (7) below.

$$CO_2 = 3.17 \cdot \sum_{i,j,k} \left[MF_k \cdot \left(\frac{V}{V_0}\right)^3 \cdot \frac{2d_{ij}}{24V} + \left(AF_{k1} \cdot \frac{2d_{ij}}{24V} + AF_{k2} \cdot t_s \right) \right] \quad (7)$$

3 Option of Operating Speed for Vessels to Minimize the Carbon Emission and When Considering Ship Owners Profit

3.1 Option of Operating Speed for Vessels to Minimize the Carbon Emission

The decrease of speed means the fewer cargo it transports in the same period of time, and that'll lead to two situations to maintain in a certain service frequency, one is a vessel has to load more cargo by increasing its loading capacity or using more efficient loading technology and packaging system to fulfill the transportation demand, another is putting more vessels in the lane to fulfill the continuing transportation demand instead of changing the sailing schedule. But the more vessels putting in this field will lead to the more CO_2 emission, which will neutralize the benefit of slowing down the operating speed.

In a short time, it is difficult to achieve the first situation, so currently we only consider the second one. In the precondition of a constant service frequency, we build a speed selection model aiming to minimize the CO_2 emission, which searching for a balance point between the decrease of carbon emission per vessel caused by slowing down the operating speed and the increase of carbon emission caused by the more vessels being put into service. Let's take an example.

Supposing there're n vessels in a particular line, dispatch interval is represented by t_i, t_r stands for the round voyage time which depends on the time for a round voyage and the loading and unloading time at two terminal ports.

$$t_r = \frac{2d_{ij}}{24V} + t_s \quad (8)$$

As we have mentioned the meaning of d_{ij}, V, t_s, the vessel number in a line, n can be represented by the equation below.

$$n = \frac{t_r}{t_i} = \frac{\frac{2d_{ij}}{24V} + t_s}{t_i} \quad (9)$$

It can be seen that if we keep the dispatch interval unchanged, the vessel number will go up when the operating speed goes down.

As is shown above, CO_2 emission per vessel per day is as follows.

$$Q_0 = 3.17 \cdot \frac{MF_k \cdot \left(\frac{V}{V_0}\right)^3 \cdot \frac{2d_{ij}}{24V} + \left(AF_{k1} \cdot \frac{2d_{ij}}{24V} + AF_{k2} \cdot t_s\right)}{t_r} \tag{10}$$

So the daily CO_2 emission of n vessels per day is,

$$Q_n = 3.17 \times \frac{MF_k \cdot \left(\frac{V}{V_0}\right)^3 \cdot \frac{2d_{ij}}{24V} + \left(AF_{k1} \cdot \frac{2d_{ij}}{24V} + AF_{k2} \cdot t_s\right)}{t_r} \times \frac{t_r}{t_i} \tag{11}$$

It's obvious that if the dispatch interval stays unchangeable, Q_n is a function of speed V. simplified as,

$$Q_n = \frac{3.17}{t_i} \times \left[MF_k \cdot \left(\frac{V}{V_0}\right)^3 \cdot \frac{2d_{ij}}{24V} + \left(AF_{k1} \cdot \frac{2d_{ij}}{24V} + AF_{k2} \cdot t_s\right)\right] \tag{12}$$

The derivative of speed V is as follows,

$$\frac{dQ_n}{dv} = \frac{3.17}{t_i} \times \left(\frac{2MF_k \times V \times d_{ij}}{12V_0{}^3} - \frac{AF_{k1} \times d_{ij}}{12V^2}\right) \tag{13}$$

If $\frac{dQ_n}{dv} = 0$, so $V_{CO_2} = \sqrt[3]{\frac{AF_{k1}}{2MF_k}} \times V_0$.

We can see the operating speed of the fleet that emit the least CO_2 depends on the rated oil consumption of both main and auxiliary engines as well as design speed for the vessel.

However, the option of operating speed for vessels is also restrained by technology, so, the option of operating speed is limited, the upper limit cannot surpass the highest speed set by the rated power of the main engine, and the lower limit cannot below the speed main engine's minimum steady speed. From Eq. (1), the lower limit of the operating speed (V_S) is 67% of the design speed. If $V_{CO_2} > V_S$, V_{CO_2} should be chosen as the operating speed, if $V_{CO_2} < V_S$, we should take some technical measures.

3.2 *Option of Operating Speed for Vessels When Considering Ship Owners Profit*

To shipping managers, slowing down the operating speed will cut down the oil cost, which is the major part of the shipping cost, but in order to sustain the service frequency, operating cost of a fleet will increase. Ship owners will always choose an operating speed to minimize the cost and maximize the profit.

Supposing K_s is a vessel's daily constant cost, including capital cost, operating cost and voyage cost which excludes oil cost (in liner, if quantity of cargo don't change, these costs all could be regarded as constant costs). The daily oil cost for a vessel is represented by K_{oil}, so

$$K_{oil} = \frac{C_f \times MF_k \times \left(\frac{V}{V_0}\right)^3 \times \frac{2d_{ij}}{24V} + C_d \times \left(AF_{k1} \times \frac{2d_{ij}}{24V} + AF_{k2} \times t_s\right)}{t_r} \quad (14)$$

C_f, C_d mean the price of fuel and derv respectively, Yuan/kg, other letters are the same as above.

As to the same voyage, the vessel number is represented by n. If these ships have the same size, scale, capital and operating cost, the daily total cost of these n ships is,

$$C = \left\{ K_s + \frac{C_f \times MF_k \times \left(\frac{V}{V_0}\right)^3 \times \frac{2d_{ij}}{24V} + C_d \times \left(AF_{k1} \times \frac{2d_{ij}}{24V} + AF_{k2} \times t_s\right)}{t_r} \right\} \times \frac{t_r}{t_i} \quad (15)$$

Simplified as follows,

$$C = \frac{1}{t_i} \times \left[k_s \times \left(\frac{2d_{ij}}{24V} + t_s\right) + C_f \times MF_k \cdot \left(\frac{V}{V_0}\right)^3 \cdot \frac{2d_{ij}}{24V} + C_d \times \left(AF_{k1} \cdot \frac{2d_{ij}}{24V} + AF_{k2} \cdot t_s\right) \right] \quad (16)$$

So, total cost C is the function of speed V, the derivative of V is,

$$\frac{dC}{dV} = \frac{1}{t_i} \times \left[-\frac{k_s d_{ij}}{12V^2} + \frac{2C_f \times MF_k \times V \times d_{ij}}{12V_0^3} - \frac{C_d \times AF_{k1} \times d_{ij}}{12V^2} \right] \quad (17)$$

If $\frac{dC}{dV} = 0$, so $V_C = \sqrt[3]{\frac{C_d \times AF_{k1} + k_s}{2C_f \times MF_k}} \times V_0$

We can come to a conclusion that the maximum operating speed relates not only to the daily rated oil consumption of both main and auxiliary engines as well as design speed, but to the prices of fuel and derv, and vessel's daily constant cost.

The model above doesn't consider the influence of competence for a fleet with the lower operating speed, which affect the cargo capacity of vessel, freight income and profit of ship operators. So we should do some modifications to the model above. We use K_r to represent the average daily loss of opportunity cost because of slow steaming, and add it to the formula (13) to calculate the derivative of speed V, after modification, the maximum speed for a profitable vessel is:

$$V_c' = 3\sqrt{\frac{C_d \times AF_{k1} + k_s + k_r}{2C_f \times MF_k}} \times V_0 \quad (18)$$

4 Conclusion

Currently, slowing down the operating speed is an effective way to sharply decrease the quantity of carbon emission in a short time. This article find a balance point between the decrease of carbon emission brought by a lower operating speed and the increase of that caused by more vessels putting into service in a mathematical method. This article works out the operating speed when maximizing ship owner's profit, and it finds out the speed when the fleet's carbon emission is at the least.

Acknowledgments This study is granted and supported by Natural Science Foundation of China, the Ministry of Education of China, Shanghai Municipal Education Commission, Shanghai Science and Technology Commission, and Shanghai Maritime University (Grant number: 70541009; 11YJA630067; S30601; ZF1209; 11510501800; 20110020).

References

1. Zhihui Wang (2008) Energy saving and emission reduction. Win Future Ship Econ Trade 7:12–14
2. Buhaug, Corbett JJ, Endresen, Eyring V, Faber J, Hanayama S, Lee DS, Lee D, Lindstad H, Mjelde A, Pålsson C, Wanquing W, Winebrake JJ, Yoshida K (2009) Second IMO greenhouse gas study. International Maritime Organization, London
3. Shuang Zhang, Shuohui Zhang, Yiliang Li (2008) Overview of measures for reduction of GHG emission from international shipping. China Marit Saf 5:60–64
4. Corbett JJ, Haifeng Wang, Winebrake JJ (2009) The effectiveness and costs of speed reduction on emissions from international shipping. Transp Res Part D: Transp Environ 8(12):593–598
5. Cariou P (2011) Is slow steaming a sustainable means of reducing CO_2 emissions from container shipping. Transp Res Part D 16:260–264
6. Psaraftis H, Kontovas C (2010) Balancing the economic and environmental performance of maritime transportation. Transp Res Part D 15:458–462
7. US Environmental Protection Agency (2000) Analysis of commercial marine vessels emissions and fuel consumption data. United States Environmental Protection Agency, Washington, DC

The Research of Low-Carbon Supply Chain Design of Manufacturing Firm

Fanrong Mu and Dan Chang

Abstract To reduce the carbon emissions, and to guarantee that the social sustainable development is an important task to the Manufacturing firm. Therefore, it should provide new request for designing the chain plan in the enterprise.

Keywords Low-carbon • Supply chain • Flow reorganization • Information sharing • DEA

1 Introduction

Along with the climate changing, Humanity faced with very serious environment and resources question. In recent years, with the promotion of low-carbon concept, the green motion which takes the low-carbon as the symbol will change the human society's production method and the life style. Supply chain, which as the important economic activity, plays the important role in the development low-carbon economy's process [1].

Because our country's production technology is backward, there are many problems to realize the low-carbon production in our country, such as the shortage of talented person, the shortage of the specialized technology, and the shortage of the related policy, which bring some difficulties. But the overseas is different, they has already carried out the study about the green manufacture, the low-carbon supply chain and so on.

When design low-carbon supply, first should meet the environment requirement, next, the node of supply chain integrated. Finally, form a closed loop system.

F. Mu (✉) • D. Chang
School of Economics and Management, Beijing Jiaotong University, Beijing 100044, China
e-mail: 11125212@bjtu.edu.cn; dchang@bjtu.edu.cn

Z. Zhang et al. (eds.), *LISS 2012: Proceedings of 2nd International Conference on Logistics, Informatics and Service Science*, DOI 10.1007/978-3-642-32054-5_6,

2 Related Technical

The integrated design of the supply chain need to use a lot of techniques and methods, such as BPR [2], Sharing technologies, multimodal, and the DEA method. BPR improve the cost, quality, service and speed, and through the information sharing and the integration, to realize the supply chain's analysis, the optimization and the coordination, so to realize the seamless connection between the chain's enterprises. Through the multi-like combined transport, it can also conform the transportation resources, and reduce the railroad, waterway's idling rate, so to reduce the transportation cost effectively [3]. DEA is based on the relative efficiency concept, it proposed a more benefit method to the same type. C^2R is the most commonly model in DEA method of a model [4].

3 Manufacturing Enterprise Low Carbon Supply Chain Solution Design

3.1 Traditional Supply Chain Analysis

The traditional supply chain compares present's low-carbon supply chain is simple, Fig. 1 is the simple traditional supply chain chart, on this supply chain relations between the node is not closely, only the neighboring nodes relation close. Moreover does not have a core enterprise to control or coordinated the entire supply chain. This caused difficulty of the traditional supply chain to realize low-carbon. So adjust to the whole supply chain structure is necessary.

The structure of the traditional supply chain has created it have the following disadvantages:

1. Because the node relates is not close, the information sharing rate is low, therefore the supply chain cost is high, the availability of resources is low, limits the investment rate of return to be low.
2. Between each node is essentially independent of the individual, makes it difficult to use high-tech on the whole supply chain , the limited level of service.
3. Whole supply chain energy to disperse, core business attention is not enough, management mode is limited to [5].

3.2 Low-Carbon Supply Chain Integration Design

As the traditional supply chains have lots of shortcomings, it was pushing people to explore a new low carbon supply chain. Economic globalizations, and the rise of multinational group, around a core enterprise of one or more products, create the upstream and downstream enterprise strategic alliance [6]. In the strategic alliance,

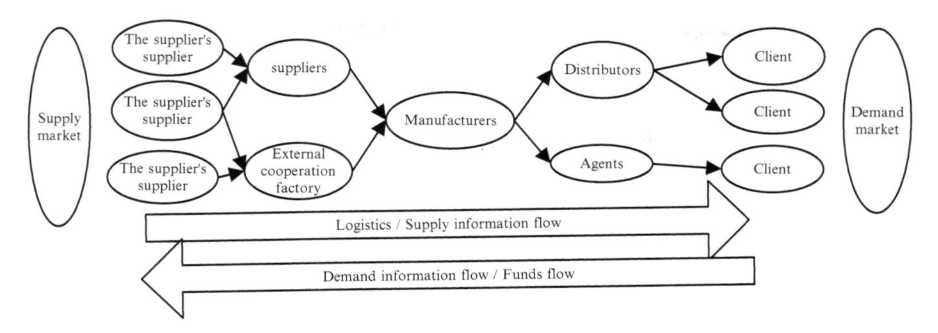

Fig. 1 Traditional supply chain chart

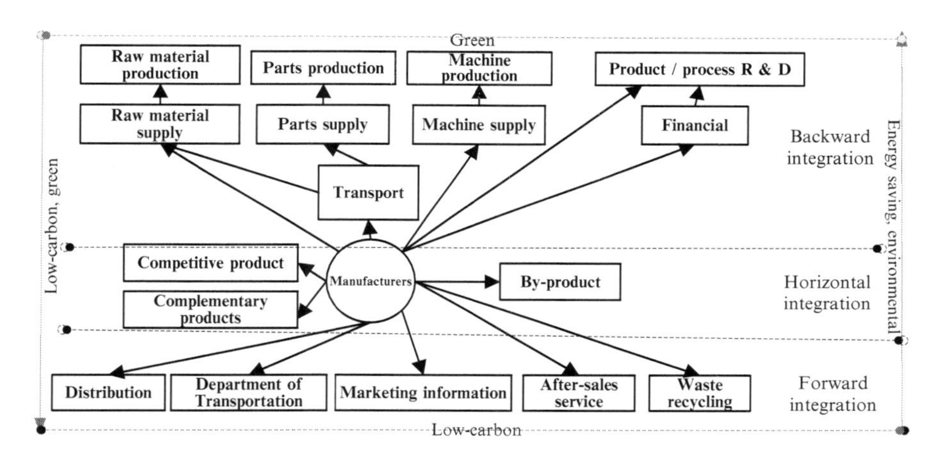

Fig. 2 Low-carbon supply chain integration

relating to the trade, logistics, information and capital formation can supply chain integration operation; It laid the foundation for the realization of low-carbon supply chain.

Figure 2 is considerate tradition supply chain's all sorts of shortcomings, Comprehensive low-carbon, green business philosophy, then designed the low-carbon supply chain integration. Supply chain integration is based on the information system, Through the information system and all over the world's transport network, and supply chain each link seamless connection form.

The realization of the supply chain integration, between enterprises and suppliers establish a set of effective material supply mechanism, reduce or even eliminate dull inventory, inhibit the supply chain' "bullwhip effect". Reduced the enterprise cost, reduce the waste of resources; make the realization of the supply chain low carbon more easy.

In the design of supply chain integration, procurement of low carbon, low-carbon production, low carbon transport, low-carbon storage, low-carbon sales, waste recycling logistics, information sharing system designed in detail.

4 The Effect of Low-Carbon Supply Chain Evaluation Model

4.1 The Basic Principle of the Model

DEA is a method that using mathematic programming model, to assess the relative effectiveness between department or units with multiple inputs, especially multiple outputs. Set x_{ij} represents the inputs of the j-th DMU i-th input indicators, the $x_{ij} > 0$; y_{rj} Represents the j-th volume of output of the r output indicators, $y_{rj} > 0$, v_i is the weight coefficient of the i-th input indicators, u_r expressed that the r-th kind delivers the target the scaling coefficient, $u_r > 0$, $(i = 1,2,\ldots,m; j = 1,2,\ldots,n; r = 1,2,\ldots,p)$; Set the input and output indicators of the weight coefficient vector respectively: $v = (v1, v2, \ldots, vm)^T, u = \left(u1, u2, \ldots, u_p\right)^T$, make $t = 1/V^T x0,\quad \omega = tv,\quad \mu = tu$. Then the C^2R linear programming model is:

$$\begin{cases} \text{Max } V_p = u^T y_0 \\ \text{s.t. } \omega^T X_i - \mu^T y_i \geq 0 \\ \omega^T X_0 = 0 \\ \omega \geq 0,\ \mu \geq 0 \end{cases} \quad (1 \leq i \leq n)$$

In order to determine DEA is more effective, simple and practical , Set ϵ non-Archimedean infinitesimal. In the domain of the generalized real number, ϵ represents a less than any number of positive and greater than zero. Considering the C^2R model with a non-Archimedean infinitesimal ϵ:

$$\begin{cases} \text{Max } \mu^T y_0 = V_p \\ \text{s.t. } \omega^T X_i - \mu^T y_i \geq 0\ (1 \leq j \leq n) \\ \omega^T X_0 = 1 \\ \omega^T \geq \epsilon \hat{e}^T,\ \mu^T >= \epsilon \hat{e}^T \end{cases} \quad (I) \qquad \begin{cases} \text{Min } \left[\theta - \epsilon\left(\hat{e}^T S^- + e^T s^+\right)\right] = V_D \\ \text{s.t } \sum X_i \lambda_i + S^- = \theta X_0 \\ \sum y_i \lambda_i - S^+ = y_0 \\ \lambda_i \geq 0,\ (1 \leq j \leq n),\ S^- \geq 0,\ S^+ \geq 0 \end{cases} \quad (II)$$

Among them, the $\hat{e}^T = (1,\ 1\ \ldots,\ \ 1)$ is the element are 1 m d vector, $e^T = (1,\ 1,\ \ldots\ 1)$ is the element are 1 p d vector. (II) is for its dual planning.

4.2 The Effectiveness of the DEA Judgment

To non-Archimedes the infinitely small quantity C^2R antithesis input model, can according to the following rules judge DEA effectiveness:

1. When $\theta_0 = 1,\ S_0^- = 0,\ S_0^+ = 0$, DMU j DEA effective, and technology and scale at the same time effective.
2. When $\theta^0 = 1$, but at least a $S_{i0}^- > 0,\ (I = 1, 2, \ldots,\ m)$ or at least $S_{r0}^+ > 0,\ (r = 1, 2, \ldots, p)$, is DMU j weak DEA effective, Not at the same technical efficiency and returns to scale the best.

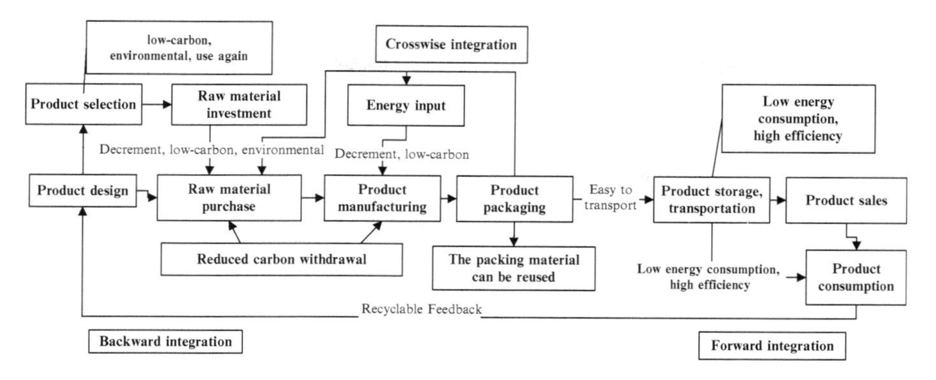

Fig. 3 M enterprise entire journey low-carbon supply chain

3. When $\theta < 0$, DMU j is not effective, and neither the DEA technical efficiency also is not the best return to scale the best.

5 Case Study

5.1 The M Enterprise Entire Journey Low-Carbon Supplies Chain's Implementation

The M enterprise is a industry which primarily with family electrical, it begin from the purchase, the research and development, the production, applies, the service and so on, as promotes the low-carbon on the supply chain, also has promoted the product, has brought this company product sales volume unceasing growth. Figure 3 is this enterprise's low-carbon supply chain design drawing.

Through the integration design, achieve a seamless connection, between supply chain nodes. In order to make low-carbon smoothly, the company also implemented a low-carbon supervision and build information sharing system.

By information system, and supply chain integration advantage, M enterprise established from the front of the supply chain core control technology to consumer demand function of the end of the integrative control. M enterprises throughout the research and development, procurement, manufacturing, logistics, installation and sale of the entire process of quality control of low-carbon supply chain management system.

5.2 Evaluation of the Implementation of the Low Carbon Supply Chain

Here are 5 M enterprise supply chain's evaluations after implemented the low carbon supply chain (Table 1).

Table 1 Input and output input and output of the low carbon supply chain

DMU	1	2	3	4	5
x1	3.34	2.23	8.21	3.4	6.79
x2	50	16	52	31	48
x3	0.62	0.2	0.4	0.54	0.15
x4	0.68	0.71	0.47	0.86	0.8
y1	10.32	20.38	3.31	17.56	4.75
y2	0.79	0.37	0.41	0.15	0.25
y3	0.67	94	0.61	0. 94	0.90
y4	3.40	10.72	12.31	3.13	5.60

Table 2 Low-carbon supply chain evaluation results

	Input				Output					
DUM	S10−	S20−	S30−	S40−	S10+	S20+	S30+	S40+	θ	K
1	0	0	0	0	0	0	0	0	1	1
2	0	0	0	0	0	0	0	0	1	1
3	0	0	0	0	0	0	0	0	1	1
4	0	7.1906	0.2609	0	2.1892	0.231	0	7.7448	0.8553	1.238
5	0	7.9709	0	0.0699	2.2221	0	0.0716	0.7473	0.7183	1.165

Explanation: X_1 (million) is environmental cost; X_2 (day) is production flexible; X_3 (%) is Scientific researchers proportion; X_4 (tons) is Unit product energy consumption quantity. Outputs evaluation index as follows: y_1 (%) is Low-carbon recognition; y_2 (%) is Return on total assets; y_3 (%) is Increase the proportion of low carbon products and services; y_4 (%) is Energy saving rate (Table 2).

1. From the table we can see DMU1, DMU2, DMU3 are on DEA = 1, and $S0 - = 0$, $S0 + = 0$, Indicated that these three low-carbon supply chain is DEA is effective, Environmental costs, production flexibility, the unit energy consumption amount of input factors to achieve the best combination of resources are fully utilized, the maximum output.
2. DMU4, DMU5 the DEA are less than 1, Indicated that these two low-carbon supply chain DEA is invalid, and the value is bigger than 1, indicated that these three low-carbon supply chain invests oversized in the production activity, the scale benefit assumes the decreasing progressively tendency.

6 Summarize and Prospect

Low carbon supply chain is in order to adapt to the needs of the time and the new concept put forward, no matter from in theory or in practice need to be further discussed. The low carbon supply chain carbon management revealed a significant reduction and a lot of economic interests between the relations. In the supply chain coordination work can win the chance to carbon emissions, enterprise change purchasing mode, use energy type, transportation mode are to create new emission reduction mode and the profit space.

Acknowledgments The research was supported by the key project of logistics management and technology lab.

References

1. Zhang Shuchun (2011) Low-carbon mode of operation of the manufacturing enterprises in Tianjin. Tianjin University of Finance and Economics, Tianjin
2. Ding Chaoxun (2010) Logistics industry in the low-carbon under the concept of ecological integration path. Guilin University of Technology, Guilin
3. Wang Guowen (2010) Low-carbon logistics and green supply chain concepts, processes and policies. Opening Herald 2(2):37–40, 53
4. Lijuan Yang, Binbin Guo (2010) Discussion of Low-carbon Supply Chain Performance evaluation based on DEA. Inquiry Into Economic Issues pp 31–35
5. Xu Hong, (2009) Chinese enterprise business process reengineering key success factors and evaluation of empirical research, Xiamen University, Xiamen, Fujian, China
6. Naichao Wang (2010) Low carbon logistics policy in advance. China storage & transport 11:37–38

Supply Chain Disruption Assessment Based on the Perspective of Trade-off in Newsvendor Model

Li Yisong, Jia Lu, and Chen Xiaofei

Abstract This paper centers on supply chain assessment, following the cost-income principle and taking the cost trade-off into account. The analysis tool is newsvendor model and its perspective—finding the critical point, which in tradition model stands for the demarcation point of profit but in this paper the is the least costs considering disruption costs and expected revenues. In this way, this paper tries to find out the optimum method to assess supply chain risks.

Keywords Supply chain disruption • Risk assessment • Newsvendor model

1 Introduction

Nowadays, facing the complicated and variable commercial environment as well as have been doing efforts in lean management for quicker response and lower cost, supply chain are tend to vulnerable and liable to affected by various risks. Supply chain risks, their impact and management are receiving much attention among practitioners and academicians alike.

The topic of risk management will continue to be important to researchers and supply management professionals. The twin areas of risk assessment and identification, and risk mitigation (approaches and theories) will continue to be of interest as the outsourcing trend continues to be a dominant strategy in firms.

Wakolbinger and Cruz T. Wakolbinger [1] summarized that the risks supply chain faced can be classified into two types: supply–demand coordination risks and

L. Yisong • J. Lu
School of Economics and Management, Beijing Jiaotong University, 100044 Beijing, People's Republic of China

C. Xiaofei (✉)
Industrial Engineering, University of Toronto, Toronto, Canada
e-mail: xiaofei.chen@utoronto.ca

Z. Zhang et al. (eds.), *LISS 2012: Proceedings of 2nd International Conference on Logistics, Informatics and Service Science*, DOI 10.1007/978-3-642-32054-5_7,

disruption risks. Moreover, the disruption risks are the most vital and most notably type because the fact that disruption can bring about huge losses and prevention cost. Zhang Song [2] emphasized disruptions and figured out that cost-income principle must be followed when build the fortification models, that is to say, the model concerned not only the cost of disruption, but also the expected cost of lost revenues.

This paper aims at supply chain disruption assess, analyzing two types of costs which possibly be related to disruption risks. The two costs respectively stand for two kinds of attitudes towards risks—risk averse and risk appetite. Based on newsvendor model and its trade-off idea, the paper establishes one model can weigh the two costs, namely the two risk attitudes, and therefore get the optimal assessment solution. In the end, by means of the conclusion, it is useful to ranking risks concerning importance.

2 Theory Review of Supply Chain Risk Management (SCRM) and Newsvendor Model

Christopher S. Tang [3] gave an integrated definitions developed by others that "the management of supply chain risks through coordination or collaboration among the supply chain partners so to ensure profitability and continuity". We can infer from the description that the objectives of SCRM fall into two aspects: the one is to enhance profit, and the other one is to lower the disruptions existing supply chain. To reduce the disruptions means to must pay much costs for precautions and controls, but on the other hand, to protect profit will ask cost-control. Contemporary researches are increasingly beginning to focus on the tradeoff between the two constraints. Based on the definition of SCRM, in general, SCRM is an issue dealing with the identification, assessment, analysis and treatment by minimizing, monitoring and controlling the probability and impact of uncertainty disruptions in order to economically effective management. Furthermore, the risk identification and assessment stage is fundamental and critical to the success of managing supply chain risks [3, 4]. The study present by Jyri P.P. Vilko and Jukka M. Hallikas [5] conducted a preliminary research concepts and findings concerning the identification and analysis of supply chain risks.

This paper aims at bring forth new approach of risks assessment by applying the newsvendor model. Taking a wide view of risks assessment theory, it revolves mainly around two aspects: (1) The probability of risk events; (2) The consequences and losses if risk events happening. After Mitchell [6] and his theory of "Risk = P(loss) *Loss"(In this equality, Risk is the assessment result; P means the probability or possibility; Loss is consequences), industry and academic circles develops the assessment theory from the two aspects: Ding Weidong et al. [7] provided a fuzzy factor technique to evaluate risks; Ericsson developed a series of tools named ERMET (Ericsson Risk Management Evaluation) [8]; Meng Kedeng [9] built a evaluation model based on grey relational analysis combining the fuzzy assessment. Most of the researches take two types variables into consideration, namely probabilities and results, so do the newsvendor model built in the paper.

Traditionally, newsvendor models is mostly assumed to be risk-neutral and insensitive to profit variations with the objective of expected profit maximization or expected cost minimization. Recently, the vulnerability and risks in the supply chains remind the managers of the tradeoff expected profit for downside protection against possible losses [10]. In this respect, Werner.Jammernegg and Peter Kischka [11–13] specially promoted this issues. They formulated the newsvendor model of the same kind that tradeoff between service level (target value) and resulting losses by the target. Yan Qin et al. [14] enriched the theory by considering the attitude of decision-maker towards the risks, devoted to the analysis newsvendor model with various risk preferences, including, but not limited to, risk-averse and risk-seeking preferences. Besides, they also reviewed and directed the future research in newsvendor problem, modeled how the buyer's risk profile moderates the newsvendor order quantity decision. Anastasios Xanthopoulos et al. [15] develop a newsvendor model for both risk neutral and risk-averse decision-makers and can be applicable for different types of disruptions related among others to the supply of raw materials, the production process, and the distribution system, as well as security breaches and natural disaster.

To sum up, the new direction of SCRM and newsvendor problem will be of tradeoff value target and losses may resulting in. Risk attitude of decision-maker is also a crucial point to be regarded.

3 Newsvendor Model for Supply Chain Disruption Assessment

The thought of traditional newsvendor model is that: when the demand is Stochastic, the managers expect to achieve profit maximization or loss minimization by optimum order quantity. In fact, the order quantity is a ratio based on a "critical point", which makes the optimal probability of target function. In addition, traditional newsvendor managers bear the risk neutral attitude towards the risks.

There is the probabilities assessment in the evaluation of supply chain risks. No matter by means of qualitative expert evaluation method or quantitative methods emerging continuously in the academic circles, it is expected to make a best assessment of supply chain risks. By virtue of thought of newsvendor model, it is a feasible approach to get the risks assessment on the foundation of an optimal "critical point ratio" weighing against costs, and the target function is to make the risks prevention cost and risks response cost minimum. Besides, the weight of two costs, at the same time, is also the weight of two attitudes—risk averse and risk appetite.

3.1 Model Foundation

This model considers risks probability as a continuous variable. To assess the probability of a risk occurring, it introduces two types of costs to weigh against. That is called opportunity cost and disruption cost in this model. The former is the cost spends on preventing risks from happening, and the latter is the loss to response the consequences after risks occur.

Assume a certain risk's probability is stochastic and obeys a known distribution. Evaluate an occurrence probability of the risk is P.

If P is higher than the actual probability. That is to say, risk averse decision overestimates the risk giving rise to an overdone prevention and emergence action, which generates opportunity cost.

If P is lower than the actual probability. That is to say, risk appetite decision underestimates the risk binging about a potential disruption point in supply chains, which a liable to generate disruption cost.

Assume a certain risk is named NI. Its actual occurrence probability is r, obeying a distribution with density function ϕ(r), namely $\int_0^r \phi(r)dr = 1$. In one risk assessment process, its calculating probability is p.

In order to get the optimal solution p*, define the two concerned cost in the first place.

1. Opportunity Cost L
   When risks are overestimated (p $\geq$ r), opportunity cost happens and the loss is: (p−r) · L, so its expectation value is $\int_0^p L \cdot (p - r)\phi(r)dr$;
2. Disruption Cost C1
   When risks are underestimated (p < r), disruption cost happens and its loss is: (r−p) · C1, so its expectation value is $\int_p^\infty C_1(r - p)\phi(r)dr$.

3.2 Optimal Decision

When a risk Ni and its calculated probability is p, combining the mentioned above (1) and (2), the total expectation losses are:

$$E[C(p)] = L\int_0^p (p - r)\phi(r)dr + C_1\int_p^\infty (r - p)\phi(r)dr$$

Target function is min E[C(P)].

Here follows the differentiation method inference process of newsvendor model:

When p is continuous variable, E[C(P)] is continuous function about p.

So, $$\frac{dE[C(p)]}{dp} = \frac{d}{dp}[L\int_0^p (p-r)\phi(r)dr + C_1\int_p^\infty (r-p)\phi(r)dr]$$
$$= L\int_0^p \phi(r)dr - C_1\int_p^\infty \phi(r)dr$$

Order, $$\frac{dE[C(p)]}{dp} = 0,$$

If $$\varphi(r) = \int_0^p \phi(r)dr,$$

Then $$L \cdot \varphi(p) - C_1 \cdot [1 - \varphi(p)] = 0,$$

And $$\varphi(p) = \frac{C_1}{L + C_1}$$

Therefore, p is solved from the arithmetic expression above, and be denoted as P*, then P* is the stationary point of E[C(P)].

And because $\frac{d^2E[C(P)]}{dp^2} = L\varphi(p) + C_1\varphi(p) > 0$, it is clear that p* is the limited minimum point of E[C(P)], minimum point of the model.

3.3 Model Analysis

It can be inferred from the optimal decision $\phi(p*) = \frac{C_1}{L+C_1}$ that the optimal assessment result towards some certain risks comprehensively affected by followings:

1. Opportunity cost L
   This part of cost in reality reflects that manager is risk averse attitude, which means the loss of potential profits.
   Because of the averse of risk, manager takes vigorous prevention and emergence measures so as to be more defensive to disruptions in supply chain. Correspondingly, the overprotection needs more cost input, so the opportunity cost is come into being.
   This kind of cost is inversely proportional to the result p*. The higher some certain kinds of risks' opportunity cost is, the lower their optimal assessment probability is.
2. Disruption cost C1
   When manager's preference towards risk is risk appetite, the cost input to prevent risk from occurring is much more than the risk averse manager. But in contrast, there are more risk events happen, and can bring more disruption cost. The optimal risk assessment will be a tradeoff between the two types of costs.

3. Density function
   This is the general rule of the occurrence of risk. And generally normal distribution is the most universal used one. Its density function can be showed as following:

$$\phi(\mathrm{r}) = \frac{1}{\sigma\sqrt{2\pi}} e^{-\frac{(r-\mu)^2}{2\sigma^2}}, \quad -\infty < r < +\infty,$$

where μ is mean value, and σ is standard deviation.

4 Conclusions

This paper builds a newsvendor model for supply chain disruption assessment, which applies the tradeoff idea of newsvendor to this model. This model considers both opportunity cost and disruption cost, between which is a cost-income principle tradeoff. When a risk's assessment is higher than optimum, disruption cost will descend whereas opportunity cost ascend; when the risk's assessment is lower than the optimal one, disruption cost will ascend while opportunity cost descend. Among which, the optimum is the "critical point" deducted by the disruption assessment in newsvendor model. The "critical point" can minimize the expectation loss of these both costs. At the same time, the two cost stand for two opposite attitudes and preferences towards risks. Opportunity cost is on behalf of risk averse; disruption cost stand for risk appetite. That can extend the traditional risk neutral newsvendor.

Simultaneously, this model can be also used to rank a series of risk events Ni (i = 1, 2, 3...) may happen in every link of a supply chain according to their importance. Pi* can represent the probability of risk Ni, whose expectation loss is the least one. Taking another look at it, the lower Pi* is, the more likely the risk event causing cost, the more attention should be paid, the more vital the risk event. In contrast, when Pi* is lower, the risk event is not that important than the former.

References

1. Wakolbinger T, Cruz JM (2011) Supply chain disruption risk management through strategic information acquisition and sharing and risk-sharing contracts. Int J Prod Res 49(13):4063–4084
2. Zhang Song (2011) Fortification models hedging disruption risks based on arborescent supply chain. Oper Res Manag Sci 20(1):186–191
3. Tang CS (2006) Perspective in supply chain risk management. Int J Prod Econ 103:451–488
4. Neiger D, Rotaru K, Churilov L (2009) Supply chain risk identification with value-focused process engineering. J Oper Manag 27:154–168

5. Vilko JPP, Hallikas JM (2011) Risk assessment in multimodal supply chains. Int J Prod Econ. doi:10.1016/j.ijpe.2011.09.010
6. Mitchell VW (1995) Organizational risk perception and reduction: a literature review. Br J Manag 6(2):115–133
7. Ding Weidong, Liu Kai, He Guoxian (2003) Study on risk of supply chain. China Saf Sci J 13(4):64–66
8. Xu Juan, Liu Zhixue (2006) Ericsson positive supply chain risk management. China Logist Purch 23:72–73
9. Meng Kedeng (2009) Research on supply chain risk with comprehensive grey fuzzy evaluation. Decis Info 58(10):178–179
10. Minghui Xu, Jianbin Li (2010) Optimal decision when balancing expected profit and conditional value-at-risk in newsvendor models. J Syst Sci Complex 23:1054–1070
11. Jammernegg W, Kischka P (2007) Risk-averse and risk-taking newsvendors: a conditional expected value approach. Rev Manag Sci 1(1):93–110
12. Jammernegg W, Kischka P (2008) A newsvendor model with service and loss constraints. Jena Research Papers in Business and Economics, Jena
13. Jammernegg W, Kischka P (2011) Risk preferences of a news vendor with service and loss constraints. Int J Prod Econ. doi:10.1016/j.ijipe.2011.10.017
14. Yan Qin, Ruoxuan Wang, Vakharia AJ et al (2011) The newsvendor problem: review and directions for future research. Eur J Oper Res 213:361–374
15. Xanthopoulos A, Vlachos D, Iakovou E (2012) Optimal newsvendor policies for dual-souring supply chains: a disruption risk management framework. Comput Oper Res 39:350–357

Multi-objective Emergency Logistics Vehicle Routing Problem: 'Road Congestion', 'Unilateralism Time Window'

Miaomiao Du and Hua Yi

Abstract According to the different characteristics of the emergency logistics compared to the general logistics system, this paper has established a mathematical model of multi-objective emergency logistics vehicle routing problem on some conditions. The conditions consider the situation that roads are congested because of roads damaging and the urgency of time in Emergency Logistics System. And in the end of the paper, we use Genetic Algorithms to solve the problem. This model has its practical significance.

Keywords Emergency logistics • Road congestion • Unilateralism time window • Vehicle routing problem • Genetic algorithms

1 Introduction

With the rapid development of China's economic and the modernization process, no matter natural disasters or public safety disasters, the occurrence and size are significantly larger than usual. It brings more and more serious affect on people's daily lives and work. This has an ever-increasing impact on people's daily life and work. And this also hinders the pace of China's sustained and healthy economic development. With the current level of technology, the occurrence of these natural disasters and public health events is inevitable and the loss it caused is immeasurable. Emergency logistics system should achieve the time benefits to maximization and the disaster to the minimization and meet the emergency logistics needs in a condition of limited time, space and resource constraints. Under the impact of the incident, the original vehicle path may no longer fit the new situation. So how to

M. Du (✉) • H. Yi
School of Economics and Management, Beijing Jiao tong University,
Beijing 100044, China
e-mail: 11120647@bjtu.edu.cn; yihua995@126.com

Z. Zhang et al. (eds.), *LISS 2012: Proceedings of 2nd International Conference on Logistics, Informatics and Service Science*, DOI 10.1007/978-3-642-32054-5_8,

elect a best path from each path program is an important work. And it worths studying.

Emergency Logistics is a special logistics activity which supplies materials, personnel and funding for the emergency support, responsing to the incident of serious natural disasters, unexpected public health events. Its aim is to pursuit the time benefits to maximization and losses caused by disasters to minimization. Emergency Logistics has some special features [1], such as uncertainty, unconventional, the urgency of time constraints, diversity of needs, government and market participation and so on.

2 The Research of Emergency Logistics Vehicle Routing Problem

In 2009, Ping Zhou established model by using operations research methods in his paper *Research on Emergency Relief Material Distribution Vehicle Routing Problem*. This model solved the problem of how to choose the path of the delivery vehicles. It had some practical significance.

In 2010, Ge Chou [2] established emergency materials vehicle scheduling model in his paper *Research on Routing Problem under the Emergency Management*. He used Genetic Algorithm to achieve it, having some certain reference significance.

In 2011, Simin Zhao [3] built a path optimization model in her paper *Network Construction and Path Optimization of Grain Emergency Logistics System* which considering road congestion and time. The aim of this model was the maximum degree of satisfaction. And this achieved good results.

3 Describing of Problem and Mathematical Model

3.1 Description of Problem

The roads may be destroyed which making the total capacity of the road network fell in some region after the occurrence of natural disasters and public emergencies. Before the road network being repaired, traffic will be concentrated to the other sections. The traffic load will increase on this sections and even leads to certain traffic congestion. In addition, the disaster itself will produce a series of traffic load, including transport and distribution of relief materials, the transfer of the wounds, visiting friends or family security and so on, which is bound to have a major impact on the entire road network [4, 5]. Therefore, the congestion constraints should be taken into account when researching the emergency vehicle routing problem. We also need consider the time window for each demand point in emergency logistics [6]. According to the characteristics of emergency logistics, we just consider the unilateral time window, meeting the latest arrival time. The upper limit of the time windows is zero [7]. It should consider the cost because of the limiting of human, material and financial. The model's main aim is the sum of the time. And the second aim is total system cost minimization. All requirements can be met according to this model.

We regard ∂ as the congestion level [8]. And ∂ is bigger, the congestion level is more serious. Road traffic only has a relation with ∂. $\partial = kQ_{\text{flow capacity}}(k > 0)$. When Link flow capacity is less than the rated flow capacity of the vehicle, that is $Q_{\text{flow capacity}} \leq Q_{\text{rated flow capacity}}$, ∂ equals to 1.

$$k = \frac{1}{Q_{\text{rated flow capacity}}} \qquad \partial = \frac{Q_{\text{flow capacity}}}{Q_{\text{rated flow capacity}}}$$

3.2 Assumption of the Model

1. There is only an emergency logistics distribution center and several emergency supplies demand points which the demand and location are known.
2. Emergency logistics centers have the same model transport vehicles and the vehicle's load capacity is known and determined. The demand for each demand point is less than the capacity of a vehicle.
3. Emergency logistics center supplies (including emergency vehicles) are adequate. There are no out of stock.
4. Each emergency logistics points need only one vehicle. One vehicle could be responsible for a number of demand points. Each vehicle starts from the emergency logistics center and returns to the emergency logistics when it completes the task.
5. We don't consider the service time of vehicles in each demand point, assuming that the loading and unloading time is 0, and only consider the travel time between the emergency demand points.
6. There is design speed limit of various sections. That is the maximum safe speed when the road conditions and environmental climatic conditions are in good case.

3.3 Explanation of Symbols

D: Emergency logistics center;

$M = \{m/m = 1, 2, \ldots, k\}$: Collection of demand point;
$V = \{v/1, 2, \ldots, n\}$: Collection of emergency vehicles;
$A = D \cup M$: A collection of all nodes;

c_n^f: The fixed costs of emergency vehicles;
c_{ijn}^t: Unit cost of the vehicle n from node i to node j;
T_{in}: Time Vehicle n arrives at the emergency demand points;
$[0, l_i]$: The time window of emergency demand point;
d_{ij}: The distance between node i and node j;
∂_{ij}: The congestion level between node i and node j;
v_{ij}: design speed between node i and node j;
Q_n: The capacity of the emergency vehicle;
q_i: Demand for emergency demand point i;
x_{in}: The Integer variable: $x_{in} = 1$, when the task of demand point is completed by the vehicle j; otherwise $x_{in} = 0$;
y_{ijn}: The Integer variable: $y_{ijn} = 1$, when Vehicle n is from node i to node j; otherwise $y_{ijn} = 0$;

3.4 Establishment of the Model

$$\min z_1 = \sum_{i \subset M} \sum_{n \subset V} T_{in} \tag{1}$$

$$\min z_2 = \sum_{n \subset V} x_n c_n^f + \sum_{i \subset A} \sum_{j \subset A} \sum_{n \subset V} c_{ijn}^t d_{ij} y_{ijn} \tag{2}$$

$$\sum_{i \subset A} \sum_{n \subset V} y_{ijn} = 1, \forall j \subset M \tag{3}$$

$$\sum_{i \subset A} y_{ijn} = x_{in}, \forall n \subset V, \forall j \subset A \tag{4}$$

$$\sum_{i \subset A} y_{jin} = x_{in}, \forall n \subset V, \forall j \subset A \tag{5}$$

$$\sum_{i \subset M} q_i x_{in} \le Q_n, \forall n \subset V \tag{6}$$

$$T_{in} \le l_i, \forall i \subset M, \forall n \subset V \tag{7}$$

$$T_{in} = T_{jn} + \frac{\partial_{ij} d_{ij}}{v_{ij}} y_{jin}, \forall i \subset M, \forall j \subset M, \forall n \subset V \tag{8}$$

$$x_{in} = \{0, 1\}, \forall i \subset M, \forall n \subset V \tag{9}$$

$$y_{ijn} = \{0, 1\}, \forall i, j \subset A, \forall n \subset V \tag{10}$$

Objective function (1) makes sure that the total time is the least. Objective function (2) is the total cost, including the fixed grid costs (gasoline consumption, vehicle depreciation, etc.) and the travel costs. Objective function (1) is the main goal. Objective function (2) is a secondary objective.

Constraint (3) refers that an emergency demand point can be serviced only by one vehicle and only service once. Constraint (4) refers that the vehicle must leave the demand points at last, to ensure the continuity of the vehicle. Constraint (6) is the capacity constraint of vehicle. Constraint (7) refers that the time which vehicles arrived at the emergency demand point must meet the requirements of the unilateral time window. Constraint (8) indicates the actual time that the vehicle n arrives at demand points i. Constraint (9) and (10) represent the integer variable constraints.

4 Solution of Model [9, 10]

In this paper, the vehicle routing problem is described as a multi-objective optimization problem and the solution to solve it is the Genetic Algorithm. Model in the paper makes sure not only time limiting, but also the minimum of the total cost of the system. And these two goals are in fact conflict with each other. We have done some processing on the target function when soluting this problem. We use the weight coefficient transformation method and give a weight to each objective function. The overall objective function is the linear weighted sum. In order to highlight the importance of time, we give relatively large weights to the time limiting function. Then the model can be seen as a general single-objective model. We can use Genetic Algorithm to solve it.

5 Conclusion

This paper takes congestion into account and uses time and cost these two factors to build a model. The model is rather perfect. And in the end of the paper, the using of Genetic Algorithms is a good way this kind problem. Generally speaking, this article has some reference significance.

References

1. Zhenya Zhao, Yiyuan Zhang, Xingkui Huang (2009) Particularity and countermeasures of emergency logistics. Res Logist Econ 03:58–61
2. Chou Ge, Chun Feng (2010) Research on VRP under emergency logistics management. Southwest Jiao tong University, Chengdu
3. Simin Zhao, Sanyou Ji (2011) The network construction and routing optimization of grain emergency logistics system. Wuhan University of Technology, Wuhan

4. Minhao Deng, Zhenhua Wen (2011) AHP-based route selection method for emergency logistics. Logist Eng Manag 33(3):93–95
5. Qian Lu, Xianliang Shi (2009) Research on road transport route choice in earthquake disaster emergency logistics system. Beijing Jiao tong University, Beijing
6. Keke (2011) Research on distribution path optimization of emergency logistics. Logist Eng Manag 33(7):63–65
7. Bing Zheng, Zujun Ma, Tao Fang (2009) Positioning – path problem of fuzzy multi-objective in emergency logistics system. Syst Eng 27(8):21–25
8. Qin Xu, Zhujun Ma, Huajun Li (2008) Research on LRP of urban public emergencies in emergency logistics, School Paper of Huazhong Science and Technology University, 1671-7023(2008)06-0036-05
9. Baker Barrie M, Ayechew MA (2003) A genetic algorithm for the vehicle routing problem. Comput Oper Res 30:787–800
10. Jianqing Yuan (2011) Optimization algorithm and reviewed of vehicle routing. Softw Guide 10(7):60–61

Analysis on Energy Conservation and Emission Reduction for Non-ferrous Metals Industry

Chunyan Cheng and Wenjia Yu

Abstract The non-ferrous metals industry is a major energy consumer and carbon emission player in China. By systematically analyzing the result of energy conservation and emission reduction in non-ferrous metals industry of China, and by looking into the development trend of the industry, this paper proposes that we enhance the energy conservation and emission reduction in the future through expanding the import of raw material, enhancing the recycling level of resources, promoting the advanced technology and so on.

Keywords Energy conservation • Emission reduction • Energy consumption • Carbon emission • Sulfur dioxide

The non-ferrous metals industry of China began to boom when the new century has come, and the industry scale jumped to the first in the world in 2010 [1]. The annual output of ten kinds of non-ferrous metals was 31.21 million tons in 2010, and the apparent consumption was 34.3 million tons, increased by 13.7 and 15.5% respectively during Eleventh Five-Year Plan; enterprises of scale and above accomplished 3,300 billion yuan sales revenue, and 219.3 billion yuan profit, increased by 29.8 and 28.1% respectively during Eleventh Five-Year Plan. With the expansion of non-ferrous metals industry, the energy consumption and waste emission also increased. In 2010, coal consumption in the whole industry reached 90.98 million tons of standard coal, the emission of carbon dioxide reached 470 million tons and emission

C. Cheng (✉)
School of the Earth Sciences and Resources, China University of Geosciences, Beijing, China

Chinese Academy of Geological Sciences, Institute of Mineral Resources, Beijing, China

School of Economics and Management, Henan Normal University, Xinxiang, China
e-mail: chengyan324@hotmail.com

W. Yu
Chinese Academy of Geological Sciences, Institute of Mineral Resources, Beijing, China

Z. Zhang et al. (eds.), *LISS 2012: Proceedings of 2nd International Conference on Logistics, Informatics and Service Science*, DOI 10.1007/978-3-642-32054-5_9,

of sulfur dioxide reached 0.92 million tons, accounting for 2.8, 5.2 and 4.2% of the domestic overall amount respectively, which indicates that the issues of high energy consumption and pollution have become the bottleneck restraining the development of non-ferrous metals industry.

In recent years, the government successively launched a number of policies and measures to strengthen the energy conservation and emission reduction in non-ferrous metals industry. Then what is the effect of these measures in the past 10 years? What are the existing problems? How will the non-ferrous metals industry develop in the next 10 years? How much is the potential of energy conservation and emission reduction? What are the key points? The research and answers to those questions are significant for the sustainable development of Chinese non-ferrous metals industry.[1]

1 Gradual Increasing of the Industry Energy Efficiency, Continuous Decreasing of the Production Energy Consumption

1.1 Gradual Decreases of Consumption Intensity, Gradual Increases of Energy Efficiency

In recent years, the boom of Chinese non-ferrous metals industry relied largely on the extensive pattern of development of increasing investment in fixed assets and expanding the industry scale. Although by promoting advanced technology and clean production, the energy consumption of unit product show decline trend, the energy consumption and waste emission rise inevitably due to the excessively growing of production [1]. In 2010, the coal consumption of the whole industry reached 90.98 million tons of standard coal, increased by 1.4 times as that in 2001, wherein the energy consumption in copper, aluminum, and lead-zinc sectors account for 6, 66 and 11% (Fig. 1); the average annual growing speed of energy consumption in 2000–2010

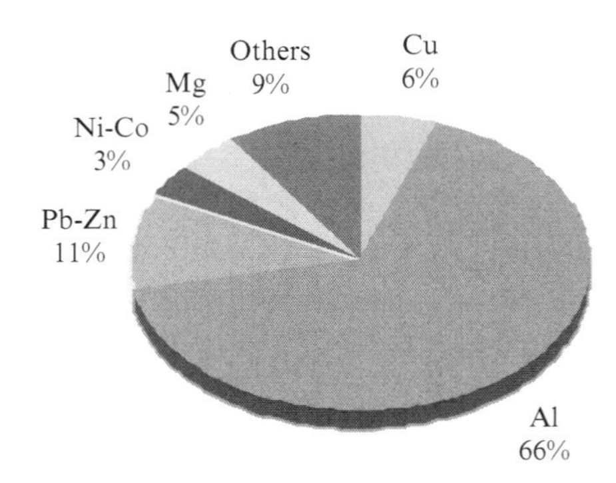

Fig. 1 Energy consumption proportion of major products in 2010 (Data Source: China Nonferrous Metals Industry Association)

[1] This article was funded by the Geological Survey Project (No.1102)

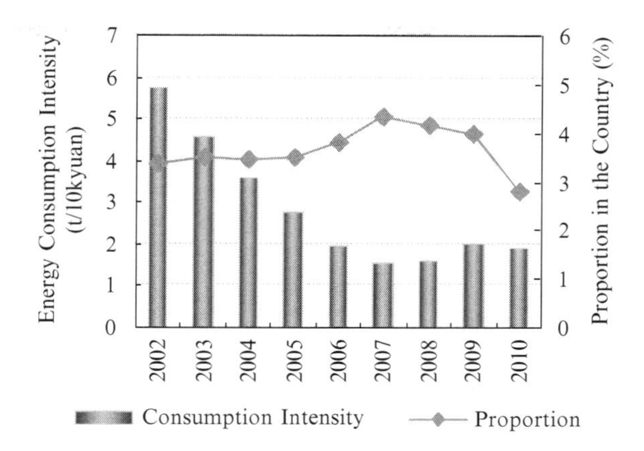

Fig. 2 Variations of intensity and proportion in the country of energy consumption during 2002–2010 (Data Source: China Statistical Yearbook, Yearbook of Nonferrous Metals Industry of China)

reached 10%, while the consumption proportion in all the country declined from 3.4 to 2.8%. Though the total amount of energy consumption of non-ferrous metals industry shows up-trend, the energy consumption intensity decreases gradually, from 5.74 to 1.88 t of standard coal/10,000 yuan during 2002–2010, with the decreasing amplitude of 67 and 13% annually on average (Fig. 2). The drop of consumption intensity indicates that the energy utilization efficiency of non-ferrous metals industry keeps rising, and unit energy consumption can generate more industrial added value. The industrial added value in non-ferrous metals industry grew by over eight times during 2002–2010.

1.2 The Energy Consumption of Unit Product Keeps Dropping

From 2000 to 2010, the comprehensive energy consumptions in smelting copper, aluminum oxide, lead and zinc are 360, 632, 454 and 947 kg/t respectively, the decreasing amplitude reached 72, 48, 37 and 59% respectively; direct current consumption of primary aluminum went down from 14,214 to 13,084 kWh/t, or by 8% (Fig. 3). Generally speaking, the non-ferrous metals industry has achieved preliminary effect in energy conservation since the new century, through heightening the technology and equipment level, adjusting the industry structure and other measures.

2 Gradual Increasing of Carbon Emission, Gradual Decreasing of Emission Intensity

In 2010, the nonferrous metals industry of China consumed electric power of 326.3 billion kWh, coal of 73.13 million tons, natural gas of 2.38 billion cubic meters, coke of 5.539 million tons. According to the IPPC CO_2 emission factor of electricity, coal, petroleum, natural gas and so on, the total emission amount of CO_2 caused by energy consumption is about 470 million tons in the nonferrous metals industry in 2010,

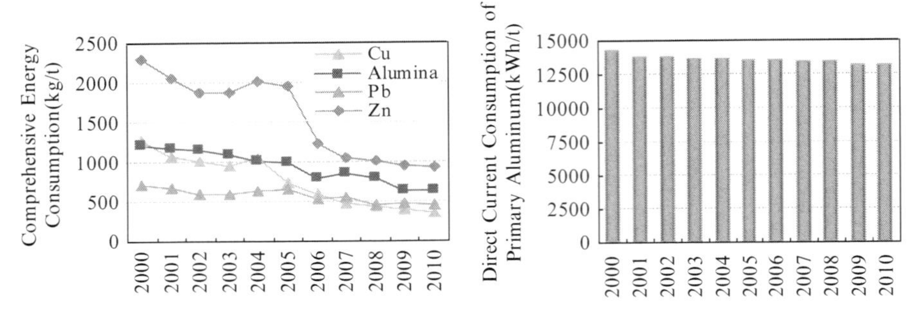

Fig. 3 Comprehensive energy consumption of the major products in nonferrous metals industry in 2000–2010 (Data Source: Yearbook of Nonferrous Metals Industry of China)

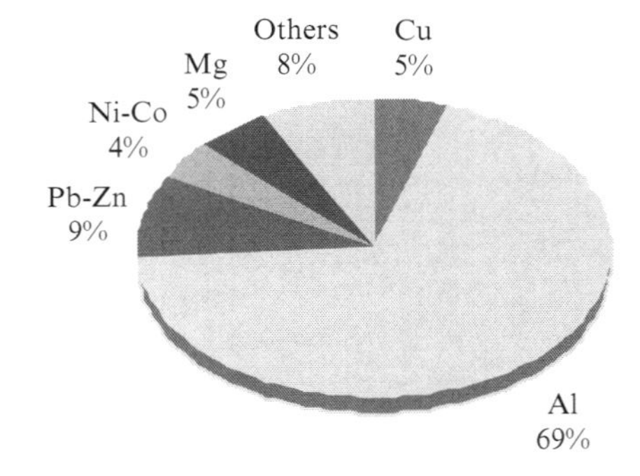

Fig. 4 Ratio of carbon emission in mining, dressing and smelting of major products (Data Source: Yearbook of Nonferrous Metals Industry of China, IPCC)

wherein the CO_2 emission of electricity and coal reached 94.9%. In 2010, the CO_2 emission of non-ferrous metals industry is mainly from the mining, dressing and smelting of aluminum, copper, lead-zinc and magnesium, accounting for 85% of the total CO_2 emission of non-ferrous metals industry, wherein, the CO_2 emission from the mining, dressing and smelting of aluminum, copper, lead-zinc accounts for 69, 5 and 9% respectively (Fig. 4). The CO_2 emission from the dressing and smelting of aluminum accounts for nearly two-thirds of the total industry, so it's the key field of energy conservation and emission reduction in non-ferrous metals industry.

The total CO_2 emissions of nonferrous metals industry kept rising in 2002–2010, from 1.2 to 4.7×10^8 t, increased by 3.1 times; whereas the emission intensity showed an inverse trend, and dropped from 17 t/10,000 yuan in 2002 to 7.3 t/10,000 yuan in 2010, or dropped by 57.4% (Fig. 5). The nonferrous metals industry has achieved remarkable results in dealing CO_2 emission.

3 Obvious Effect of Treatment of Sulfur Dioxide Pollution

At present the SO_2 emission of China ranks the first place of the world [2], and the industrial SO_2 emission accounts for 78% of the total emission of the country, including 4.2% from the nonferrous metals industry. The SO_2 emission of

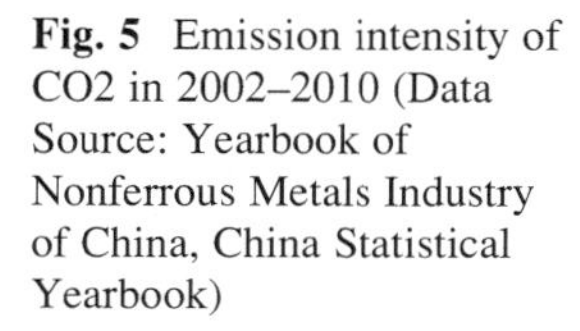

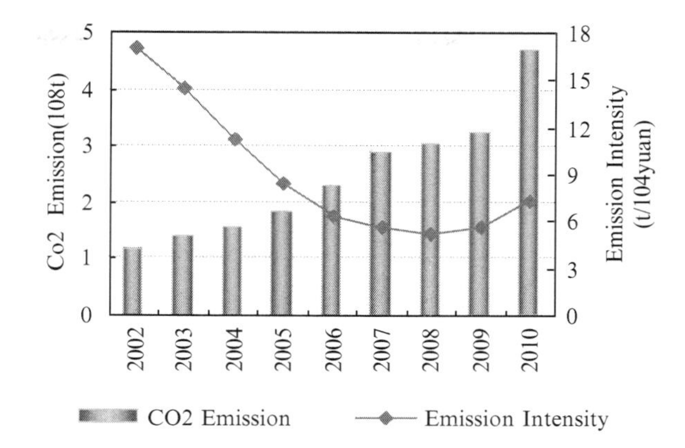

Fig. 5 Emission intensity of CO2 in 2002–2010 (Data Source: Yearbook of Nonferrous Metals Industry of China, China Statistical Yearbook)

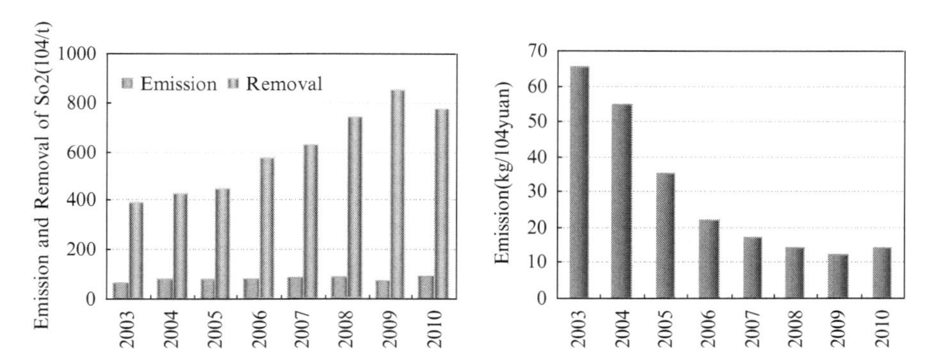

Fig. 6 The variation of SO_2 emission, removal and emission intensity during 2003–2010 (Data Source: Yearbook of Nonferrous Metals Industry of China, IPCC)

nonferrous metals industry comes mainly from combustion of sulfur-bearing fossil energy, especially the grizzle [3]. From 2003 to 2010, the SO_2 emission of the nonferrous metals industry rose from 630,000 to 920,000 t, or by 45%; the SO_2 emission share in the whole country increased from 2.9 to 4.2%; the SO_2 removal rose from 3.91 to 7.69 million tons, or nearly doubled; the emission intensity showed downtrend, during 2003–2010, SO_2 emission of unit GDP in non-ferrous metals industry declined from 65.3 to 14.2 kg, by 78.3%. The change of SO_2 removal and emission intensity indicates that SO_2 pollution treatment has gained obvious effect (Fig. 6).

4 Key Points of Energy Conservation and Emission Reduction for Chinese Non-ferrous Metals Industry

According to the prediction of Research Center for Global Mineral Resources Strategies, Chinese Academy of Geological Sciences, in 2020, the demand for copper and aluminum in China will reach 12.99 and 28 million tons respectively, increasing by 75 and 77% of that of 2010 respectively; it is predicted that by then, domestic production of refined copper and primary aluminum will increase to 7.4 million tons and 23 million respectively, by 63 and 42% of that in 2010. In the next 10 years, the sustainable development of the nonferrous metals industry will certainly result in the continuous growing of energy demand, so the situation of energy conservation and emission reduction will be more severe. From the view of sustainable development of the industry, the key points of energy conservation and emission reduction lie in the following aspects:

Firstly, strictly restrain the blind expansion of domestic industries of high energy consumption, and enlarge the import scale of products such as primary aluminum.

In recent years, large amount of capitals has rushed in and lead to the rapid expansion of smelting industry, especially the blind expansion and repeated construction of primary aluminum and aluminum oxide capacity. Up to 2010, the capacity of aluminum oxide and primary aluminum had reached 43.9 and 23 million tons respectively, with idle capacity of 34 and 29% respectively. Currently, exactly according to the national standard, prebaked aluminum electrolysis tank under 100 kA and lagging recycled aluminum capacity should be removed, with the new projects in the industry of high energy consumption strictly restrained. At the same time, changes should be adopted to the situation of importing great amount of bauxite and aluminum oxide while enlarging the import of primary aluminum. Thus, the energy consumption and waste emission in the non-ferrous metals industry can be effectively controlled.

Secondly, improve the resource recycling level and promote the energy conservation and emission reduction.

Compared to primary aluminum, the secondary aluminum per ton can save energy of 3.4 t of standard coal, water 22 m^3, and reduce solid waste by 20 t. Improving the using of secondary resource can relieve pressure of supplying as well as decrease energy consumption and pollution.

Thirdly, advance the technical progress, raise the technological equipment level, improve the energy efficiency, and reduce the energy consumption of unit products.

It is revealed by China Nonferrous Metals Industry Association that, in the technology innovation of aluminum electrolysis, developed by China, Special shaped tank technology with new cathode structure for aluminum electrolysis, Diversion tank technology with new structure for aluminum electrolysis, and Optimization and control technology of intelligent polycyclic synergy in aluminum electrolysis with efficient energy conservation which can largely increase the anodic current density as well as obviously decrease the tank voltage, have gained great breakthrough. The direct current consumption of aluminum per ton is between 12,043 and 12,400 kWh, with the saved electricity over 1,000 kWh compared to 2010. China ENFI and other institutes have also achieved key breakthrough in the technology of copper smelting with oxygen bottom blowing. It's indicated by the operation that, energy consumption of copper per ton decreased to 320 kg standard coal (including anode copper), and 131 kg standard coal less than that of foreign technology. Henan Yuguang Gold and Lead Co., Ltd has developed the technology of lead smelting in direct reduction method with high-lead-bearing liquid slag bottom blowing furnace. It has realized the continuity of lead smelting, and decreased the comprehensive energy consumption of lead bullion to below 300 kg standard coal, with energy saved over 30%. New progress has also been achieved in the technology application of zinc smelting in direct leaching method with atmospheric/pressurized enriched oxygen which effectively decreased the energy consumption of zinc smelting [4]. The popularization and application promotion of those technologies will largely reduce the energy consumption of unit product for companies in the non-ferrous metals industry, and thus reduces the pollutant emission of CO_2 and SO_2.

5 Conclusions

Based on the analysis of the energy conservation and emission reduction history in the past 10 years and the key points and solutions for the future non-ferrous metals industry, we can draw the following conclusions:

1. In the past 10 years, the industry energy efficiency improved gradually, and production energy consumption dropped constantly. The energy consumption intensity reduced by 67%, unit consumption of copper and aluminum oxide decreased by 72 and 48% respectively, and direct current consumption of primary aluminum declined by 8%.
2. The total carbon emission rose, while the emission intensity declined continuously. During 2002–2010, CO_2 emission caused by energy consumption increased by 3.1 times in the non-ferrous metals industry, while the emission intensity declined by 57%.
3. The treatment of SO_2 pollution achieved obvious effect. During 2003–2010, the SO_2 emission grew by 45%, the removal increased by 100%, and the emission intensity dropped by 78.3%.
4. Strictly Control should be exercised over the blind expansion of high energy consumption industries of aluminum, primary aluminum and so on, while enlarging the import scale of primary aluminum and increasing the utilization rate of secondary resource to promote energy conservation and emission reduction. In addition, developing domestic independent research and bringing in international technology to reduce the energy consumption of unit product, and remove the lagging technology step by step, improve the energy efficiency, decrease the pollutant emission, and thus promote the sustainable development of the non-ferrous metals industry.

References

1. Zhao W (2007) Energy conservation and emission reduction is the develop priority of nonferrous metals industry. Chin J Nonferrous Metals 2:24–25
2. Qie J (2011) Carbon dioxide emission of China ranking the top of world. Legal Daily 11–10(2)
3. RuiPing Li etc. (2010) Factor analysis of SO2 emission trend in typical industrialized countries and its revelation to China. J. Acta Geosci Sin 31(5):749–757
4. Hongguo Zhang (2010) Independent innovation obtaining major breakthrough in nonferrous metals industry of 2009. The yearbook of nonferrous metals industry of China. China Nonferrous Metals Industry Yearbook Agency, Beijing, China, pp 46–54

Decision Model for the Subsides to Low-Carbon Production by the Government Under the Emission Trading Scheme

Sheng Qu, Xianliang Shi, and Guowei Hua

Abstract In the view of supply chain aspect, reducing carbon footprint becomes to the recent business trends. We build a model which includes a re-manufactory which can reduce the carbon emissions by using the recycled products participating in the production process. To support the environmental mode of production, national governments offer subsidies for the re-manufactory to support its production, but not to reduce the total profit of the whole supply chain. We build a bi-objective optimization model to analyze the optimal strategy to reduce the emissions without serious pernicious influence to the industry profit. The outcome proposes significant instructions for the decision making of government and manufactory, when they are under supply chain management.

Keywords Low carbon production • Government subsidies • Carbon trading • Multi-objective optimization

1 Introduction

By the reason of huge consumption of fossil energy and accumulative emissions of greenhouse gases, global warming becomes worse. Thus, a series of policies have been introduced by governments in worldwide to encourage the public to use the low-carbon productions and support enterprises to reduce the emission in producing, marketing, and recycling process [1–3]. Therefore many scholars focus on design green supply chain which can effectively reduce carbon emissions [4–6] and they seek to achieve the optimal state of the environmental and economic

S. Qu (✉) • X. Shi • G. Hua
School of Economics and Management, Beijing Jiaotong University, Shangyuancun 3, District Haidian, 100044 Beijing, People's Republic of China
e-mail: 11120659@bjtu.edu.cn; xlshi@bjtu.edu.cn; gwhua@bjtu.edu.cn

Z. Zhang et al. (eds.), *LISS 2012: Proceedings of 2nd International Conference on Logistics, Informatics and Service Science*, DOI 10.1007/978-3-642-32054-5_10,

indicators [14]. In this paper, we assume a kind of environmental manufacturer participating in production progress, we call it remanufacturer which produces by using the recycled materials like Nagurney et al. [15]. And it has a monopolistic competition relationship with the original manufacturer which produces in a normal way in the market. As we said below, the remanufacturer increases the cost recycling, classifying and sorting raw materials. Besides, remanufacturer produces less profit than the manufacturer due to the materials recycled and the lower perceived value of customers. To encourage these enterprises and avoid heavy polluted commercial activity, several policies of emission reduction are enacted and have effective results, such as emission tax, emission trade and government subsidies.

However, there are many restrictions in the policies and t and methods [7, 8]. For instance, emission trade is available to the resource-intensive enterprises and government subsidies are suitable for the renewable resources enterprises. Based on actual situation of enterprises [7], we design a combination measures that can offer appropriate emission reduction policy for various enterprises and the enterprises could accept and implement it better. That means that under a fixed transaction price of emission trading, the government should decide how much subsidies to the remanufacturer to reduce the total emission without affect the total profits of the whole industry. We build an optimal model including maximizing the interests and minimizing environmental impact as the targets. Finally the outcome of this paper offer useful suggestions about the strategy of government subsidies.

2 Research Methodology

2.1 Model Description

A remanufacturer which produces by recycling waste and sold products can greatly reduce the emissions in the process of production. But in competitive market, the remanufacturer's production cost is higher than the general production activities in both technically and economically. Besides, for the traditional consumers, the perceived value of remanufactured goods may become lowers. Generally, the government usually makes use of the subsidy policy to foster the remanufacturing enterprises in order to achieve reduction. Our research assumes that government affects the market share by subsidies. Obviously, if the manufacturer's markets share increase, the environmental indicators improve, while the economic benefits of the industry as a whole will be seriously affected. We focus on reducing the emission without affecting the economic benefit seriously. At the same time, based on the emission trading, it can transfer the emission target into profit function [9]. In the model, it offers the suitable subsidy to get the largest profit.

2.2 Variables Descriptions

P_m, C_m, L_m, E_m: per new unit product's selling price, cost, profit and carbon emission;
P_r, C_r, L_r, E_r : per remanufactured unit product's selling price, cost, profit and carbon emission;
γ: the market share of remanufacturer;
E_l: the emission limit for whole industry;
P_s, P_b: selling and buying price of per unit emission;
$\lambda \in [0, C_r - C_m]$: subsidies for per remanufactured product;
Λ: total profit of whole industry;
E: total emission of whole industry

2.3 Model Assumptions

Several complex conditions will be simplified without changing the nature of the problem. The simplifications of the model are as follows:

1. We roughly assume that both the manufacturer's market share is inversely with their prices. And the market share of manufacturer is $\frac{P_r}{P_r+P_m}$, and the remanufacturer's is $\gamma = (1 - \frac{P_r}{P_r+P_m})$.
2. We assume that the total market demand is 1, and the prices of both kinds of products are denoted on the interval $(0, 1)$ [10]. Based on assumption (1), γ represents the demand of the remanufactured products.
3. We use the carbon emission factor of the electronic and communications industry, which is the amount of carbon dioxide emissions per unit of economic output as the carbon emissions [11]. We assume that the emission limit is 10.
4. The subsidy can compensate for high cost, and the cost becomes $C_r - \lambda$. We assume that government subsidies are just used to expand the market share of the remanufacturer. If so, the market share becomes $\gamma = \frac{P_m}{P_m+(P_r-\lambda)}$.
5. We assume that the ability of selling and buying carbon emission is limitless. So the price of selling or buying is constant. And the resource for recycling is limitless, so the remanufacturer doesn't have to consider the ability of product.

2.4 Model Building and Optimal Solutions

$$\max \Lambda = [(P_r - \lambda) - (C_r - \lambda)]\gamma + (P_m - C_m)(1 - \gamma) \tag{1}$$

$$\min E = E_m(1-\gamma) + E_r \cdot \gamma \tag{2}$$

Subject to

$$0 \le \lambda \le C_r - C_m$$

$$E_m > E_l > E_r$$

We can combine two objective functions into one profit function based on emission trading [12, 13].

$$F(\lambda) = \begin{cases} \Lambda + (E_l - E)P_s & \left(\lambda \ge P_r - \frac{E_l - E_r}{E_m - E_l} P_m\right) \\ \Lambda - (E - E_l)P_b & \left(\lambda \le P_r - \frac{E_l - E_r}{E_m - E_l} P_m\right) \end{cases} \tag{3}$$

In the following, we just consider the realistic situation: $\frac{P_r - C_r + C_m}{P_m} \le \frac{E_l - E_r}{E_m - E_l} \le \frac{P_r}{P_m}$

1. If $E_m - E_r \ge \frac{L_m - L_r}{P_s}$, namely emission saved cost of producing one more unit of manufactured products more than the profit of producing one more unit of new products. When we let subsidies $\lambda = C_r - C_m$ we can get the optimal profit function as follow:

$$F_1 = L_m - (E_m - E_l)P_s + \frac{P_m[L_r - L_m + (E_m - E_r)P_s]}{P_m + P_r - (C_r - C_m)} \tag{4}$$

2. If $E_m - E_r \le \frac{L_m - L_r}{P_s}$, when the subsidies denoted $\lambda = P_r - \frac{E_l - E_r}{E_m - E_l} P_m$ we can get the optimal profit function:

$$F_2 = L_m + \frac{(E_m - E_l)[L_r - L_m]}{E_m - E_r} \tag{5}$$

3. If $E_m - E_r \ge \frac{L_m - L_r}{P_b}$, namely buying permits cost of producing one more unit of 'new' product is more than the profit of producing one more unit of new products. When we let subsidies $\lambda = P_r - \frac{E_l - E_r}{E_m - E_l} P_m$, we can get the optimal profit function as follow:

$$F_3 = L_m + \frac{(E_m - E_l)[L_r - L_m]}{E_m - E_r} \tag{6}$$

4. If $E_m - E_r \le \frac{L_m - L_r}{P_b}$, when the subsidies denoted $\lambda = 0$, we can get the optimal profit function:

$$F_4 = L_m - (E_m - E_l)P_b + \frac{P_m[L_r - L_m + (E_m - E_r)P_b]}{P_m + P_r} \tag{7}$$

Table 1 Variable assignment

E_l	P_m	C_m	P_r	C_r	P_s	P_b
10	1	0.5	0.9	0.7	0.08	0.15

Now, we give an overall strategy about the government subsidies for the remanufacturing under the emission trading as follow:

$$F(\lambda) = \begin{cases} F_4 = F(0) \\ F_{2,3} = F\left(P_r - \frac{E_l - E_r}{E_m - E_l} P_m\right) \\ F_1 = F(C_r - C_m) \end{cases} \tag{8}$$

Then, our numerical experiments simulate the decision-making process and analyze the relationship between strategy and emissions.

3 Numerical Experiments

In this sector, we rely on a numerical analysis to compare the pros and cons of several emission reduction approach through the assignment of variables in the model. And then, we analysis the sensitivity of total profit to (E_r, λ).

3.1 Comparison of Emission Reduction Efficiency

It is seen from that (8) the best subsidies are influenced by the emission of remanufacturer. Based on our assumption below, we have the result of best subsidy policies and the best profit by the assignment of the variable in Table 1.

The current emission reduction will make the emission reduction efficiency (economic losses of reducing per unit emission) lower. When we just consider the government subsidy under the assignment and the assumption, our profit ranges from $F_{\max} = 0.342$ to $F_{\min} = 0.323$, and the emission range from $E_{\max} = 10.15$ to $E_{\min} = 9.94$, when the subsidies range from 0 to $C_r - C_m$. But when we consider the subsidy under the emission trading, the profit rang from $F_{\max} = 0.342$ to $F_{2,3} = 0.328$, and the emission range from $E_{\max} = 10.15$ to $E = 10$. It will easily discover that the new emission reduction way can decrease more emission with the same economics loss.

3.2 Sensitivity of Total Profit to $(\boldsymbol{E_r}, \boldsymbol{\lambda})$

With the consideration that the emission from (re)manufacturer can also affect the most optimal subsidies in (8), we have conducted a set of four sensitivity analyses of the total profit with respect to (E_r, λ), where the remanufacturer's production

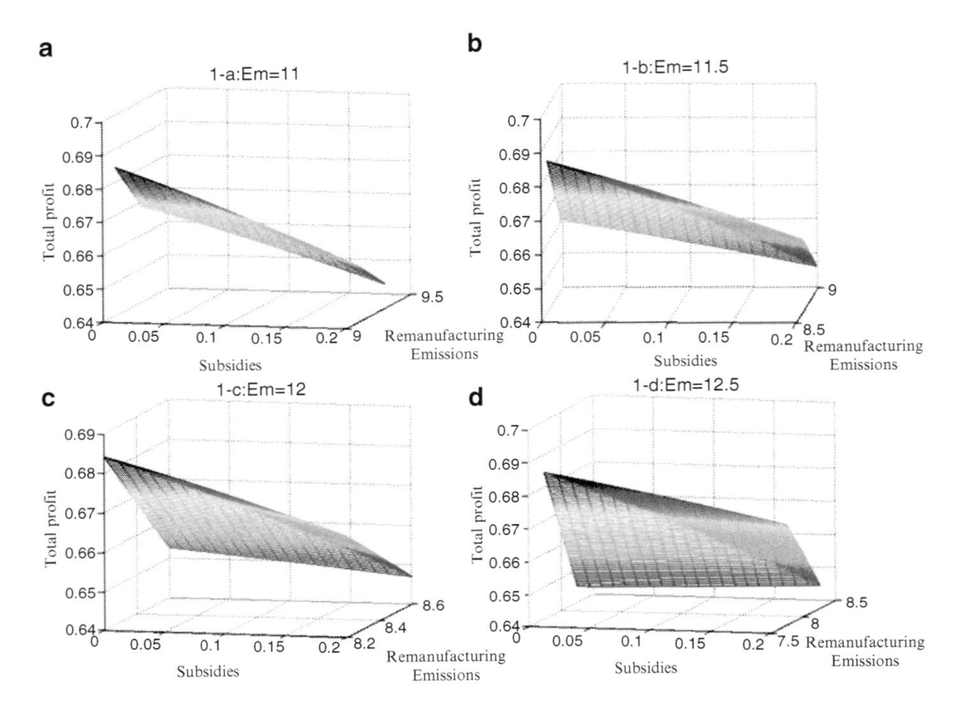

Fig. 1 The images of profit function respect to (E_r, λ) the z axis represents the total profit, and the x axis represents the government subsidies for the manufacturers, and the y axis represents the remanufacturer's emission. (**a**) The total profit function when Em = 11 (**b**) The total profit function when Em = 11.5 (**c**) The total profit function when Em = 12 (**d**) The total profit function when Em = 12.5

emission E_r have to be subjected to $\frac{P_r - C_r + C_m}{P_m} \leq \frac{E_l - E_r}{E_m - E_l} \leq \frac{P_r}{P_m}$ and the manufacturer's emission $E_m = 11, 11.5, 12, 12.5$. The results are illustrated in Fig. 1a–d.

As we seen from the Fig. 1, the optimal policy of the government subsidies is influenced by the two types of manufacturer's emissions. First, when the optimal total profit comes to the flex point, the government need supply the more subsidies, with the E_r increasing. Secondly, if E_m is in higher position, emission reduction efficiency of government subsidies increases.

We can find that the (re)manufacturer emission can determine the optimal strategy of government subsidies. When the two kind of manufacturers' emission decrease, the subsidies can have more choices to balance the emission reduction and the total profit. On the contrary, one unit emission reduction can affect the economic benefits.

4 Discussions and Conclusions

In this paper, we established a multi-objective optimization model to discuss the relationship between the economic benefit and the emission reduction. And we analyzed the optimal government subsidies in different situation that different emission of two kinds of manufacturers. And we gave the optimal strategy of subsidies for the government. And we can summarize several suggestions. Firstly, the combination of a variety of emission reduction policies can bring higher emission reduction efficiency and less economic loss than single original policy. Secondly, if the remanufacturer's emission is low, the government is able to balance the economic and environmental benefit by adjusting the subsidies. Thirdly, if manufacturer's emission is high, the efficiency of subsidy will decrease. These recommendations can guide the government to make the optimal subsidy policy.

Acknowledgments This paper was supported by "the Fundamental Research Funds for the Central Universities" (2011JBM234) and "Program for New Century Excellent Talents in University" (NCET-11-0567).

References

1. de Brito MP, Carbone V, Meunier Blanquart C (2008) Towards a sustainable fashion retail supply chain in Europe: organization and performance [J]. Int J Prod Econ 114:534–553
2. Sarkis J (2006) Greening the supply chain. Springer, Berlin
3. Zhu Qinghua, Dou Yijie (2007) An evolutionary model between governments and core-enterprises in green supply chains. Syst Eng: Theory Pract 27:85–89
4. Hall J (2000) Environmental supply chain dynamics. J Clean Prod 8:455–471
5. Koplin J, Seuring S, Mesterharm M (2007) Incorporating sustainability into supply management in the automotive industry: the case of the Volkswagen AG. J Clean Prod 15:1053–1062
6. Vachon S, Klassen RD (2008) Environmental management and manufacturing performance: the role of collaboration in the supply chain. Int J Prod Econ 111:299–315
7. Liu Xiaochuan and Wang zengtao (2009) Carbon dioxide emissions policy and our optimization options. J Shanghai Univ Financ Econ 11:3–80
8. Shrum T (2007) Greenhouse gas emissions: policy and economics. http://kec.kansas.gov/reports/GHG_Review_FINAL.pdf
9. Chaabane A, Ramudhin A, Paquet M (2012) Design of sustainable supply chains under the emission trading scheme. Int J Prod Econ 135:37–49
10. Mitra S, Webster S (2008) Competition in remanufacturing and the effects of government subsidies. Int J Prod Econ 111:287–298
11. Jin Ke-Di, Chu Chun-Li, Wang Yuan-Sheng (2011) Chinese high-tech industrial carbon emission trends and influencing factors analysis. Jianghuai Tribune 3:16–20
12. Ellerman AD, Buchner BK (2007) The European Union emissions trading scheme: origins, allocation, and early results. Rev Environ Econ Policy 1:66–87
13. Farahani RZ, Elahipanah M (2008) A genetic algorithm to optimize the total cost and service level for just-in-time distribution in a supply chain. Int J Prod Econ 111:229–243
14. Guillen-Gosalbez G, Grossmann (2009) Optimal design and planning of sustainable chemical supply chains under uncertainty [J]. AIChE J 55:99–121
15. Anna Nagurney (2006) On the relationship between supply chain and transportation network equilibria: a super network equivalence with computations. Transp Res E Logist Transp Rev [J] 42:293–316

Game Theory Application in Government's Guide on the Implementation of Enterprises Recovery Reverse Logistics

Chundi Liu

Abstract In order to pursue the operation objective of biggest profit, the enterprise is not willing to pay the cost for the implementation recovery reverse logistics. The government as the macroeconomic regulation and control department is necessary to guide it. Using the game theory, this article analyzes the government and the enterprise about the implementation of recovery reverse logistics, and proposes the most superior strategy of the government guiding on the implementation of enterprise recovery reverse logistics.

Keywords Strategy • Game theory • Recovery reverse logistics

1 Introduction

With the rapid development of economy, enterprise logistics activities produce a large number of industry wastes, which caused global warming, greenhouse effect and environment pollution. However, every enterprise all pursues the operation objective of biggest profit [1]. Whether enterprises can implement recovery reverse logistics actively depends on government's attitude. Chinese Government is now caring more and more about the implementation of recovery reverse logistics. As we all know that production is like people's arteries, recovery is like people's vein, which is difficult to drive relying on interest. Therefore, we need for government's environment protection policy, tax policy [2], as well as a series of co-operation mechanisms punishment.

Wherever recovery Reverse Logistics will take waste materials held by the ultimate customers to each node in the supply chain. It includes five kinds of material

C. Liu (✉)
Advanced Vocational Technical College, Shanghai University of Engineering Science, 200434 Shanghai, People's Republic of China
e-mail: lousia466@126.com

Z. Zhang et al. (eds.), *LISS 2012: Proceedings of 2nd International Conference on Logistics, Informatics and Service Science*, DOI 10.1007/978-3-642-32054-5_11,

flow: Direct re-sold product flow (recovery → test → distribution), re-processing product flow (recovery → test → reprocessing), components processing flow (recovery → test → split → reprocessing), scrapping product flow (recovery → test → treatment), end-of-life components and parts flow (recovery → test → split → treatment). Government and enterprises play different role in implementation of recovery reverse logistics. So they have different point of view on recovery reverse logistics [3].

Enterprises implementing the recovery reverse logistics will improve their resource utilization, enhance corporate image and win the trust and support from consumers, which will promote their sustainable development so as to obtain long-term profits.

However, enterprises must pay some cost. After comparison with benefits and costs, enterprises will have a final decision on whether to implement the recovery reverse logistics.

Government, as the macroeconomic regulation and control department, has the responsibility to maintain society stability and promote sustainable economic development and improve people's quality of life. These responsibilities promote government to be involved in recovery reverse logistics in order to achieve rational use of resources, protect for ecology and environment.

2 Model Assumptions

In this paper, we assume that government and enterprises are risk-neutral. On one hand, government firstly sets a recovery reverse logistics standard G, which is a standard vector including enterprises' ability of renewable resources [4], waste disposing capacity and the ability of Environment Protection. On the other hand, enterprises decide whether achieve the standards set by Government in the implementation of recovery reverse logistics.

The extent of Enterprises implanting recovery reverse logistics is T, which is variable. In general, there is inequality $T \leq G$, the cost that enterprises needed in the implementation of recovery reverse logistics is $C_1 = C_1(T)$, at the same time, there is $\frac{\partial C_1}{\partial T} > 0$. The enterprise benefits from recovery reverse logistics is R, so there is $R = R(T)$,at the same time $\frac{\partial R}{\partial T} > 0$, which means the more actively enterprises implement recovery reverse logistics, the more benefits they will obtain.

If enterprises implement recovery reverse logistics without achieving the standard, there will be some side effect. The cost government burdened is C_3, $C_3 = C_3(T)$ and $\frac{\partial C_3}{\partial T} < 0$, which means that the more actively enterprises implement recovery reverse logistics, the less cost government burdened is.

When enterprises achieve the standard set by government, that is $T = G$, at this time, private costs and social costs are equal, we have $C_3(T) = 0$.

3 Model Analysis

Government hopes that enterprises will achieve the standard [5]. The probability of government's inspection to enterprises is α, the cost of its inspection is C_2. When government finds T, it will take the punishment. Penalty function is as follows,

$$F = F(G - T) \quad \frac{\partial F}{\partial (G - T)} > 0$$

When $T = G$, there is $F(0) = 0$

The probability of enterprises not achieving the standard in the implementation of recovery reverse logistics is β. Based on the above assumptions, the game process between government and enterprises are shown below,

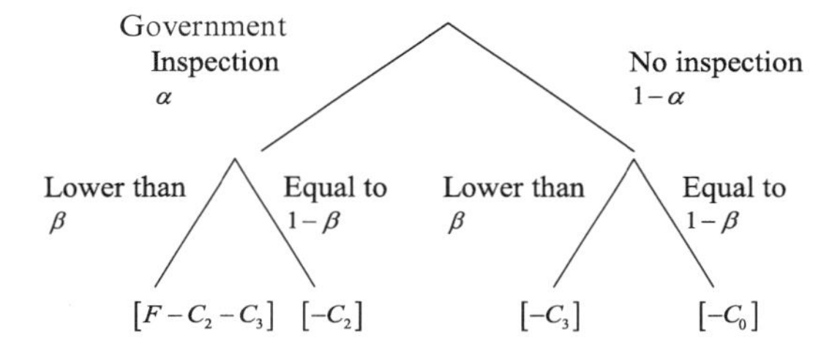

We can obtain enterprises' biggest profits is as follows,

$$\begin{aligned} A = {} & \alpha\beta[R(T) - F - C_1(T)] + \alpha(1-\beta)[R(G) - C_1(G)] \\ & + \beta(1-\alpha)[R(T) - C_1(T)] + (1-\alpha)(1-\beta)[R(G) - C_1(G)] \end{aligned}$$

After calculating, we can get

$$A = R(G) + \beta C_1(G) + \beta R(T) - \alpha\beta F - C_1(G) - \beta R(G) - \beta R(T)$$

We can obtain government's biggest profits is as follows,

$$B = \alpha\beta(F - C_2 - C_3) + \alpha(1-\beta)(-C_2) + \alpha(1-\beta)(-C_3)$$

After calculating, we can get

$$B = \alpha\beta F - \alpha C_2 - \beta C_3$$

As for enterprises, when $\frac{\partial A}{\partial \beta} = 0$, they can have the biggest profits. Then,

$$C_1(G) + R(T) - \alpha F - R(G) - R(T) = 0$$

We can obtain,

$$\alpha = \frac{C_1(G) - R(G)}{F}$$

When enterprises obtain their biggest profits, the most superior strategy of the government is to inspect and compare with the benefits from achieving the standard and not achieving the standard. If the profits are too high, it is necessary for government to inspect. At the same time, if government punish too heavy, the probability of government's inspection will decrease.

Whether to inspect all depends on the government's profits;

$$\frac{\partial B}{\partial \beta} = \alpha F - C_2 = 0$$

At last, we get $\beta = \frac{C_2}{F}$. When government has its biggest profits, we can find that the lower government's inspection cost is, the more actively enterprises achieve the standard in the implementation of recovery reverse logistics.

If the punishment taken by government is too high, enterprises will achieve the standard more actively.

The most superior strategy of enterprises is

$$T = G$$

The inspection government taken to enterprises includes two aspects. One is the appropriate probability, the other one is the punishment degree. These two aspects can be replaced each other. We have obtained $F \geq \frac{R(T)-C_1(T)-R(G)+C_1(G)}{\alpha}$. When government inspect enterprises, that is to say, when $\alpha = 1$, $F = R(T) - C_1(T) - R(G) + C_1(G)$. But if government can't inspect enterprises, the punishment degree should add to,

$$\begin{aligned}\Delta F &= \frac{R(T) - C_1(T) - R(G) + C_1(G)}{\alpha} - [R(T) - C_1(T) - R(G) + C_1(G)] \\ &= \frac{1-\alpha}{\alpha}[R(T) - C_1(T) - R(G) + C_1(G)]\end{aligned}$$

If government's inspection has difficulty, the punishment degree should increase. But if the probability of inspection is zero, the punishment degree has no meaning. Therefore, on one hand, government's inspection should exist; on the other hand, the punishment to enterprises should be appropriate. Then, good results are coming.

4 Conclusion

It is urgent for our society and government to find the effective measures for recovery reverse logistics [6–7]. Based on the above analysis, it needs the co-operation of government, enterprises and customers.

References

1. Stock JR (1992) Reverse logistics. Oak Brook Council of Logistics Management, Oak Brook, pp 58–61
2. Kopicky RJ, Berg MJ, Legg L et al (1993) Reuse and recycling: reverse logistics opportunities. Council of Logistics Management, Oak Brook, pp 44–47
3. Jia Xiao-Mei (2005) The game analysis of reverse logistics. Bus Econ Manag 160(2):9–12
4. Tanyanhua (2006) Business activities of reverse logistics game analysis. Green Economy, 11:77–79
5. Wang Qiong (2004) Green logistics. Chemical Industry Press, Beijing, pp 119–225
6. Wang Shu-yun (2002) Modern logistics. People's Traffic Press, Beijing, pp 68–70
7. Wu Zhong, Liu Hai, Zhou (2005) The implementation of reverse logistics management research. Mod Logist, 11:23–28

Empirical Analysis on Network Structure of the Inter-Regional Oil Railway Transportation

Wang Zhe and Wang Wei

Abstract By applying the complex network method to defining the inter-regional oil railway transportation network, the paper suggests the evaluation indexes of inter-regional oil railway transportation network. According to the relevant data, the paper makes empirical analysis on the structure of inter-regional oil railway transportation network in China. The research shows that Chinese inter-regional oil railway transportation network mainly aggregates in north and southwest China, which has characteristics of complex networks, such as small world, scale-free, group structure, etc.

Keywords Oil railway transportation • Network structure • Weight • Clustering coefficient • Cluster structure

1 Introduction

Inter-regional oil railway transportation network has an evident characteristic of weighted network. The regions constitute its nodes and the lines represent the inter-regional oil railway transportation route. The node weight is the total weight of transporting in and transporting out and the edge weight is the transporting weight among provinces and regions. The present paper is based on the above and discusses

W. Zhe (✉)
Research Institute of Transportation and Urban Planning and Designing, China Railway, Aryan Engineering Group Co. LTD, 610031 Chengdu, People's Republic of China
e-mail: flybird81@sohu.com

W. Wei
School of Architecture, Southwest Jiao tong University, 610031 Chengdu, People's Republic of China

School of Art and Communication, Southwest Jiao tong University, 610031 Chengdu, People's Republic of China

Z. Zhang et al. (eds.), *LISS 2012: Proceedings of 2nd International Conference on Logistics, Informatics and Service Science*, DOI 10.1007/978-3-642-32054-5_12,

the topological structure features of the inter-regional oil railway transportation network. The researcher also carries out an empirical analysis to find out the basic principles, trends and changes of the inter-regional oil railway transportation by analyzing the topological structure of the inter-regional oil railway transportation network. The results can provide basis for the planning and optimization of the inter-regional oil railway transportation.

2 Description of the Inter-Regional Oil Railway Transportation Network

Based on the description method of complicated network, the researcher applies the figures in math to describe the inter-regional oil railway transportation network. The network G is composed of a vertex set V and an edge set E, i.e. $G = (V,E)$. For the vertex set V, $V(G) = 1,2,\ldots,N$, the vertexes represent provinces, autonomous regions and municipalities directly under the central government (provinces and municipalities for short), the total number of which is 31. For the edge set E, $E(G) = (i_1,j_1),(i_2,j_2),\ldots,(i_E,j_E),E$ represents the total number of edges. The transporting of oil includes transporting in and transporting out, so oil transporting is a directional figure, of which the direction of in represents transporting in oil and the direction of out represents transporting out oil. That is to say, (i_k, j_k) does not equal (j_k, i_k). However, the range of edge set is 0 or the positive natural number, which is a typical character of weighted network. In a complicated edge weight, there are categories of difference weight and similarity weight. The former refers to that the more weight there is, the more distant the relationship between two nodes is or the farther the distance between two points is. It is opposite for the similarity weight that the more weight there is, the closer the relationship between two points is or the closer the distance between two points is. Inter-regional oil railway transportation network is a similarity weight. The line represents the railway oil transporting amount between two regions. The more transporting amount there is, the closer and stronger the relationship between regions is.

It can be seen from in a micro sense that the inter-regional oil railway transportation network reflects the source place, destination, transporting route and transporting amount. The lines connecting nodes show that there is a relationship of oil transporting among provinces and cities. The thicker the line is (the more weight there is), the more transporting amount there is. It can be seen in a macro sense that inter-regional oil railway transportation network reflects the whole situation of national oil railway transporting regulation and the basic situation of regional oil demand and supply. The safe and efficient operation of the inter-regional oil railway transportation network can realize the benefit of participative subjects in a micro sense and can realize the guarantee function of the oil supply in a macro sense. If there is something wrong with the nodes or route, the function of the inter-regional oil railway transportation network will be influenced.

3 Basic Statistical Indices of the Structure of the Inter-Regional Oil Railway Transportation Network

According to the analysis of complicated network, the structure of the inter-regional oil railway transportation network can be described by degree, degree distribution, vertex weight distribution, edge weight distribution, average shortest route, clustering coefficient and cluster structure, etc. [1–5].

4 Empirical Analysis of the Structure in the Inter-Regional Oil Railway Transportation Network

With the communicating data of regional oil in the National Railway Statistical Yearbook (the total amount of oil transported in or transported out by national railway from one region to the other one, the transporting data of the amount of oil transported by the highway, waterway and other transportation modes not included) and the statistical indices about network structure, the empirical analysis on the structure of the inter-regional oil railway transportation network is carried out.

4.1 Vertex Weight, Vertex Weight Distribution, Edge Weight Distribution and Unit Weight

The vertex weight data of the inter-regional oil railway transportation network from 2003 to 2006 are calculated. Compared with the amount of oil transported from regions to regions by railway in 2003, the amounts of oil transported in most provinces and cities in 2006 increased. The increasing rate of the amount of oil transported by railway in Inner Mongolia and Ningxia is more than 100%. The increasing rate of the amount of oil transported by railway in Shanxi, Hei Longjiang, Shanghai, Zhejiang, Anhui, Sichuan, Shaanxi, Gansu, Xinjiang and other provinces and cities is more than 20%. However, there is a decrease in the amount of oil transported by railway in Jilin, Guangdong, Guangxi, Chongqing, Yunnan, Guizhou and other provinces and cities. Besides, there is a big difference in the total amount of oil transported by railway in the provinces and cities. The amount ranges from the maximum of 18,870,000 t in Xinjiang to 500,000 t in Fujian. The amounts differ from each other in a level of two magnitudes.

The edge weight distribution curve and the vertex weight distribution weight of the inter-regional oil railway transportation network in 2006 are respectively calculated. It can be seen from the results that the edge weight and the vertex weight of the inter-regional oil railway transportation network are featured with the similar nature of power-law distribution. The edge weight distribution curve and the vertex weight distribution weight from 2003 to 2006 are fit. The fitting result shows that the vertex weight of the inter-regional oil railway transportation network accords with the power-law distribution. The maximum power exponent is 1.058 in 2006 and the minimum is 0.953 in 2004. The edge weight also accords with the power-law distribution. The maximum power exponent is 1.585 in 2006 and the minimum is 1.518 in 2003.

The unit weight is calculated. The unit weight can reflect the average transporting amount in different provinces and cities. Similar with the vertex weight, there is a big difference between the unit weights in different nodes. Take the year of 2006 as an example, the unit weight of Inner Mongolia in 2006 is 704,000 t while the unit weight in Chongqing is 42,000 t which is the smallest. The unit weights in the provinces and cities in the west China, northeast and northwest are relatively big. The unit weights in the provinces and cities in east China, middle south and southwest are relatively small. The results show that there is a relatively big average transporting amount in the provinces and cities connected by the nodes in the north China and northeast and there is a relatively small average transporting amount in the provinces and cities connected by the nodes in the east China, middle south and the southwest. The result accords with the fact that the national oil transporting center lies in the north China, the northeast and the northwest.

As discussed above, the difference of edge weight distribution connected by nodes can be measured as Yi. The difference of edge weight distribution connected by nodes reflects the balance of transporting amount between the node and the one connected with it, or the degree of dependence on other nodes. For the provinces and cities out of which the oil is transported, the difference reflects the generalization of selling, and for the provinces and cities in which the oil is transported, the difference reflects the generalization of the source of the oil. The bigger Yi is, the more uneven the edge weight distribution is. It means that the difference of edge weight connected with the node is small. It can be seen from results that the node of the transporting-out and transferring provinces and cities always has a higher difference of node edge weight, such as Qinghai, Gansu and Xinjiang. There is also a higher node edge weight in Anhui, Hei Longjiang and Inner Mongolia. The reason is that the oil transferred in Gansu is mainly transported from Xinjiang and the oil in Anhui is mainly transported from Jiangsu and Shandong and also the oil transferred in Hei Longjiang is mainly transported from Inner Mongolia.

4.2 Clustering Coefficient

Using the weighted clustering coefficient formula, the weighted clustering coefficients from 2003 to 2006 are calculated. It can be seen from results that the oil railway transportation network focuses on the north China, north east of China, northwest of China and southwest of China. There is a fluctuation in the clustering coefficients from 2003 to 2006.

4.3 Clustering Structure

According to the division of six regions in our country (North China: Beijing, Tianjin, Hebei, Shanxi, and Inner Mongolia; the Northeast of China: Liaoning, Jilin, and Hei Longjiang; East China: Shanghai, Jiangsu, Zhejiang, Anhui, Fujian, Jiangxi, and Shandong; the Middle South of China: Henan, Hubei, Guangdong, Guangxi, and Hainan; the Southwest of China: Chongqing, Sichuan, Guizhou, Yunnan, and Tibet; the Northwest of China: Shaanxi, Gansu, Qinghai, Ningxia, and Xinjiang), the clustering structure of the inter-regional oil railway transportation network is analyzed. It can be seen from results that the total amount of oil transported from the inner-region takes up a big proportion. Provinces and cities related with the most amount of oil transported may not be the provinces and cities in the same region. The clustering structure of the inter-regional oil railway transportation network does not exactly coincide with the division of the six regions. Besides, the actual structure of the inter-regional oil railway transportation network is not the same with the administrative division. For Example, the inter-regional oil railway transporting in the southwest takes up a low proportion. The oil in Chongqing, Sichuan and Yunnan is mainly transported from Gansu and Xinjiang in the northwest of China. And the oil in Guizhou is mainly transported from the middle south of China. Although Inner Mongolia belongs to the north China, it is the biggest transportation-related province of Hei Longjiang. Shandong belongs to the region of the east China and it is closely related with the transportation regions of Henan and Hebei. Although Henan belongs to the middle south of China, most of its oil is transported from Shaanxi in the northwest. Considered from the transportation relationship, the south of Inner Mongolia can be included by the transportation region of north China; the east of Inner Mongolia can be included by the transportation region of the northwest of China; Chongqing, Sichuan and Yunnan in the southwest of China can be included by the northwest transportation region; Guizhou can be included by the middle south transportation region and the production and planning of transportation ability can be carried out there. From the aspect of the whole country, the Northwest of China is the transporting center, which plays an important role in the supply of oil in the region and is also the main source area of oil transporting in the southwest of China. The northeast of China is the transporting center, which plays an important role in the supply of oil in the region and is also the main source area of oil transporting in the north China. Besides, there are also some regional centers of oil railway transportation. Hebei, Jiangsu, Guangdong, Sichuan are respectively the centers of the north China region, the east China region, the middle south region and the southwest region.

5 Conclusion

The inter-regional oil railway transportation network has the features of small world and no scale. Its vertex weight and edge weight are featured with the power-law distribution and reflects the key role of partial provinces and cities and partial transporting lines in the oil railway transportation.

There is a feature of non-balance in the inter-regional oil railway transportation network in our country. That is to say, it is very dense in the north area and sparse in the south area. According to the result of the clustering coefficient, there is a high degree of clustering in some provinces and cities in the northeast, north China, the northwest and the southwest. There is some degree of clustering in these areas.

Although there is a character of clustering to some degree in the inter-regional oil railway transportation network in our country, the main provinces and cities and those that with which there is a transporting relationship may not be in the same region. Due to the clustering feature of the transporting network, there are two major national transporting centers, respectively the northeast of China and the northwest of China. There are also some regional centers, such as Hebei, Jiangsu, Guangdong, and Sichuan, etc.

References

1. Lv Tao, Cao Yongrong (2009) Empirical analysis on the structure of inter-regional coal transportation network. Railw Transp Econ 31:1–7
2. Fang Aili, Gao Qisheng, Zhang Siying (2009) Clustering of the industrial network and the relative analysis. Syst Eng—Theory Prac 29:178–183
3. Liu Hongkun, Zhou Tao (2007) Empirical research and analysis on urban aerial network. Chinese Physics 56:106–112
4. Guo Lei, Xu.Xiaoming (2006) Complicated network. Shanghai Technology Education Press, Shanghai
5. Wang Xiaofan, Li Xiang, Chen Guanrong (2006) The theory and application of complicated network. Tsinghua University Press, Beijing

Inventory Rationing Based on Different Shortage Cost: A Stable Inventory Demand Case

Qifan Chen and Jie Xu

Abstract This paper discusses an inventory rationing system with a threshold in a relatively stable inventory demand environment. There are two different demand classes in this inventory system, prior demand and secondary demand, distinguished by different shortage cost. The inventory system functions as follow. When the inventory level is above the threshold, demands come from both classes are filled. Once the inventory level falls below the threshold, all demands come from secondary demand class are backordered, only prior demand would be served. Heuristic approach was used to solve the model in this paper, and the optimal order quantity, single inventory cycle time and inventory threshold were given. Compare between inventory system with and without inventory threshold shows the cost advantage of the first system. Sensitivity analysis was performed to give a better understanding of two crucial parameter, inventory threshold and cost saving rate.

Keywords Inventory replenishment • Inventory rationing • Shortage cost

1 Introduction

As an important component of the cost of a company, inventory cost has long been the mainly focus of the management. Both business and academic world have come up with many different practice and ideas in inventory management field in decades. Among them, inventory rationing, which differentiated replenishment for different customers based on their own attributions, has brought in increasing attention. Differentiating replenishment refers to issuing the inventory to satisfy some customers, while rejecting or delaying other customers' needs.

Q. Chen (✉) • J. Xu
School of Economics and Management, Beijing Jiao tong University,
Haidian District, Beijing 100044, People's Republic of China
e-mail: 09120752@bjtu.edu.cn; jxu@bjtu.edu.cn

Z. Zhang et al. (eds.), *LISS 2012: Proceedings of 2nd International Conference on Logistics, Informatics and Service Science*, DOI 10.1007/978-3-642-32054-5_13,

In the modern marketing, the practice of providing different standards of service to different customers according to their special needs is a common business strategy. Industries like airline industry, hotel industry and car rental industry have already provided differentiated service (for example, airline companies provides first class, business class and economic class seats), and they mainly depend on price that charged from different customers, as well as possible rationing levels such as booking limits to indentify different customer priorities [1]. It is worth mentioning that, from the perspective of inventory, services provided by above industries can be seen as timely inventory resources with limited supply, while no need to consider the inventory replenishment strategy. Thus, it is a special inventory management case.

As for single product, multiple demand classes inventory rationing research, Veinott [2] is known to be one of the first person to investigate in this field. Under a multi-period, non-stationary inventory environment, he got a solution of the replenishment strategy of the inventory system (when to replenish and order quantity). Based on Veinott's work, Topkis [3] further discussed how to allocate inventory among two different demand classes within one single inventory circle in a periodic review inventory system. Topkis' method of distinguishing demand classes by their different shortage cost is widely adopted in many research, Vinayak et al. [4] also use different shortage cost to describe demand classes that requested 90% fulfillment rate and 95% fulfillment rate in the research of U.S. military spare part logistics system. Other criteria to separate demand class such as Elliot [5], when he developed an inventory rationing strategy for a company that possesses both traditional physical distribution and internet distribution channel, he distinguished different classes by assuming the company would choose to satisfy the physical distribution channel first.

The assumption of inventory demand distribution is another key point in inventory management research. Assuming inventory demand subject to Poisson distribution, Nahimas and Demmy [6] discussed the fulfillment rate of two demand classes for given rationing and replenishment levels in a continuous review (Q, r) system. Other examples of Poisson demand assumption include Ha [7]'s research of a single product, make to stock inventory system, and Erhan and Mohit [8]'s research of a time-based service target, two-echelon service part distribution system. Anteneh et al. [9] assumed inventory demand to be triangle distributed, then discussed an internet retailer inventory system under such assumption. The sensitivity analysis they did also indicated best inventory threshold values and optimal profits in different cases.

The interest of this paper stemmed from a company's inventory system. This company holds inventory to provide materials for both the railway system (prior demand) and to the open market (secondary demand). The railway system is a major client for the company. As a result, fails to supply would cost larger shortage cost that includes penalty, loss of creditability and profit, even social and political effects. While the shortage of secondary inventory demand contents much less penalty and creditability loss, also loss of profit. Hence, inventory rationing by different demand class should be a good way for the company to cope with its inventory management problem.

2 Basic Model

2.1 Model Assumption

Following most research on inventory rationing, this paper also adopts shortage cost as the criteria to separate different demand classes. The shortage cost per time unit of prior and secondary demand classes are denoted by c_1 and c_2 respectively, therefore, the weighted shortage cost per time is denoted by $c = (R_1c_1 + R_2c_2)/(R_1 + R_2)$. The prior demand comes from railway system is relatively stable, so we assume this demand is linear distributed with demand rate R_1 per time unit. Compare to Poisson demand assumption, linear demand is more reasonable when meeting with more stable demand such as supplying for projects with well planned time schedule, or long term contract supply. The company has both long term major clients with stable demand and temporary clients that create demand much more like Poisson distributed in the open market. In order to simplify our discussion, we also assume secondary demand is linear distributed with demand rate R_2 per time unit. Therefore, total demand rate per time unit is $R = R_1 + R_2$. Other parameters are explained as follow:

s: Reorder point;
h: Inventory holding cost per time unit;
c_3: Fixed setup cost;
Q: Order quantity;
L: Lead time;
K: Threshold, only prior demand would be served once inventory level falls below K, secondary demand would be backordered.

2.2 Inventory Model Without Rationing

Suppose the company doesn't differentiate the two demand classes, then the system issues stocks to both classes equally, despite the inventory level (Fig. 1). This inventory system performs as Fig. 2.

The inventory holding cost and backorder cost are:

$$H_1 = \frac{1}{2}h\frac{Q^2}{R} \tag{1}$$

$$B_1 = \frac{1}{2}cR(t - \frac{Q}{R})^2 \tag{2}$$

The inventory cost per time unit of this inventory system is:

$$C_1 = \frac{H_1 + B_1 + c_3}{t} \tag{3}$$

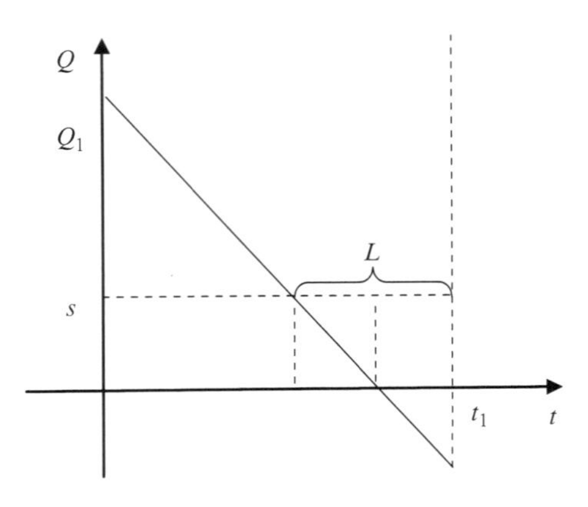

Fig. 1 Inventory system without rationing

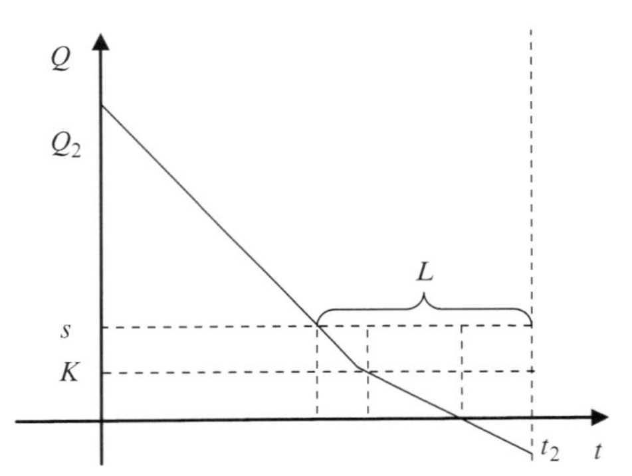

Fig. 2 Inventory system with rationing policy

The first order differential conditions to minimize C_1 with respect to Q and t are:

$$\frac{\partial C_1}{\partial Q} = \frac{1}{t}[h\frac{Q}{R} - c(t - \frac{Q}{R})] = 0 \tag{4}$$

$$\frac{\partial C_1}{\partial t} = -\frac{1}{t^2}[h\frac{Q}{R} - c(t - \frac{Q}{R})] + \frac{1}{t}cR(t - \frac{Q}{R}) = 0 \tag{5}$$

Because $t \neq 0$, we solve Eqs. (4) and (5), the optimal order quantity and single inventory circle time

$$t_1{}^* = \sqrt{\frac{hQ^2 + cQ^2 + 2c_3R}{cR^2}} \tag{6}$$

$$Q_1{}^* = \sqrt{\frac{2cc_3R}{h(h+c)}} \tag{7}$$

Note, by using lead time L, Eqs. (6) and (7), we can easily calculate two parameters that much more practice, reorder point and reorder time.

3 Rationing Model

Our rationing model separates two demand classes by their different shortage costs. The rationing policy functions like this: when the inventory level is above threshold K, stocks would be issued for both demand classes; once the inventory level drops below threshold K, stocks would be issued just for prior demand class, requests come from secondary demand class would not be responded until replenishment. The rationing inventory system works as Fig. 2.

The inventory holding cost and backorder cost of rationing model are:

$$H_2 = \frac{1}{2}h\frac{(Q-K)^2}{R} + \frac{1}{2}hK[\frac{2(Q-K)}{R} + \frac{K}{R_1}] \tag{8}$$

$$B_2 = \frac{1}{2}c_1R_1(t - \frac{Q-K}{R} - \frac{K}{R_1})^2 + \frac{1}{2}c_2R_2(t - \frac{Q-K}{R})^2 \tag{9}$$

The inventory cost per time unit of this rationing inventory system is:

$$C_2 = \frac{H_2 + B_2 + c_3}{t} \tag{10}$$

The first order differential conditions to minimize C_2 with respect to Q is:

$$\frac{\partial C_2}{\partial Q} = \frac{1}{Rt}\{[h(Q-K) + hK)] - [c_1R_1(t - \frac{Q-K}{R} - \frac{K}{R_1}) + c_2R_2(t - \frac{Q-K}{R})]\} = 0 \tag{11}$$

Because $t \neq 0$, we can get an equation of t and Q:

$$t = \left[\frac{hR + c_1R_1 + c_2R_2}{R(c_1R_1 + c_2R_2)}\right]Q - \frac{c_1R_1K + c_2R_2K - c_1RK}{R(c_1R_1 + c_2R_2)} \tag{12}$$

The first order differential conditions for C_2 with respect to t is not very easy to get. Even if we got it, the accurate value for the three variables, Q, K, t is still unknown because there would be only to equations. Here, we use a heuristic approach to solve this problem.

Note that when the inventory level falls below threshold K, demand comes from secondary demand classes was backordered in order to be sure more demands come from prior demand class would be satisfied, this inventory system is expected to have a lower backorder cost. Thus, it is reasonable to conclude that $Q_2 \leq Q_1^*$, as the rationing system doesn't need to prepare as much inventory as Q_1^* to lower the risk of backorder because of threshold K.

Based on this conclusion, we set up an approach as follow:

Step 1: let $Q_2 = Q_2^{'} = Q_1{}^*$, $K = 0$, we get $C_2 = C_1$ as the initial value;

Step 2: while $K^{'} = 0, 1, 2 \cdots Q_2^{'}$, calculate $t_2^{'}$ and $C_2^{'}$, if $C_2^{'} < C_2$, then $C_2 = C_2^{'}$, $Q_2 = Q_2^{'}$, $K = K^{'}$, $t_2 = t_2^{'}$;

Step 3: let $Q_2^{'} = Q_2^{'} - 1$, if $Q_2^{'} > 0$, repeat step 2, else, stop the approach.

Now, we let c_1 =50, c_2 =10, R_1 =30, R_2 =20, h =5, L =1.5 as a numerical example. By our approach, results for the basic model are minimum inventory cost C_1 =660.23 (correct to two decimal), optimal order quantity $Q_1{}^*$ =132 (correct to the nearest integer), optimal inventory cycle time $t_1{}^*$ =3.03 (correct to two decimal), reorder point s_1 =6 (correct to the nearest integer); results for the rationing model are C_2 =637.23 (correct to two decimal), optimal order quantity $Q_2{}^*$ =127 (correct to the nearest integer), optimal inventory cycle time $t_2{}^*$ =4.27 (correct to two decimal), reorder point s_2= 2 (correct to the nearest integer), inventory threshold K= 23.

4 Sensitivity Analysis

In order to further discuss the inventory model we built, we performed a sensitivity analysis for the key parameter, thresholds K, and a critical indicator, inventory cost saving rate r ($r = \frac{C1-C2}{C1} \times 100\%$). One of the major focuses of this paper is how to set up the inventory threshold K, thus, the sensitivity analysis for K are showed in Figs. 3, 4, 5 and 6 below (parameters do not show in the figures are the same as previous section).

From Fig. 3, inventory threshold K increase as the increase of prior demand class shortage cost per time unit c_1, which is also very reasonable that if the shortage cost of prior demand increase, more inventory should be kept to ensure the supply. Figure 4 shows that, when fix setup cost c_3 increase, the threshold K increase correspondently. We believe the reason for K to increase is that setup cost extended the inventory period, therefore, more inventory should be kept for the prior demand class. In Fig. 5, when inventory holding cost h is relatively small, K increase as h increase. This is because the inventory threshold functions as that the additional inventory holding cost is less than the reduced shortage cost, when h increases, more shortage cost should be reduced in order to "offset" the increase of holding cost. However, when h is relatively big, K decrease as h increase, the reason is that optimal order quantity Q gets smaller with h increase, which then "compress" K. In Fig. 6, K increase as prior class demand rate per time unit R_1 increase, indicating a strong positive correlation between these two parameters.

Another focus in this section is inventory cost saving rate for the rationing model we set up, which is also a major motivation for companies to adopt new inventory management policy. Sensitivity analysis for r are showing as follows (parameters do not show in the figures are the same as previous section).

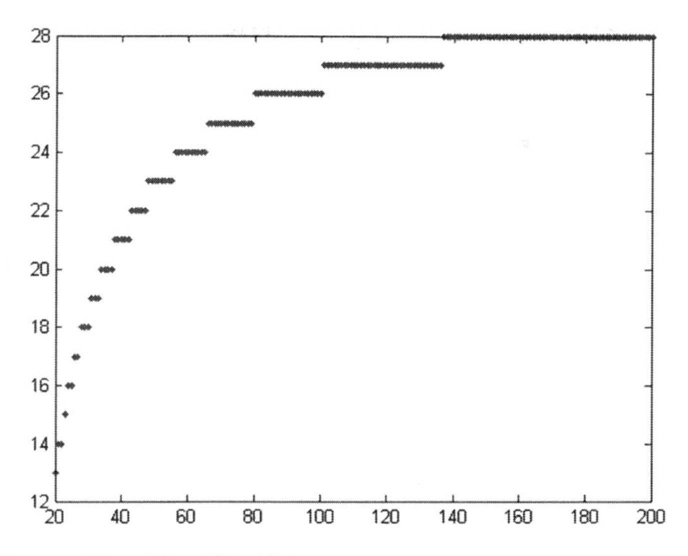

Fig. 3 C_1 increased from 20 to 200 with 1 per step

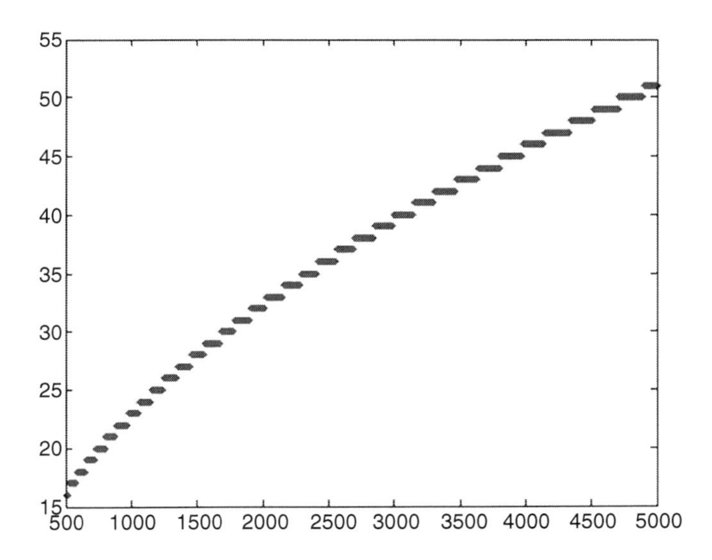

Fig. 4 C_3 increased from 500 to 5,000 with 10 per step

From Fig. 7, the inventory cost saving rate r increase as prior demand shortage cost increase, however, after $r > 6\%$, the increasing rate becomes very slowly. We further point out that as c_1 becomes infinitely great (which is similar to not shortage is allowed for prior demand), $r \approx 7\%$, indicating that the maximum inventory cost

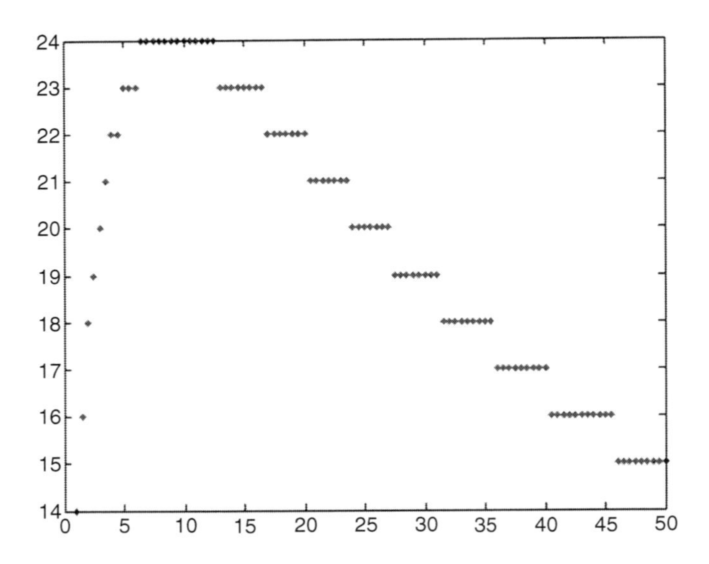

Fig. 5 h increased from 1 to 50 with 0.5 per step

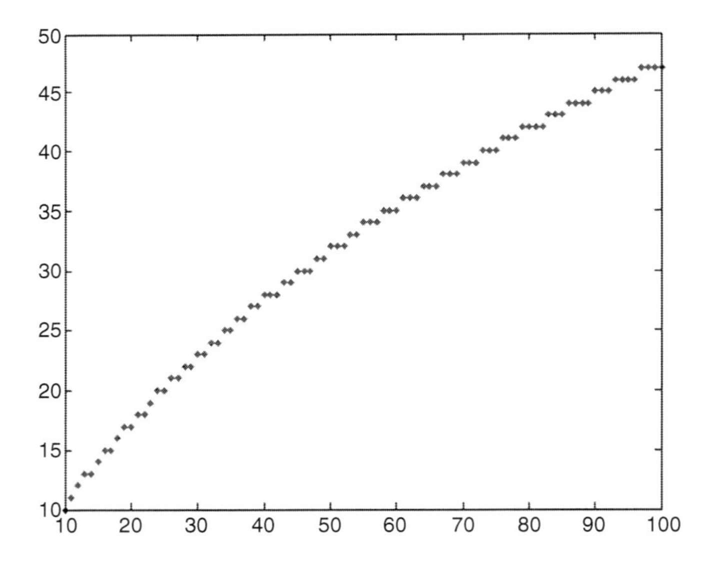

Fig. 6 R_1 increased from 10 to 100 with 1 per step

saving rate is 7%. The relation between inventory saving rate and inventory holding cost per time unit is showed in Fig. 8, the highest saving rate is obtained when inventory holding cost per time unit is 20.

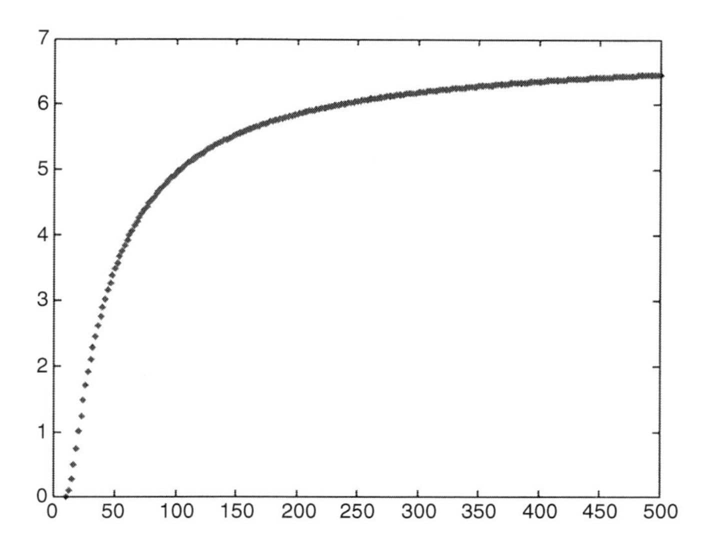

Fig. 7 C_1 increased from 10 to 500 with 2 per step

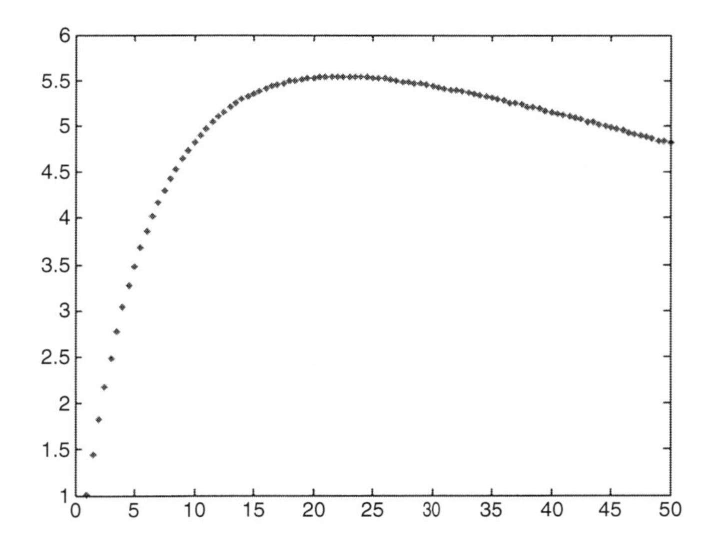

Fig. 8 h increased from 1 to 50 with 0.1 per step

5 Conclusions and Extensions

In this paper, we discussed an inventory rationing system with two demand classes in a relatively stable demand rate environment. The two demand classes were distinguished by different shortage cost. The rationing policy operates under the following rules. Demands come from both classes are filled when the inventory level is above threshold K, once the inventory level drops below K, all remaining stocks would be reserved for prior demand class, secondary demand is backordered. Heuristic approach used in this paper gave the optimal order quantity, single inventory circle time and threshold. Results showed that compare to inventory system without rationing, our rationing system with threshold is fairly cost efficient. Besides, the approach we designed for the model is easy to be utilized and solved. Sensitivity analysis performed in this paper provided us a better understanding of our two major interests, inventory threshold K and inventory saving rate r.

One shortcoming of our model is that we assume the secondary demand rate to be fixed for simplification. In reality, this demand rate is more like a random number, which would be better described by random distribution such as Poisson distribution. Our target function could also be extended to calculate profit rather than just inventory cost, which is the price of stocks multiplies quantity, then minus total inventory cost.

Acknowledgments This research was supported by the National Natural Science Foundation of China #71140007, the Fundamental Research Funds for the Central Universities # B11JB00410 and the key project of logistics management and technology lab.

References

1. Kimes SE (1989) Yield management: a tool for capacity constrained service firms. J Oper Manag 8:348–363
2. Veinott AE (1965) Optimal policy in a dynamic single product non-stationary inventory model with several demand classes. Oper Res 13:761–778
3. Topkis DM (1968) Optimal ordering and rationing policies in a non-stationary dynamic inventory model with n demand classes. Manag Sci 15(3):160–176
4. Deshpande V, Cohen MA, Donohue K (2003) A threshold inventory rationing policy for service-differentia demand classes. Manag Sci 49(6):683–703
5. Bendoly E (2004) Integrated inventory pooling for rms servicing both on-line and store demand. Comput Oper Res 32:1465–1480
6. Nahmias S, Demmy WS (1981) Operating characteristics of an inventory system with rationing. Manag Sci 27(11):1236–1245
7. Ha AY (1997) Stock rationing policy for a make-to-stock production system with two priority classes and backordering. Nav Res Logist 44:458–472
8. Kutanoglu E, Mahajan M (2009) An inventory sharing and allocation method for a multi-location service parts logistics network with time-based service levels. Eur J Oper Res 194:728–742
9. Ayanso A, Diaby M, Nair SK (2006) Inventory rationing via drop-shipping in Internet retailing: a sensitivity analysis. Eur J Oper Res 171:135–152

Absolute Stability Decision Models of Supplier Selection Problem Under Uncertain Demand

Jianhua Ma and Yong Fang

Abstract Supplier selection is one key problem of logistics and supply chain management. Uncertainty of demand is the main challenges to the supplier selection. In this paper the stability of supplier selection was considered. Firstly the optimization model of supplier selection was given, then the absolute stability decision model of supplier selection was put forward, and finally an example was given.

Keywords Supplier selection problem • Stability decision • Robust optimization • Optimal model

1 Introduction

Supplier selection is one key problem of logistics and supply chain management. The main research methods of supplier selection problem are comprehensive evaluation, cost analysis, mathematical programming [1–3].

The uncertainty of demand is the main challenges of supplier selection decision, general processing method is based on the average demand forecast. But it is difficult to give a precise prediction, the longer the time the greater of the prediction error. Sometime it can only give the range of demand, so need to study the supplier selection problem with the uncertainty demand.

J. Ma (✉)
School of Management Science and Engineering, Shandong University of Finance and Economics, 250014 Jinan, People's Republic of China
e-mail: jianhuama@126.com

Y. Fang
Academy of Mathematics and Systems Science, Chinese Academy of Sciences, 100190 Beijing, People's Republic of China

Z. Zhang et al. (eds.), *LISS 2012: Proceedings of 2nd International Conference on Logistics, Informatics and Service Science*, DOI 10.1007/978-3-642-32054-5_14,

The supplier selection problems with fuzzy demand and stochastic demand have been studied [4–6]. Fuzzy demand and stochastic demand need to know the probability distribution function or the fuzzy membership function. But many cases it is difficult to get them. Thus supplier selection problem with completely uncertain demand should be considered.

The range programming [7] and robust optimization method [8] are main methods to solve completely uncertain decision-making problem. These methods required decision to meet all the parameters of the possible changes in the constraints, or make a choice based on regret value of worst-case. To extreme cases the decision-making optimality will be relatively poor. We have given stability decision [9] to overcome the shortcomings. In this paper we will solve supplier selection problem under completely uncertain demand by stability decision method.

2 The Supplier Selection Problem

Consider one enterprise selects raw material suppliers from m candidate enterprises. Each enterprise's maximum production capacity is $s_i^+, i = 1, 2, \ldots, m$, the minimum contract order quantity is $s_i^-, i = 1, 2, \ldots, m$, the unit transportation cost is $c_i, i = 1, 2, \ldots, m$, and demand of the enterprise is d.

The variable $x_i, i = 1, 2, \ldots, m$ indicates whether the ith candidate enterprise is selected, $x_i = 1$ if it is selected, otherwise $x_i = 0$. The variable $y_i, i = 1, 2, \ldots, m$ is contract order quantity of the ith candidate enterprise. If $x_l = 0$ then $y_i = 0, i = 1, 2, \ldots, m$. If the demand is certain, there are constraints:

$$\begin{gathered} s_i^- x_i \le y_i \le s_i^+ x_i, i = 1, 2, \ldots, m \\ \sum_{i=1}^{m} y_i = d \end{gathered} \tag{1}$$

The total transportation cost is $\sum_{i=1}^{m} c_i y_i$, its programming model is:

$$\begin{aligned} &\min \sum_{i=1}^{m} c_i y_i \\ &s.t. \begin{cases} s_i^- x_i \le y_i \le s_i^+ x_i, i = 1, 2, \ldots, m \\ \sum_{i=1}^{m} y_i = d \\ x_i = 0, 1, y_i \ge 0, i = 1, 2, \ldots, m \end{cases} \end{aligned} \tag{2}$$

If demand is uncertain, the contract order quantity may be not equal to the actual demand. If $\sum_{i=1}^{m} y_i < d$, the enterprise needs to buy raw material from the market, the cost higher than the supplier is r_1. If $\sum_{i=1}^{m} y_i > d$, the excess part should be stored, increasing storage costs is r_2.

The additional expenditure cost is determined by the demand, let z^+ is the part that the demand exceeds the contract order and z^- is the part that the demand is lower than the contract order. Apparently z^+ and z^- can not be greater than 0, so there have:

$$\begin{aligned} &\sum_{i=1}^{m} y_i + z^+ - z^- = d \\ &z^+ z^- = 0 \end{aligned} \tag{3}$$

The additional expenditure cost is $r_1 y + r_2 z$. According to the principles of goal programming, when the additional expenditure cost is the minimum, z^+ and z^- should not simultaneously be greater than zero, so the programming is:

$$\min \sum_{i=1}^{m} c_i y_i + r_1 z^+ + r_2 z^-$$
$$s.t. \begin{cases} s_i^- x_i \le y_i \le s_i^+ x_i^-, i = 1, 2, \ldots, m \\ \sum_{i=1}^{m} y_i + z^+ - z^- = d \\ x_i = 0, 1, y_i \ge 0, z^+, z^- \ge 0, i = 1, 2, \ldots, m \end{cases} \tag{4}$$

The programming is a parameter programming, and demand d is its parameters.

3 Absolute Stability Decision Model of Supplier Selection Problem

If demand is uncertain, the supplier selection model (10) is a parameter programming. Its optimal solutions under different demand are different. If the demand is a continuous parameter, it is to calculate the optimal solution for all parameters. Here assumed the demand is a discrete parameter, its value may be $d_1, d_2, \ldots, d_n$. For each demand value, the programming (10)'s optimal value is $v_1, v_2, \ldots, v_n$.

For a given decision (x, y, z), it is stable to demand d_l if its objective function value is less than or equal to α times the optimal value under demand d_l. And demand d_l is called as its stability parameter. Absolute stability decision is a solution which stable range or the number of stable parameters is maximum [9].

In order to establish the mathematical model of absolute stability decision, it should identify the stability parameters of every solution. Lets the variable $u_l, l = 1,2,\ldots,n$ indicates whether the parameters d_l is stability parameter, if it is $u_l = 1$, else $u_l = 0$. So the total number of stability parameters is $\sum_{i=1}^{n} u_i$, and there:

$$u_l = \begin{cases} 1, if \sum_{i=1}^{m} c_i y_i + r_1 z_l^+ + r_2 z_l^- \leq \alpha v_l \\ 0, else \end{cases} \quad l = 1,2,\ldots,n \tag{5}$$

It is equivalent to:

$$\sum_{i=1}^{m} c_i y_i + r_1 z_l^+ + r_2 z_l^- - \alpha v_l\text{-}M(1 - u_l) \leq 0, l = 1,2,\ldots,n \tag{6}$$

Where M is a large enough positive number. The absolute stability model of supplier selection is:

$$\max \sum_{l=1}^{n} u_l$$
$$s.t. \begin{cases} s_i^- x_i \leq y_i \leq s_i^+ x_i^-, i = 1,2,\ldots,m \\ \sum_{i=1}^{m} y_i + z_l^+ - z_l^- = d_l, l = 1,2,\ldots,n \\ \sum_{i=1}^{m} c_i y_i + r_1 z_l^+ + r_2 z_l^- - \alpha v_l\text{-}M(1 - u_l) \leq 0, l = 1,2,\ldots,n \\ x_i = 0,1, y_i \geq 0, i = 1,2,\ldots,m, z^+, z^- \geq 0 \end{cases} \tag{7}$$

This programming is an integer linear programming, it can be solved by branch and bound algorithm. α is its parameter, with its increase the number of stability parameters will increase.

4 Example

Consider there have eight candidate enterprises, the maximum production capacity, minimum contract order quantity and unit transportation costs of each candidate enterprise are shown as Table 1.

Table 1 Production capacity of enterprises

Enterprise	1	2	3	4	5	6	7	8
Maximum production capacity	4	5	4	3	6	5	6	4
Minimum contract order quantity	1.5	2	1.5	1	2	2	2	1.5
unit transportation cost	1.2	2.1	1.6	1.8	1.4	2.5	2.2	1.3

Table 2 Optimal value of demand

Demand	8	9	10	11	12	13	14	15
Optimal values	10	11.5	12.8	14.2	15.6	17	18.4	20.15

$r_1 = 1.75$, $r_2 = 1.2$, and its the optimization model is:

$$\min 1.2y_1 + 2.1y_2 + 1.6y_3 + 1.8y_4 + 1.4y_5 + 2.5y_6 + 2.2y_7 + 1.3y_8 + 1.75z^+ + 1.2z^-$$
$$s.t.\begin{cases} y_1 + y_2 + y_3 + y_4 + y_5 + y_6 + y_7 + y_8 + z^+ - z^- = d \\ y_1 - 4x_1 \le 0, y_1 - 1.5x_1 \ge 0, y_2 - 5x_2 \le 0, y_2 - 2x_2 \ge 0 \\ y_3 - 4x_3 \le 0, y_3 - 1.5x_3 \ge 0, y_4 - 3x_4 \le 0, y_4 - x_4 \ge 0 \\ y_5 - 6x_5 \le 0, y_5 - 2x_5 \ge 0, y_6 - 5x_6 \le 0, y_6 - 2x_6 \ge 0 \\ y_7 - 6x_7 \le 0, y_7 - 2x_7 \ge 0, y_8 - 4x_8 \le 0, y_8 - 1.5x_8 \ge 0 \\ x_i = 0, 1, y_i, z^+, z^- \ge 0, i = 1, 2, \ldots, 8 \end{cases} \quad (8)$$

The demand d is its parameter, its value may be integer from 8 to 15. For all demand values, the programming's optimal values are shown as Table 2.

Lets $\alpha = 1.056$ and $M = 30$, the absolutely stable model is:

$$\max \sum_{l=1}^{8} u_l$$
$$s.t.\begin{cases} y_1 + y_2 + y_3 + y_4 + y_5 + y_6 + y_7 + y_8 + z_l^+ - z_l^- = d_l, l = 1, 2, \ldots, 8 \\ y_1 - 4x_1 \le 0, y_1 - 1.5x_1 \ge 0, y_2 - 5x_2 \le 0, y_2 - 2x_2 \ge 0 \\ y_3 - 4x_3 \le 0, y_3 - 1.5x_3 \ge 0, y_4 - 3x_4 \le 0, y_4 - x_4 \ge 0 \\ y_5 - 6x_5 \le 0, y_5 - 2x_5 \ge 0, y_6 - 5x_6 \le 0, y_6 - 2x_6 \ge 0 \\ y_7 - 6x_7 \le 0, y_7 - 2x_7 \ge 0, y_8 - 4x_8 \le 0, y_8 - 1.5x_8 \ge 0 \\ 1.2y_1 + 2.1y_2 + 1.6y_3 + 1.8y_4 + 1.4y_5 + 2.5y_6 + 2.2y_7 + 1.3y_8 \\ \quad + 1.75z_l^+ + 1.2z_l^- - 1.056v_l\text{-}30(1 - u_l) \le 0, l = 1, 2, \ldots, 8 \\ x_i = 0, 1, y_i, z_l^+, z_l^- \ge 0, i = 1, 2, \ldots, 8, l = 1, 2, \ldots, 8 \end{cases} \quad (9)$$

The optimal solution of the program is to select Enterprise 1, 5 and 8 for suppliers, and it is stable on 7 cases of all potential demands.

5 Conclusion

The absolute stability decision model of supplier selection problem with uncertain demand was given. The model strikes a balance between optimality and stability of decision. It can be widely used in supply chain design and logistics management. The parameters of these stability decision models are discrete parameters. For continuous parameter, it can be solved by discretization of continuous demand.

References

1. Lin Yong, Ma Shi-hua (2000) Research on supplier comprehensive evaluation in supply chain management. Log Technol 5:30–32
2. Liu Xiao, Li Hai-yue, Wang Cheng-en, Chu Cheng-bin (2004) A survey of supplier selection models and approaches. Chin J Manag Sci 1(12):139–148
3. Chen Qijie, Qi Fei (2009) A survey of supplier selection, Foreign Economics & Management 31(5):30–37
4. Wang TY, Yang YH (2009) A fuzzy model for supplier selection in quantity discount environments. Expert Syst Appl 36(6):12179–12187
5. Awasthi A et al (2009) Supplier selection problem for a single manufacturing unit under stochastic demand. International J Prod Econ 117(1):229–233
6. Li SL et al (2009) Selection of contract suppliers under price and demand uncertainty in a dynamic market. Eur J Oper Res 198(3):830–847
7. Guo Zixue, Jiang Hongxin, Qi Meiran (2011) Method for supplier selection based on interval-valued intuitionist fuzzy sets. Comput Eng Appl 47(33):216–218
8. Ben-Tal A, Nemirovski A (2002) Robust optimization: methodology and applications. Math Progr 92(3):453–480
9. Jianhua Ma (2011) Models and application of stability decision under uncertain environment research. J Appl Sci Eng Tech 3(9):986–992

On Optimization of Logistics Operation Models for Warehouse & Port Enterprises

Junyang Li and Xiaomin Zhu

Abstract Warehouse logistics business is a combination of warehousing services and terminal services which is provided by port logistics through the existing port infrastructure on the basis of the port. However, if there is little internal contact among the warehouses & ports in one company, then unnecessary competition will rise up. In order to deal with the competitive relationship among them, this paper carries out an optimization on warehouse & port logistics operation models based on game theory and accordingly solving the optimized model using Genetic Algorithm. We set Sinotrans Guangdong Company as an example to illustrate the newly proposed method. Finally, this paper gives some responses and suggestions for the logistics operation in Sinotrans Guangdong warehouse & port for its future development.

Keywords Warehouse & port • Logistics operation • Optimized model • Game theory • Genetic algorithm

1 Introduction

Warehouse logistics business is a combination of warehousing services and terminal services which is provided by port logistics through the existing port infrastructure on the basis of the port. Modern logistics refers to the whole process of raw materials, finished goods from start to finish and the efficient flow of information. It is a combination of the transportation, storage, handling, processing, finishing, distribution, information, and other aspects. It forms a complete supply chain which can provide multi-functional, integrated and comprehensive services to the users [1].

J. Li (✉) • X. Zhu
School of Mechanical, Electronic and Control Engineering, Beijing Jiaotong University, Shangyuancun 3, 100044 Beijing, People's Republic of China
e-mail: 11121401@bjtu.edu.cn

Z. Zhang et al. (eds.), *LISS 2012: Proceedings of 2nd International Conference on Logistics, Informatics and Service Science*, DOI 10.1007/978-3-642-32054-5_15,

Modern logistics, as an advanced organization and technology on efficient planning, management and delivery during the period of the product from production to consumption, has become "the third profit source" for companies and society [2].

At present, there are many studies on port logistics operation models.

(a) Port planning theory. Xin Zhang [3] analyses the conflict between the port and the ship in the development process of the port.
(b) Port transportation network optimization decisions. Chuanxu Wang and Liangkui Jiang [4] optimize the inland transportation network on the regional port.
(c) Modeling and simulation of logistics systems. Wei Zhao, Chunlin Tian, Zijian Bai's [5] modeling and simulation of real container reflects the high performance of the professional software WITNESS in terminal logistics operations.

The emphases of these algorithms are different. Therefore they can solve different kind of problems. This paper aims at carrying out an optimization on warehouse & port logistics operation models based on game theory and accordingly solving the optimized model using Genetic Algorithm.

2 Optimization Model of Warehouse & Port Logistics Operation Based on Game Theory

2.1 Assumptions

We make the following assumptions to make the problem solvable. There're little difference between these assumptions and the actual situation. So these assumptions are reasonable.

(a) Regard the selection and use of the catchment, the transit center, warehouse & port and the target port as a unity.
(b) The layout of all the catchment, the transit center, warehouse & port and the target port have been identified, regardless of location problems.
(c) The transportation of containers from the catchment to the target port is road transport, rail transport and sea transport, regardless of other modes of transportation.
(d) Taking the two-way operations into account, assume that the container can be transported two-way between the catchment and the target port.
(e) Assume that the traffic of road transport, rail transport and water transport is within the designed capacity in the selected route. There are no security problem in the process of transport and storage and the goods can be delivered in time.

2.2 Network Model of Container Transportation

According to different conditions, it can be transported in land route such as highways and railways, or it can be transported via branch route from Spoke port

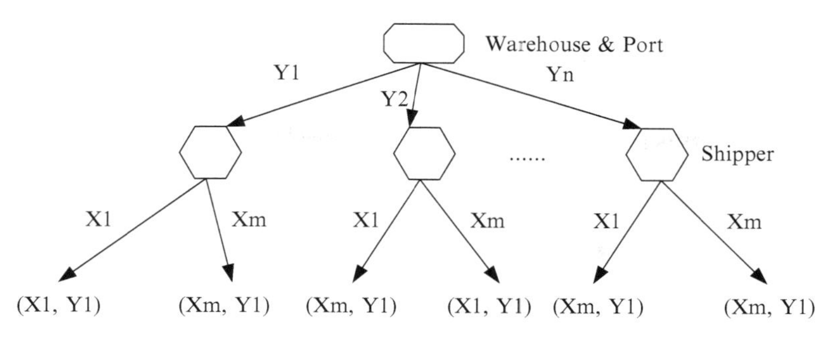

Fig. 1 Game tree

S1 to Hub port H1, then main route from Hub port H1 to Hub port H3, and the hinterland transport the same as Hub port H1. In this transport mode, the choice of transport mode and transport path is not just to select the warehouse & port, but the choice of the entire transport chain [6].

2.3 *Game Theory Model*

According to the assumptions above, we can separately build the network model of the owner and the carrier. This problem can be regarded as a complete information dynamic non-cooperative game problem, considering the carrier's route set and the owner's volume choice. Carriers and owners' game process is as follows: carriers carry out a variety of transportation services program, i.e., arrange transportation routes and frequency of services in their own transportation network and obtain optimal strategy by estimating the owner's reaction to various transport programs.

The dynamic game can be described as follows with expansion-type: (a) Collection of participants: two interest groups – companies and owners; (b) Sequence of actions for participants: the companies have decision-making priority; owners make their decisions after that; (c) Action space for participants: companies – transport path and frequency of services(Y); owners – transport volume(X); (d) Payment functions for participants: two sub-models above. Figure 1 is a strategy-type game tree.

3 The Genetic Algorithm Modeling and Analysis for the Optimization Model

Section 2 described a complete information dynamic Stackelgerg problem [7]. Its decision-making mechanism is that: the company has decision-making priority. The company selects a route program to their advantage. According to the decision, the company selects a proper transport volume to optimize the objective function which in turn affects the company's decision-making. Therefore, this problem is a joint decision-making problem by two decision-makers who are in grade levels. It is usually described as linear bi-level programming in mathematical programming [8].

Simplify the linear bi-level Stackelberg programming problem into a single-stage problem with penalty function using the theory of penalty function. And carry out a global optimal algorithm using genetic algorithms [9]. After Calculation and simplifying, the fitness function of GA is as follows:

$$\begin{aligned} f_1 = 1/[&\sum_r \sum_i \sum_{j \neq i} \sum_{s \neq r} (c_{ri} + c_{js}) X_{rijs} + \sum_r \sum_i \sum_{j \neq i} \sum_{s \neq r} P_{rs} i_0 d y_{ij}] \\ &+ \sum_i \sum_j f_{ij} \sum_r \sum_s X_{rijs} \\ &+ (ct)^2 \sum_i \sum_j \min\{0, 1/(\sum_i fr^t_{ij} v l^t Y^t_{ij} - \sum_r \sum_s X_{rijs})\} \\ &- (ct)^2 \sum_r \sum_s (\sum_i \sum_j X_{rijs} - OD_{rs})^2 \\ f_2 = Z_1 + &(ct)^2 (\phi_1 + \phi_2) \end{aligned} \tag{1}$$

4 Case Study

Sinotrans Guangdong Company has eight self-owned container terminal at Huangpu, Dongjiang, Jiaoxin, Zhongshan, Jiangmen, Foshan, Zhanjiang and Guangxi Wuzhou in Guangzhou. Among them, Jiaoxin and Guangxi Wuzhou are newly added warehouse & ports. In general, eight ports are operating independently currently. All ports are lack of fluid exchanging of information and it is unknown which port is most suitable for a certain kind of business. This situation leads to the cost increasing inevitably. The operation mode of Sinotrans Guangdong Company needs to be adjusted in order to decrease the cost.

After optimizing the logistics operation mode in the warehouse & port of Sinotrans Guangdong Company according to the process proposed in this paper, we can find that from the 895th generation, the result has already reached the stop condition of GA. According to the result, flow volume of goods and service frequency for each warehouse & port can be obtained. Choose two maximum values of the volume and frequency, and then we can obtain the logistics map, shown in Fig. 2. Huangpu and Foshan have the potential to be the hub port when measured by throughput, route number, freight per month and other indicators. The company should theoretically develop Huangpu and Foshan.

According to the results and the current economic situation, we put forward the following recommendations to Sinotrans Guangdong Company warehouse & port

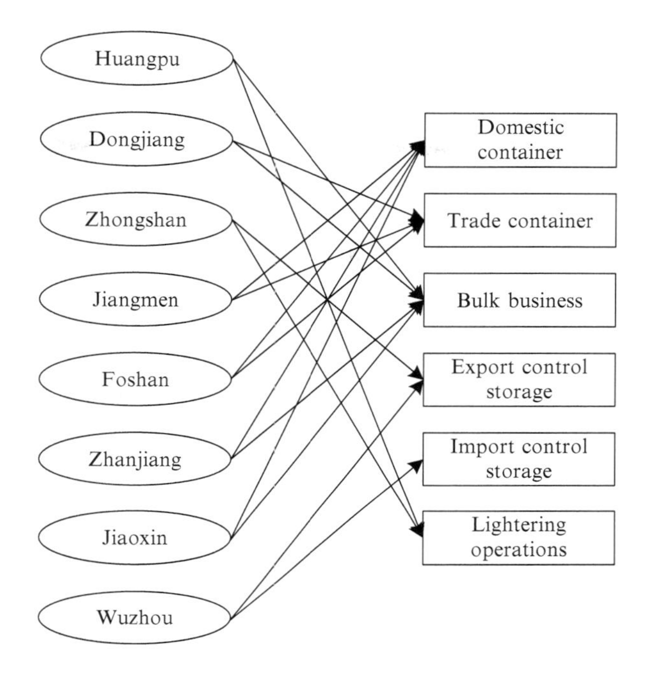

Fig. 2 Logistics business of warehouse & port

logistics business market positioning for future development: (a) Develop the expertise advantages respectively; (b) Integrate the resources among the warehouses & ports; (c) Explore the hinterlands by water-rail transportation; (d) Improve the level of information.

5 Conclusions

This paper simulates the decision-making process of the company and the owner by network optimization and game theory based on the inherent economic laws in water economy and establishes a network optimization model based on game theory. GA is used to solve the combinatorial optimization problems. Then the suggestion that the warehouse & port should make adjustment is raised according to the result. Some other advises are also made to the further development of the company.

After the study, the following conclusions can be made:

(a) The inherent economic laws in water economy are that must be obeyed for warehouse & port logistics business operation;
(b) The simulation of the decision-making process using game theory in economics totally matches the real container transport process.
(c) Genetic algorithms can solve complex combinatorial optimization problems quickly and efficiently, especially for such a large-scale optimization problem in this paper which contains hundreds of variables.

References

1. Gao X, Lin J, Zhang W (2005) Supply chain simulation and modeling design and implementation based on agent. Comput Eng Appl 32:183–192 (In Chinese)
2. China Federation of Logistics & Purchasing: China Logistics Development Report (2001), China Logistics Publishing House. (2002) 46–53 (In Chinese)
3. Zhang X (2008) Port berth planning study based on dynamic programming. J Guangdong Ocean University 28:45–52 (In Chinese)
4. Wang CX, Jiang LK (2008) Inland port transportation network optimization decisions based on bi-level programming. J Industrial Eng Eng Manag 4:35–42 (In Chinese)
5. Zhao W, Tian CL, Bai ZJ, Qian F (2009) Simulation and modeling of internal logistics operating systems in container terminal based on WITNESS. Commun Stand 1:35–42 (In Chinese)
6. Nijkamp P, Reggiani A, Tsang WF (2004) Comparative modeling of interregional transport flows: application stimuli modal European freight transport. Eur J Oper Res 155:584–602
7. Zhang ZL, Wang YF (2011) Land acquisition based on incomplete information dynamic game. China Land Sci 25:49–53
8. Steenbrink, P. A.(1978) Optimization of transport networks, Wiley, No. 4. 22–28
9. Zhou GH, Wang R, Jiang YP, Zhang GH (2010) Non-cooperative game model for job shop scheduling with hybrid adaptive genetic algorithm. J Xi'an Jiaotong University 44:35–39, 70

The Competitive Model and Simulation Research of Logistics Enterprises' Resource Acquisition Process

Yunlong Hou, Zhenji Zhang, Qianqin Qu, and Xiaolan Guan

Abstract In this paper, according to the theory of ecology logistic equation, we gave and analyzed the competitive model of logistics enterprises' customer resources requiring process, and did the system dynamics simulation analysis based on Anylogic software platform, and finally analyzed the enterprises' customer resources acquiring process in different initial business scale under the influence of different cooperation and competition level. The simulation model and simulation results have been great significant to the research of logistics enterprises' resources acquiring process; the realization of logistics enterprises' resource allocation and integration and the enterprises' strategic decision in reality.

Keywords Logistics enterprise • Resources acquisition • Ecological competition • Logistic equation • Anylogic • System dynamics • Simulation

1 Introduction

Logistic resources integration among enterprises is achieved through the process of resources acquisition and redistribution, therefore the study of the logistics enterprises' resources earning process has great significance for the achievement of logistics enterprises resources integration. Zhang Rui et al. [1] presented a competitive model of enterprises with lower critical based on ecological theory and gave the stable point analysis; Li Zhaolei, Wu Qunqi, Zhang Yaqi [2] established a

Y. Hou (✉) • Z. Zhang • Q. Qu
School of Economics and Management, Beijing Jiaotong University, No.3 Shang Yuan Cun, Hai Dian District, Beijing 100044, People's Republic of China
e-mail: 11125177@ bjtu.edu.cn; zhjzhang@ bjtu.edu.cn; 11125183@ bjtu.edu.cn

X. Guan
Beijing Institute of Graphic Communication, Beijing 102600, People's Republic of China
e-mail: 08113101@ bjtu.edu.cn

Z. Zhang et al. (eds.), *LISS 2012: Proceedings of 2nd International Conference on Logistics, Informatics and Service Science*, DOI 10.1007/978-3-642-32054-5_16,

model of regional logistic system, with the Logistic growth curve, and analyzed the effects of logistics resources integration to the evolution cycle of a regional logistics system; Zhang Wenbin [3] analyzed the competition and cooperation relationship and self-organization evolution process of enterprises in the supply chain based on organization theory. But there are so few studies of logistics enterprises' resource acquisition process, especially the effect of enterprises' competition and cooperation to it, and the studies based on ecological competitive theory and system dynamics simulation are fewer, so, in this paper we will do some research of logistics enterprises' resource acquisition process based on ecological competitive theory and system dynamics simulation.

2 The Resources Acquisition Process of Logistics Enterprises

Because of logistics enterprises' resource types are numerous, a lot of resources would be difficult to do quantitative analysis, while logistics enterprise customer resources is of scarcity, convenient quantitative, and is also the important scrambling for logistics enterprise resources that directly affect the enterprise's survive and development, so in this paper we will focus on the analysis of competition between logistics enterprises during their customer resources acquisition process.

In this paper we assume that the enterprise can establish a long-term relationship to customers, and they won't be taken by other enterprises. So we will only analysis the competition process among enterprises during their new customer resources acquiring process. Logistics enterprise new customer resources acquiring process is mainly realized through the competition and cooperation between enterprises, such as the reorganization merger between enterprises, cooperation agreement alliance, order outsourcing, lease custody etc. For logistics enterprise, it shall, according to the logistics management operation of technical and economic features, select different competition and cooperation strategy combined with its own logistics service ability and customer demand.

3 The Construction of Foundation Model

3.1 Hypotheses of Model

Hypothesis 1: The letter 'n' stands for the number of logistics enterprises in the same area, and which x_i ($i = 1, 2, \ldots, n$, similarly hereinafter) stands for the logistics enterprise's business ability. 'x_i' is the function of 't' which stands for time. Because enterprise business ability can also be influenced by the factors such as technology, information, equipment and trading cost, but these factors in the model may be simply regarded as the function of 't'. So the letter of 't' here has a broader meaning, that is, it also represents the influence of technology, information, equipment and other factors on enterprise's business ability [4].

Hypothesis 2: We assume that for the logistics enterprise of number i, the existing and potential demand for logistics business from customer resources is represented by N_i. While without the competition from other logistics enterprises which can provide the same service, the enterprise's business ability increasing rate will increase by the coefficient of r_i (r_i has a positive correlation with the scale and strength of the enterprise, so 'r_i' can also be regarded as the metrics of the enterprise's scale and strength), and will finally achieve to the maximum logistics demand of N_i.

Hypothesis 3: Because every logistics enterprise has its own characteristics in reality, for different characteristics and different scale logistics enterprises, they can coexist in competition and cooperation [5], in order to reflect this situation, now we use formula '$\alpha_{ij}*(N_i - x_i)$' ($i \neq j$, $j = 1, 2,\ldots, n$) to represent the influence from logistics enterprise j to logistics enterprise i during the cooperation process between them, of which 'α_{ij}' is the coefficient of influence from logistics enterprise j to enterprise i during their cooperation process; $\beta_{ij}x_j$ ($i \neq j, j = 1, 2,\ldots, n$) to represent the influence from logistics enterprise j to enterprise i during the competition process between them, of which 'β_{ij}' is the coefficient of influence from logistics enterprise j to enterprise i during their competition process ($0 < a_{ij} < 1, 0 < \beta_{ij} < 1$).

3.2 The Construction of Model

Based on the above hypotheses and the logistic equation, we get the competition model of the logistics enterprises' customer resources acquiring process, as is shown in formula (1).

$$\frac{dx_i}{dt} = r_i x_i \left(1 - \frac{x_i}{N_i} + \frac{\alpha_{ij}(N_i - x_i)}{N_i} - \frac{\beta_{ij} x_i}{N_j}\right) \tag{1}$$

Formula (1) is the general model for the competition of logistics enterprises during their customer resources acquiring process. Here we assume that there are two logistics enterprises, enterprise 1 and enterprise 2, in one region, and then we get the model of two logistics enterprises' resources acquiring competition process, as is shown in formula (2):

$$\frac{dx_1}{dt} = r_1 x_1 \left(1 - \frac{x_1}{N_1} + \frac{\alpha_{12}(N_1 - x_1)}{N_1} - \frac{\beta_{12} x_2}{N_2}\right)$$
$$\frac{dx_2}{dt} = r_2 x_2 \left(1 - \frac{x_2}{N_2} + \frac{\alpha_{21}(N_2 - x_2)}{N_2} - \frac{\beta_{21} x_1}{N_1}\right)$$

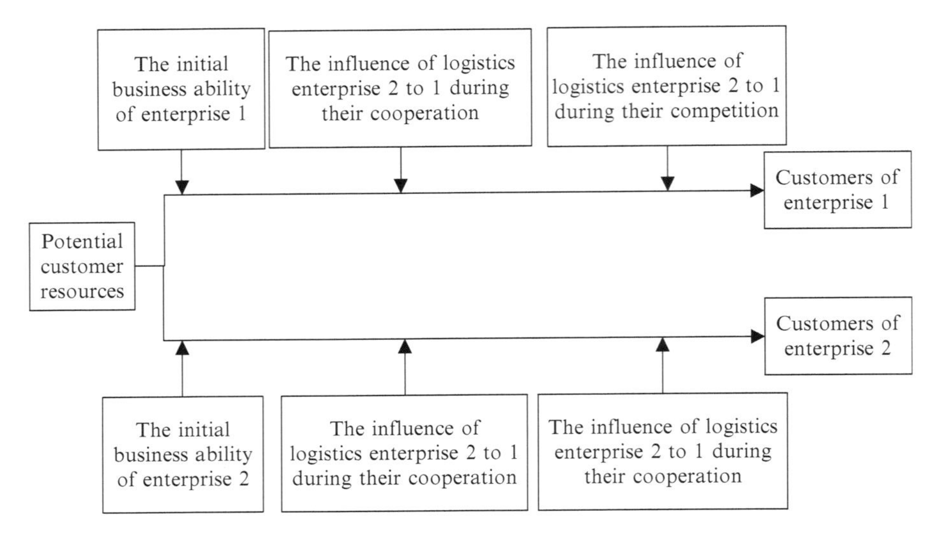

Fig. 1 The conceptual model of customer resources competition process

In formula (2), 'N_1', 'N_2' represent the maximum customer number of logistics enterprise 1 and 2, and they are equivalent to the enterprises' biggest market demand; 'r_1', 'r_2' represent the initial business ability increasing rate of logistics enterprise 1 and 2, without considering other enterprises competition and cooperation; 'a_{12}', 'a_{21}' represent the coefficient of influence between logistics enterprise 1 and enterprise 2 during their cooperation process; 'β_{12}', 'β_{21}' represent the influence between logistics enterprise 1 and enterprise 2 during their competition process.

4 Simulation Research

4.1 Conceptual Model

Here the enterprise's business ability is considered to be the major measurement for the customers' choices, and the increasing rate of existing customers is decided by the enterprise's inherent ability of business. When it exists that enterprise 1 and 2 can cooperate or compete with each other during the competition or cooperation process such as consolidation, lease custody, information sharing or facilities and sharing, it could affect the enterprise's business ability directly or indirectly, and will finally influence enterprise's customers number's changing curve, from which we get the conceptual model of competition process between logistics enterprise 1 and 2, as is shown in Fig. 1.

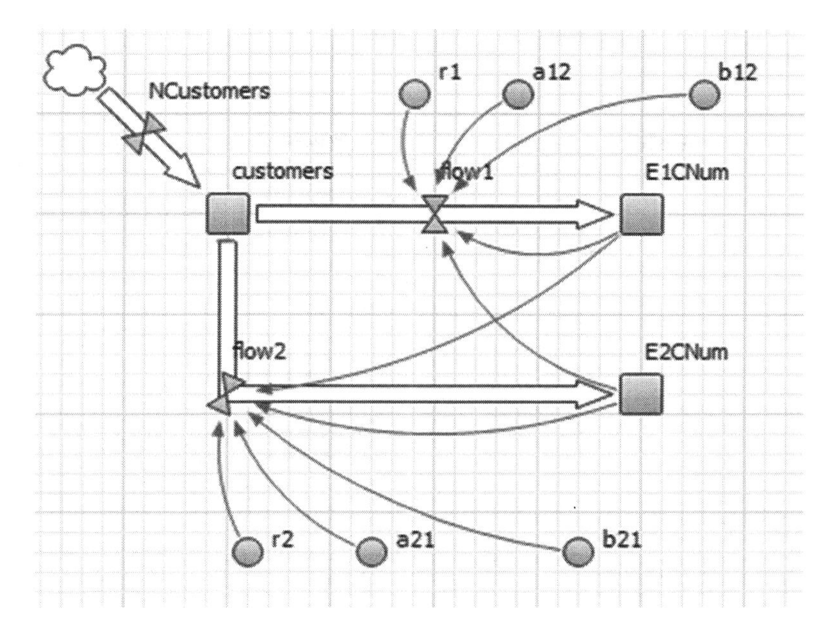

Fig. 2 Simulation model of logistics enterprises customer resources acquiring process

4.2 Computer Simulation

Now according to the foundation model and hypotheses, we will give the system dynamics model and analysis the competition between two logistics enterprise during the customer resources acquiring process based on Anylogic software platform, as is shown in Fig. 2, and the main parameters' settings are shown as Table 1.

4.3 Simulation Analysis

Hypothesis 1: We assume that $r_1 = 0.1$, $r_2 = 0.09$, $a_{12} = a_{21} = 0.1$, $b_{12} = b_{21} = 0.1$. This situation represent that there exist two logistics enterprises, and their initial business ability increasing rate are similar, and there is no obvious cooperation and competition between them, then we get the changing law of the two enterprises' business ability increasing rate and number of customers, as is shown in Figs. 3 and 4.

Table 1 Model parameters setting and interpretation

Parameter	Initial value	Interpretation
E1CNum	1	The number of customers logistics enterprise 1 earned
E2CNum	1	The number of customers logistics enterprise 2 earned
a12	0.1	The influence from logistics enterprise 2 to enterprise 1 during the cooperation process between them
a21	0.1	The influence from logistics enterprise 1 to enterprise 2 during the cooperation process between them
b12	0.1	The influence from logistics enterprise 2 to enterprise 1 during the competition process between them
b21	0.1	The influence from logistics enterprise 1 to enterprise 2 during the competition process between them
Ncustomers	0	New customer resources
Customers	1,000	Existing customer resources
r1	0.1	Initial business ability increasing rate of logistics enterprise 1
r2	0.1	Initial business ability increasing rate of logistics enterprise 2

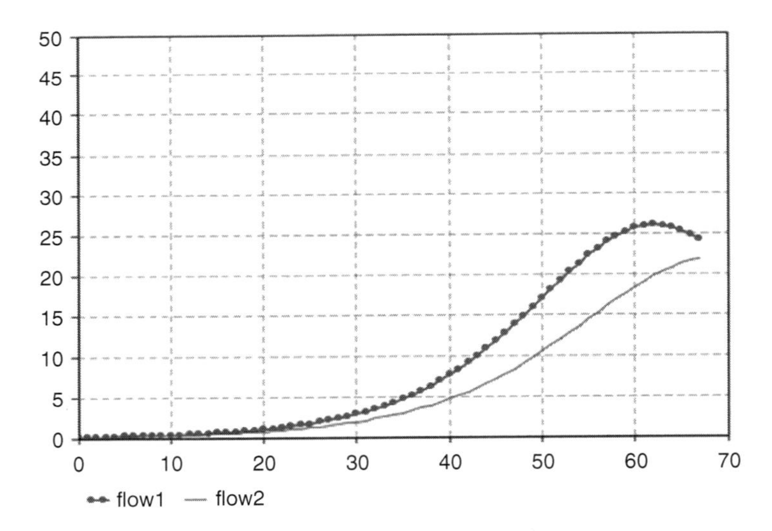

Fig. 3 Business ability increasing rate changing curves of two similar scale enterprise without obvious cooperation and competition

Hypothesis 2: We assume that $r_1 = 0.1$, $r_2 = 0.09$, $a_{12} = a_{21} = 0.8$, $b_{12} = b_{21} = 0.1$. This situation represent that there exist two logistics enterprises, and their initial business ability increasing rate are similar, but there is greater cooperation between them, and the competition between them is not very fierce, then we get the changing law of the two enterprises' business ability increasing rate and number of customers, as is shown in Figs. 5 and 6. Contrast with the simulation result of hypothesis 2, we find that the in this state the simulation time is significantly shortened, and that the cooperation between two enterprises can increase their business ability increasing rate greatly.

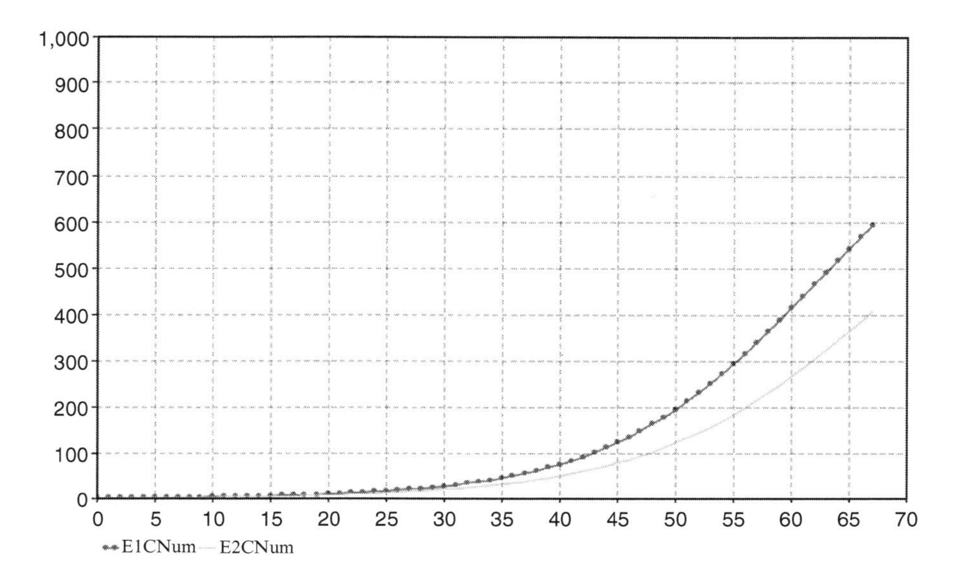

Fig. 4 The changing curves of the customers number of two similar scale enterprises without obvious cooperation and competition

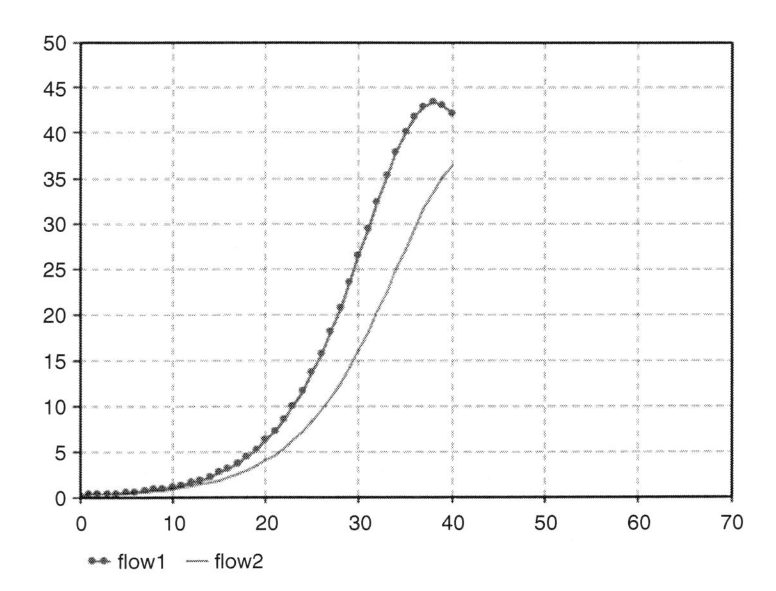

Fig. 5 Business ability increasing rate changing curves of two similar scale enterprises with good cooperation

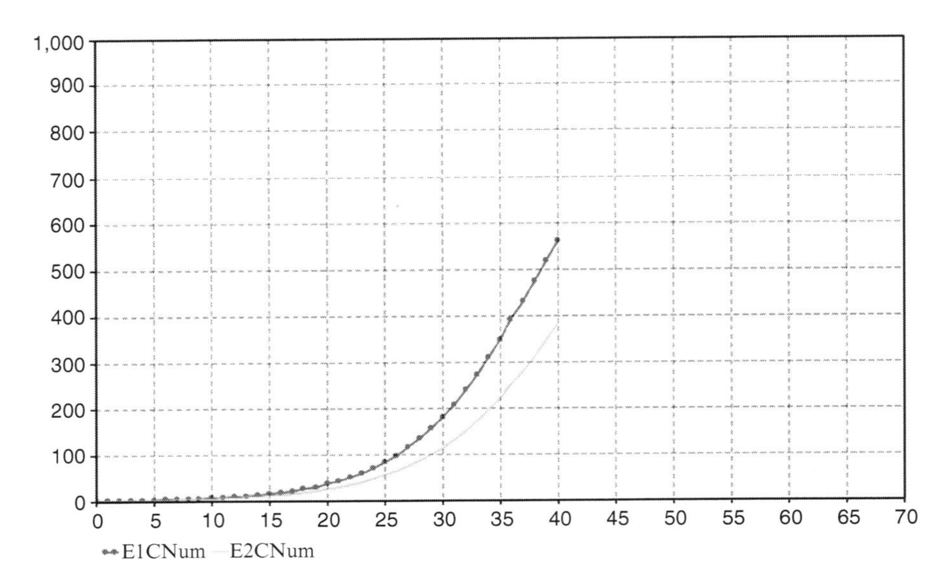

Fig. 6 The changing curves of the customers number of two similar scale enterprises with good cooperation

Hypothesis 3: We assume that $r_1 = 0.1$, $r_2 = 0.05$, $a_{12} = a_{21} = 0.1$, $b_{12} = 0.1$, $b_{21} = 0.2$. This situation represent that there exist two logistics enterprises, and the initial scale and strength of enterprise 1 is great larger than enterprise 2, and because of this the coefficient of influence from logistics enterprise 1 to enterprise 2 is larger than that from logistics enterprise 2 to enterprise 1, but there is no obvious cooperation between them, then we get the changing law of the two enterprises' business ability increasing rate and number of customers, as is shown in Figs. 7 and 8. According to the simulation result we find that because of the inherent superiority of scale and strength, logistics enterprise 1 will acquire almost all customer resources, but in the reality considering the factor of costs and other cases, logistics enterprise 2 obviously can't survive.

Hypothesis 4: We assume that $r_1 = 0.1$, $r_2 = 0.05$, $a_{12} = 0.1$, $a_{21} = 0.8$, $b_{12} = 0.1$, $b_{21} = 0.2$. This situation represent that there exist two logistics enterprises, and the initial scale and strength of enterprise 1 is great larger than enterprise 2, and because of this the coefficient of influence from logistics enterprise 1 to enterprise 2 is larger than that from logistics enterprise 2 to enterprise 1 during their competition process, but there is great cooperation between them, and the coefficient of influence from logistics enterprise 1 to enterprise 2 is great larger than that from logistics enterprise 2 to enterprise 1 during their cooperation process, then we get the changing law of the two enterprises' business ability increasing rate and number of customers, as is shown in Figs. 9 and 10. According to the simulation result and contrast with the result of hypothesis 4, we find that only in the case of that logistics enterprise 2 can cooperate with enterprise 1 actively could it coexist with enterprise 1, and this is more in line with the reality of the actual situation.

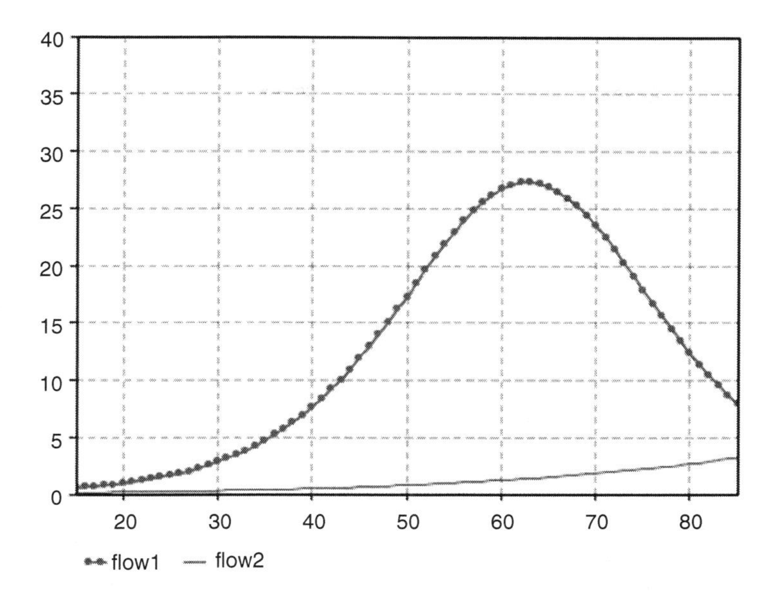

Fig. 7 Business ability increasing rate changing curves of two enterprises in different initial scale without obvious cooperation

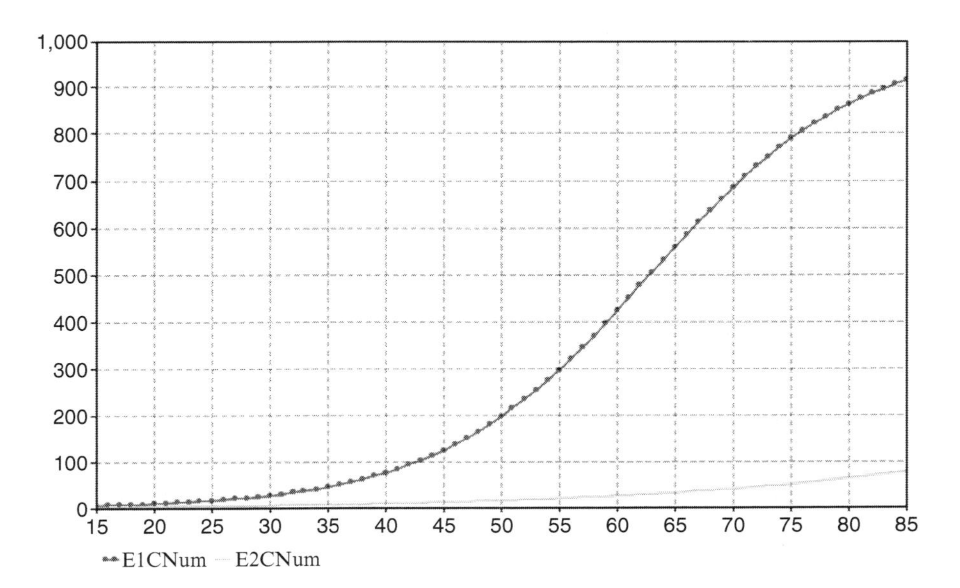

Fig. 8 The changing curves of the customers number of two enterprises in different initial scale without obvious cooperation

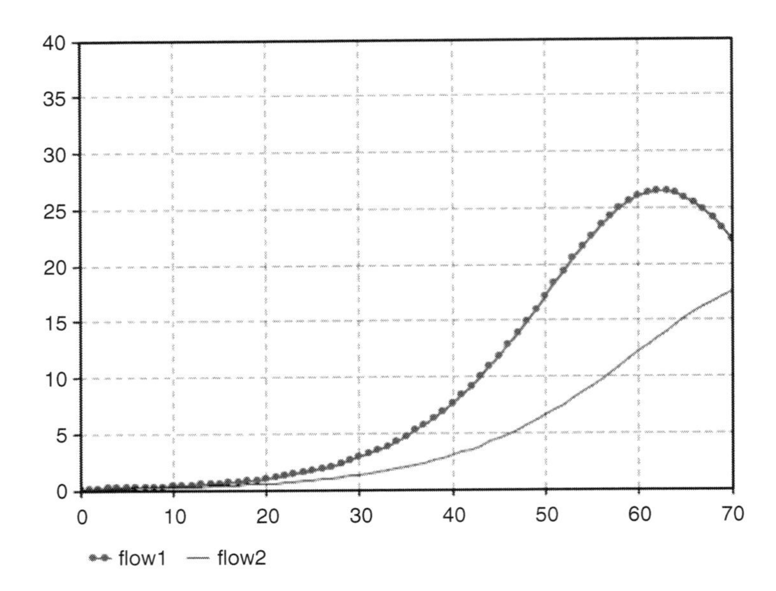

Fig. 9 Business ability increasing rate changing curves of two enterprises in different initial scale with good cooperation

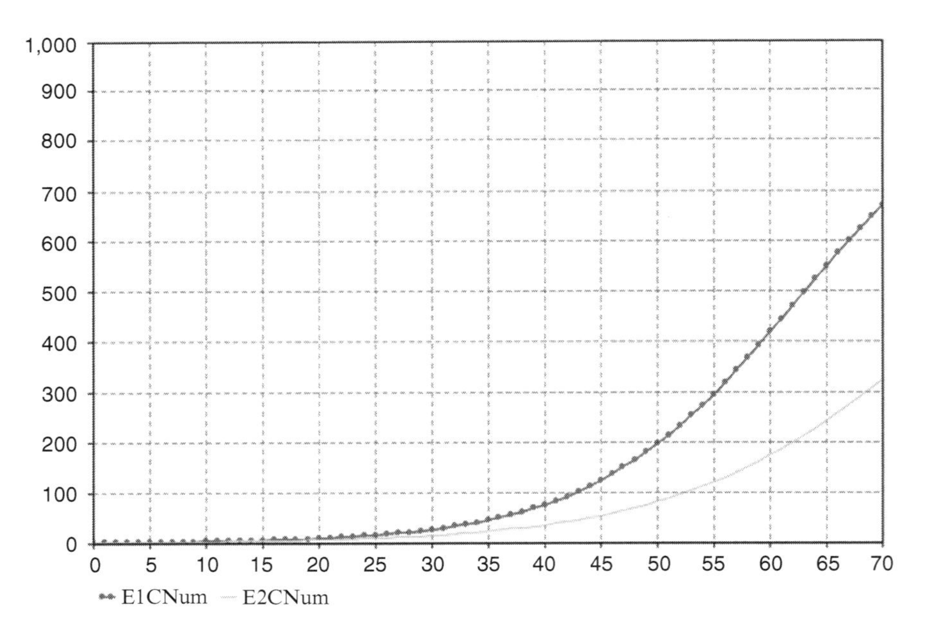

Fig. 10 The changing curves of the customers number of two enterprises in different scale with good cooperation

5 Conclusion

In this paper we constructed the corresponding logistic model and the competition model of customer resources acquiring process of two logistics enterprises in the same link of supply chain, and gave the system dynamics simulation analysis based on the Anylogic software platform. Then we analyzed the influence of different competition and cooperation level to the customer resources acquiring process of two logistics enterprises in different initial scale and got some conclusions as follows: (1) The good cooperation between two logistics enterprises in similar initial scale at the same link of supply chain can greatly improve the enterprises customer resources acquiring rate; (2) If the gap of the initial scale between two logistics enterprises at the same link of supply chain is too large, the smaller enterprise will not be able to get enough customer resources to maintain its operation without effective cooperation with the bigger enterprise, and almost all the customer resources will be acquired by the stronger enterprise exclusively. But if the small logistics enterprise can seek for cooperation with its competitor actively according to its characteristics, the number of customers that the small enterprise acquired will increase effectively, so the small enterprise can coexist with its competitors.

Acknowledgments This paper is supported by National Natural Science Foundation of China. Project id is: 71132008.

References

1. Zhang Rui, Qian Xingsan, Gao Zhen (2008) A competitive model of enterprises based on ecology theory. Syst Eng 26(2):116–119
2. Li Zhaolei, Wu Qunqi, Zhang Yaqi (2010) Evolution mechanism of regional logistic system based on dissipation structure theory. J Chang Univ 12(4):33–37
3. Zhang Wenbin (2005) The competition and cooperation relationship between supply chain numbers based on self-organization theory. Xi'nan Jiaotong University, ChengDu
4. Zhou Hao (2003) Enterprises cluster co-existence model and stability analysis. Syst Eng 21(1):32–37
5. Liu Rui (2005) The analysis of logistics enterprises' conformity of resource based on economic theory. ShangHai Maritime University, ShangHai

Integrated Logistics Network Design in Hybrid Manufacturing/Remanufacturing System Under Low-Carbon Restriction

Yacan Wang, Tao Lu, and Chunhui Zhang

Abstract This paper examines the eco-efficiency of different forward/reverse logistics network integration models in manufacturing/remanufacturing closed-loop supply chain under low-carbon restriction. A multi-objective mixed linger programming is established to optimize the site selection and flow allocation. In the objective function, three minimum targets are set: economic cost, energy consumption and waste generation. We compare the cost and environment efficiency of independent design and integrated design as well as sequential integration and simultaneous integration respectively, and find the optimal network integration option based on eco-efficiency to meet the requirements from low-carbon economy.

Keywords Hybrid manufacturing/remanufacturing system • Integration • Logistic network design • Low-carbon economy • Eco-efficiency

1 Introduction

Remanufacturing has been regarded as the optimum way of terminal resource reutilization to deal with waste machinery and electric appliance products, which are reaching the summit of scrapping globally [1]. Remanufacturing is a typical manufacturing/remanufacturing hybrid system, consisting of not only the forward logistics network for manufacturing and distribution but also the reverse logistics network for remanufacturing [2]. However, for a long time, manufacturers stay disintegrated in both forward and reverse logistics network, which make it very difficult to coordinate the designing, planning and decision of the two systems

Y. Wang (✉) • T. Lu • C. Zhang
School of Economics and Management Beijing, Jiaotong University, Xizhi Men Wai, Shangyuan Road, Haidian District, Beijing, People's Republic of China
e-mail: yacan.wang@gmail.com; tluaa@ust.hk; progresszhang@gmail.com

Z. Zhang et al. (eds.), *LISS 2012: Proceedings of 2nd International Conference on Logistics, Informatics and Service Science*, DOI 10.1007/978-3-642-32054-5_17,

effectively [3, 4]. Meanwhile, low-carbon economy proposes new requirements for the integration of forward/reverse logistics network design. In essence, it is the pursuit of eco-efficiency, which is a win-win situation between economic and environmental efficiency [5]. Therefore, it is necessary to integrate the designing of forward/reverse logistics network in pursuit of the optimal eco-efficiency [6–8].

This paper adds to the existing literature by considering both the integrated facilities and transportation routes and eco-efficiency in manufacturing/remanufacturing closed-loop supply chain, with three targets of economic cost, energy consumption and waste generation. Section 2 put forward the design model of manufacturing/remanufacturing logistics network integration. Section 3 sets up the mathematical model. Section 4 does the numeric study on an example of refrigerator industry and does comparative studies of eco-efficiency among different integration models, attempting to find the optimal network integration option under low-carbon restriction. The paper closes with Sect. 5.

2 The Integration Model of Manufacturing/Remanufacturing Logistics Network

2.1 Description of the Problem

In this paper, the design of manufacturing/remanufacturing integrated logistics network is supported by mixed-integer linear programming (MILP). This model contributes to determine the optimal locations of factory, distribution center, recycling center and distribution/recycling center, as well as the allocations of both forward and reverse material flows in the closed-loop supply chain.

This model considers both the structure design of the product and the logistics network design. Product design adopts three eco-friendly structure designs, which are reuse, maintenance and comprehensive recycling [9, 10]. Structure tree below describes the connectivity relationships among products, models and components under three different design methods (Fig. 1).

There are two ways of designing manufacturing/remanufacturing hybrid logistics networks: integration approach and independent approach. Integration design can be divided into sequential integration and simultaneous integration. No matter which method we choose, we still need to consider integration on two levels: the integration of facilities and the integration of transportation. This model considers both of these levels of integration. The design model of the closed-loop supply chain is illustrated in Fig. 2. We assume a sample manufacturing/remanufacturing logistics network, and the topology structure of the model is illustrated in Fig. 3.

2.2 Basic Assumptions

First, we concern the basic assumptions in this model, as those are in Table 1.

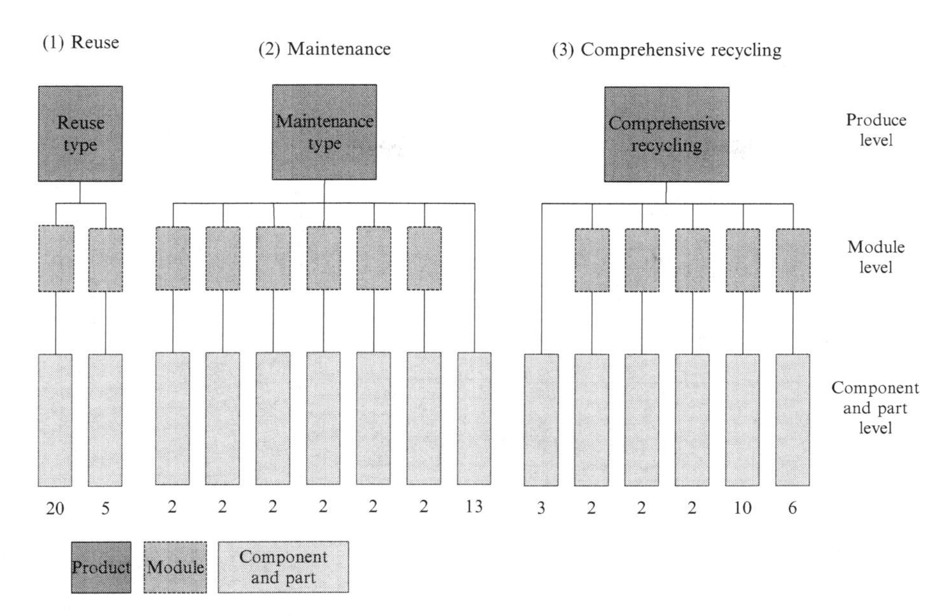

Fig. 1 Structure tree of three product structure design

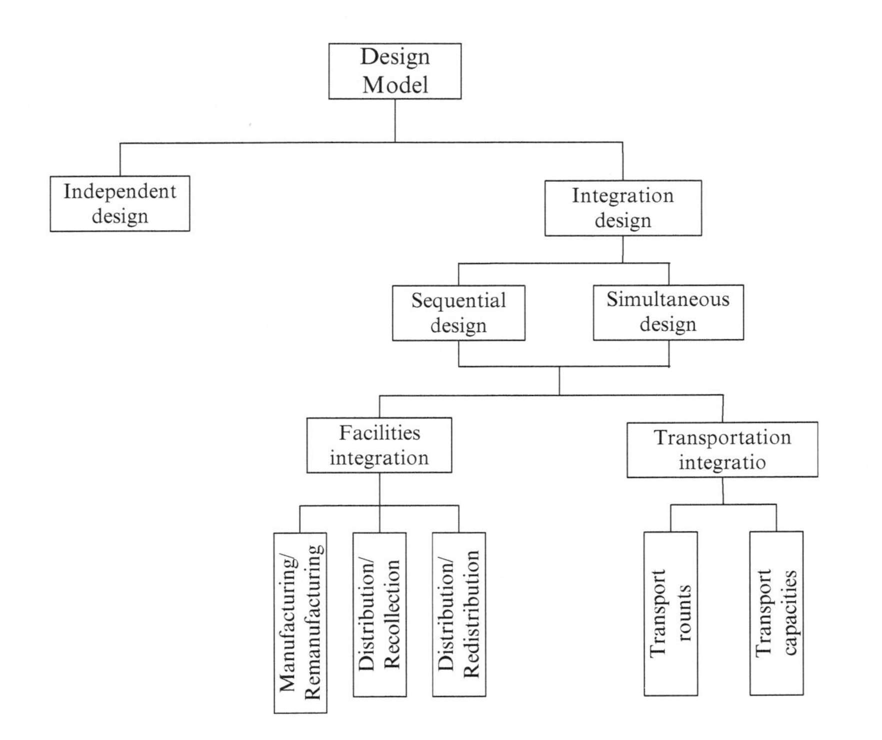

Fig. 2 Design model

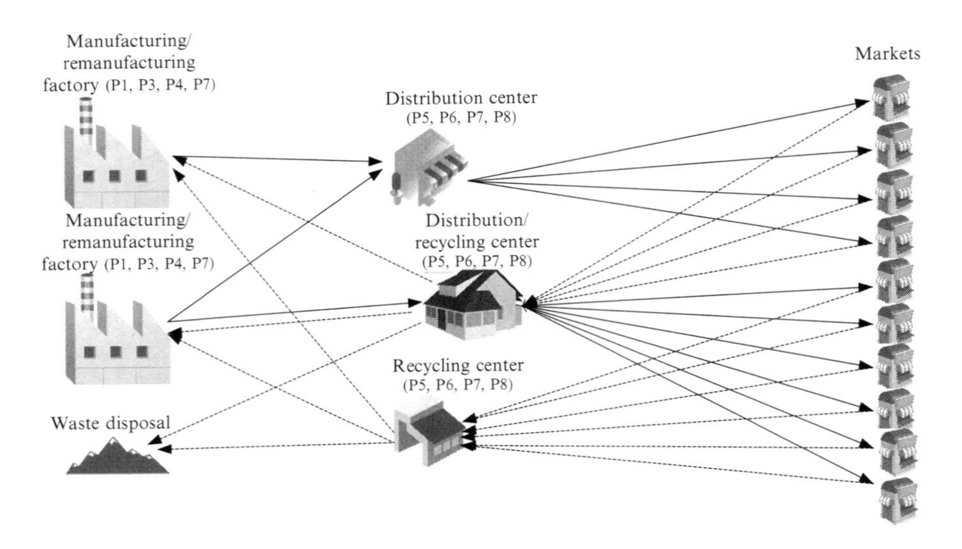

Fig. 3 Logistics network topology structure

Table 1 Assumed conditions of the model

No.	Assumptions
1	Factories can both manufacture and remanufacture. Both the new products and remanufactured products are sold through recycling centers (or integrated distribution/ recycling centers) to meet the market demand. The distribution operation cost per unit should equal to the price, and the remanufactured products as well as the new products could equally meet the demand of the market
2	The quantity of collection could be calculated by regions, and is directly related to the consumption of the region, with the recollection rate known
3	The recycling centers examine and categorize the products, and repair the products then put them into market. Those cannot be repaired can be dissembled into modules and parts, then all transported to manufacture/remanufacture factories. Those products which are unavailable for remanufacturing could be further dissembled in smaller parts and be discarded
4	The storage is completed nearby of inside the facilities, so the transportation between the store and related facilities could be ignored

3 Formulations

To answer the call of low-carbon economy, in this model, we set the optimization goals as: (1). Minimum total cost; (2). Minimum environmental impact. In this paper, we adopt waste generation and energy consumption for an approximation to evaluate environmental impact [4, 5]. The optimization objectives are illustrated in Fig. 4.

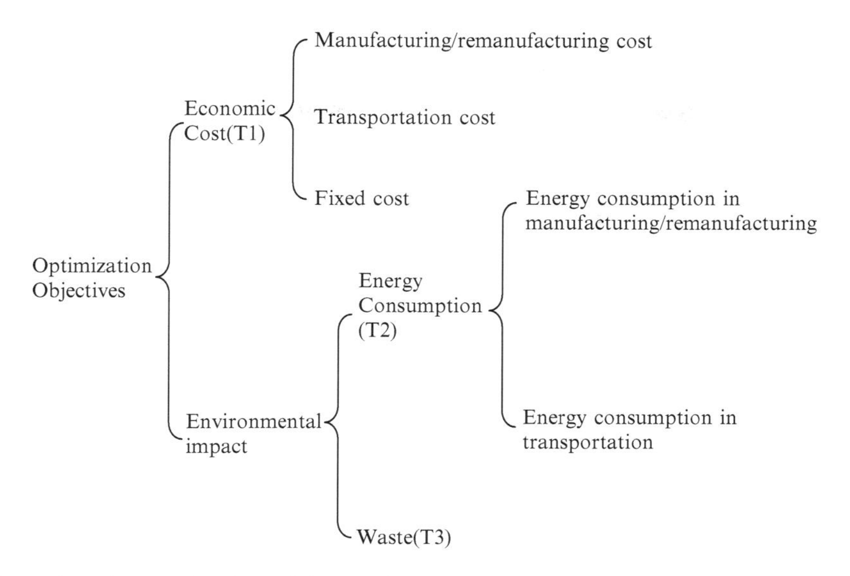

Fig. 4 Optimization objectives of the model

The first optimization objective is minimum economic cost:

$$\min\ T_1 = FC + TC + MC \tag{1}$$

Which include fixed costs of setting up factories, distribution centers and collection points.

$$FC = \sum_m F_m^f Y_m^f + \sum_i Y_i^w F_i^w + \sum_j Y_j^r F_j^r - \sum_n Y_n^s F_n^s$$

Transportation cost consists of that in forward logistics and in reverse logistics; and the integrated centers use transportation team in reverse logistics to transport parts and modules to save cost; the last cost generated in transporting products back to collection centers.

$$\begin{aligned} TC = & \sum_i \sum_m C_{mi} X_{mi}^o + \sum_i \sum_c C_{ic} X_{ic}^o + \sum_c \sum_j C_{cj} X_{cj}^o \\ & + \sum_j \sum_m \left(\sum_a C_{jm}^a X_{jm}^a + \sum_t C_{jm}^t X_{jm}^t\right) + \sum_j \sum_d \sum_a C_{jd} X_{jd}^a \\ & - \sum_c \sum_n C_{cn} X_{cn}^o - \sum_n \sum_m \left(\sum_a C_{nm}^a X_{nm}^a + \sum_t C_{nm}^t X_{nm}^t\right) + \sum_c \sum_j C_{jc} X_{jc}^o \end{aligned}$$

Cost of manufacturing/remanufacturing: cost of manufacturing minus cost saved by reproducing modules and parts

$$MC = \sum_{m}\sum_{i} C_o X^o_{mi} - \sum_{j}\sum_{m}\sum_{t} C_t X^t_{jm} - \sum_{j}\sum_{m}\sum_{a} C_a X^a_{jm}$$

The second optimization objective is minimum energy consumption

$$\min\ T_2 = TE + ME \tag{2}$$

The calculation of the energy consumption during transportation is similar to that in (1), which includes the energy consumption in transportation.

$$\begin{aligned} TE = & \sum_{i}\sum_{m} E_{mi} X^o_{mi} + \sum_{i}\sum_{c} E_{ic} X^o_{ic} + \sum_{c}\sum_{j} E_{cj} X^o_{cj} \\ & + \sum_{j}\sum_{m}\Big(\sum_{a} E^a_{jm} X^a_{jm} + \sum_{t} E^t_{jm} X^t_{jm}\Big) + \sum_{j}\sum_{d}\sum_{a} E_{jd} X^a_{jd} \\ & - \sum_{c}\sum_{n} E_{cn} X^o_{cn} - \sum_{n}\sum_{m}\Big(\sum_{a} E^a_{nm} X^a_{nm} + \sum_{t} E^t_{nm} X^t_{nm}\Big) + \sum_{c}\sum_{j} E_{jc} X^o_{jc} \end{aligned}$$

The energy consumption in manufacturing/remanufacturing

$$ME = \sum_{m}\sum_{i} E_o X^o_{mi} - \sum_{j}\sum_{m}\sum_{t} E_t X^t_{jm} - \sum_{j}\sum_{m}\sum_{a} E_a X^a_{jm}$$

The third optimization objective is minimum waste

$$T_3 = \sum_{a}\sum_{d}\sum_{j}\left[R_{oa} U^a_d (1 - U_a)\delta^a \sum_{c} X^o_{cj} - X^a_{jd}\right] \tag{3}$$

$$X^o_{mi}, X^o_{ic}, X^o_{cj} \ldots \geq 0 \tag{4}$$

$$Y^f_m, Y^w_i, Y^r_j, Y^s_n = 0, 1 \tag{5}$$

(4) represents the materials flow allocation of each route is nonnegative;

(5) represents the location decision variable is constrained by 0–1. 1 represents the correspondent facility is under construction and 0 represents the opposite.

$$\sum_i X^o_{ic} \le A^f_m Y^f_m, \forall m \tag{6}$$

$$\sum_c X^o_{ic} \le A^w_i Y^w_i, \forall i \tag{7}$$

$$\sum_c X^o_{cj} \le A^r_i Y^r_i, \forall j \tag{8}$$

(6), (7), and (8) represent the capacity constraints of factory, distribution center and collection point respectively

$$\sum_j X^o_{jc} + \sum_i X^o_{ic} \ge D_c, \forall c \tag{9}$$

(9) represents products and the reproduced products delivered from the distribution center to the market must meet its demand

$$\sum_m X^o_{mn} + \sum_c X^o_{nc} \le MY^s_n, \forall n, \quad m \text{ is big enough} \tag{10}$$

(10) represents only when distribution/recollection center is in location n, then the flow there could not be zero

$$X^o_{cn} \le X^o_{nc}, \forall c, n \tag{11}$$

$$\sum_a X^a_{nm} + \sum_t X^t_{nm} \le X^o_{mn}, \forall m, n \tag{12}$$

(11) represents the flow in forward logistics from distribution/collection center n to market c is more than that in reverse logistics and this is because the transportation team also loads products in reverse routes.

$$\sum_n X^o_{nc} \le \sum_i X^o_{ic} + \sum_j X^o_{jc}, \forall c \tag{13}$$

$$\sum_n X^o_{cn} \le \sum_j X^o_{cj}, \forall c \tag{14}$$

$$\sum_n X^a_{nm} \le \sum_j X^a_{jm}, \forall m, a \tag{15}$$

$$\sum_n X^t_{nm} \le \sum_j X^t_{jm}, \forall m, t \tag{16}$$

$$\sum_{n} X_{mn}^{o} \leq \sum_{i} X_{mi}^{o}, \forall m \tag{17}$$

(13), (14), (15), (16), and (17) represents the total flow of a distribution/collection center should not be more than the total flow of a collection center

$$Y_{n}^{s} \leq Y_{i}^{w}, \text{ when } \quad n = i \tag{18}$$

$$Y_{n}^{s} \leq Y_{j}^{r}, \text{ when } \quad n = j \tag{19}$$

(18) and (19) represents only when distribution and collection center is in n, the squalor quantity of the correspondent distribution/collection centers could be 1.

$$\sum_{m} X_{mi}^{o} \leq \sum_{c} X_{ic}^{o}, \forall i \tag{20}$$

(20) represents the logistics balance among the distribution centers.

$$\delta_{c} \sum_{i} X_{ic}^{o} = \sum_{j} X_{cj}^{o}, \forall c \tag{21}$$

(21) represents the products collected from the market by the collection point equals to the amount of products sold in the market times to return rate

$$R_{oa} U_{d}^{a} (1 - U_{a}) \delta^{a} \sum_{c} X_{cj}^{o} \geq X_{jd}^{a}, \forall d, j, a \tag{22}$$

$$R_{oa} U_{a} \delta^{a} \sum_{c} X_{cj}^{o} \geq \sum_{m} X_{jm}^{a}, \forall j, a \tag{23}$$

$$R_{oa} U_{t} \delta^{t} \sum_{c} X_{cj}^{o} \geq \sum_{m} X_{jm}^{t}, \forall j, t \tag{24}$$

$$\delta^{0} \sum_{c} X_{cj}^{o} \geq \sum_{c} X_{jc}^{o}, \forall j \tag{25}$$

(22), (23), (24), and (25) represent after the products are examined and categorized in the collection point, the logistics balance on the parts and modules level.

4 Algorithm and Numeric Study

We do the numeric study on an example of refrigerator industry. We use comprehensive solvers, like cplex and Lingo to solve mixed integer programming.

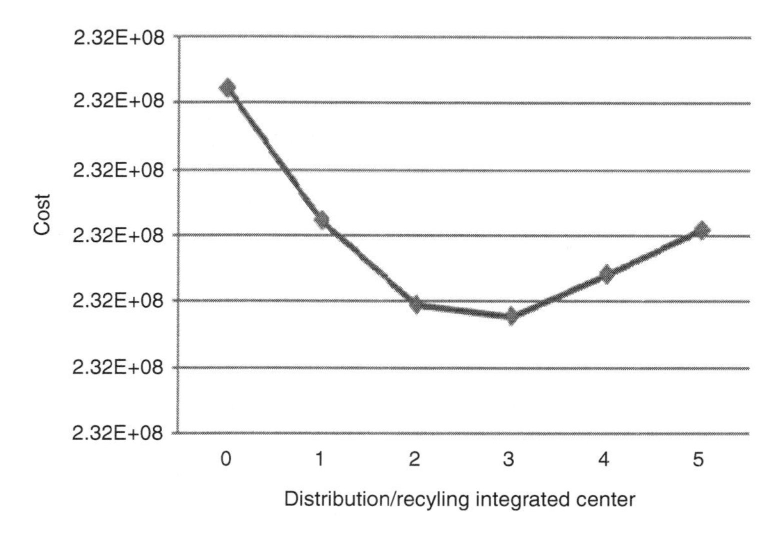

Fig. 5 Trend of economic cost

4.1 *Integrated Design vs Individual Design*

By managing the number of distribution/collection centers, we can analyze how the number of integrated facilities influences the energy consumption of the system. The quantitative relationship between economic cost (Yuan) and number of integrated centers and relationship between energy consumption (Joule) and number of integrated centers are shown in Figs. 5 and 6 respectively. They are shown that by integrating forward/reverse logistics network, we can reduce the economic cost and energy consumption of the network. If only waste generation is concerned, the change of the number of integrated centers has very minor effect on waste generation.

4.2 *Sequential Integration vs Simultaneous Integration*

Compare the optimal economic cost and energy consumption got by the two methods, although the location solution is different, the difference between optimal cost (unit: Yuan) and optimal energy consumption is minor, and the two curves almost coincident (as that in Figs. 7 and 8). What's more, simultaneous integration in method two gets smaller optimal economic cost.

In addition, because the waste generation relates mainly to the recollection and recovery rate of the discarded products but has no relationship to the number of integrated centers, so waste submission of the two integration models is the same.

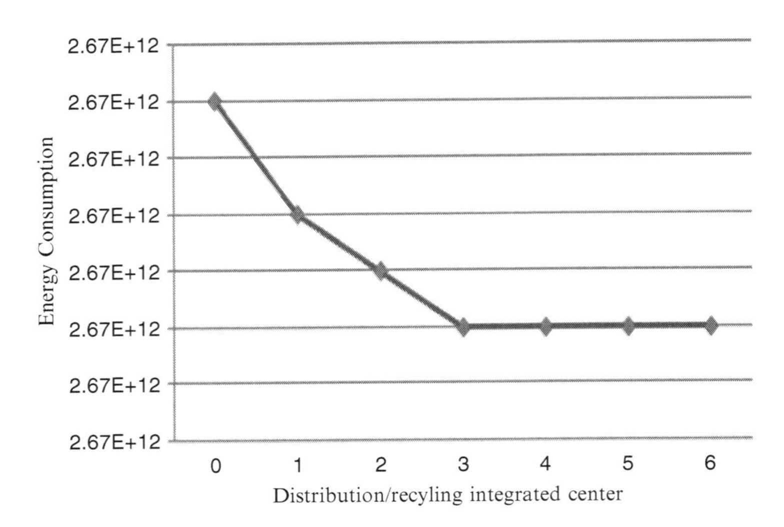

Fig. 6 Trend of energy consumption

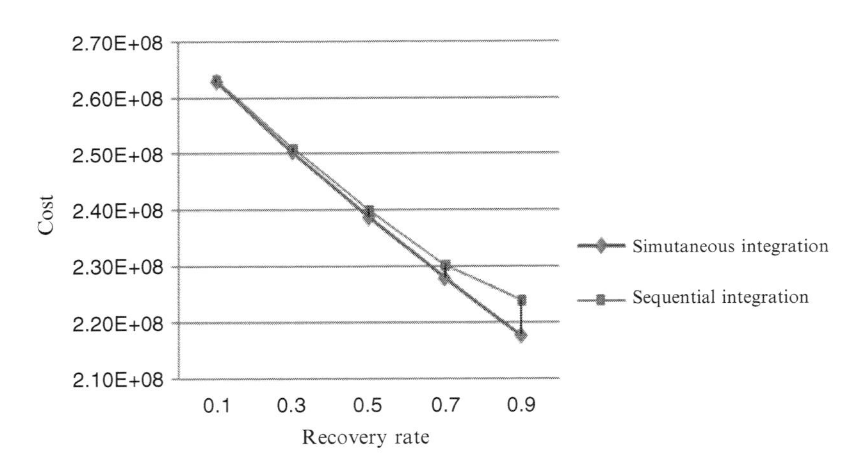

Fig. 7 Comparison of cost

5 Conclusion and Future Work

This paper examines the eco-efficiency of different forward/reverse logistics network integration models in manufacturing/remanufacturing closed-loop supply chain under low-carbon restriction. According to the results of the numeric experiments, forward/reverse network integration can reduce cost and energy consumption in operation. Since there are only very little cost and energy consumption differences between sequential integration and simultaneous integration, companies do not need reform their previous forward logistics network facilities when setting up reverse logistic network.

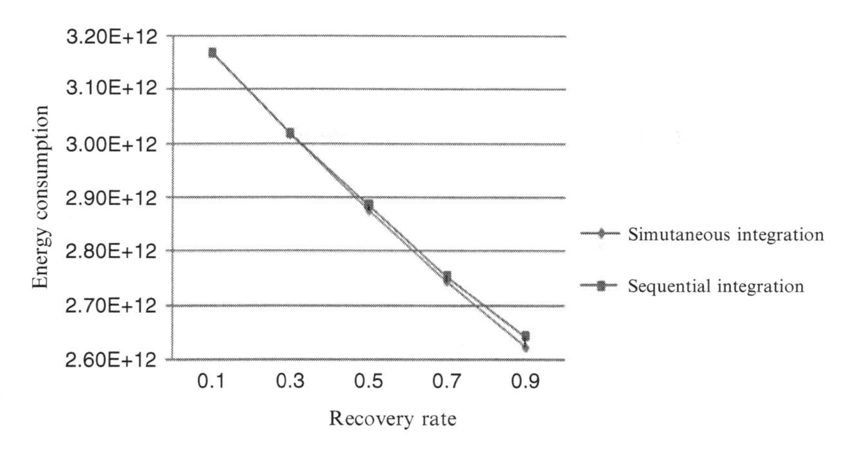

Fig. 8 Comparison of energy consumption

The current study includes several limitations that offer opportunities for future research. First, we can consider the extension of this mode and fit it to multi-product and multi-loop situations to better reflect the real situations. Second, we can discuss the design with variant parameters. Last but not least, there is still possibility to study their real statistics to verify this model in further research.

References

1. Xu BS (2009) Current, situation and policy recommendation of remanufacturing industry in China. Guangxi Jieneng 3:12–12
2. Ma ZJ, Dai Y (2005) Integrated network optimization design model in manufacturing/remanufacturing hybrid system. Comput Integr Manuf Syst 11:1151–1157
3. Sasse H, Karl U, Renz O (1999) Cost efficient and ecological design of cross-company recycling systems applied to sewage sludge re-integration. In Despotis DK, Zouponidis C (eds), Proceedings of DSI conference, vol. 3–4. New Technologies Publications, Athens, Greece, pp 1418–1420
4. Mutha A, Pokharel S (2009) Strategic network design for reverse logistics and remanufacturing using new and old product modules. Comput Ind Eng 56:334–346
5. Fang SJ (2010) Low-carbon development in the context of the green economy. Chin J Popul Res Environ 20(4):8–11
6. Francasa D, Minner S (2009) Manufacturing network configuration in supply chains with product recovery. Omega 37:757–769
7. Fleischmann M, Krikke HR, Dekker R, Flapper SDP (2000) A characterisation of logistics networks for product recovery. Omega 28(6):653–666
8. Easwarana G, Usterb H (2010) A closed-loop supply chain network design problem with integrated forward and reverse channel decisions. IIE Trans 42(11):779–792
9. Umeda Y, Nonomura A, Tomiyama T (2000) Study on life-cycle design for the post mass production paradigm. AIEDAM 14:149–161
10. Krikke H, Blemhof Ruwaard J, Wassenhove LN (2003) Concurrent product and closed-loop supply chain design with an application to refrigerators. Int J Prod Econ 41(16):3689–3719

Classification and Application on Food Cold Chain Collaborative Distribution Models

Hongjie Lan, Hailin Xue, and Yongbin Tian

Abstract Food cold chain (FCC) collaborative distribution (co-distribution) is an efficient way to solve the FCC distribution system infrastructure weak, high cost and low efficiency. In this paper cold chain distribution modes were classified and the applicable conditions of each collaborative model were also analysed to help the enterprise select collaborative models.

Keywords Food cold chain • Collaborative distribution • Models • Classification • Application

1 Introduction

The FCC distribution can reduce food product corruption and ensure the "last mile" foods safety, meeting people's urgently requirements for high quality and safe food. However, merely increasing facilities and equipments to change city's backward status of cold chain distribution in a short period of time is unrealistic. The crucial way is carrying out co-distribution. Co-distribution is meaning that several distribution centers united by ways of sharing resources and logistics facilities to deliver foods with common users on a specific region in order to have complementary advantages and improve the utilization on finance, equipments, personal and times [1]. It's the advanced stage of the city cold chain development and also the first and critical choice to intensive the integration on resource [2].

H. Lan (✉) • H. Xue • Y. Tian
School of Economics and Management, Beijing Jiao Tong University,
Beijing 100044, China
e-mail: hjlan@bjtu.edu.cn; snowseaforest@sina.com; tyb1518@126.com

Z. Zhang et al. (eds.), *LISS 2012: Proceedings of 2nd International Conference on Logistics, Informatics and Service Science*, DOI 10.1007/978-3-642-32054-5_18,

2 Cold Chain Collaborative Distribution Models Classification

Collaborative delivery models can be classified by various views. Hall [3] divided them into the same industry and different industry collaborative distribution on the perspective of industrial attribution. Hokey Min [4] gave the main classification method from the cooperation between horizontal collaboration and vertical coordination. Li Ding [5] put forward the subject of the logistics industry and goods owner industry classification but it did not consider the emerging public network distribution platform and collaborative distribution between industry organizations. Some scholar have also classified by the length of the collaboration time.

Combined with the previous studies, this paper presents nine kinds of co-distribution modes on FCC. We classified models to food owner inter-enterprise co-distribution, FCC logistics inter-enterprise co-distribution, social co-distribution by the collaborative subject. The horizontal and vertical relationship and the collaborative time were also considered for the further classification. What's more we give some existing examples on Beijing. The detail is shown in Table 1.

For the more concise expression, we called subjects like farmers, producers, wholesalers, distribution centers, retailers the good owner enterprise. They can also be divided into supplier enterprises which means the goods outflow and receiver enterprise which means the goods receiving. Cold chain logistics enterprise is the third part enterprise. It's only responsible for food distribution.

3 Description of Cold Chain Co-distribution Model

In order to tersely describe the nine models, Fig. 1 was given to show how the different enterprises that involved in models collaborative their FCC distribution business.

4 Application Analysis of Cold Chain Co-distribution Model

To help companies select the appropriate co-distribution model, the characteristics and suitable use situation were analysed for each models.

Model 1: Dynamic Co-Distribution between the FCC Logistics Enterprises
This model is suitable for the different cold chain logistics enterprises with period complementary. Most of the food production and consumption are seasonal. When making dynamic cooperation with others cold chain logistics enterprises with complementary distribution business, partners can entrusting their distribution business to achieve "clipping valley filling" effect and improve resource utilization rate.

Table 1 The classification of cold chain co-distribution models

Classification		No	Models	Cases
FCC logistics enterprises collaboration	Horizontal collaborative	1	Dynamic Co-Distribution between the FCC Logistics Enterprises	Express Channel Food Logistics & Beijing for Logistics
		2	Stable Co-Distribution between the FCC Logistics Enterprises	Beijing for Logistics & Shanghai Jiaorong Cold Chain Distribution
Food owner enterprise collaboration		3	Large-scale Supplier Enterprise Oriented Co-Distribution	Beijing Er Shang Group
		4	Supplier Enterprises Co-Distribution Union	Japan 7-11 Distribution Center
		5	Co-Distribution within the Group System	Wal-Mart, Wu-Mart, Metro, Jing Ke Long
		6	Supplier and Receiver Formed New Co-Distribution Organization	Japan Aichi-ken Co-distribution
Social collaboration	Vertical collaborative	7	Outsourcing Distribution to Cold Chain Logistics Enterprise	HAVI Logistics, The Cold Fresh Logistics, Shun Xin Green Logistics
		8	Cold Chain Logistics and Owner Formed New Co-Distribution Organization	The Beijing Chaopi Trade Co., Ltd.
		9	Co-Distribution Based on the Public Logistics Information Platform	Chengdu Logistics Public Information Platform

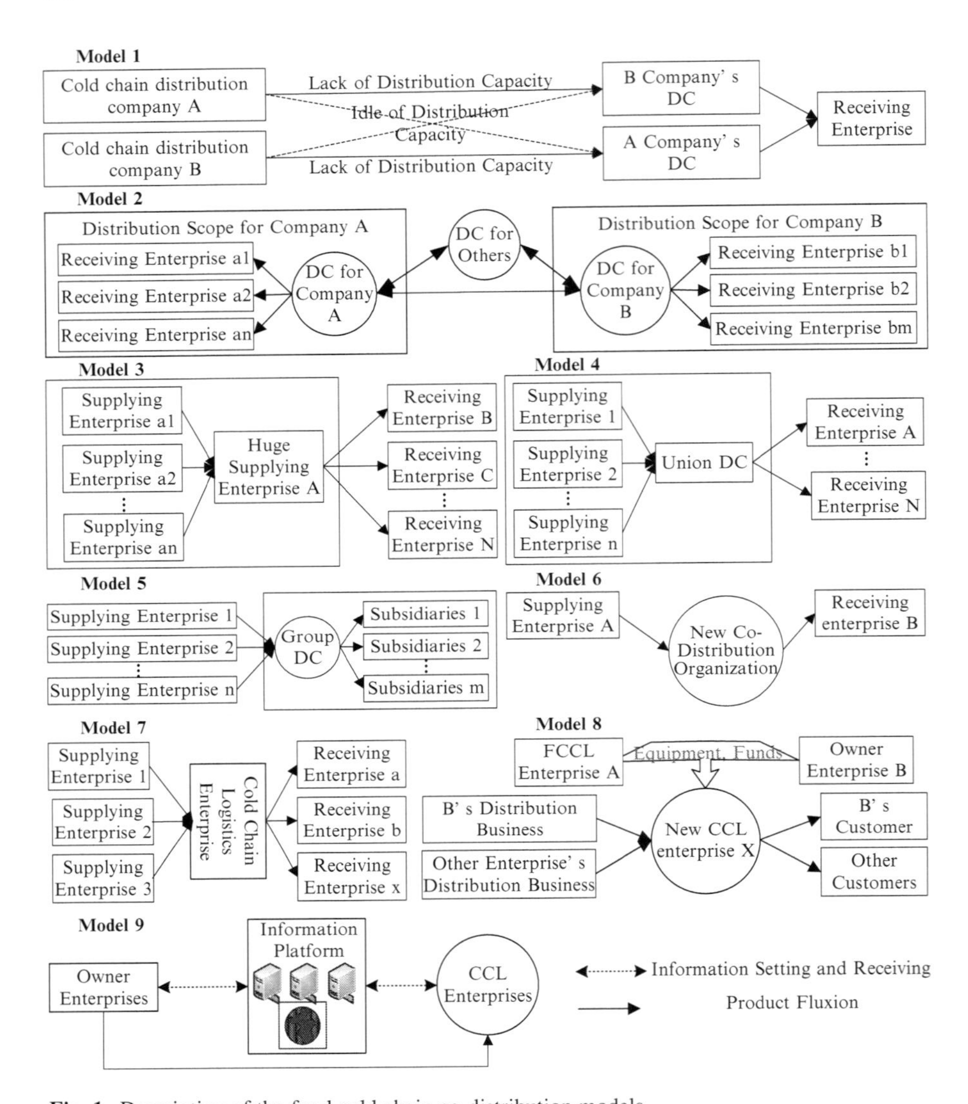

Fig. 1 Description of the food cold chain co-distribution models

Model 2: Stable Co-Distribution between the FCC Logistics Enterprises
This model is suitable for the cold chain logistics enterprises with reasonable economic distribution areas. But beyond this region, due to lack of cargoes and distribution capabilities it will add high cost for enterprises. This limits the market expansion especially when enterprises are facing shortage of funds. However, different cold chain logistics enterprises can share distribution networks by contracts and establish stable cooperation relations. This promotes the market expansion and gets the economies scale within their distribution region. By focusing on distribution business in their own region, it will also improve the enterprises' service capabilities.

Model 3: Large-scale Supplier Enterprise Oriented Co-Distribution
On the circulation of food, there are some larger enterprises, it has its own food distribution sector, or even set up food distribution subsidiary, with strong cold chain logistics capabilities and the great distribution system. At he same time, there was some small and medium food owner enterprises, which lack of cold chain logistics capabilities. These enterprises can delegate its distribution business to the larger enterprises, and joint distribution under its auspices.

Model 4: Supplier Enterprises Co-Distribution Union
This model is suitable for the situation that more than one supplying enterprises face the same receiving enterprise and their cold chain distributions ability are limited. These enterprise set up a co-distribution union to harmonize cold chain resource and their distribution business will be arranged by the DC.

Model 5: Co-Distribution within the Group System
This model has three available premises. First, food Enterprise Group has a certain number of subsidiaries. Second, the Group has sufficient demand for cold chain logistics. Third, the Group has strong ability of cold chain distribution. Different supply companies transport food products to the DC of the group, and then the DC develop scientific delivery plan to prove distribution efficiency.

Model 6: Supplier and Receiver Formed New Co-Distribution Organization
This model is suitable for the situation that there is a stable trade relationship between supplying enterprise and receiving enterprise. Both of then have certain distribution capabilities but didn't meet the cold chain distribution business need. In order to meet the distribution needs of the business, adding new logistics equipments and building up a new organization to integrate logistics resources of both sides is feasible. The new organization is in charge of enterprise business.

Model 7: Outsourcing Distribution to Cold Chain Logistics Enterprise
This mode applies to the food owner enterprise which has loss or weak cold-chain delivery ability. What's more, the cold chain distribution business is not the enterprise's key business or the cost of the food cold chain distribution by owners is higher than outsourcing. The enterprise outsources the cold chain distribution business to a professional third-party cold-chain logistics enterprise which will realize the centralized distribution of the goods.

Model 8: FCC Logistics and Owner Formed New Co-Distribution Organization
This mode is used in large scale owner enterprise. The enterprise has bigger cold chain cargoes and has a strictly requirement on cold chain distribution. In order to better serving customers who have larger cold chain logistics quantity, according to the business needs of the owner, cold chain logistics enterprise integrates cold chain logistics resources of the owner enterprise and then builds new cold chain distribution center or distribution spot with customer.

Model 9: Co-Distribution Based on the Public Logistics Information Platform
It is suitable for the case that there was a perfect logistics public information platform had been established. By utilizing the logistics public information platform, the food owner enterprise can release distribution demand information and

the cold-chain logistics enterprise can releases information of distribution service ability. So, with the information sharing, both of the owners and the cold chain logistics enterprise can choose their suitable collaborative partners. It can effectively integrate cold chain distribution resources within same regions.

5 Conclusion

This paper classified the collaborative models and described the suitable using case s. It's the groundwork for the further study of the cold chain collaborative distribution. These modes can also be used to refer on the other industry. The next step in this study is using quantitative analysis methods to choice the collaborative models.

Acknowledgement This research has been supported by the Beijing Municipal Science & Technology Commission's project (Z111105000111010-3) and the Foundation of Beijing Jiaotong University (2011JBM233).

References

1. Minghua Wang, Yi Zheng (2007) Collaborative distribution cost allocation model research based on virtual enterprise. Sci Technol Ind 17(12):71–73
2. Jiani Liu, Li Zhou, Yiu Li (2011) Develop the suitable cold chain distribution model in China. Market Cond 44:33–33
3. Hall RW (1987) Consolidation strategy: inventory, vehicles and terminals. J Bus Logist 8(2):57–72
4. Hokey Min (1996) Consolidation terminal-allocation and consolidated routing problems. J Bus Logist 17(2):235–238
5. Li Ding, Huang Yuanxin (2011) Research on collaborative distribution based on virtual enterprise, Southwest Jiaotong University, Xi'an, china

Optimization of the Vehicle Route of Express Company with Multiple-Dynamic Saving Algorithm

Jun-chao Liu and Wei Liu

Abstract As the special transportation requirements of express companies, optimized the vehicle route base on Saving Algorithm, and then use multiple-dynamic Saving Algorithm, consider the timeliness requirements of the express company to achieve a balance cost and timeliness. In the actual problems of SF for the example, has carried on the analysis of the example. It has a great practical significance to the transportation network and path optimization of express companies.

Keywords Saving algorithm • Multi-dynamic • Limitation

1 Introduction

The express industry in China is developing rapidly. Current research on the logistics vehicle routing optimization is considered from the perspective of cost savings, such as Cao Jian-dong et al. [1]. Liu Wusheng and Liu Jun [2] studied the vehicle distribution routing problem and proposed an improved saving criteria based on the basic principles of traditional conservation method, which had living examples to demonstrate the improved Saving Algorithm which can reduce the transport distance and cost compared to the typical one.

This article considers the special transport requirement of express enterprises, optimizing the vehicle routing with Saving Algorithm, and then use the multiple dynamic Saving Algorithm, considering the timeliness requirements, to make a balance between the cost and the limitation. It has the significance to the transport network path optimization in express enterprise.

J.-c. Liu (✉) • W. Liu
College of Transport and Communications, Shanghai Maritime University,
Shanghai 201306, People's Republic of China
e-mail: liujc0723@163.com

Z. Zhang et al. (eds.), *LISS 2012: Proceedings of 2nd International Conference on Logistics, Informatics and Service Science*, DOI 10.1007/978-3-642-32054-5_19,

2 Saving Algorithm

2.1 Model Parameters and Established.

$q_i(i = 1, 2, \ldots, \quad n)$: Signify the demand of customer i
$Q_k(k = 1, 2, \ldots, \quad K)$: Signify the load capacity of truck k
G : Signify the set of all customers, $G = \{1, \ldots n\}$;
$G_0 = G \cup \{0\}$, $\{0\}$ stands for distribution center
G_k : Signify the customer set serviced by truck k
D_k : Signify the maximum driving distance of truck k
C_{ij} : Signify all transport costs of the truck from customer i to customer j.
n : Signify the total number of customers who need service

$$Y_{ijk} = \begin{cases} 1, & \text{Traffic from i to j} \\ 0, & \text{else.} \end{cases}$$

$$X_{ik} = \begin{cases} 1, & \text{Customer accomplished by truck i} \\ 0, & \text{else.} \end{cases}$$

W_1, W_2 : The weight value of each target
μ_{ij} : The distance from customer i to j
δ_k : The no-load driving cost coefficient of k-truck

Objection:

$$Z = w_1 \min \sum_{k=1}^{K} \sum_{i=1}^{n} \sum_{j=1}^{n} C_{ij}\, y_{ijk} + w_2 \min \sum_{k=1}^{K} \sum_{i=1}^{n} -\delta_k q_i x_{ki}$$

s.t.

$$\sum_{i-1}^{n} q_i\, x_{ik} \leq Q_k, \ \forall\, k \in K \tag{1}$$

$$\sum_{j-1}^{n} y_{ijk} = x_{jk}, \ \ \forall j \in G_0, \ \ k \in K \tag{2}$$

$$\sum_{i-1}^{n} y_{ijk} = x_{ik}, \ \forall i \in G_0, \ \ k \in K \tag{3}$$

$$\sum_{k-1}^{K} x_{ik} = 1, \ i = 1, 2 \ldots, \quad n \tag{4}$$

$$0 \leq \sum_{i-1}^{n} x_{ik} \leq n \tag{5}$$

$$\sum_{i,j} \mu_{ij} \, y_{ijk} \leq D_k \tag{6}$$

$$\sum_{i} \sum_{k} x_{ik} = n \tag{7}$$

$$\sum_{k=1}^{K} x_{ik} = \begin{cases} m, & \mathrm{i} = 0 \\ 1, & \mathrm{i} = 1, \ \ldots, \quad n. \end{cases} \tag{8}$$

$$y_{ijk}, \ x_{ik} \in \{0, 1\} \tag{9}$$

$$\mathrm{i} = 0, \ C_{ij} = C_{ok} + C_{1k}\mu_{ij}; \ \mathrm{i} \neq 0, \ \ C_{ij} = C_{1k}\mu_{ij}$$

The objective function means the optimal total cost. Constraints: Function (1) constrain the capacity of transport vehicles. Functions (2) and (3) limit the number of vehicle arriving and departing a distribution point is only one. Function (4) means that a distribution point completed only by one truck. Function (5) means that each vehicle with the distribution points cannot exceed the number of all distribution points. Function (6) means that driving distance of each vehicle cannot exceed the maximum travel distance. Function (7) means that all distribution points are loaded. Function (8) ensures that each delivery point is only completed by a truck, but also all distribution points completed by the m vehicles.

2.2 *To Solve the Saving Algorithm*

Saving Algorithm has used the geometry principle of two sides's length must be longer than the third side [3, 4].

In Fig. 1, the delivery distance is 2 × (a + b), in Fig. 2, the distribution distance is a + b + c. Obviously, the former need more mileage. Therefore, the core of Saving Algorithm is making two circuits combing into one loop, which minimizes the total transportation distance, until reaching a car loaded limit, and then optimize the next one [5].

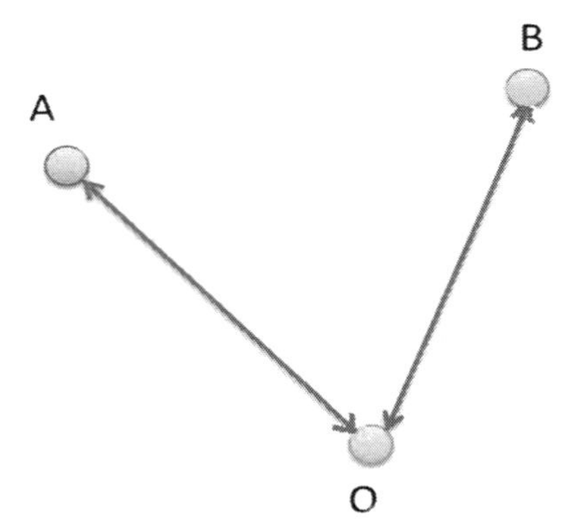

Fig. 1 Line one

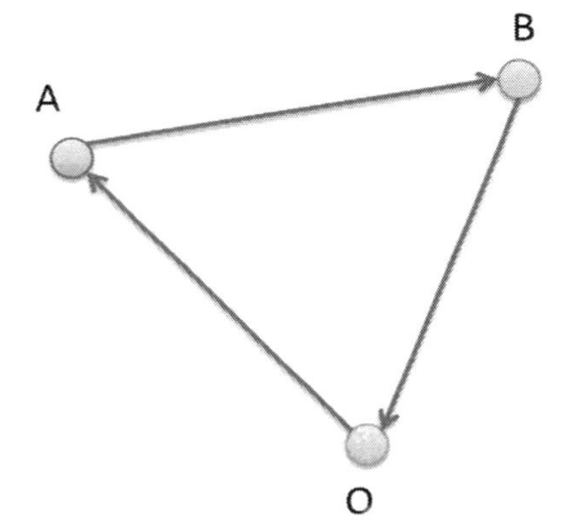

Fig. 2 Line two

3 Apply the Saving Algorithm

In order to testify the effectiveness of the method, randomly selected seven points from the East China of SF company. First distribution Hangzhou station is the origin point, assuming that the demand for each sorting station is equivalent to the sorting station handling express number of votes. We get the following data (Tables 1 and 2).

The paper estimates the per kilometer operating costs: 7.1 t and 11.2 t operating costs are 2.09 yuan/km, 2.55 yuan/km.

First, calculate the Distance of saving in same route (Table 3):

The second step, to determine transport routes and transport costs of the initial program, now arranged six 7.1 t trucks delivery to each sorting station, transport routes and their costs in the table below (Table 4):

According to the basic principles of Saving Algorithm, get the following results of final optimization of the routes (Fig. 3a and Table 5):

By comparing the initial proposal and the final proposal we can get that by optimizing the rout we can save about 252.2 km and 527.1 yuan.

Table 1 Express demand for some cities (unit: million votes per day)

Customer(i)	Suzhou	Wuxi	Jiaxing	Shaoxing	Ningbo	Wenzhou
Freight volume (q_i)	2.8	2.3	1.8	2.4	2.3	3.3

Table 2 Road distance between some cities (unit: km)

	Hangzhou	Suzhou	Wuxi	Jiaxing	Shaoxing	Ningbo	Wenzhou
Hangzhou	0	–	–	–	–	–	–
Suzhou	166	0	–	–	–	–	–
Wuxi	208	50.8	0	–	–	–	–
Jiaxing	90.9	80.8	121	0	–	–	–
Shaoxing	64.2	197	238	123	0	–	–
Ningbo	155	230	272	156	117	0	–
Wenzhou	364	493	532	419	312	269	0

Table 3 The distance of saving in same route

	Suzhou					
Wuxi	323.2	Wuxi				
Jiaxing	176.1	177.9	Jiaxing			
Shaoxing	33.2	34.2	32.1	Shaoxing		
Ningbo	91	91	89.9	102.2	Ningbo	
Wenzhou	37	40	35.9	116.2	250	Wenzhou

Table 4 Transport routes and transport costs

Transit route	Truck	Distance (km)	Unit cost (Yuan/km)	Freight (Yuan)
Hangzhou – Suzhou	7.1 t	166	2.09	346.94
Hangzhou – Wuxi	7.1 t	208	2.09	434.72
Hangzhou – Jiaxing	7.1 t	90.9	2.09	189.98
Hangzhou – Shaoxing	7.1 t	64.2	2.09	134.18
Hangzhou – Ningbo	7.1 t	155	2.09	323.95
Hangzhou – Wenzhou	7.1 t	364	2.09	760.76
Total		1,048.1		2,190.53

4 Multi-dynamic Saving Algorithm

Single Saving Algorithm is not without flaws. It obviously that saving more cost means sacrificing more time [6, 7].

To compensate for the time loss caused by Saving Algorithm, we apply the Saving Algorithm in every distribution center, forming a "multi-dynamic" type of Saving Algorithm to ensure the unity of cost savings and the whole time in SF company. When regard second distribution of Ningbo stations as the originating point, we can draw a new independent Saving Algorithm route, as shown (Figs. 3b and 4):

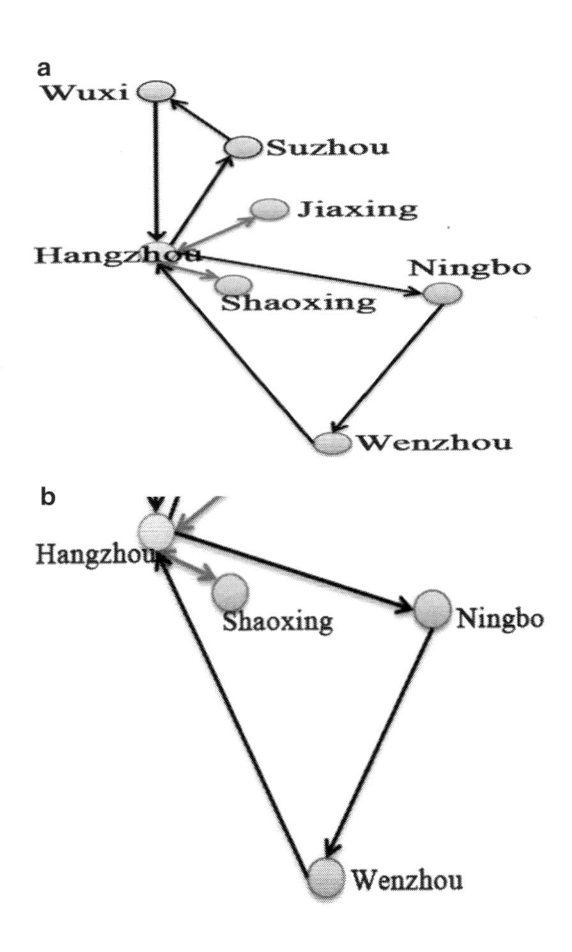

Fig. 3 The final optimization results (**a**) and (**b**)

Table 5 Final scheduling program and cost

Transit route	Truck	Distance (km)	Unit cost (Yuan/km)	Freight (Yuan)
Hangzhou – Suzhou – Wuxi	7.1 t	216.8	2.09	453.11
Hangzhou – Jiaxing	7.1 t	90.9	2.09	189.98
Hangzhou – Shaoxing	7.1 t	64.2	2.09	134.18
Hangzhou – Ningbo – Wenzhou	7.1 t	424	2.09	886.16
Total		795.9		1,663.43

Each individual mileage saving is responsible only for the goods of demand points, and no other waste motion, such as cargo undertaking and exchanging. Such arrangement has the following two advantages: (1) it saves the Lay time in sorting points; (2) vehicles do not have to wait for the delay in shipment, ensuring the timeliness of the system as a whole [8].

Fig. 4 The optimization results of Ningbo origin

5 Conclusion

As for saving algorithm, there can be a balance improvement with cost and timeliness. The mileage saving algorithm saves was the trend of diminishing marginal, that is, the more close to the cost-optimal, limitation loss is more. So it can be considered in combination with timeliness, the optimal solution can best meet the cost and timeliness to a certain extent. Although the return path will be no-load, compared to ensure the continuity and timeliness of the overall system, the expense of some sections of the timeliness and cost is acceptable.

Acknowledgments This study is granted and supported by Natural Science Foundation of China, the Ministry of Education of China, Shanghai Municipal Education Commission, Shanghai Science and Technology Commission, and Shanghai Maritime University (Grant number: 70541009; 11YJA630067; S30601; 11510501800; 20110020).

References

1. Cao Jian-dong, Zheng Si-fa, Li Bing, Yang Yang, Lian Xiao-min (2008) Optimization of urban pickup and delivery costs with one way multi load and unload. J Syst Simul 20(1):29–32
2. Liu Wusheng, Liu Jun (2007) Application of an improved economical method in delivery route optimization. Mod Transport Technol 4(6):72–74,78
3. Dror M, Trudeau P (1986) Stochastic vehicle routing with modified savings algorithm. Eur J Oper Res 23(2):228–235
4. David Simchi-Levi, Philip Kaminsky, Edith Simchi-Levi (2002) Designing and managing the supply chain: concepts, strategies, and cases. McGraw-Hill/Irwin, New York
5. Dong Mingfeng (2010) Study on the vehicle routing problems of TFGL company. South China University of Technology, Guangzhou
6. Potvin J, Duhamel C, Guertin F (1996) A genetic algorithms for vehicle routing problem with backhauling. Appl Intell (S0924-669X) 6:345–355
7. Wade AC, Sal Hi S (2002) An investigation a new class of vehicle routing problem with backhauls. Omega (S0305-0483) 30(6):479–487
8. Montemanni R, Gambardella LM, Rizzoli AE, Donati AV (2002) A new algorithm for a dynamic vehicle routing problem based on ant colony system. Technical report IDSIA-23-02, IDSIA, Manno, Nov 2002, pp 164–168

A Novel Method on Layout Planning for Internal Functional Areas of Logistic Parks

Xiaomin Zhu and Qian Zhang

Abstract This paper proposes a novel method on layout planning for internal functional areas of logistics parks by combining SLP and Fuzzy Cluster Analysis, where the layout planning problem can be studied in two angles of quantity and quality. Based on this new method, the reasonable correlation is achieved between the various functional areas of logistics parks. Meanwhile, close links of the logistics parks and external transport facilities can also be achieved.

Keywords Logistics parks • Layout planning • SLP • Fuzzy cluster

1 Introduction

Layout planning for logistics parks is becoming more and more important in the construction of logistics system. A rational layout has a direct impact on the operational efficiency of the logistics park, and it's the critical process for the implementation of the logistics park's strategy [1].

Richard Muther [2] put forward the theory of Systematic Layout Planning (SLP) in 1961. Wuhan University of Technology [3] put forward application of Fuzzy Clustering for logistics park general layout planning in 2010.

Existing methods on layout planning of functional areas of the logistics park mainly includes two aspects: qualitative and quantitative research.

Quantitative methods, such as Systematic Layout Planning, mainly emphasis on the relevance of business processes. The basis and foundation of this kind of method is the flow of goods between the logistics facilities or degree of closeness.

X. Zhu (✉) • Q. Zhang
School of Mechanical, Electronic and Control Engineering, Beijing Jiaotong University, Shangyuancun 3, 100044 Beijing, People's Republic of China
e-mail: xmzhu@bjtu.edu.cn; 11121411@bjtu.edu.cn

Z. Zhang et al. (eds.), *LISS 2012: Proceedings of 2nd International Conference on Logistics, Informatics and Service Science*, DOI 10.1007/978-3-642-32054-5_20,

But the exchange of external logistics relations and non-logistics relationship is rarely considered.

Fuzzy cluster method clusters functional areas that have similar properties or strong correlations from a qualitative point of view. The layout of functional areas is carried out on this basis. Although service capabilities and dynamic of the park is considered more, relatively more subjective factors are pulled in.

This paper aims at layout planning for internal functional areas of logistics parks using SLP and Fuzzy Cluster Analysis. The layout planning problem can be described in two angles of quantitative and qualitative. Based on this new method, the reasonable correlation is achieved between the various functional areas of logistics parks. Meanwhile, close links of the logistics parks and external transport facilities can also be achieved.

2 Layout Method

The layout method for internal functional areas of logistic park in this paper is a combination of SLP and Fuzzy Cluster Analysis. This layout method mainly contains these following steps:

The first step, Preliminary Layout of Logistic Park with SLP Method.

Firstly, logistic park should be analyzed according to the actual situation and planning requirements. And then the logistic park could be divided into the best classified scheme. According to the nature of logistics activity, the internal function areas of logistic park can be divided into the following two categories, shown in Fig. 1 [4]:

Then, the logistic park should be implemented comprehensive quantitative analysis. Furthermore, the integrated closely related degree R_{ij} of each functional area and control point could be worked out. And then the comprehensive quantitative correlation graph is obtained, shown in Fig. 2.

According to the integrated closely related degree, the distance between each functional area can be determined. After being estimated of the space, each functional area's location would be implemented. Then the preliminary layout of logistic park is worked out.

The second step, Preliminary Layout of Logistic Park with Fuzzy Cluster Analysis Method.

Firstly, the functional areas should be fuzzy clustering analyzed. The functional areas having similar property and strong correlation should be clustered in order to decrease the handing expense and avoid the bypass of People flowing and goods flowing between the functional areas.

There have many factors, such as geographical conditions, domain area, surrounding traffic conditions and land character, should be analyzed, shown in Fig. 3.

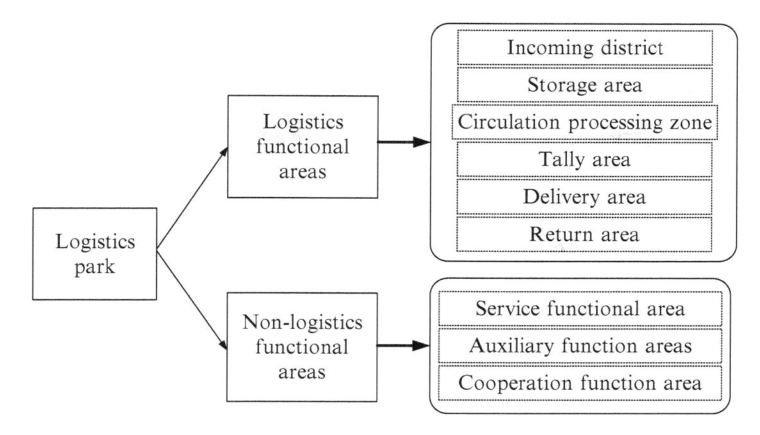

Fig. 1 Generally division of the functional areas of logistics center

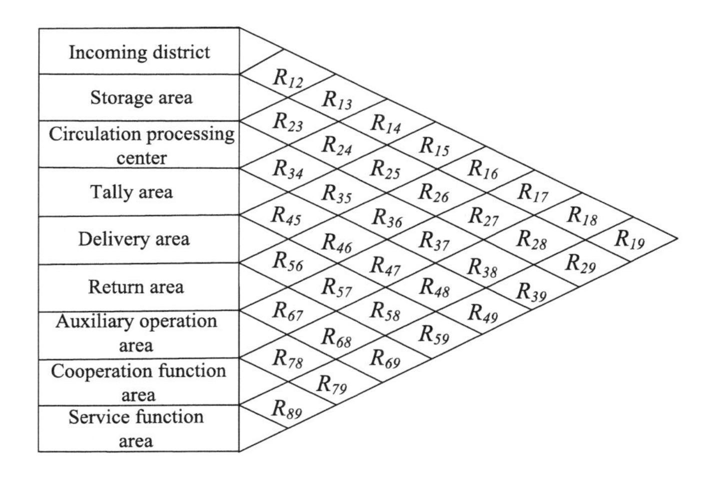

Fig. 2 Comprehensive correlation graph of functional area

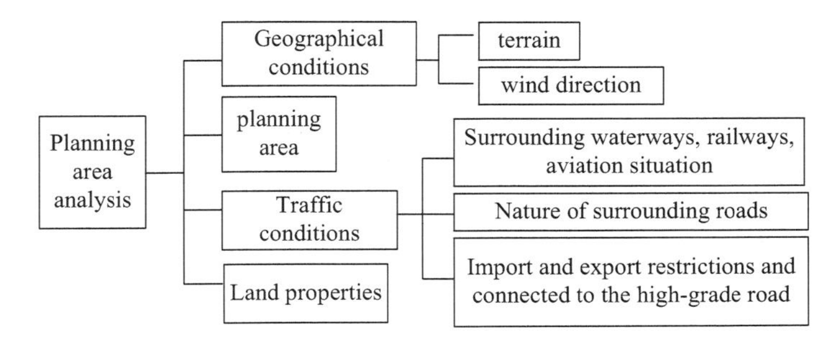

Fig. 3 Analysis of the planning area

The clustering functional areas should be analyzed from three points, such as transportation, goods and environment. The planned area also should be analyzed from three points, such as geographical conditions, domain area, surrounding traffic conditions and land character. Taking all these factors, the functional area of logistic park can be planned and arranged. Then the preliminary layout of logistic park is worked out.

The third step, Simulation, Optimization and The Final Layout scheme.

The previous two stages have applied SLP and fuzzy cluster analysis method to layout the logistic park and obtained two kinds of layout scheme. In order to test the dynamic service capabilities of logistic park and achieve the smooth flow of goods, we can use Arena [5] software to model and simulate the two preliminary layout schemes. The average queuing time of the two preliminary layout schemes should be compared. On the basis of above, the two layout scheme should be optimized and adjusted.

On the basis of optimized preliminary layout, non-logistics functional areas should be arranged in the reasonable position. Then the final layout scheme is worked out.

3 Case Analysis

In order to verify the practicability and superiority of the method elaborated in this paper, we will use this new method to layout the internal functional area of one logistic park.

According to the preliminary planning, this logistics park will achieve the function of warehousing, transshipment, circulation and processing. According to this, the logistic park can be divided into many functional areas, such as warehousing logistics center, Freight transportation center, professional logistics operations center, counter-trade center, exhibition center, comprehensive office service center.

These functional areas should be layout with SLP and Fuzzy Cluster Analysis method. Then the two preliminary layout schemes will be obtained and they should be modeled and simulated. Comparing the average queuing time of it, the two schemes should be optimized and combined. Then, the final layout will be worked out, shown in Fig. 4.

4 Conclusion

In the future, we can layout the internal functional areas of logistics parks by combining SLP and Fuzzy Cluster Analysis. This method is more reasonable.

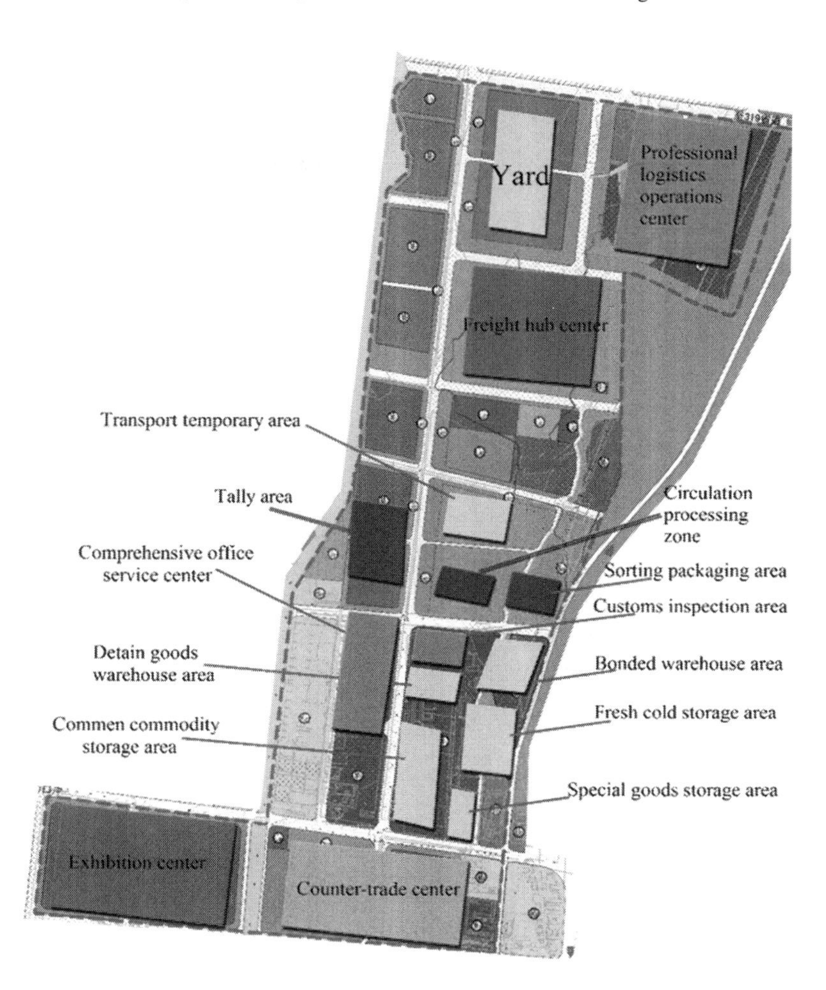

Fig. 4 Final layout scheme of logistic park

References

1. Huang PP, Ge YX (2010) The analysis of the problem of Chinese logistics Park's development. Logist Sci-Tech 11:29–31
2. Lin QL, Liu L, Li P (2009) Application of systematic layout planning to operation rooms in a hospital. Ind Eng J 23:111–14,125
3. Pan NN, Zhang PL, Yang CB (2010) Application of fuzzy clustering for logistics park general layout planning. Ports Waterway Eng 11:9–11
4. Wang SQ, Liu W (2008) Layout planning with a controlling structure to logistics parks. In: 2008 I.E. international conference on automation and logistics, Shanghai, China. J Shanghai Sec Polytech Univ 26:2039–2043
5. Kelton WD (2002) Simulation with Arena. The McGraw-Hill Company, New York

Research on Supply Chain Disruption Risk and Its Hedge Based on Reliability Theory

Fu Shaochuan and Han Qingming

Abstract This article analyses the disruption risk of the supply chain. By introducing the supply chain reliability theory, it establishes a supply chain model with reliability constraints based on the reliability theory. The article simplified the model and is proposed to use *Microsoft Excel* Solver to solve the 0–1 integer programming problem. This method is simple and convenient, laying the foundation for the application of this theory in practice.

Keywords Reliability • Supply chain disruption risk • 0–1 integer programming

1 Introduction

Supply chain disruptions are unanticipated events that disrupt the normal flow of goods and materials within a supply chain [1]. Supply chain disruptions could lead to many problems, such as extending the lead time, out of stock, not satisfactory customers. Such as "9 • 11" terrorist attacks, which led to the closure of U.S. airspace, earthquakes in Japan and the SARS outbreak in 2003, all of them have paralyzed the supply chain.

These pressures are driving intense effort and initiatives to reduce exposure to risk. So it is important to have insight into network construction to avoid disruptions in the supply chain. Therefore, to prevent the supply chain disruption risk and improve the reliability of the supply chain has theoretical and practical significance.

F. Shaochuan • H. Qingming (✉)
School of Economics and Management, Beijing Jiaotong University,
Beijing, Haidian District, People's Republic of China
e-mail: hanqingming@foxmail.com

Z. Zhang et al. (eds.), *LISS 2012: Proceedings of 2nd International Conference on Logistics, Informatics and Service Science*, DOI 10.1007/978-3-642-32054-5_21,

2 Literature Review

2.1 Supply Chain Disruptions Management

Snyder and Shen [2] develop simulation models for several networks comparing stochastic demand and supply disruptions. They conclude the two different sources of stochasticity have very different impacts on optimal supply chain design. Hopp and Yin [3] examine an assembly system with the potential for disruptions in capacity.

While among the research, sophisticated quantitative analysis continues to appear, but in general it is not extensive. In view of this, this article analyses the disruption risk of the supply chain by introducing supply chain reliability theory.

2.2 Classification of Risks in a Supply Chain

Classification of risks in a supply chain is also a growing research trend. For example, Deleris and Erhun [4] present a Monte Carlo simulation that they use to evaluate risk levels in the supply chain. According to the risk source, supply chain disruption risk fall into three broad categories [5]: External risks can be driven by events either upstream or downstream in the supply chain. Internal risks provide better opportunities for mitigation because they are within your business's control. Risks between enterprises in the supply chain make influences on all of the members in the supply chain.

2.3 Supply Chain Reliability

The concept of supply chain reliability is related to network reliability theory [6], which is concerned with calculating or maximizing the probability that a graph remains connected after random failures due to congestion, disruptions, or blockages.

There are three structural models of supply chain, the series connection structure, parallel structure and series–parallel structure.

Series structure is a system constructed in series, in which with the increase of nodes, the reliability of supply chain will gradually decline. Parallel structure is a system constructed in parallel, in which the entire system can run as long as a node is in the normal operation. Series–parallel structure is a system constructed both in series and parallel. The reliability of the system depends on not only the series structure but also the parallel structure, and its reliability analysis is more difficult, but it can be a better solution to the practical problem.

3 Modeling the Supply Chain Structure Network

According to the above analysis, a network model of series–parallel structure of supply chain will be made and optimized, building a reliable supply chain model.

3.1 Analysis Reliability Function of Network Structure of Supply Chain

Supply chain network model is constructed in a series–parallel structure in reliability analysis, whose reliability block diagram is shown as following Fig. 1. The reliability of the j node of the subsystem is $R_{ij}(t)$, whose operational lifetime is T_{ij}. And the operational life of the system is set as t.

$$\begin{aligned}R_i(t) &= P(T_i > t)\\ &= P\{\max[T_{i1}, T_{i2}, \ldots, T_{im}] > t\}\\ &= 1 - P\{T_{i1} \le t, T_{i2} \le t, \ldots, T_{im} \le t\}\\ &= 1 - \prod_{j=1}^{m} [1 - R_{ij}(t)]\end{aligned} \tag{1}$$

Assume that the distribution of the operational life of each node within each subsystem is independent of each other. $R_i(t)$ is strictly increasing which means that with m increased, the reliability of node i , $R_i(t)$, will increase.

$$\begin{aligned}R(t) &= P(T > t) = P\{\min[T_1, T_2, \ldots, T_n] > t\}\\ &= P\{T_1 > t, T_2 > t, \ldots, T_n > t\}\\ &= \prod_{i=1}^{n} R_i(t)\end{aligned} \tag{2}$$

Assume that the distribution of the operational life of each node is independent of each other. $R(t)$ is strictly decreasing which means that with n increased, the

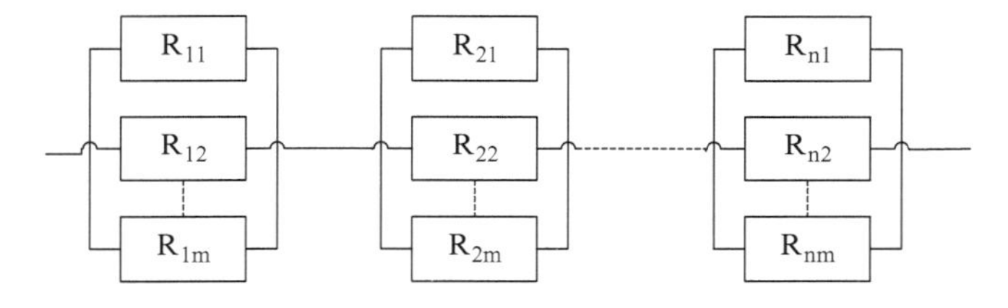

Fig. 1 Reliability diagram of the supply chain network model

reliability of system $R(t)$ will decrease. In summary for the whole supply chain, with the increase of subsystem the system reliability will gradually decline.

3.2 Mathematical Model and Resolution

Supply chain model consists of n sub-systems. The cost of child nodes in each sub-system is C_{ij}. The sub-system of supply chain is composed of m nodes.

The objective of the supply chain model is the reliability of the series system which consisting of subsystem to meet the condition, $R \geq R_0$. And the reliability of each subsystem meets the condition $R_i \geq R_0(i)$, and spends the minimum total cost of C.

$$\sum C = \min\left\{\sum_{i=1}^{n} \sum_{j=1}^{m} C_{ij} \cdot X_{ij}\right\} \tag{3}$$

$$R = \prod_{i=1}^{n} R_i \cdot X_i \geq R_0 \tag{4}$$

$$R_i \cdot X_i \geq R_0(i) \tag{5}$$

$$R_i = 1 - \prod_{j=1}^{m} [1 - X_{ij} \cdot R_{ij}] \tag{6}$$

$X_i = 1$, subsystems i should be included in the supply chain model, or $X_i = 0$
$X_{ij} = 1$, node j should be included in the subsystem i, or $X_{ij} = 0$

According to the definition of the model, we need to determine which subsys-tem-s should be included in the whole supply chain, and which node enterprises s-should be included within each subsystem in order to find a program with the minimum cost based on the reliability.

3.3 Solving Ideas

This model is an 0–1 integer programming model. Usually the exhaustive method would be used to solve the 0–1 integer programming problem to find the optimal solution. Using Microsoft Office Excel to solve such problems is fast and accurate, and you can get a multiplier effect. In accordance with practi-cal requirements, the solution ideas are given as following:

(1) Entry the known cost, reliability data, increase the variable region; (2) Calcu-late the cost of each subsystem; (3) Calculate the reliability of each subsystem; (4) Set constraints; (5) Calculate the total cost of the supply chain system.

9783642320538

9 783642 320538

4 Numerical Studies

4.1 Conditions and Assumptions

To illustrate more convenience, the numerical example was simply adjusted, assuming that there are four alternative subsystems in the supply chain, and each of the subsystem has four alternative nodes. The reliability of the supply chain system requires 0.95. In addition, each subsystem should include at least two nodes to make sure the stability of the system. The cost and reliability of each node in each subsystem are shown in Tables 1 and 2.

4.2 Modeling Resolving

Set the corresponding parameters in the Solver function in Excel and select the target cell, constraints and the variable cell. This results show in a graphical representation of Fig. 2. The supply chain consists of four subsystems, the first subsystem consists of 1,3 node, the second consists of the 1,4 node, the third consists of the 1,2 node, the fourth consists of the 2,3 node. The optimal cost is

Table 1 The cost of each subsystem node units: 10,000 yuan

	Subsystem 1	Subsystem 2	Subsystem 3	Subsystem 4
Node 1	1.90	2.00	1.60	1.90
Node 2	2.00	2.20	2.00	2.30
Node 3	1.80	2.40	1.80	1.80
Node 4	1.50	1.80	2.40	1.50

Table 2 The reliability of each subsystem of each node

	Subsystem 1	Subsystem 2	Subsystem 3	Subsystem 4
Node 1	0.90	0.85	0.80	0.75
Node 2	0.80	0.70	0.80	0.85
Node 3	0.85	0.80	0.90	0.90
Node 4	0.70	0.90	0.70	0.70

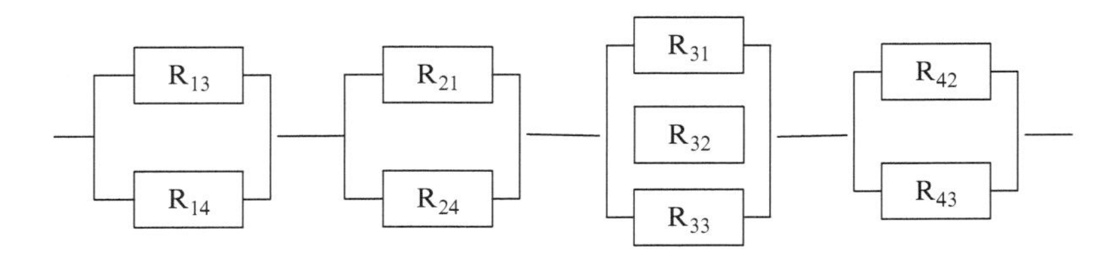

Fig. 2 Solver results

17 million dollars. The reliabilities of the four subsystems are 0.99, 0.99, 1.0 and 0.99. The reliability of the entire supply chain is 0.95 which means that it meets the reliability requirement.

5 Conclusions

This paper analyses the disruption risk of the supply chain. By introduction of supply chain reliability theory, establishing a supply chain model with reliability constraints based on the reliability theory, taking into account the conditions of reliability and resource and the minimization of the cost of the supply chain system. The numerical study show that the use of the model and the algorithm could provide a reference to the supply chain network determination with reliability constraints, which can not only reduce the overall cost of the supply chain to ensure rapid response capability, but also improve the supply chain to resist interruption risks.

References

1. Hendricks KB, Singhal VR (2003) The effect of supply chain glitches on shareholder wealth. J Oper Manag 21:501–523
2. Snyder LV, Shen ZJM (2006) Supply and demand uncertainty in multi-echelon supply chains. Working paper, P.C. Rossin College of Engineering and Applied Sciences, Lehigh University, Bethlehem, PA
3. Hopp WJ, Yin Z (2006) Protecting supply chain networks against catastrophic failures. Working paper, Department of Industrial Engineering and Management Science, Northwestern University, Evanston, IL
4. Deleris LA, Erhun F (2005) Risk management in supply networks using Monte-Carlo simulation. In: Kuhl ME, Steiger NM, Armstrong FB, Joines JA (eds) Proceedings of the 2005 winter simulation conference, Association for Computing Machinery, Piscataway, pp 1643–1649.
5. Oke A, Gopalakrishnana M (2009) Managing disruptions in supply chains: a case study of a retail supply chain. Int J Prod Econ 118(1):168–174
6. Colbourn C (1987) The combinatorics of network reliability. Oxford University Press, New York

The Location-Routing Problem in the Food Waste Reverse Logistics System

Shisen Li, Li Wang, Haiping Wu, and Shiqiang Bai

Abstract According to the characteristics of food waste reverse logistics system, we studied the recovery point location-routing problem. Using the 0–1 programming and the minimal cost-maximal flow graph theory, we established a mathematical model of the food waste reverse logistics system, and find out the location of the recovery point and the route by LINGO language program.

Keywords Food waste • Reverse logistics • 0–1 programming • Minimal cost-maximal flow

1 Introduction

With the development of socio-economic and the improvement of people's living standards, massive amount of food wastes are generated by residents' daily diet consumptions. If these food waste collections are not transported timely or handled properly, they will result in a huge waste of resources and serious environmental pollution.

As you know, food waste has two basic characteristics [1], environmental pollution and resource regeneration. Firstly, food waste and the surrounding environment will cause adverse effects without timely collection and handling properly, such as the spread of pathogens and infection, and even endangering human health. On the other hand, the food waste is rich in nutrients, which represents a good resource recovery value, in addition the organic nutrient content showed an

---

Hebei Social Science Fund Project (HB12GL069): Research on Incentive Mechanism of Common Food Waste Resources in Reverse Logistics

S. Li (✉) • L. Wang • H. Wu • S. Bai
Department of Economics and Management, Shijiazhuang Institute of Railway Technology, 050041 Hebei, People's Republic of China
e-mail: forests_li@126.com; wangqiuxi999@126.om; zu_zhi@126.com; bsq1234xx@126.com

Z. Zhang et al. (eds.), *LISS 2012: Proceedings of 2nd International Conference on Logistics, Informatics and Service Science*, DOI 10.1007/978-3-642-32054-5_22,

increasing trend with the improvement of people's living standards. Therefore, the food waste recycling has become a major problem that plagued the city for sustainable development. It is urgent to establish the food waste recycling system to meet the requirements of the increasing public awareness on environmental protection.

2 Food Waste Reverse Logistics System Location-Path Analysis

The most important parts of the food waste recycling system are the collection and the transportation. In fact, the collection and transportation costs can account for up to 50% of the total costs of handling food wastes [2]. Therefore, the choice of the optimal transport path and the recovery position of food waste become the critical factors in establishing an effective food waste logistics system. It is more advanced to separate the short distance collection and haul transportation by a transfer station in order to save transportation costs [3]. This recovery system will certain create the cost of the construction and operation of transfer station. Therefore, the number and the location of transfer station, as well as the transport path become the two critical issues, which will be discussed in the following paragraphs.

Because of the dispersion of the food waste origin [4], we take the third-party repo structure to descript the recyclable procedure. First, the food wastes are transported from the food waste original points to the professional collection points, and then transported to a recycling center (which is transfer station). In this structure, the main problems are where and how many collection points should be built, and how to arrange the route.

3 System Modeling

During the operation of the system, the costs are mainly generated from two ways: (1) the cost of transportation, which is depended on the cost of transport distance and unit freight; (2) the cost of building a recycling point.

3.1 Parameters and Variables Explain

(1) $h_i, i = 1, 2, \cdots, m$ – the amount of the food waste original points.

$s_j, j = 1, 2, \cdots n$ – the capacity of the collection points.

(2) (x_i, y_i) – The coordinates of the food waste original points.
$(\bar{X}_j, \bar{Y}_j)$ – The coordinates of the total recycling center.
(X_j, Y_j) – The coordinates of the collection points.

(3) $l_{ij} = \sqrt{(x_i - X_j)^2 + (y_i - Y_j)^2}$ – the distance from food waste original points to the collection points.

$\bar{l}_j = \sqrt{(x_i - X_j)^2 + (y_i - Y_j)^2}$ – the distance from the collection points to the total recycling center.

(4) c_{ij} – the unit freight from food waste original points to the collection points.

$\bar{c}_j$ – the unit freight from the collection points to the total recycling center.

(5) r – the fixed costs of the construction and operation of collection points.

(6) R – the capacity of the waste recycling center.

(7) $x_i = \begin{cases} 1, & \text{Build a recycling point} \\ 0, & \text{else.} \end{cases}$

3.2 Model Construction

The objective function:

$$\frac{\min z}{\text{objection}} = \frac{\sum_{j=1}^{n}\sum_{i=1}^{m}(l_{ij} \times c_{ij} \times t_{ij})}{(1)} + \frac{\sum_{j=1}^{n}\bar{l}_j \times \bar{c}_{ij} \times \left(\sum_{i=1}^{m} t_{ij}\right)}{(2)} + \frac{\sum_{j=1}^{n} x_j \times r_j}{(3)}.$$

The objective function is obtaining the minimum total cost to run the system. The part (1) of the objective function is the cost of food waste transported from the point of origin to collection points. The part (2) of the objective function is the cost of food waste transported from collection points to the recycling center. The part (3) of the objective function is the costs of the construction and operation of collection points. The objective function needs to satisfy the following five constraints:

$$\sum_{i=1}^{m} t_{ij} \leq x_j s_j \tag{1}$$

$$\sum_{j=1}^{n} l_{ij} = h_i \tag{2}$$

$$\sum_{i=1}^{m} t_{ij} = \bar{t}_j \tag{3}$$

$$\sum_{j=1}^{n}\sum_{i=1}^{m} t_{ij} \leq R \tag{4}$$

The formula (1) is to ensure each collection points does not exceed its capacity limit. The formula (2) is to ensure all the food waste which created from the original points is shipped to the collection points, without surplus. The formula (3) is to ensure that the food wastes from collection points are all shipped to a recycling center, without surplus. The formula (4) is to ensure that the processing capacity of the recycling center is not less than the amount of food waste generated. The last one is all variables are non-negative to ensure the result obtained is of practical significance.

3.3 Algorithm Description

We use the thought of 0–1 program to solve the minimum cost maximum flow problem in this model. So, the algorithm should also take the idea of solving the 0–1 program and the minimum cost maximum flow problem. Minimum cost maximum flow problem in operations research has a more mature algorithm, we can use the following steps to solve the model:

Step 1: Take the food waste original points, the collection points, the recycling center as the node. Set the path from food waste original points to the collection points and the path from the collection points to the total recycling center as a directed arc, set the transported amount of waste to food waste as flow. Set the transportation unit price as cost. We could make a directed graph.

Step 2: Assuming that there is a starting point, the distance and the unit freight of the point to the food waste original points are zero. The total flow of this point to all nodes must be equal to the total amount of all the food waste original points.

Step 3: Define the distance from v_i to $\mathrm{V_j}$ as $v_i - V_j = \begin{cases} l_{ij} & if\ x_j = 1 \\ \infty & if\ x_j = 0 \end{cases}$, which means if the collection point is not built, there will be no food waste shipped to this collection point.

Step 4: Add the cost of building collection points to the objective function of the minimal cost-maximal flow problem, and use the Dijkstra's algorithm to solve the model.

4 Case Study

According to a case studied in Hebei Province, we got 19 food waste original points coordinates and the amount generated per day through the gravity method (see Table 1). In our research, we selected 5 appropriate coordinates to address the collection points in the city. The coordinates of each collection points and

Table 1 Food waste original points coordinates and the amount generated per day

Order number	Horizontal coordinate	Vertical coordinate	Daily output (100 kg)	Order number	Horizontal coordinate	Vertical coordinate	Daily output (100 kg)
1	5.23	1.27	3.93	11	4.27	13.9	5.27
2	6.22	4.76	3.8	12	5.26	15.6	3.13
3	2.48	3.76	2.47	13	11.17	5.57	3.93
4	9.54	6.67	4.6	14	13.46	3.02	1.8
5	8.29	9.8	2.87	15	13.17	4.64	3.4
6	12.44	9.08	2.47	16	13.09	14.3	4.87
7	15.46	12.12	4.47	17	13.56	16.13	2.87
8	4.98	10.87	3.93	18	10.2	12.82	3.27
9	3.26	13.5	5.93	19	9.48	13.78	3.8
10	6.2	10.66	1.67				

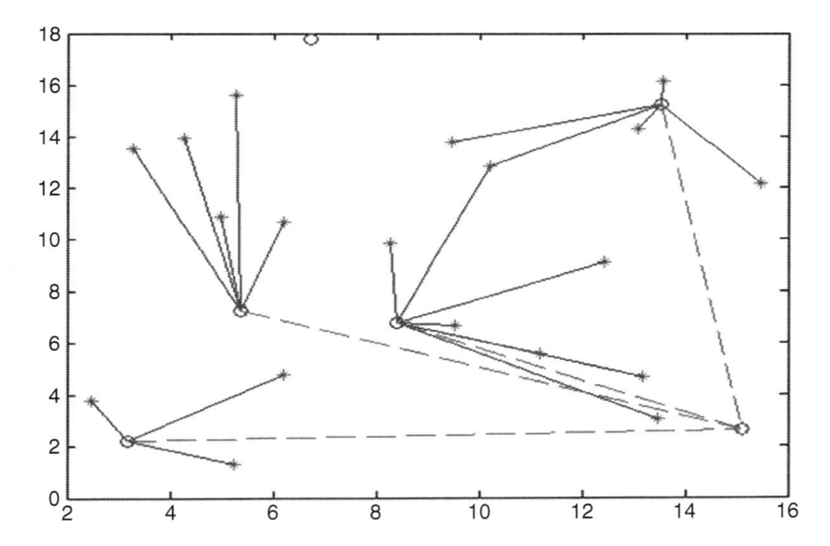

Fig. 1 The optimal result of the location-routing problem

corresponding daily handling capacity are (3.19, 2.17, 20), (6.71, 17.74, 10), (13.52, 15.22, 30), (8.41, 6.75, 20), (5.36, 7.24, 20), values in brackets represent the horizontal coordinate, the vertical coordinate and the daily handling capacity (100 kg). The coordinates of the recycling center is (15.1, 2.62).

Using LINGO11.0 to run the program, we could get the optimal result: 364 million yuan, and the coordinates of collection points should be (3.19, 2.17), (13.52, 15.22), (8.41, 6.75) and (5.36, 7.24). The respective handling amount of each collection point is 10.2, 18.35, 20 and 19.93 (100 kg). The recovery paths are shown in Fig. 1. In this figure the blue solid line is from the food waste original points to the collection points, and the red dashed line is from the collection points to the recycling center.

5 Conclusion

With the rapid growth of Chinese economy, food waste production in each city has grown. Therefore, how to handle and recycle these wastes is becoming an important issue the government faced. This paper mainly studied the location of the urban food waste recycle location-routing problem and taken into account both the distance and transportation costs. In addition, it considered the cost of building collection station. Further research includes the following aspects:

(1) This article does not consider food waste during transport, means of transport load capacity constraints and the choice of means of transport routes. Therefore, the actual means of transport constraints is taken into account to establish a more realistic model. (2) To consider the uncertainties, such as every food waste generated waste generated by the point (hotel) is not fixed, so you should consider the establishment of a probability model. (3) Considering the negative effects of waste recycling on people's lives, we could establish the multiobjective programming model.

References

1. Li Wang, Yingzong Liu (2009) Food waste recycling research based on the reverse logistics. J Xi'an Electron Sci Univ 19:62–67
2. Xinfu Lv, Linning Cai, Zhiwei Qu (2005) The location-routing problem in the waste recycling logistics. Syst Eng Theory Pract 5:89–94
3. Fuhua Xu, Lihua Huang (2004) The utilization of food waste in Shanghai. China Environ Prot Ind 4:42–43
4. Aiguo Guan (2006) The establishment of modern renewable resource recycling system to promote the sustainable development of social economy in China. Res Renew Res 2:1–4

Application of Quality Cost and Quality Loss Function in Food Supply Chain Systems Modeling

Tianyuan Zheng and Michael Wang

Abstract Today, the food supply chains not only fulfill the human daily needs, but also contribute to the economical development of domestic and in many cases, off-shore economics. In this paper, we propose a basic model for evaluating and optimizing the performance of the retailers in food supply chain systems. The basic model can be extended to different scenarios such as (1) producer with retailer functions, (2) producer with no retailer functions, and (3) retailer with no production functions. Within each model, the performance for different stakeholder is evaluated based on the food quality of nutrition value, physical sense, and the opportunity cost of food product risks.

Keywords Food supply chain • Quality cost models • Food quality loss

1 Introduction and Review

Worldwide, the food industry has recently drawn much attention due to issues related to human health and safety. The performance of such a food supply system is heavily based upon the interactive activities between other business entities. It follows the similar chain structure as the manufacturing industries but the structure is more complex and the variety of the food products is much more diverse. The value of food products includes two parts, the nutritional value, and the physical senses value. Previous models include zero-order reaction kinetics, first-order reaction kinetics, fractional conversion kinetics, the Bigelow model, and non-linear microbiological death model [1,2]. Zero-order reaction kinetics is the traditional model, with simple calculation but with larger errors in estimation. First-order

T. Zheng (✉) • M. Wang
Department of Industrial and Manufacturing Systems Engineering,
University of Windsor, Windsor, ON, Canada
e-mail: zhengs@uwindsor.ca; wang5@uwindsor.ca

Z. Zhang et al. (eds.), *LISS 2012: Proceedings of 2nd International Conference on Logistics, Informatics and Service Science*, DOI 10.1007/978-3-642-32054-5_23,

reaction kinetics and fractional conversion kinetics are models based on the experiments by changing the content during certain stages for food storage. The Bigelow model and non-linear microbiological death model are models that have been used to illustrate the changes of nutrients within the food product during a more complex situation, including the effects of changes in temperature during food cooking and after [3]. In most literatures, the costs due to the loss of product value is considered as result of food deterioration, and is normally modeled with linear or exponential deterioration rate to illustrate the cost of such loss. Their assumption is that, the reduction of inventory level is a result of joint operation of demand and deterioration. Two models were developed by Fujiwara [4] with linear deterioration rate, and Chung and Huang [5] with an exponential deterioration rate. Quantification of risk is part of the risk assessment process. Huss et al. [6] developed a semi-quantitative assessment system to evaluate the risk of seafood products. Van Gerwen et al. [7] developed a SIEFE system, and by setting different scale of risk factors on each risk level, the overall quantitative risk could be obtained. Ross and Summers [8] developed a model with nine risk input values. The most important part of their model is the concept of comparative risk. This risk contains the evaluation of probability of illness over all servings, annual exposures per person in a daily basis, and the hazard severity factor.

2 Design and Methodology

Notation and Assumptions:

D, d Annual (D) and daily demand (d) for a food product at the retailer, $d = D/365$.
Q Order quantity, $Q = d*n$, where n is days for each ordering period.
C_o Ordering cost, all cost associated with the placement of an order.
C_s Setup cost, all cost associated with the setup of product for each batch
Q_p Size for each production batch.
S_r Retail price for food product at the retailer.
S_p Price for food product at the producer.
y Producer's quality level (in Taguchi Quality Loss function concept).
h Holding cost rate at the retailer, during the storage and on shelf period before being purchased by consumers.
Q_{10} Food deterioration parameter, used in the model for loss of nutrition.
F_1 Food product life labelled, which is based on the storage temperature of T_1.
F_2 Food product real shelf life, which is based on the storage temperature of T_2.
k Food product deterioration rate or quality loss rate.
k_1 Deterioration rate from micro-organisms' activities.
k_2 Deterioration rate from enzymes' activities.
k_{NL} Food product nutrition quality loss.
k_{PS} Food product physical senses quality loss.
L_{NL} Nutrition loss.

L_{PS} Physical senses loss.
P_p Production cost per unit of food product.
P_q Cost of quality per unit, a sum of internal failure cost, prevention cost and appraisal cost.
P_t Cost of transportation per unit.
Q_i Probability of occurrence for a certain risk related activity i.

To facilitate the modeling of the food chains, we made the following assumptions:

1. Replenishment rate is infinite and lead time is zero. And shortage is not allowed.
2. The food products follow the general form of deterioration. Deterioration process starts when retailers receive product, and no deterioration during transportation.
3. All cost parameters are known in advance.
4. Demand for a food item is assumed to be deterministic. No seasonal effect.
5. During storage, transportation, and on shelf period, the environment, such as temperature, lighting, packaging quality, are assumed to be steady and unchanged.

2.1 *Cost of Quality, and Food Quality Loss*

According to research literature in food science as discussed in the review section, quality factors for food could be divided into two major categories: (1) nutrition and energy supply, and (2) quality related to physical senses. Nutrition and energy supply serves the basic function of food: to supply energy and bio-chemical needs to maintain the survival and functions of human body. Quality senses includes the physical senses customer received from the food product. Such senses include sight, touch, smell, taste, and even hearing. In general, these senses are grouped into appearance factors, textural factors, and flavor.

2.1.1 Nutrition and Energy Supply Loss

To quantify the food nutrition loss over the time factor, a widely used method of Q_{10} in food science is adopted in our model. The equation of Q_{10} is used to describe the duration of storage to reach the same nutrition level under different temperature.

$$f_2 = f_1 * Q_{10}^{\frac{\Delta}{10}} \tag{1}$$

In Eq. 1, f_1 is the reference duration at reference temperature T_1, f_2 is the duration of the targeting temperature T_2. Δ is the difference between targeting temperature and sample temperature of. Notice that this 'potential' nutrition loss for the customers is in proportion to the food retail price and the ratio of actual shelf period over the labeled shelf life. For example, a food item has been labeled for a shelf life of F_2 days, when this item has been purchased before on F_1 ($F_1 < F_2$), a certain portion (F_1/F_2) of nutrition loss can be expressed as NL_{F1}.

$$NL_{F_1} = a \cdot \frac{S_r}{F_2} \cdot F_1 \tag{2}$$

2.1.2 Physical Senses Loss

According to food science, there are (1) appearance factors, (2) textural factors, and (3) the flavor factors, that are considered physical senses by consumers on food products. Appearance factors usually include size, shape, wholeness, color, consistency for liquid, and so on. Textural factors include hand feel and mouth feel of firmness, softness, juiciness, chewiness, grittiness. Flavor factors include both taste and odor. The loss of physical senses for food products is normally the result of food deterioration. Major causes for food deterioration include (1) Growth and activities of microorganisms, such as bacteria, yeast and so on; (2) Activities of natural food enzymes; (3) Insects, parasites, and rodents; (4) Temperature; moisture and dryness; air (particularly oxygen); and light; and (5) Time duration.

Growth and Activities of Microorganisms

According to food science, if the original number of microorganisms in food product is A, then the number of microorganisms in food after time t will be:

$$N = A \cdot t^2 \tag{3}$$

And the quality loss can be formulated as in proportional to the number of microorganisms in the food. If the loss rate is k, the loss from microorganisms is:

$$L_{micro} = k \cdot N = k \cdot A \cdot t^2 = k_1 \cdot t^2 \tag{4}$$

Activities of Natural Food Enzymes

According to Potter [3], bacteria or microorganisms are the greatest factors in food deterioration, and the activity of enzyme is the second greatest. We can define the food quality loss due to enzymes as the following:

$$L_{enzyme} = k_2 \cdot t^2 \tag{5}$$

Insects, Parasites, Rodents, Temperature, Moisture, Dryness, Oxygen, and Light

With proper packing and storage of food items in modern food retailer facilities, these factors are not considered in our models.

Time

It is clear that one of the most important factors for food quality loss calculation due to food storage and food shelf life is dominated by the time factor. The quality loss due to physical senses for one product unit can be expressed as:

$$L_{PS} = L_{micro} + L_{enzyme} = k_1 \cdot t^2 + k_2 \cdot t^2 = k_{PS} \cdot t^2 \quad (6)$$

2.2 *Quality Loss Functions and Food Quality Loss Due to Physical Senses*

Taguchi's quality loss function is originally designed for the manufacturing industry. In our case, the longer a product has been stored, the greater the loss customer will be suffering. Equation 6 can be modified according to the Taguchi format as:

$$L_{PS} = k_{PS} \cdot t^2 = \frac{S_r}{F_1^2} \cdot t^2 \quad (7)$$

Within each ordering period, food items may be stored on shelf from *0* to *n* days, waiting to be purchased. The Total Quality Loss due to physical senses as:

$$L_{PS} = \int_0^n (d)\cdot(k_{PS} \cdot t^2)dt = \frac{d \cdot k_{PS} \cdot n^3}{3} = \frac{d \cdot S_r \cdot Q^3}{3F_1^2\, d^3} = \frac{S_r \cdot Q^3}{3\, F_1^2 d^2} \quad (8)$$

2.3 *Food Risks Associated with Time*

One of the main drawback in Ross and Summers model [8] is the assumption of the constant risk levels for food items, regardless of the time spent on the shelf. In our model, we combine all the risk factors in Ross and Summers model and eliminate the frequency of consumption, market share, as well as population and exposure distribution. The overall risk for a certain period of *t* is:

$$\bar{R} = \int k_{fr} t^2\, dt = \frac{1}{3} k_{fr} n^3 = \frac{1}{3} k_{fr} \frac{Q^3}{d^3} \quad (9)$$

2.4 Objective Function for the Food Supply Chain Model

The Total Cost including all direct costs (production, transportation, shortage, inventory holding, etc.), quality loss cost(nutrition loss and physical sense loss), as well as cost associated with food risk factors is summarized as:

Total Cost = Production cost + Setup cost + Yield and Defective cost + Transportation cost + Ordering cost + Inventory holding cost + Nutrition loss cost + Physical senses loss cost + Food risk loss (converted cost)

$$\text{Total Cost} = P_p \cdot D + C_s \cdot \frac{Q}{y \cdot Q_F} + P_Q \cdot D \cdot \frac{(1-y)}{y} + P_t \cdot D + C_o \frac{D}{Q} + h \cdot S_p \cdot \frac{Q}{2} + \frac{Q \cdot S_r \cdot D \cdot Q}{2F_2 \cdot d} + \frac{b \cdot S_r \cdot D}{3F_1^2 \cdot d^2} \cdot Q^2 + \frac{k_{fr} Q^3}{3d^3} \tag{10}$$

Similar to the traditional EOQ model, optimization technique is then applied to find the optimal ordering quantity that minimizes the Total Cost. Due to the limit of paper length. We will not present the solution procedure here.

3 Conclusion

In this paper, we present a model for food supply chain procurement decision making by considering the traditional EOQ factors as well as additional factors such as food nutrition loss, physical senses quality loss, and food risk financial loss. Preliminary results on four different food items, not presented in the paper, demonstrated that the feasibility of adding the additional factors in a food supply chain decision making process. Our future work will be to refine the food risk financial loss factor and to expand the model for different scenarios such as including distributors, and producers.

References

1. Manuel Angel Palazón et al (2009) Determination of shelf-life of homogenized apple-based beikost storage at different temperatures using Weibull hazard model. LWT- Food Sci Technol 42:319–326
2. Martins RC (2006) Simple finite volumes and finite elements procedures for food quality and safety simulations. J Food Eng 73:327–338
3. Potter NN (1986) Food science, 4th edn. The AVI Publishing Company Inc, Westport
4. Fujiwara O (1993) EOQ models for continuously deteriorating products using linear and exponential penalty costs. Eur J Oper Res 70:104–114
5. Chung K-J, Huang T-S (2007) The optimal retailer's ordering policies for deteriorating items with limited storage capacity under trade credit financing. Int J Prod Econ 106:127–145

6. Huss H, Reilly A, Embarek P (2009) Prevention and control of hazards in seafood. Food Control 11:149–156
7. van Gerwen S, te Giffel M, van t Riet K, Beumer R, Zwietering M (2000) Stepwise quantitative risk assessment as a tool for characterization of microbiological food safety. J Appl Microbiol 88:938–951
8. Ross T, Sumner J (2002) A simple, spread-sheet based food safety risk assessment tool. J Food Microbiol 77:39–53

A Comparative Study on Increasing Efficiency of Chinese and Korean Major Container Terminals

Bo Lu and Xiao Lin Wang

Abstract As the competition among East Asia container terminals has become increasingly fierce, every port is striving to increase its investments constantly to maintain the competitive edge. The unreasoning behavior, however, has induced that substantial waste and inefficiency exists in production. From this perspective, data envelopment analysis provides a more appropriate benchmark. By applying three kinds of DEA models, this study acquires a variety of analytical results on operational efficiency of the 31 major container terminals. Firstly, this study finds the reason of inefficiency. It is followed by identification of the potential areas of improvement for inefficient terminals by applying slack variable method. Furthermore, return to scale approach is used to assess whether each terminal is in a state of increasing, decreasing, or constant return to scale. The results of this study can provide container terminal managers with insights into resource allocation and optimization of the operating efficiency.

Keywords Efficiency • Container terminal • Data envelopment analysis

1 Introduction

With rapid expansion of global business and international trade, the distinctive feature is that competition among container terminals is more intensive than previously. To maintain its competitiveness in such competitive condition,

---

B. Lu (✉)
Institute of Electronic Commerce and Modern Logistics, Dalian University, Dalian, People's Republic of China
e-mail: lubo_documents@hotmail.com

X.L. Wang
College of Tourism, Dalian University, Dalian, People's Republic of China
e-mail: wangxiaolin@dlu.edu.cn

Z. Zhang et al. (eds.), *LISS 2012: Proceedings of 2nd International Conference on Logistics, Informatics and Service Science*, DOI 10.1007/978-3-642-32054-5_24,

Cullinane et al. [1] claimed that container terminals have to invest heavily in sophisticated equipments or in dredging channels to accommodate the most advanced and largest container ships in order to facilitate cost reductions for the container shipping industry. However, that pure physical expansion is constrained by a limited supply of available land, especially for urban centre terminals, and escalating environmental concerns. In addition, the excessive and inappropriate investment also can induce the phenomenon of inefficiency and wasting of resources. In this context, improving the productive efficiency of container terminal [2] appears to be the viable solution.

From this perspective, data envelopment analysis model provides a more appropriate benchmark for the container terminal [3]. The aim of this study is assumed to be the minimization of the use of input(s) and maximization of the output(s), by applying with DEA-CCR, DEA-BCC, and DEA-Super-Efficiency, three models, to acquire a variety of analytical results about the productivity efficiency for the 31 Chinese and Korean major container terminals. According to efficiency value analysis, this study firstly identifies efficient container terminals and ranks the sequence of them, then finds the reason of inefficiency ones. It is followed by identification of the potential areas of improvement for inefficient terminals by applying slack variable method. Return to scale approach is used to assess whether each terminal is in a state of increasing, decreasing, or constant return to scale. Finally, by comparing the efficiency scores between Chinese and Korean container terminals, the study can identify which input or output variables are more critical to the models, and would more impact the efficiency of terminals.

The paper is structured as follows: after the introductory section of Sect. 1, the required definition of input/output variables and the data collection have been described in Sect. 2. Estimates of the efficiency of a sample of container terminals are derived in Sect. 3. Finally, conclusions are drawn in Sect. 4.

2 Data Collection and Definitions of Variables

For doing a typical analysis, the data sample comprises the 14 Chinese and 17 Korean major container terminals. In order to gain the accurate performance, this study defines the variables of each terminal at the level of per berth by dividing variables by berth number.

Container throughput is the most important and widely accepted indicator of container terminal output [4]. Most importantly, it also forms the basis for the revenue generation of a container terminal. Another consideration is that container throughput is the most appropriate and analytically tractable indicator of the effectiveness of the production of a container terminal. Synthesizing the former research, in this study, the terminal productivity indicator is defined as the per berth handling capacity.

In order to determine the input variables, the used factors for variables in the study are discovered through an abundant literature review and discussion with experts. As far as the process of production is concerned, a container terminal depends crucially on the efficient use of infrastructures and facilities [5]. On the basis of that, yard area per berth, the quantities of quay crane, yard crane, yard tractor per berth, water depth and berth length have been deemed to be the most suitable factors to be incorporated into the models as input variables.

3 Efficiency Analysis and Implication

The efficiency analytical results for container terminals are summarized in Table 1. DEA-CCR model yields lower average efficiency estimates than the DEA-BCC model, with respective average values of 0.783 and 0.939, where an index value of 1.000 equates to perfect (or maximum) efficiency. This result is reasonable since a DEA model with an assumption of constant returns to scale provides information on pure technical and scale efficiency taken together, while a DEA model with the assumption of variable returns to scale identifies technical efficiency alone. DEA-Super-Efficiency model which removes an efficient DMU, and then estimates the production frontier again and provides a new efficiency value that can be greater than 1. Therefore, the average efficiency value of Super-Efficiency model, 0.815 is greater than CCR model.

By using of efficiency value analysis, slack variable approach and return to scale method can be summarized as:

Firstly, the aggregate efficiency value acquired from the CCR model of Waigaoqian phase-2, HIT, COSCO, BICT, DPI, MTL, Shekou, PCTC, NBCT, CS-4 and INTERGIS terminals were all equal to 1. The efficiency values of other terminals in that year were less than 1, which indicated that they were relatively inefficient terminals. The ‘pure technical efficiency value’ obtained from the BCC model represented the efficiency in terms of the usage of input resources. All of the pure technical efficiency values of the Waigaoqiao phase-2, HIT, COSCO, DPI, MTL, Shekou, PCTC, NBCT, CS-4, INTERGIS, HBCT, KX3-1, KBCT, Hanjin, ICT, JUCT, UTC, KIT2-2 and SGCT terminals were equal to 1. The technical efficiency values of other terminals were less than 1, thus indicating that they would need to improve their usage of resources. Among these, GICT phase-1 terminal had the least pure technical efficiency value.

Then, the DEA-Super-efficiency model is utilized to reinforce the discriminatory power of the CCR model. Waigaoqian phase-2 has the best performance among these 31 container terminals. HIT and COSCO ranked as the second and third best in model, respectively. However, SGCT has the lowest score which was 0.100.

The slack variable analysis, showed that HIT, COSCO, MTL, DPI, Shekou, NBCT, CS-4, Waigaoqian phase-2, INTERGIS, BICT and PCTC terminals were relatively efficient; their ratios of input variables to output variable were appropriate, and they were capable of applying their input resources effectively to achieve enhanced efficiency. In contrast, the terminals of ACT, Yantian, Chiwan, Nansha, NBSCT, YS-1&2, HGCT and GICT 1 terminals were relatively inefficient as a result

Table 1 Efficiency under three DEA models

	Efficiency			Reasons of inefficiency		
	Score					
Models Terminals	CCR efficiency	Super efficiency	Rank	BCC efficiency	Scale efficiency	Return to scale
WQ-2(C)	1.000	1.343	1	1.000	1.000	Constant
HIT(C)	1.000	1.211	2	1.000	1.000	Constant
COSCO(C)	1.000	1.130	3	1.000	1.000	Constant
BICT(K)	1.000	1.091	4	1.000	1.000	Constant
DPI(C)	1.000	1.088	5	1.000	1.000	Constant
MTL(C)	1.000	1.031	6	1.000	1.000	Constant
Shekou(C)	1.000	1.030	7	1.000	1.000	Constant
PCTC(K)	1.000	1.028	8	1.000	1.000	Constant
NBCT(C)	1.000	1.027	9	1.000	1.000	Constant
CS-4(C)	1.000	1.012	10	1.000	1.000	Constant
INTERGIS(K)	1.000	1.005	11	1.000	1.000	Constant
HBCT(K)	0.981	0.981	12	1.000	0.981	Increasing
KX3-1(K)	0.933	0.933	13	0.936	0.997	Increasing
Chiwan(C)	0.907	0.907	14	0.908	1.000	Increasing
KBCT(K)	0.903	0.903	15	0.909	0.994	Increasing
ACT(C)	0.901	0.901	16	0.915	0.985	Increasing
HGCT(K)	0.893	0.893	17	0.901	0.992	Increasing
DPCT(K)	0.805	0.802	18	0.819	0.979	Increasing
YS-1&2(C)	0.800	0.799	19	0.831	0.962	Increasing
Hanjin(K)	0.759	0.750	20	1.000	0.750	Increasing
NBSCT(C)	0.746	0.746	21	0.860	0.867	Increasing
Yantian(C)	0.743	0.743	22	0.800	0.929	Increasing
ICT(K)	0.663	0.663	23	1.000	0.663	Increasing
Nansha(C)	0.656	0.657	24	0.729	0.900	Increasing
JUCT(K)	0.652	0.652	25	1.000	0.652	Increasing
UTC(K)	0.538	0.538	26	1.000	0.538	Increasing
DBE2-1(K)	0.482	0.482	27	0.922	0.522	Increasing
KIT2-2(K)	0.347	0.347	28	1.000	0.347	Increasing
HKTL(K)	0.304	0.304	29	0.830	0.366	Increasing
GICT1(K)	0.182	0.182	30	0.733	0.248	Increasing
SGCT(K)	0.100	0.100	31	1.000	0.100	Increasing
Average	0.783	0.815		0.939	0.831	

of inappropriate application of input resources. KBCT, HBCT, Hanjin, DPCT, UTC, KIT2-2, KX3-1, HKTL, DBE2-1, ICT, SGCT and JUCT terminals were also relatively inefficient; however, in these cases, an inappropriate production scale was the cause of the inefficiency. The results indicated that Nansha, YS-1&2, KBCT, KIT2-2, KX3-1, HKTL, GICT1, DBE2-1, ICT, SGCT and JUCT terminals had adjusted their yard area of container base. Nansha, HBCT, UTC, KIT2-2, KX3-1, ICT and JUCT terminals had adjusted their number of quay crane. YS-1&2 and JUCT terminals had adjusted their number of terminal crane. Yantian and HKTL terminals had adjusted their number of yard tractor. ACT, Yantian, Chiwan, Nansha, YS-1&2, HBCT, Hanjin, DPCT, UTC, KIT2-2, KX3-1, HKTL, DBE2-1, ICT and JUCT terminals had adjusted the length of their container berth. Nansha, YS-1&2, KBCT, HBCT, Hanjin, DPCT, KIT2-2, KX3-1, HKTL, DBE2-1, ICT and JUCT had adjusted the deep-water of piers. In addition to adjusting and improving the input variables, each inefficient terminal had increased their loading/unloading volumes if they were to reach a relatively efficient state.

3.1 Implication of Efficiency Value Analysis

For making a concrete analysis for the integral empirical results, an estimate of the scale efficiency and, based on this, the returns to scale classification of each terminal. Among those large terminals (classified as having annual container throughput per berth of more than 0.5 million TEU), 11 of 16 show constant return to scale, other large terminals show increasing return to scale. On the other hand, all of the small terminals, except PCTC terminal, having annual container throughput of less than 0.5 million TEU, exhibit an increasing returns to scale. These results do suggest an association between large terminals and constant returns to scale and between small terminals and increasing returns to scale. On the other hand, the terminals that exhibit constant returns to scale are only large terminals.

With respect to comparing analysis, the reason of aggregate efficiency values of Chinese terminals is higher than Korean terminals can be summarized by:

An overwhelming majority of this increasing international trade is conducted by sea transportation; therefore, the huge investments of equipments have been put into the container terminals production. In addition, taking geographic advantage of huge area and respective cheap cost, and the rapid development of international container and intermodal transportation of Chinese container terminal production has drastically changed the market structure, and then attracted more customers and the cargo.

4 Conclusions

According to efficiency analysis of container terminals, empirical results reveal that substantial waste exists in the production process of the container terminals in the sample. For instance, the average efficiency of container terminals using the DEA-CCR model amounts to 0.783. This indicates that, on average, the terminals under this study can dramatically increase the level of their outputs by 1.28 times as much as their current level while using the same inputs. Empirical results also reveal that the terminals in the study were found to exhibit a mix of increasing and constant returns to scale at current levels of output. Such information is particularly useful for terminals managers or policy makers to decide on the scale of production.

Moreover, the reason why aggregate efficiency values of Chinese terminals are higher than Korean terminals can be summarized that the huge investments of equipments have been put into the Chinese container terminals production, geographic advantage of huge area and respective cheap cost. However, the pure technical efficiency values of Korean terminals are more than Chinese terminals, thus indicating that the most Korean terminals handle application of input resources better.

References

1. Cullinane K, Wang T-F, Song D-W (2006) The technical efficiency of container ports: comparing data envelopment analysis and stochastic frontier analysis. Trans Res Part A 40:354–374
2. Hanh Dam, Melissa M, (2006), Container terminal productivity: experiences at the ports of Los Angeles and Long Beach. In: The proceeding of National urban freight conference, Feb 1-3 2006, The West Long Beach, Long Beach, CA, 1–3 Feb 2006. pp. 1–21
3. Cheon SH, David E, Song D-W (2010) Evaluating impacts of institutional reforms on port efficiency changes: ownership, corporate structure, and total factor productivity changes of world container ports. Trans Res Part E 46:546–561
4. Lin LC, Tseng CC (2007) Operational performance evaluation of major container ports in the Asia-Pacific region. Marit Policy Manag 34(6):535–551
5. Park NK, Bo LU (2010) A study on productivity factors of Chinese container terminals. J Korean Nav Port Res 34(7):559–566

Evaluation on Logistics Enterprise Normalization Degree Based on Fuzzy Analysis

Lina Ma, Zhenji Zhang, and Xiao Xiao

Abstract The fuzzy mathematical theory is applied to conduct the comprehensive assessment model with a combination of qualitative and quantitative features on logistics enterprise informatization degree, based on theoretical analysis, according to characteristics of logistics industry, the questionnaire is designed, the factorial analysis is selected to handle data obtained by means of questionnaire to establish the index system of evaluation on logistics industry informatization degree, and this system and fuzzy analysis method are adopted to conduct evaluation on the informatization degree of the logistics industry of our country.

Keywords Fuzzy analysis • International competitiveness • Logistics enterprise

1 Introduction

In recent years, along with development of e-commerce of China as well as sustainable growth of state economy, no matter the gross of the logistics enterprise or the sum involving in logistics, they increase substantially year by year, however, the logistics enterprise of our country as a whole is in the underdeveloped state, the service cost of the enterprise is always much higher than that of other developed countries. In 2010, the all-in cost of the logistics enterprise of our country always takes up 17.8% of GDP, which keeps considerable gap between the 8% cost in other developed countries [1]. While the logistics cost nowadays in China remains high for the ground that the logistics management as well as logistics technical equipment suffer a low level, especially the national logistics informatization technology remains low [2]. For this reason, the national logistics enterprise needs to strengthen

---

L. Ma (✉) • Z. Zhang • X. Xiao
School of Economics and Management, Beijing Jiaotong University,
Beijing, People's Republic of China

Z. Zhang et al. (eds.), *LISS 2012: Proceedings of 2nd International Conference on Logistics, Informatics and Service Science*, DOI 10.1007/978-3-642-32054-5_25,

logistics informatization construction so as to boost the comprehensive competitiveness of the logistics enterprise.

The paper attempts designing the questionnaire, and applies factor analysis and other methods to analyze data of questionnaire to generate the evaluation index system of informatization of the logistics industry of our country, and the fuzzy method is applied to evaluate modern informatization to evaluate national logistics industry.

2 Design of the Normalization Assessment Index System

2.1 Design of Questionnaire

For the aim at apply the fuzzy analytical method to effectively evaluate informatization of national logistics enterprises, to begin with, it must rely on the combination of theoretical analysis with mathematics analysis to design the informatization evaluation index system of logistics enterprises. By means of some retrospect, conclusion and analysis on previous relevant literature, this text considers the factors influencing informatization of logistics industry mainly consist of Quality of employees in logistics enterprises, S&T innovative factors in logistics enterprises and Internal capacity of enterprises [3].

The questionnaire design of informatization facing logistics enterprises contains the mentioned variables, and each influencing factor sets five grades that influence informatization of logistics enterprises: (1) Very low; (2) Low; (3) Middle; (4) High; (5) Very high (Table 1).

2.2 Issue of Questionnaire

This research sends out 320 pieces of questionnaire mainly in allusion to division managers and administrative leaders in logistics enterprises with 198 being recovered including 185 effective ones. All testers have at least the junior college education and have long-term experience in logistics enterprises, all testers make commitment that all information filling in questionnaire is authentic and effective. Enterprises of testers are logistics enterprises with assets exceeding RMB 100 million, and specially, listed companies occupy 62.24%. This text uses SPSS19 software to conduct statistical analysis. According to the analytical result on questionnaire validity, the coefficient of Cronbach5 is 0.874, which indicates that key enterprise informatization influencing factors scale and various sub-scale possess favorable internal consistency, and basically fits reliability requirement.

Table 1 Normalization variables in logistics enterprises

Name	Normalization variables
Quality of employees in logistics enterprises	Advanced technology
	Technical force
	System integration
	IT support
	Equipment quality
	Technical maturity
	Technology application
S&T innovative factors in logistics enterprises	Infrastructure
	Corollary equipment
	Information platform
	Overall arrangement of infrastructure
	Technology primacy
Internal capacity of enterprises	Capability support
	Investment operation
	Flow integration
	Information sharing
	Knowledge learning
	Cost control
	Enterprise cognition

3 Evaluation on Logistics Enterprise Normalization Degree

FCE, the fuzzy comprehensive evaluation, can be better used for some researches involving multiple fuzzy factors so as to conduct comprehensive evaluation [4]. Logistics enterprise informatization in itself is a kind of fuzzy concept possibly, for this reason; influencing factors of enterprise informatization possess certain fuzziness. This research utilizes traditional scoring method-based questionnaire to conduct discussion and analysis on the entire informatization of logistics enterprises so as to establish the fuzzy evaluation method of informatization of logistics enterprises to conduct the quantitative analysis and confirm level of informatization of logistics enterprises.

3.1 Establishment of the Model

In order to adopt the fuzzy evaluation method to evaluate the logistics enterprise informatization, it should start rely on mathematical method to establish the model, that is, the fuzzy evaluation model of logistics enterprise informatization [5]. This text sets $\Omega = \{\Omega 1, \Omega 2, \ldots, \Omega n\}$ to indicate the set of influencing factors of logistics enterprise informatization, n indicates the number of the first-class evaluation indexes, $\Omega i = \{vi1, vi2, \cdots, vim\}$, m indicates the number of the second-class indexes below various first-class evaluation indexes, $i = 1, 2, \cdots, n$; $j = 1, 2, \cdots, m$.

Suppose $\Pi = \{\Pi1, \Pi2, \cdots, \Pi p\}$ indicates the comment set, specifically, Πk is the evaluation result, $k = 1, 2, \cdots, p$, p indicates the number of the order of evaluation. By means of confirming the membership of various enterprise informatization on various evaluation grades, that is, a fuzzy mapping from the set Ω to Π f: $\Omega \rightarrow \Pi$ is adopted to obtain the fuzzy comprehensive evaluation matrix Ti of Ωi.

3.2 Membership Matrix of Computering First-Class Index

There into, the mentioned fuzzy evaluation matrix Ti of the secondary index can be expressed:

$$\begin{aligned} \Gamma i = (\alpha i\ j\ k\ )\ m \times p = \\ \alpha i11\ \alpha i12 \cdot s\ \alpha i1\ p \\ \cdots\ \cdots\ \cdots \\ \alpha im1\ \alpha im2 \cdot s\ \alpha im\ p \end{aligned} \tag{1}$$

αi j k is the membership of Πk of v i j in k-grade comment.

With regard to the index vi j, if ti j1 Π1-grade comment exists, there will be twoΠ2-grade comments, . . ., and there will be ti j m Πm-grade comments, then, the value of αi j k can be confirmed in accordance with the following formula:

$$\alpha i\ j\ k = ti\ j\ k\ /\ \Sigma mk = 1ti\ j\ k \tag{2}$$

ti j 1 can be calculated according to data statistics of questionnaire.

By means of weight confirmation, the first-class index membership matrix can be obtained through Дi = KiΓi = (βi1, βi2, ···, βi p), thereinto, βi j indicates the membership of the first-class evaluation index on comment Πk.

3.3 Judgment of Fuzzy Comprehensive Evaluation Results

The weight coefficient in predecessors' literature is adopted in this text to confirm the method, according to results obtained in the questionnaire survey, the text calculates the first-grade and the secondary index weight value.

Take 3 first-grade evaluation indexes and 19 secondary evaluation indexes to form one factor set and confirm the comment set as $\Pi = \{1, 2, 3, 4, 5\}$, specifically, 1, 2, 3, 4 and 5 represent five grades of informatization degree of logistics enterprises, namely, very low, low, medium, high and very high. The results of vague judgment matrixΓ1, Γ2 and Γ3 built by the secondary international competitiveness evaluation index are followed as follows:

$$\Gamma 1 = \begin{matrix} 0.0272 & 0.0024 & 0.3120 & 0.1431 & 0.0080 \\ 0.0216 & 0.2118 & 0.5250 & 0.1503 & 0.0452 \\ 0.1240 & 0.0350 & 0.3317 & 0.0654 & 0.0246 \\ 0.0714 & 0.2313 & 0.2390 & 0.2072 & 0.0734 \\ 0.0612 & 0.1109 & 0.4450 & 0.1897 & 0.0174 \\ 0.2023 & 0.1433 & 0.4121 & 0.3072 & 0.1042 \\ 0.3200 & 0.1674 & 0.3213 & 0.4031 & 0.1032 \end{matrix}$$

$$\Gamma 2 = \begin{matrix} 0.0210 & 0.2393 & 0.2675 & 0.3607 & 0.0518 \\ 0.3015 & 0.3841 & 0.4205 & 0.2335 & 0.0326 \\ 0.0241 & 0.1436 & 0.3147 & 0.4051 & 0.0317 \\ 0.4153 & 0.0577 & 0.3243 & 0.0781 & 0.0075 \\ 0.0183 & 0.1082 & 0.4260 & 0.1722 & 0.2414 \\ 0.0233 & 0.1224 & 0.4102 & 0.2016 & 0.0188 \end{matrix}$$

$$\Gamma 3 = \begin{matrix} 0.0223 & 0.2414 & 0.1183 & 0.4117 & 0.1036 \\ 0.0036 & 0.1005 & 0.3216 & 0.5120 & 0.0042 \\ 0.3020 & 0.1325 & 0.4518 & 0.3245 & 0.0053 \\ 0.0076 & 0.2562 & 0.2478 & 0.3230 & 0.0734 \\ 0.1022 & 0.2148 & 0.3263 & 0.2021 & 0.3116 \\ 0.3200 & 0.1674 & 0.3213 & 0.4031 & 0.1032 \\ 0.0118 & 0.1137 & 0.4350 & 0.3270 & 0.0236 \end{matrix}$$

Suppose the secondary index weight in this text is expressed as Ki = (Ki1, Ki2,. . ., Kim), and the result can be obtained through Д = Ki × Γi = (βi1, βi2 ,. . ., βip)

$$R = \begin{matrix} 0.0346 & 0.4893 & 0.3347 & 0.0451 & 0.2147 \\ 0.0614 & 0.2125 & 0.3670 & 0.2133 & 0.0612 \\ 0.0330 & 0.3913 & 0.1424 & 0.3028 & 0.0580 \end{matrix}$$

Suppose K = (K1, K2, K3,. . ., K8) as the vector of the first-order weight so as to calculate the membership of informatization level in national logistics enterprises on comments of different levels:

$$Д = K \times R = (0.4235, 0.3166, 0.2913)$$

In accordance with β*i = max (βi1, βi2,. . ., βip), judge results of international competitiveness factors of national logistics industry to obtain:β*1 =0.4893, β*2 = 0.3670 and β*3 = 0.3913.

In accordance with calculated results, judge employees' quality of logistics enterprises in the logistics industry, internal capacity of enterprises, technological innovation factors of logistics enterprises and influencing degree of informatization in logistics enterprises. Specifically, the quality of employees in logistics enterprises is the biggest, while the technological innovation factor in logistics enterprises is the smallest.

4 Conclusion

Based on mentioned demonstration results, it can be known that national logistics industry informatization suffers a low level at present, and national logistics industry informatization level suffers poor stability at present. For this reason, we can concentrate on various factors influencing enterprise informatization to increase informatization of national logistics industry. Construction of logistics informatization will enhance logistics competitiveness [6]. For this reason, enterprises must establish the intense awareness of informatization, seriously research and study the application of modern logistics information technology to get rid of various difficulties existing in the informatization construction, vigorously boost the informatization construction so as to shorten the gap between informatization and logistics enterprises of developed countries as soon as possible.

References

1. Wang Han (2010) Research on IT application in management of modern logistics enterprises. Chin Logist Purchase 17:72–73
2. Li Jingyu (2011) Logistics informatization enters into the innovative era. Chin Storage Transport 04:39–40
3. Chen Yan (2011) Research on measurements of informatization management implemented by logistics enterprises. Value Eng 01:143–144
4. Zhu Guangyu, Lei Li, Pu Yanjun (2007) Relative competitiveness evaluation model of the 3rd logistics enterprises. Logist Technol 4:36239
5. Na Baoguo, Liang Jingguo (2008) International competitiveness evaluation research of national telecommunications industry based on fuzzy gray comprehensive method. Modern Manag 4:32233–32279
6. Li Li, Dong Hong, Liu Henan (2007) Constructing research on international competitiveness generation model of modern logistics industry. Logist Technol 8:7211–7259

Analysis of the Impact of Return Price on Competing Supply Chains

Jian Liu and Haiyan Wang

Abstract In this paper, we build the model of two competing supply chains in the presence of customer returns. The competition exists on retail price and return price. We examine the optimal pricing and ordering strategy of two competing supply chains in the Bertrand-Nash equilibrium and Stackelberg equilibrium respectively, and then analyze the impact of return price on supply chains. We conclude that the optimal price, order quantity and profits in the Stackelberg equilibrium are affected more greatly by return price than that in the Bertrand-Nash equilibrium.

Keywords Competing supply chain • Return policy • Bertrand-Nash equilibrium • Stackelberg equilibrium

1 Introduction

Accepting customer returns has been an important strategy for retailers to attract customers and stimulate demand, in the increasingly competitive market environment. Customer returns policies can enhance customers' confidence in purchasing goods, stimulating the demand and possibly increasing the retailer's market share. However, it is also bound to increase the retailers' and the manufacturers' processing cost. Hence, from an operational standpoint, a natural question emerges: how should two effects of customer returns be traded off to yield greater profits?

Customer returns impacts the pricing and ordering strategy of firms. In general, firms can use full refund policy, which is a 100% money-back-guarantee (MBG) offered to ensure consumer satisfaction, and sometimes they also offer partial

J. Liu • H. Wang (✉)
School of Economics and Management, Institute of Systems Engineering,
Southeast University, Nanjing 210096, People's Republic of China
e-mail: liujane1124@126.com; hywang@seu.edu.cn

Z. Zhang et al. (eds.), *LISS 2012: Proceedings of 2nd International Conference on Logistics, Informatics and Service Science*, DOI 10.1007/978-3-642-32054-5_26,

refund policy for customers. Mukhopadhyay and Setoputro [1, 2] held that the market demand was the linear function of retail price and return price, where it would decrease with retail price, and increase with return price. Then they studied the manufacturer's optimal pricing and return policy. Chen and Bell [3] addressed the simultaneous determination of price and inventory replenishment under the full refund policy. However, they ignored the market competition.

The work is also related with literature on competing supply chains. There are several factors which can cause the competition, such as price, quantity, fulfill rate, quality, warranty period and so on. In the work of Choi [4], they studied that two manufacturers faced the same retailer competing on price. Bernstein and Fedegruen [5] examined a supply chain system competing on price and fulfill rate, and demonstrated the existence of Nash equilibrium. Moorthy [6] and Banker et al. [7] studied a model with two identical firms competing on product quality and price. Recently, Chen et al. [8] investigated the optimal decision of one manufacturer and two competing retailers when the demand was dependent on warranty period.

This paper is different with the above literature. Firstly, we initially consider that the demand is related with the retail price and return price of both the supply chain itself and its competing chain. Secondly, the return price considered in this paper is not limited to the full refund, but it can change from 0 to the retail price.

2 Problem Statement and Model

There are two competing supply chains in the same market, where the manufacturer sales the substitutable and perishable product directly to customers in one single sale period, and accept the customer returns. Such policy can improve the manufacturer's profit through stimulating demand, but it can also increase the manufacturer's processing cost because of the high customer return. The problem defined in this paper is that the manufacturers should make which kind of decision to maximize their profits in the presence of customer returns.

We assume that the retailer's demand $D_i(p, r), (i = 1, 2)$ is related with both retail price p_i $(i = 1, 2)$ and return price r_i $(i = 1, 2)$ in two supply chains, where $r_i \leq p_i$ $(i = 1, 2)$, and it is the linear function of them. That is

$$D_i(p, r) = a - bp_i + cp_j + \xi r_i - \eta r_j, j = 3 - i. \tag{1}$$

Because the retail price limits the range of the return price, we can assume $r_1 = \beta_1 p_1$, $r_2 = \beta_2 p_2$, where $0 \leq \beta_1$, $\beta_2 \leq 1$. $\beta_1 = \beta_2 = 0$ denotes neither of supply chains chooses customer returns policy. $\beta_1 = \beta_2 = 1$ denotes both supply chains choose the full refund policy. Generally, $0 < \beta_1$, $\beta_2 < 1$ denotes that both supply chains choose the partial refund policy. So, the ith chain's demand function becomes

$$D_i(p, r) = a - (b - \xi\beta_i)p_i + (c - \eta\beta_j)p_j. \tag{2}$$

Let the parameters $a, b>0, c \geq 0, \xi \geq 0, \eta \geq 0$. a is the initial market demand, which reflects the whole developing level of the products. b denotes the demand responsiveness to the supply chain's retail price, while c denotes the demand responsiveness to its rival's retail price. ξ denotes the demand responsiveness to the supply chain's return price, while η denotes the demand responsiveness to its competitor's return price. The parameters β_1, β_2 are called return factors, and reflect the size of return price. We require $\xi < b$, $\eta < c$, $b - \xi\beta_1 > c - \eta\beta_2$ and $b - \xi\beta_2 > c - \eta\beta_1$.

In this paper, the production cost is not considered. We assume the inventory quantity is adequate, and the return rate is H_i, $i = 1, 2$, respectively.

Then, the ith chain's profit is

$$\Pi_i = (1 - \beta_i H_i)\, p_i \left[a - (b - \xi\beta_i)p_i + (c - \eta\beta_j)p_j\right]. \tag{3}$$

3 Supply Chain Competition Equilibrium Results

In this section, we analyze equilibrium results in Bertrand-Nash game and Stackelberg game respectively.

3.1 The Equilibrium Results in Bertrand-Nash Game

In Bertrand-Nash game, each supply chain decides its retail price conditional on that of its competing supply chain, which develops a duopoly market. Therefore, the ith chain's equilibrium retail price is

$$p_{iB} = [(c - \eta\beta_j) + 2(b - \xi\beta_j)]Ua. \tag{4}$$

Where $U = [4(b - \xi\beta_1)(b - \xi\beta_2) - (c - \eta\beta_1)(c - \eta\beta_2)]^{-1}$.

The ith chain's equilibrium demand becomes

$$D_{iB} = (b - \xi\beta_i)[2(b - \xi\beta_j) + (c - \eta\beta_j)]Ua. \tag{5}$$

The ith chain's equilibrium profit is

$$\Pi_{iB} = (1 - \beta_i H_i)a^2 U^2 (b - \xi\beta_i)[2(b - \xi\beta_j) + (c - \eta\beta_j)]^2. \tag{6}$$

3.2 The Equilibrium Results in Stackelberg Game

In Stackelberg game, without loss of generality, we assume that supply chain 1 is a leader and supply chain 2 is a follower. Then, the decision sequence is that supply chain 1 firstly set the retail price p_1, and then supply chain 2 chooses the optimal retail price to make its profit $\Pi_2(p_1, p_2)$ maximize. Then, the ith chain's optimal retail price is

$$p_{iS} = a[(b - \xi\beta_j) + (c - \eta\beta_j)]U_0. \tag{7}$$

The ith chain's equilibrium demand is

$$D_{iS} = U_0 a(b - \xi\beta_1)(b - \xi\beta_2). \tag{8}$$

Where $U_0 = [2(b - \xi\beta_1)(b - \xi\beta_2) - (c - \eta\xi_1)(c - \eta\beta_2)]^{-1}$. Therefore, in Stackelberg game, the ith chain's equilibrium profit is

$$\Pi_{iS} = (1 - \beta_i H_i)U_0^2 a^2[(b - \xi\beta_j) + (c - \eta\beta_j)](b - \xi\beta_i)(b - \xi\beta_j). \tag{9}$$

4 Analyze the Impact of Return Price on Retail Price, Order Quantity and Profits

In this section, we analyze the impact of return price on the optimal retail price, order quantity and profits, and we also compare the optimal price, order quantity and profits in Stackelberg game with that in Bertrand-Nash game. We get some conclusions as follows.

Proposition 1. *In Bertrand-Nash game and Stackelberg game, the optimal retail price* p_{1B}, p_{1S} *increase with* β_1, *and the optimal retail price* p_{2B}, p_{2S} *increase with* β_2. *When* $c\xi\eta$, *the optimal retail price* p_{1B}, p_{1S} *increase with* β_2, *and the optimal retail price* p_{2B}, p_{2S} *increase with* β_1. *When* $c\xi\eta$, *the optimal retail price* p_{1B}, p_{1S} *decrease with* β_2, *and the optimal retail price* p_{2B}, p_{2S} *decrease with* β_1.

Proposition 2. $p_{1S} > p_{1B}$, $p_{2S} > p_{2B}$ *Furthermore, if* $\beta_2 < \beta_1$, *then* $p_{2S} < p_{1S}$ *and* $p_{1B} < p_{2B}$. *Otherwise, if* $\beta_2 > \beta_1$, *then* $p_{2S} > p_{1S}$ *and* $p_{2B} > p_{1B}$.

Proposition 2 suggests that in Stackelberg game, the retail price in two supply chains is higher than corresponding that in Bertrand-Nash game. In addition, both in Stackelberg game and Bertrand-Nash game, when the return price in supply chain 1 is higher (lower) than that in supply chain 2, the corresponding retail price is also higher (lower).

Proposition 3. *When* $c\xi\eta$, *the optimal order quantity* D_{1B}, D_{2B}, D_{1S} *and* D_{2S} *increase with* β_1 *and* β_2. *When* $c\xi < b\eta$, D_{1B}, D_{2B}, D_{1S} *and* D_{2S} *decrease with* β_1 *and* β_2.

Proposition 4. *$D_{1S} < D_{1B}$, $D_{2S} < D_{2B}$. Furthermore, if $\beta_2 < \beta_1$, then $D_{1B} < D_{2B}$. Otherwise, if $\beta_2 > \beta_1$, then $D_{1B} > D_{2B}$.*

From Proposition 4, we know that in Stackelberg game, the order quantity in two supply chains is higher than corresponding that in Bertrand-Nash game. In Bertrand-Nash game, the change trend of optimal order quantity is opposite with that of return price.

Proposition 5. *When $c\xi > b\eta$, the optimal profit Π_{1B}, Π_{1S} increase with β_2, and the optimal profit Π_{2B}, Π_{2S} increase with β_1. When $c\xi < b\eta$, the optimal profit Π_{1B}, Π_{1S} decrease with β_2, the optimal profit Π_{2B}, Π_{2S} decrease with β_1.*

Proposition 6. *$\Pi_{1S} > \Pi_{1B}$, $\Pi_{2S} > \Pi_{2B}$. Furthermore, in the case of $c\xi > b\eta$, if $\beta_2 < \beta_1$, then $\Pi_{2S} > \Pi_{1S}$ and $\Pi_{2B} > \Pi_{1B}$. Otherwise, if $\beta_2 > \beta_1$, then $\Pi_{2S} < \Pi_{1S}$ and $\Pi_{2B} < \Pi_{1B}$. In the case of $c\xi < b\eta$, the optimal profits satisfy $\Pi_{2S} < \Pi_{1S}$ and $\Pi_{2B} < \Pi_{1B}$.*

Proposition 6 suggests that the profits of two competing supply chains in Stackelberg game are larger than corresponding that in Bertrand-Nash game. Furthermore, in the case of $c\xi > b\eta$, when the return price of the competing supply chain is smaller (larger) than that of its own supply chain, the profits of competing supply chain are larger (smaller) than that of its own supply chain. In the case of $c\xi < b\eta$, it has an opposite trend.

5 Conclusion

Customer returns policy is an important decision for firms to strive and develop in the fiercely competitive market. In this paper, we analyze the impact of return price on the competing supply chains. From our analysis, we can get that the retail price, order quantity and profits are affected by return price both in Bertrand-Nash game and in Stackelberg game. In addition, retail price, order quantity and profits of two competing supply chains in Stackelberg game are affected more greatly by the return price than that in Bertrand-Nash game.

In the future work, we will further consider the manufacturers' optimal pricing and ordering decision when the demand is stochastic. Furthermore, our model also can be extended to incorporate the customer behavior.

Acknowledgments This work is supported by the grant from National Natural Science Foundation of China (Project No. 71171049).

References

1. Mukhopadhyay SK, Setoputro R (2004) Reverse logistics in e-business: optimal price and return policy. Int J Phys Distrib Logist Manag 34(1):70–88
2. Mukhopadhyay SK, Setoputro R (2005) Optimal return policy and modular design for build-to-order products. J Oper Manag 23:496–506

3. Chen J, Bell PC (2009) The impact of customer returns on pricing and order decisions. Eur J Oper Res 195:280–295
4. Choi SC (1991) Price competition in a channel structure with a common retailer. Mark Sci 10 (4):271–296
5. Bernstein F, Fedegruen A (2004) A general equilibrium model for industries with price and service competition. Oper Res 52(6):868–886
6. Moorthy KS (1988) Product and price competition in a duopoly. Mark Sci 7(2):141–168
7. Banker RD, Khosla I, Sinha KK (1998) Quality and competition. Manag Sci 44(9):1179–1192
8. Chen X, Li L, Zhou M (2012) Manufacturer's pricing strategy for supply chain with warranty period-dependent demand. Omega 40:807–816

Studies on the Supply Chain Risk Management Using Complex Network

Yi Cong-qin, Meng Shao-dong, and Zhang Da-min

Abstract The complexity of supply chain networks become increasingly vulnerable to the invasion of various types of risks and thus the supply chain risk management (SCRM) is widespread concerned in academia and industry. Our contribution reveal the source of the supply chain risk through a complex supply chain network modeling evolution, identify the key nodes in the supply chain network, and improve the ability to resist risks in supply chain network by working closely with key nodes and increasing network resilience.

Keywords Complex network • Supply chain risk • Key nodes • Network resilience

1 Introduction

Several incidents such as China SARS and 2011 Japanese tsunami indicated that once risks occurred the supply chain would result inevitably damage and huge losses. SCRM thus was widespread concerned in academia and industry. Complex network theory can reveal the formation mechanism of different complex systems. With the help of complex network theory the risk root of supply chain could be determined under a new perspective view. Objective of this contribution was to enhance the ability of supply chain to resist risks using complex network.

Y. Cong-qin
College of Economics & Management, Shanghai Ocean University, Shanghai 201306, China

College of Computer Science & Information, GuiZhou University, Guiyang 550000, China

M. Shao-dong (✉)
Customs Management Department of Shanghai Customs College, Shanghai 201204, China
e-mail: sdmeng@customs.gov.cn

Z. Da-min
College of Computer Science & Information, GuiZhou University, Guiyang 550000, China

Z. Zhang et al. (eds.), *LISS 2012: Proceedings of 2nd International Conference on Logistics, Informatics and Service Science*, DOI 10.1007/978-3-642-32054-5_27,

2 Current Developments on the Supply Chain Risk Management

Some serious events such as the 911 event, China SARS, the earthquake in Taiwan, tsunami and earthquake in Japan and other such events present supply chain management a huge challenge. With internal and external environment increasing complicated the supply chain was more vulnerable to diverse invasion of risks. At present most of the studies on SCRM are superficial and major focused on the source of the risks, risks evaluation and control. Uncertainty in the supply chain was considered to be the mainly reason of risk [1].

The supply chain itself is a complex network. Using complex theory can trace origin of supply chain risk from a holistic view and thus related measures could adopt direct work on the source, reduce the losses coming from the risk. Some scholars researched part of SCRM by using complex networks theory. Study of Dirk Helbing et al. indicated that the bullwhip effect in supply chain management, i.e. the information amplifying effect, was correlated with the nature of the supply chain network topology [2]. Siemieniuch et al. thinked of the complexity of the supply chain network was the unpredictable behavior between the business entities [3]. Goh et al. provided a stochastic model of the global supply chain network by analyzed global supply chain provision, demands, trade and distribution [4]. Yanyan et al. used complex network to research test methods of cascade effects in the supply chain [5]. Zhu Bing-xin et al. applied the complex network theory to the urgent management research in the supply chain [6]. These studies researched supply chain risk from different aspects of supply chain but not from the perspective of whole supply chain network.

3 Modeling Supply Chain Complex Network

3.1 Basic Theory of Complex Network

Complex network represents a wide range of real-world systems occurred in nature. A complex network can be represented as a graph $G = (V, E)$ with a set of vertexes (V) connected by a set of edges (E). For a weighted network, suppose w is the correlation between two vertexes, and w_{ij} represents the correlation between node i and node j. The weight of edges represent the intensity of the connection between the two vertexes and higher w_{ij} demonstrate more intensive connection. Suppose the distance between node i and j is $d_{ij} = 1/w_{ij}$. Then a higher weight of the edge joining node i and j indicates a shorter distance between them. The strength of vertex i is the sum of the weights of the edges indexed on i:

$$s_i = \sum_{j \in \Gamma_i} w_{ij}$$

Where Γ_i is the neighbor vertexes of node i. Point intensity indicates that the node has a point of strengths probability distribution P (s).

3.2 Modeling Supply Chain Complex Network

In a supply chain network, nodes represent different subjects, which could be manufacturers, suppliers, retailers, and customers, while edges represent the collaboration and competitive relations among these subjects. The weights of edges could be the trading volume. The strength of a node indicates the importance of a subject in the supply chain. In supply chain with complex networks, manufacturers are hubs, and other companies join the network. The connections among vertexes are generated based on their intentions, regardless of the physical distance. The model is a dynamical network in that edge weights change as the collaboration intensities change, which in turn refine the edge weight and affect the overall topological structure of the network.

Model constructed as follows:

1. at $t = 0$, the network has m_0 vertexes and e_0 edges
2. at every time interval, a new node N joins the network and connects with the nodes w in the following ways, where $w \leq m_0$ and depends on N.

   ① If the node N represents a supplier and has a probability α, the probability of node N connects with node i is a function of the weights of Si, i.e.,

$$\prod_{n \to i} = \frac{s_i}{\sum_j s_j} \tag{1}$$

   where $s_i = \sum_{j \in V(i)} w_{ij}$ is the sum of all edges indicted on node i.

   ② Node N represents a retailer and has the probability of $1-\alpha$.
   Suppose δ is the attraction factor of a node. δ reflects the quality of products, exchange time, and service quality. At each time interval, following the joining of a new node N to the network, every node i (including N) in the network selects m other nodes. The probability of node i selects node j is

   ③ $\prod_{i \to j} \frac{s_j+\delta}{\sum_{k \neq i} s_k+\delta}$. In case that node i selects j but not vice versa, the connection between the two will not be established or the edge weight remains unchanged if there is already a connection between them. Otherwise, if two nodes select each other simultaneously, then the edge weight increases by 1, i.e.,

$$w_{ij} = w_{ij} + 1 \tag{2}$$

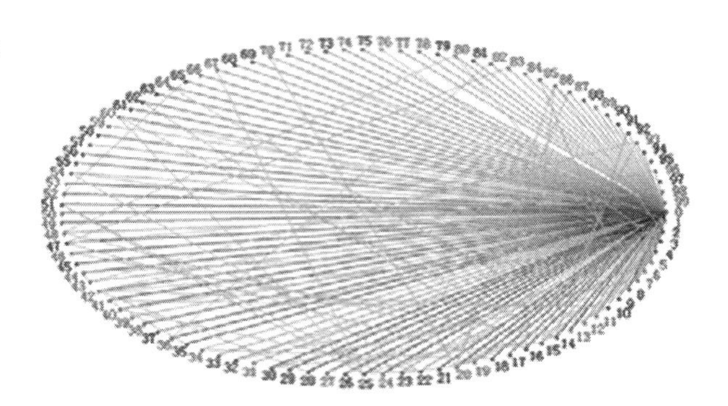

Fig. 1 Model of supply chain with the manufacture as the core

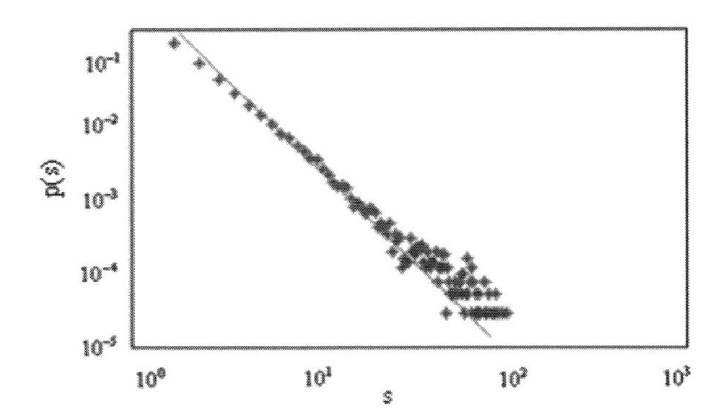

Fig. 2 Intensity distribution of supply chain network model (S: strength; P(s): strengths probability distribution)

Figure 1 demonstrates a supply chain network, whose weight distribution is shown in Fig. 2. As shown by the fitted curve, the weight in a supply chain follows the power law distribution. That is, a node tends to interact with nodes with higher weights. With the expansion of the network, these nodes will finally have more connections than the other nodes and are the "hubs". But weightier node would not passive interaction with lighter nodes and core enterprise would take all factors into consideration of suppliers' quality of merchandise, services and price etc.

4 Analysis of Supply Chain Based on Complex Network

4.1 Key Nodes in the Complex Network

Complex network improve the ability to resist risks mainly by protect important nodes and temporary separation them. In order to analyze the risk-resisting ability of the supply chain network, we must first find an important node in the entire network, and take relevant measures on important nodes. From the local and global aspects to measure the importance of the nodes in the supply chain network, in order to find the key nodes in the supply chain network. Right to the edge of the node represents the trading volume between nodes enterprises, the value of the node side of the right and the greater the more closely to the node with other nodes in

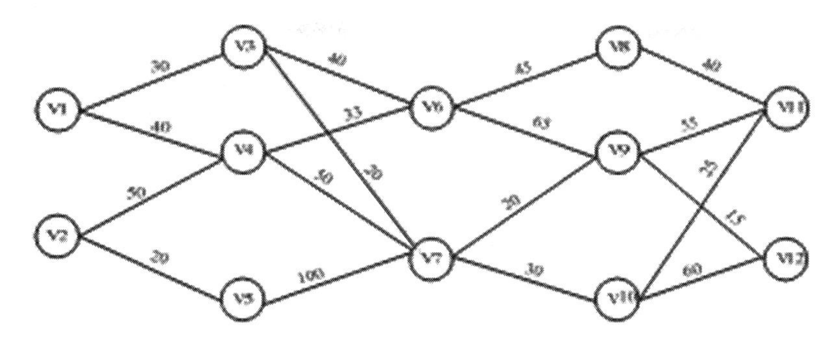

Fig. 3 Topology structure of supply chain network

contact, so that the node is important. Position has also identified the importance of the nodes in the entire network, more by the shortest path to the node, the node is more important; the connectivity of the network has a crucial role.

Take manufacturers as core, suppliers and customers as the network edge, the topology of the supply chain is shown in Fig. 3. To find the key nodes in the network, let we first make the following assumptions:

Assumption 1. Nodes in the network are independent. Sum of the edges' weights $s_i = \sum_{j \in N_l} \frac{1}{w_{ij}}$ and $s_i \in (0, \infty)$, that is to say the more the weight of the edge, the closer connection between the node and the other nodes, and if the node was removed then heavier influence will put on ambient supplier chain enterprise.

If just consider the weights of factors, the larger the node weights, the node is more important. Weight of node V7 is the most and of node V1, V2 is the least in Fig. 3.

Assumption 2. Let us consider the importance of nodes from a whole point of view. Importance of nodes used change of mean network shortest path, put value of 1 to all edges connected to node v_i and others are constant, compare the change of mean network shortest path then the importance could express quantized.

Take supply chain model in Fig. 3 as example, consider the change of mean shortest path, the quantized results was shown in Table 1. Inspection the whole supply chain globally or locally node V7 is the most important in the supply chain, in other words, node V7 is key node in the whole supply chain and play key role in the network.

4.2 Response Measures Based on the Key Node in the Supply Chain Risk

There are two measures to improve network survivability in complex network: one is to take protective measures of the key nodes; the other is to intend to weaken the scale-free network key node and change its topology to a certain extent, thus to

Table 1 Assessment of key nodes important

Node	Quantized mean shortest path	Node	Quantized mean shortest path
V7	0.5979	V3	0.2279
V4	0.5449	V11	0.2294
V6	0.4373	V2	0.2277
V9	0.3217	V8	0.1787
V5	0.2841	V1	0.0952
V10	0.2401	V12	0.2279

enhance anti-attack capability of the network, that is to say split an important node into multiple nodes and node ability to resist risks improved.

Supply chain network can design in optimization under the foundation of understanding the topology and key nodes of the supply chain network. On one hand to maintain close contact with key suppliers, to learn the supplier's supply capacity and inadequate, to reduce interrupt the probability of risk occurring by taking cooperation, information sharing, strengthen the awareness of risk management measures; to made various enterprises can more clearly understand the changes in the supply chain risk through information sharing and exchange of supply chain. On the other hand, to enhance the elasticity of supply chain networks in the risk of downtime occurs by select key node in multi-source suppliers and safety stock strategy. In order to improve the elasticity of the supply chain network should at least have a backup supplier to maintain a reasonable level of stock.

5 Conclusion

Through the supply chain network modeling and simulation we know that the supply chain network was scale-free, from the mechanism point of view of the evolution of complex networks, the source of the supply chain risk mainly came from exogenous risk caused by inner-power-driven endogenous risks between node enterprises and the external environment. Thus in the management of supply chain risk the ability of resist risks of supply chain could be improved by protecting and splitting key nodes, that is, supply chain risk could managed in working closely with key node enterprises and improving network resilience.

Acknowledgement This research was funded by the International Cooperation Project of Guizhou Province (no. [Guizhou Province International Cooperation G(2011) 7007]).

References

1. Spekman RE, Davis EW (2004) Risky business: expanding the discussion on risk and the extended enterprise. Int J Phys Distrib Logist Manag 34(5):414–433
2. Helbing D et al (2006) Information and material flows in complex networks. Phys A: Stat Mech Appl 363(1):xi–xvi

3. Siemieniuch C, Sinclair M (2002) On complexity, process ownership and organisational learning in manufacturing organisations, from an ergonomics perspective. Appl Ergon 33 (5):449–462
4. Goh M, Lim J, Meng F (2007) A stochastic model for risk management in global supply chain networks. Euro J Oper Res 182(1):164–173
5. Yan Y, Xiao L, Xin-tian Z (2010) Cascading failure model and method of supply chain based on complex network. J Shanghai Jiaotong Univ 3:322–325
6. Bing-xin Z, Yi-hong H (2008) Study on supply chain disruption management based on complex network theory. Logist Technol 26(11):147–150

Research on the Mode and Countermeasure of Closed Supply Chain Management about Green Agricultural Products Logistics System

Xiaolin Zhang and Xiaodong Zhang

Abstract The closed supply chain system is an advanced form of the supply chain management. It is an effective means to promote circulation modernization and guarantee food safety. The paper analyzes the concepts about green supply chain management, puts forward the technological means to implement green supply chain of agricultural products. It also presents countermeasures for the green supply chain management of agricultural products from strengthening infrastructure construction, improving technique level of agricultural production logistics, developing the leading enterprise, and so on.

Keywords Closed supply chain • Agricultural products logistics • Green agricultural products

1 Introduction

Agricultural products are closely linked with people's life, and thus people pay close attention to the "green" issues of agricultural products. There is a huge risk in food safety as the cold chain facilities is lacking and logistics management of agricultural products lags behind, which result in big loss of primary agricultural products, especially the perishable food. The backwardness of agricultural products logistics also directly leads to the weak anti-risk ability of agriculture. In China, the development of green food and green food industry needs the strong support of green logistics technologies. Therefore, to construct the agricultural products green logistics system based on the ecological agriculture and green circulation has

X. Zhang (✉)
Department of Economy and Management, Tianjin Agriculture University, Tianjin, People's Republic of China
e-mail: tszhangxl@126.com

X. Zhang
Department of Library, Tianjin Agriculture University, Tianjin, People's Republic of China

Z. Zhang et al. (eds.), *LISS 2012: Proceedings of 2nd International Conference on Logistics, Informatics and Service Science*, DOI 10.1007/978-3-642-32054-5_28,

become the top priority of Chinese agricultural industry. The construction of agricultural green logistics system must form the integration of the cold chain and technological system from the transportation, preservation, package standards, processing storage, distribution and retail. Closed supply chain management is an important research subject on the agricultural products logistics mode in recent years. It means that the operation of the entire supply chain is placed in a closed system, which is not interfered and invaded by the outside world. With uniform practice and technical standards, closed supply chain implements strict access management system on member companies of supply chain. It makes the supply chain system traceable by carrying out the real-time monitoring and dynamic tracking for the whole process of the supply chain [1, 2]. It plays an important role in meeting the green consuming demands and enhancing the competitiveness of agricultural products.

2 Construction on Closed Supply Chain System of Green Agricultural Products

2.1 The Main Mode of Closed Supply Chain of Green Agricultural Products

Lacking in coordination and management, the traditional agricultural products supply chain is not the supply chain mode in the real sense. Closed supply chain is a centralized supply chain with close organization, with the core enterprises playing an important role in organization and coordination and bearing the corresponding security liabilities. Therefore, the core enterprise of agricultural products should be selected first in the design of modern agricultural supply chain. Generally speaking, there are two types of core enterprises leading or dominating closed supply chain of the agricultural products. The one is the distributors as the leaders of closed supply chain, which mainly refers to the large supermarkets, the large distribution centers, and other distribution enterprises. The characteristics of this mode contain three aspects. Firstly, distributors gain the strong integrating and controlling ability in the whole process from procurement, transportation, sale, distribution to quality monitoring of the agricultural products. Secondly, distributors possess extensive business network and the ability to control the market. Thirdly, the distributors have the ability to control suppliers (Fig. 1).

The other one is the manufacturers as the leaders of agricultural products closed supply chain, which mainly refers to the large-scale agricultural processing enterprises leading the supply chain. This mode has its own characteristics like the former one. Firstly, manufacturers gain the strong integrating and controlling ability concerning the whole supply chain. Secondly, manufacturers can control the products channels and own a large market share. Thirdly, manufacturers are able to control suppliers.

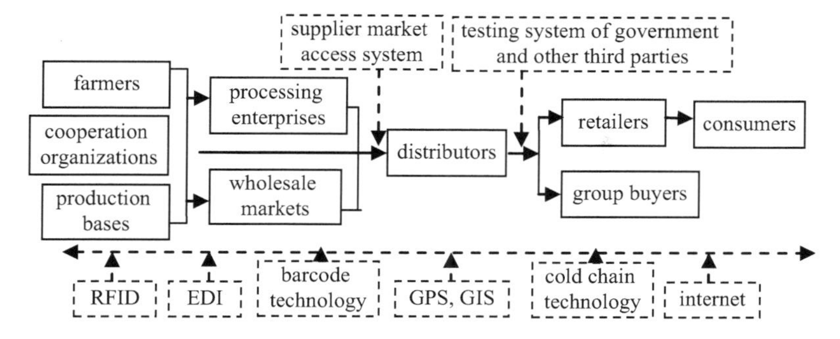

Fig. 1 Closed supply chain mode of distributors as the leaders

2.2 The Technical Means of Green Agricultural Products Closed Supply Chain

The development of supply chain management mode is the results of enterprises' internal development demands and also the promoting from advanced management techniques. The more centralized is supply chain system, the more advanced management level it needs. Specifically, the advanced management techniques to support closed system in different agricultural products logistics links are as follows:

1. Purchasing stage: on-time purchasing management technology, packaging and transportation technology based on cold chain, inventory management technology, on-time delivery management technology, continuous replenishment technique, virtual logistics technology, third-party logistics management technology [3].
2. Distribution stage: distribution technology based on JIT (just in time), RFID and GPS tracking technology, HACCP monitoring technology.
3. Processing stage: pre-cooling and preservation technology, cold chain packaging and transportation technology.
4. Sale stage: electronic commerce technology and information traceability technology (Fig. 2).

3 Countermeasures on the Closed Supply Chain Management of Green Agricultural Products

3.1 Strengthening Infrastructure Construction

Infrastructure is the foundation of supporting efficient and stable logistics. First, it is necessary to set up a number of professional agricultural products wholesale markets that own powerful distribution functions in the main producing areas of grain, vegetables, fruits, eggs and aquatic products. Secondly, The construction of highway

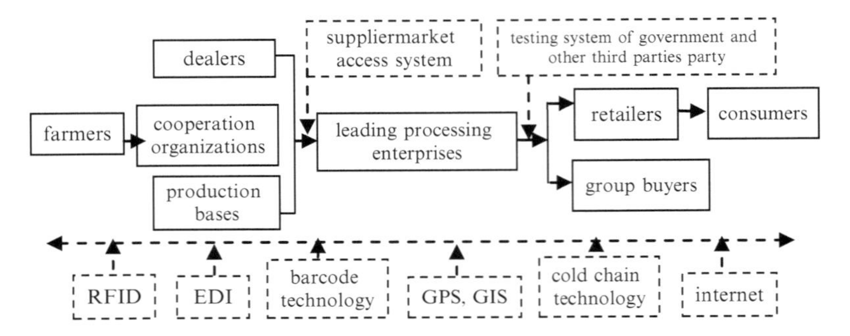

Fig. 2 Closed supply chain mode of processing enterprises as the leaders

infrastructure need to be strengthened to ensure transportation of agricultural products smoothly, speed up the special logistics facilities research and development of cold chain equipment, temperature control equipment and moisture-proof equipment, and improve the inspection system of agricultural products quality including product classification-testing, pesticide residue testing inspection equipments. Thirdly, the construction of logistics base is supposed to be accelerated to adopt the pollution-free production technology and standardized procedures to meet requirements of green agricultural products.

3.2 Improving Agricultural Products Logistics Technology

Technological innovation is the driving force for development of supply chain management. It is advisable to put the whole process into the standardization track to achieve the high quality of agricultural products from seedlings, fertilizers, pesticides, cultivation, breeding to processing, packaging, quality inspection and all the other sections of the agricultural production. Research and applications of key logistics technology must be also strengthened, for example, tracking and locating system, RFID, logistics information platform, intelligent transportation, logistics management software and so on. It is also important to promote the application of internet in the logistics field and implementation of logistics standards, improve the modern level of logistics equipment, especially the cold-chain logistics equipment technology, accelerate innovation of agricultural preservation technology, promote the resource sharing of logistics information to improve the efficiency and management level of logistics services [4].

3.3 Developing the Leading Enterprises of Agricultural Products

Core enterprise is the key for the closed agricultural supply chain to operate effectively. Firstly, enterprises are expected to play a leading role in promoting the "contract farming". It is beneficial to establish the mechanism of mutual risk-sharing and benefit sharing between the enterprises and farmers. Secondly, it is admirable to implement the brand strategy vigorously to form a number of strong brands of agricultural products, expand the market share by establishing brand images and enhancing brand recognition. Thirdly, it is also expected to promote the transformation of the traditional logistics enterprises to third-party logistics enterprises. Fourthly, the development of green agricultural products industry should be promoted by the implementation of preferential policies and encouraging the production and sales of green agricultural products.

3.4 Improving the Agricultural Logistics Information System

Firstly, it is vital to take the "Agricultural Information Network" as a platform, and gradually build the information network of connecting farmers and wholesale markets and industrialized leading enterprises as well as improve the consulting services level of market information in order to provide the prospective and guided market information for farmers and leading enterprises. Secondly, As Logistics speed will be improved and the circulation and market will be expanded by the use of modern information networks and online trading, it is essential to establish the agricultural product logistics information management system and apply bar code technology widely in the whole process of agricultural supply chain. Thirdly, it is desirable to build the network on manufacturing information platform of green supply chain, and establish the green supply chain management database and information system.

3.5 Developing Farmers' Cooperative Organization

The farmers are the starting point of agricultural products supply chain, however, their organization stays at a low level at present. Enhancing the farmers' organizing degree and forming effective interest allocation system rely on developing the farmers' cooperative organization. Various forms of professional cooperative economic organization need to be set up based on the model of "Leading enterprises + professional cooperatives + farmers" to play the association role in providing full service for farmers, agriculture and rural areas. In the meantime, it is also crucial to accelerate direct connecting between the cooperative organization and supermarkets to reduce the circulation cost, thus improving the agricultural products quality and safety [5].

4 Conclusion

The operation and management of closed supply chain of green agricultural products is helpful to improve the agricultural products quality, guarantee the consumption safety of agricultural products, improve the circulation efficiency, and reduce the market terminal price of green agricultural products. The paper analyzes the mode and characteristics of closed supply chain management of green agricultural products, puts forward the main technical means for the efficient operation of the closed supply chain, presents the main measures to promote its development from strengthening infrastructure construction, improving agricultural production logistics technique level, developing the leading enterprise, and so on. Of course, for such a new topic, many significant problems need to be further researched. How to reform the traditional agricultural products supply chain based on the green supply chain is an important and urgent issue to research in future.

References

1. Liu Weihua, Xiao Jianhua, Jiao Zhilun (2009) A study on typical operation model and its cost control in agricultural product closed supply chain. Soft Sci 11:58–63
2. Li Huiliang, Wen Xiawei (2011) Research and application of fresh agricultural products supply chain safety. Sci Technol Manag Res 1:119–212
3. Liu Weihua1, Liu Yanping, Liu Binglian (2010) The method and application of the closed reconstruction of green agricultural product supply chain. Soft Sci 4:49–52
4. Wu Qinggang (2011) The current situation and the countermeasures of China's cold chain logistics development. China Bus Mark 2:24–28
5. Zhang Xiaolin, Luo Yongtai (2011) Re-construction and optimization of agricultural products logistics system. J Bus Econ 11:5–10

Channel Selection and Co-op Advertising in Ecommerce Age

Yongmei Liu and Yuhua Sun

Abstract The focus of this paper is channel selection and co-op advertising strategies in Ecommerce Age. We use Stackelberg game to develop two co-op advertising models under different channels. Then we compare the two models to select optimal channel and co-op advertising strategies. Furthermore, we analyze the impact of product web-fit on optimal strategies and illustrate by some numeral examples. Based on our results, we provide some significant and managerial insights, and derive some probable paths of future research.

Keywords Channel selection • Co-op advertising • Stackelberg game

1 Introduction

The surge in the Internet and Third part logistics has significantly changed supply chain channel structure, and many manufacturers adopt direct channel [1]. Comparing with any single channel, dual channel can better realize market penetration, but it leads to channel conflict [2]. Moreover, because co-op advertising can increase channel demand, it partly alleviates channel conflict. Therefore, channel selection and co-op advertising strategies in Ecommerce age are important and interesting.

Most multi-channel studies devote to determining whether manufacturers should set up direct channel, such as Chiang et al. [3], Tsay and Agrawal [2], Chiang et al. [4], Arya et al. [5]. In general, direct channel breaks the supply chain balance, and reasonable channel structure can alleviate channel conflict. However, little discussion has been given to co-op advertising in their studies, even through co-op

Y. Liu (✉) • Y. Sun
Department of Management Science and Engineering, Business School,
Central South University, Changsha 410083, People's Republic of China
e-mail: liuyongmeicn@yahoo.com.cn; syhyxt@126.com

Z. Zhang et al. (eds.), *LISS 2012: Proceedings of 2nd International Conference on Logistics, Informatics and Service Science*, DOI 10.1007/978-3-642-32054-5_29,

advertising has been used by many industries [6,7]. More recently, about $50 billion was spent on co-op advertising of U.S. companies (http://advertising suite101.com article cfm.coop_advertising programs). Co-op advertising literatures mainly focus on game theoretical to explore, such as Huang and Li [6], Huang et al. [7], and Karry and Zaccour [8], but these literatures little involve dual channel. Yan et al. [9] makes up the defect, but this paper doesn't make co-op advertising as decision variable and co-op advertising doesn't increase total demand. In fact, co-op advertising offers customers product information and thus it can increase total demand. Moreover, the characteristics of two channels are different, so the impacts of co-op advertising on two channels are different, too.

In our study, we attempt to address these limitations by using Stackelberg game to develop and analyze two co-op advertising models under different channels. Then we obtain channel selection and co-op advertising strategies by comparing the two models. Furthermore, we focus on the impact of product web-fit on these optimal strategies and identify probable paths of future research.

2 Model Framework

We study two scenarios(R and RD) as illustrated in Fig. 1. We assume only one kind of product for sale. D_d and D_r represent direct channel and traditional channel demand, respectively. As in Huang et al. [10], we assume that each channel demand is influenced by co-op advertising in different ways. Thus in Scenario R, we have

$$D_{r-R} = a + k_r\sqrt{A_R}. \tag{1}$$

$$D_{d-R} = 0. \tag{2}$$

In Scenario RD, we have

$$D_{r-RD} = (1-\theta)a + k_r\sqrt{A_{RD}}. \tag{3}$$

$$D_{d-RD} = \theta a + k_d\sqrt{A_{RD}}. \tag{4}$$

where a represents base demand; k_r and k_d represent the impact factors of co-op advertising on traditional channel and direct channel; θ represents product web-fit,

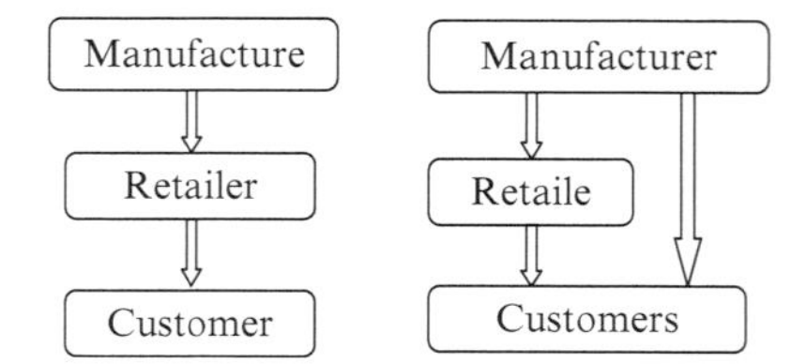

Fig. 1 The supply chain structure of Scenarios R, RD

which is the compatibility of the product with web direct channel, $0 \leq \theta \leq 1$, where 0 means no compatibility with web direct channel and 1 means complete compatibility with web direct channel; A represents co-op advertising expenditure.

The manufacturer's unit marginal profits in traditional channel and direct channel are ρ_{m1} and ρ_{m2}, $0 < \rho_{m1} < \rho_{m2}$; the retailer's unit marginal profit is ρ_r; the manufacturer's co-op advertising fraction is t.

In Scenario R, the manufacturer's, the retailer's and supply chain's profits:

$$\pi_{m-R} = \rho_{m1} D_{r-R} + \rho_{m2} D_{d-R} - t_R A_R. \tag{5}$$

$$\pi_{r-R} = \rho_r D_{r-R} - (1 - t_R) A_R. \tag{6}$$

$$\pi_{t-R} = \pi_{m-R} + \pi_{r-R} = \rho_{m1} D_{r-R} + \rho_{m2} D_{d-R} + \rho_r D_{r-R} - A_R. \tag{7}$$

In Scenario RD, the manufacturer's, the retailer's and supply chain's profits:

$$\pi_{m-RD} = \rho_{m1} D_{r-RD} + \rho_{m2} D_{d-RD} - t_{RD} A_{RD}. \tag{8}$$

$$\pi_{r-RD} = \rho_r D_{r-RD} - (1 - t_{RD}) A_{RD}. \tag{9}$$

$$\pi_{t-RD} = \pi_{m-RD} + \pi_{r-RD} = \rho_{m1} D_{r-RD} + \rho_{m2} D_{d-RD} + \rho_r D_{r-RD} - A_{RD}. \tag{10}$$

3 The Models

3.1 Co-op Advertising Model in Scenario R

The Stackelberg model leader is the manufacturer, who acts as the first mover by choosing co-op advertising fraction to maximize its profit. The retailer, acting as the follower, chooses optimal co-op advertising to maximize its profit. In Scenario R, we obtain the optimal strategies in Table 1. From Table 1, we obtain Property 1.

Table 1 Optimal strategies in the co-op advertising model in Scenario R

Parameters	Base values
Co-op advertising expenditure, A_R	$\frac{k_r(2\rho_{m1} + \rho_1)}{4}$
Manufacturer's optimal co-op advertising fraction, t_R	$\frac{2\rho_{m1} - \rho_r}{2\rho_{m1} + \rho_r}$
Retailer's profit, π_{r-R}	$\rho_r a + \frac{1}{8} k_r^2 \rho_r (2\rho_{m1} + \rho_r)$
Manufacturer's profit, π_{m-R}	$\rho_r a + \frac{1}{16} k_r^2 (2\rho_{m1} + \rho_r)^2$

Table 2 Optimal strategies in the co-op advertising model in Scenario RD

Parameters	Base values
Co-op advertising expenditure, A_{RD}	$\left(\frac{2\rho_{m1}k_r + 2\rho_{m2}k_d + \rho_r k_r}{4}\right)^2$
Manufacturer's optimal co-op advertising fraction, t_{RD}	$\frac{2\rho_{m1}k_r + 2\rho_{m2}k_d - \rho_r k_r}{2\rho_{m1}k_r + 2\rho_{m2}k_d + \rho_r k_r}$
Retailer's profit, π_{r-RD}	$\rho_r(1-\theta)a + \frac{1}{8}\rho_r k_r(2\rho_{m1}k_r + \rho_{m2}k_d + \rho_r k_r)$
Manufacturer's profit, π_{m-RD}	$\rho_{m1}(1-\theta)a + \rho_{m2}\theta a + \frac{1}{16}(2\rho_{m1}k_r + 2\rho_{m2}k_d + \rho_r k_r)^2$

Property 1. In Scenario R, (a) $\frac{dA_R}{dk_r} > 0$, $\frac{dA_R}{d\rho_{m1}} > 0$, $\frac{dA_R}{d\rho_r} > 0$; (b) $\frac{dt_R}{d\rho_{m1}} > 0$, $\frac{dt_R}{d\rho_{m1}} < 0$; (c) $\frac{d\pi_{r-R}}{dk_r} > 0$; (d) $\frac{d\pi_{m-R}}{dk_r} > 0$.

Property 1(a) is intuitive that when the retailer's co-op advertising efficiency of traditional channel and the retailer's unit marginal profit increase, the retailer's profit is high. The retailer would like to invest more into co-op advertising. Moreover, if the manufacturer's unit marginal profit in traditional channel is more, the manufacturer's profit is more and he will share more which means Property 1(b), so the retailer will invest more. Property 1(b) also reveals that if the retailer's unit marginal profit is less, the manufacturer will share more to encourage the retailer. Property 1(c) and (d) mean that when co-op advertising efficiency of traditional channel is more, traditional channel demand is higher and the retailer's and the manufacturer's profits is more.

3.2 Co-op Advertising Model in Scenario RD

In Scenario RD, co-op advertising can increase two channels demand. In addition, the impact of co-op advertising on direct channel is different from traditional channel. Thus, we can obtain the optimal strategies summarized in Table 2.

From Table 2, we can find that most results in Scenario R are hold in Scenario RD. In addition, the manufacturer's direct channel unit marginal profit and product web-fit have impact on the optimal results in Scenario RD. First, when the manufacturer's direct channel unit marginal profit and the impact of co-op advertising on direct channel are high, the manufacturer's profit is high and he share more and the retailer will invest more. Second, when the impact of co-op advertising on traditional channel is high, the retailer's profit is high and the manufacturer needn't share more to encourage the retailer. Third, when product web-fit is more, the manufacturer's profit is more but the retailer's profit is less.

3.3 *Analysis of the Two Models*

In this section, we discuss which channel is better for the manufacturer and the retailer and analyze the differences between the optimal strategies in two channel scenarios. We obtain the following results through comparing of the two models.

Theorem 1. *(a) For the manufacturer,* $\pi_{m-RD} > \pi_{m-R}$*; for the retailer,* $\pi_{r-RD} > \pi_{r-R}$*, if* $\theta < \frac{\rho_{m2}k_r k_d}{4a}$*; (b)* $t_{RD} > t_R$*; (c)* $A_{RD} > A_R$.

Theorem 1(a) reveals that the manufacturer always better in Scenario RD and the retailer is the same case only when product web-fit is smaller than some threshold. It is intuitive that the existing of direct channel increases total demand and it generates more profits for the manufacturer. However, when product web-fit is low, traditional channel demand is high, the retailer can benefit more from Scenario RD. Theorem 1(b) indicates that the manufacturer share more in Scenario RD. As showed in Theorem 1(a), the retailer is not always better in Scenario RD. In order to remit channel conflict, the manufacturer will share more with the retailer. So the retailer invests more co-op advertising to attract potential customers, as showed in Theorem 1(c).

4 Numerical Examples

In this section, some numerical examples are given to illustrate the effect of changes in product web-fit on the manufacturer's profit, the retailer's profit and profit of the supply chain. For our numerical examples, parameters settings are shown as follows: $A = 200$, $\rho_{m1} = 2$, $\rho_{m2} = 8$, $\rho_r = 5$, $k_r = 1.2$, $k_d = 1.8$.

Figure 2 reveals that profit of the supply chain in Scenario RD is positively related to product web-fit, but profit of the supply chain in Scenario R is not related

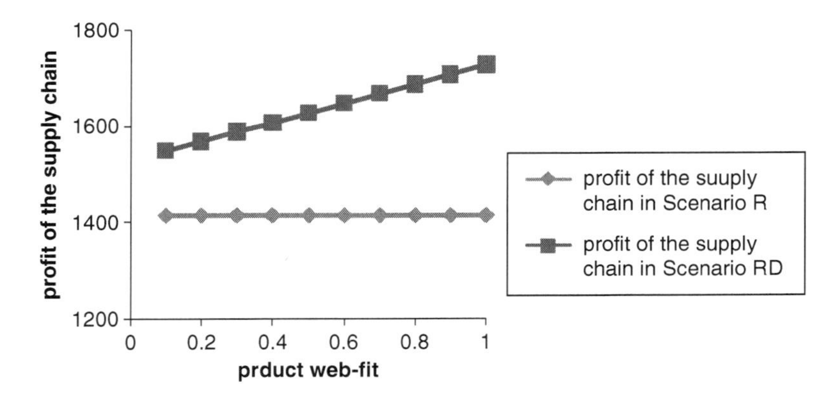

Fig. 2 Profit of supply chain under Scenario R and RD

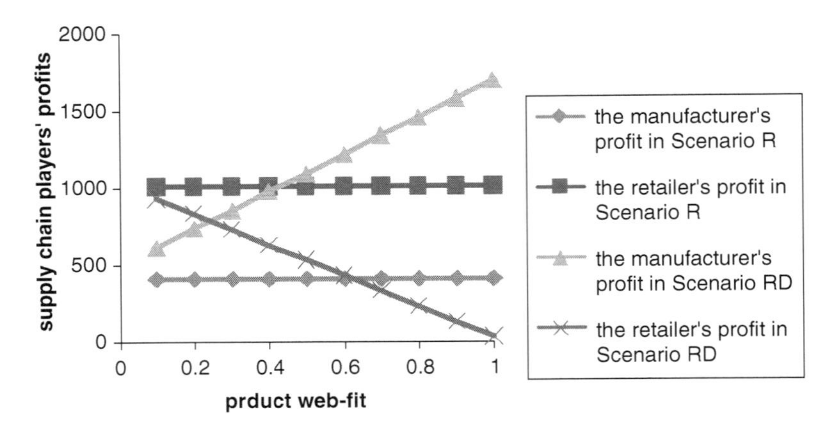

Fig. 3 Supply chain players' profits under Scenario R and RD

to product web-fit. Because traditional channel doesn't have web direct channel, traditional channel demand is not related to product web-fit. Also, we observed that profit of the supply chain in Scenario RD is higher than in Scenario R. Therefore, for the supply chain, it is better to adopt dual channel.

Figure 3 indicates that the manufacturer's profit and the retailer's profit in Scenario R are not related to product web-fit, too. The manufacturer's profit in Scenario RD increases with product web-fit and the retailer's profit in Scenario RD decreases with product web-fit. Note worthily, when product web-fit is less than 0.25, the retailer's profit in Scenario RD is larger than the manufacturer's profit; but when product web-fit is larger than 0.25, the case is opposite.

5 Concluding Remarks

In this paper, we provide a framework for researching optimal co-op advertising strategies in a two-level supply chain. First, we discuss co-op adverting model under traditional channel and dual channel based on Stackelberg game, and we derive optimal co-op advertising strategies. Next, comparisons of these two models are discussed and we find that the manufacturer always benefits from dual channel, but the retailer benefits from dual channel under certain conditions. According to these results, we explore some important theories and managerial insights. Furthermore, our numerical examples illustrate the impact of product web-fit on these optimal market strategies. More specifically, our study presents some managerial implications for managers. If the manufacturer and the retailer know that the impacts of co-op adverting on different channels, both would like to choose reasonable strategies to improve the channel coordination. Therefore, it would be best if managers conduct market survey before they start their co-op advertising campaign.

This paper has its limitations. First, we focus on two channel structures, so a further comparison with other structures can be explored. Second, we ignore some factors that influence demand, such as service and price. We can do some researches from the point of these factors.

References

1. Dumrongsiri A, Fan M, Jain A et al (2008) A supply chain model with direct and retail channels. Eur J Oper Res 187(3):691–718
2. Tsay AA, Agrawal M (2004) Channel conflict and coordination in the e-commerce age. Prod Oper Manag 13(1):93–110
3. Chiang WK, Chhajed D, Hess JD (2003) Direct marketing, indirect profits: a strategic analysis of dual-channel supply chain design. Manag Sci 49(1):1–20
4. Chiang WK, Monahan G (2005) Managing inventories in a two-echelon dual-channel supply chain. Eur J Oper Res 62(2):325–341
5. Arya A, Mittendorf B, Sappington DE (2007) The bright side of supplier encroachment. Market Sci 26(5):651–659
6. Huang Z, Li SX (2001) Co-op advertising models in a manufacturer-retailer supply chain: a game theory approach. Eur J Oper Res 135(3):527–544
7. Huang ZM, Li SX, Mahajan V (2002) An analysis of manufacturer-retailer supply chain coordination in cooperative advertising. Decis Sci 33(3):1–20
8. Karry S, Zaccour G (2006) Could co-op advertising be a manufacturer's counterstrategy to store brands? J Bus Res 59(9):1008–1015
9. Yan R, Ghose S, Bhatnagar A (2006) Cooperative advertising in a dual channel supply chain. Int J Electron Market Retailing 1(2):99–114
10. Huang S, Yang C, Zhang X (2011) Pricing and cooperative advertising decision models in dual-channel supply chain. Comput Integr Manuf Syst 17(12):2683–2692

The Extension of Third-Party Logistics Value-Added Services in the Financial Field

Chengkai Lv and Xuedong Chen

Abstract The paper gave a research on the third-party logistics value-added services in the financial field, proposing two services the third-party logistics can expand and giving a comparative analysis of the advantages and disadvantages of the two services based on two cases. The study has some practical significance for the third-party logistics on how to utilize its advantages in developing high value-added services in supply chain financing and improve the competitiveness of the supply chain.

Keywords Third-party logistics • Value-added services • Supply chain financing • Agent supervision service • Gent purchasing service

1 Introduction

This paper tries to make a discussion about the third-party logistics value-added services at the end of the supply chain financing, for further analysis in the traditional model of operation of third-party logistics. The discussion is based on the cases of value-added service model for third-party logistics companies to expand, offering some suggestions for third-party logistics enterprises in China on the value-added model innovation. It has a certain practical significance for the third-party logistics enterprises to improve the efficiency of supply chain management.

C. Lv (✉) • X. Chen (✉)
School of Economics and Management, Beijing Jiaotong University,
Beijing, People's Republic of China
e-mail: 11120698@bjtu.edu.cn; xdchen@bjtu.edu.cn

Z. Zhang et al. (eds.), *LISS 2012: Proceedings of 2nd International Conference on Logistics, Informatics and Service Science*, DOI 10.1007/978-3-642-32054-5_30,

2 The Status of the Third-Party Logistics Value-Added Services in the Financial Field

According to the survey, internationally, a number of large multinational logistics enterprises have begun to increase the financial value-added services in logistics services, such as using inventory and accounts receivable for finance, mortgages, global trade finance, agent collection and payment and cash on delivery. Representative of the enterprises is the UPS. Several large multinational third-party logistics companies including UPS, have made the financial and logistics services which are included in the entire service system [1] an important strategy to fighting for customers.

In China, there have been few third-party logistics enterprises involved in the service of financial aspects in logistics, such as Zhongchu Group. It has been engaged in logistics and finance-related part of the business since 1999, and gets new opportunities for development. The research of the domestic supply chain financing services is also relatively small at the present stage, most of the discussion is limited to the specific business of the warehouse.

3 LNCS Online

In traditional model third-party logistics enterprises only provide the support of logistics provider to make the enterprises in the supply chain be more focused on their core business [1]. It is difficult for the small and medium enterprises to obtain financing service from commercial banks directly in this model due to the lower credit grades.

3.1 Agent Supervision Service

In agent supervision service, third-party logistics plays a supporting role in the bank financing, assists the bank to supervise the management of the supply chain and provides credit guarantee for the enterprise which needs financing indirectly. It is an operation model that bank and the third-party logistics operate jointly with each other.

The model brings the following values to the underfunded supply chain: it makes the bank reduce its financial risk, and expand the range of its business. It also makes small logistics enterprise get financing service more easily. Enterprise parties can then realize a scientific, reasonable and efficient financing, and achieve their optimal decision making [2]. This kind of financing business under the agent of the third-party logistics enterprises builds a industrial ecology among banks, enterprises and commodity chains which is mutually beneficial, sustainable developing and positively interacting [3].

3.2 Agent Purchasing Service

In agent purchasing service, third-party logistics provides SMEs with agent purchase using its own capital or financial advantages, it solves the financing problem of small and medium enterprises, and makes an incentive effect on the order quantity of the SMEs at the same time [4].

What's different from the agent supervising service is that in the agent purchasing service the third party logistics enterprise plays a leading role in the underfunded supply chain, controlling the entire supply chain operations [5]. For instance, some large logistics enterprises can take advantage of its fund advantage to make large scale of purchases and then distribute to the underfunded retailers, thereby it serves as the retailer's direct suppliers of raw materials which provide agent purchases and distribution support [6]. At the same time, The third party provides financing services for retailers indirectly which offers deferred payment credit contracts for underfunded retailers.

4 The Case Studies of Third-Party Logistics Financial Value-Added Services

4.1 Cheng Tong Group

Cheng tong Group successfully used the financing service through the pledge of warehouse receipt, a service provided by the third-party logistics. It effectively solved the difficult situation in the difficulty of recycle of the loan due to the lack of credit of some enterprises. Its business operational mode is shown in Fig. 1.

Cheng tong Group's third-party storage company's warehouse receipts pledge financing business conformed to the third-party logistics agent supervising service model discussed above, in which the third-party logistics enterprises have the advantage of risk-averse. Since Cheng tong has a long-term cooperative relationship with industrial and commercial enterprise customers, it has a deep grasp of the credit of its customers and has a deep understanding of the market of their customers' products. It chooses the goods that have high quality and have small price fluctuations for pledge financing so that effectively avoid the risks of selecting pledged goods. It also has a strong distribution capacity, making use of the large sales channels to dispose the pledged goods which have a big price fluctuation to reduce the risk of bank financing [7].

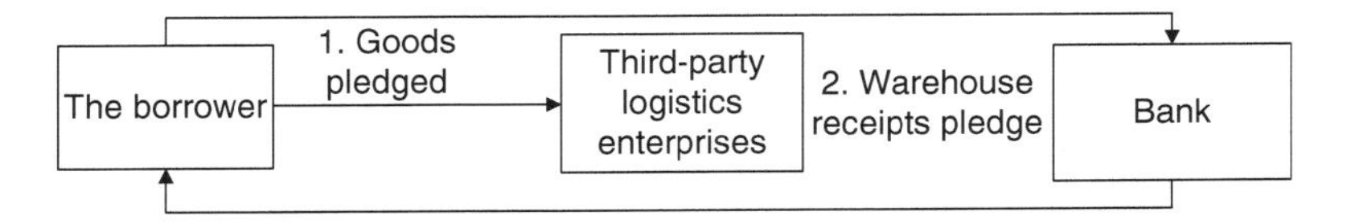

Fig. 1 The operational model of Cheng tong Group's third-party logistics

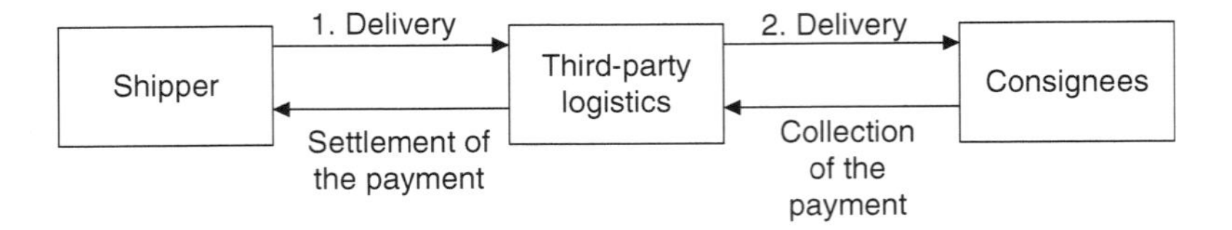

Fig. 2 The operation model of the Meiqi logistics platform

However, many disadvantages of this model exposed. The third-party logistics can't estimate very accurately about the situation of the sales and the cash converting of goods. And it is in an inferior position for the financial institutions to obtain the business information in the entire course of the operation.

4.2 Meiqi Logistics

Meiqi Group is founded as China's first "network of logistics business platform". It can be understood as a third-party logistics providers taking part in the purchase of enterprise cluster, finance and distribution. The network platform is established for small commercial enterprises in China, designed for low scale benefits especially. Its operation model is shown in Fig. 2.

In this model, the Meiqi logistics platform served as a pool of financing. It centralized the retailer's small claims of fund to the center of the company and bank settlement center and formed large funds, so that the individual credit turned to the group credit, and the group credit turned into the bank credit. Meiqi then obtained large fund credit from banks, thereby it provided agent purchase to suppliers and achieved economies of scale, distributed to retailers on-demand and thus complete the efficient operation of the entire process. This model reduced the bank credit risk and solved SMEs' financing problem, and its economies scale made the whole transaction process intensive based on the platform resources.

It also has some limitations; both the scale of most domestic third party logistics and some restrictions of the environment can't meet the needs of this model.

5 Conclusion

Obviously, the traditional services can't create new values for the underfunded supply chain. However in agent supervising services the third party logistics acted as an important connecting bridge providing the special supervising service for banks and enterprises, making the supply chain achieve an integration service of logistics and funds flow and creating new value. In agent purchasing services, the third-party logistics ensured a greater effectiveness, it integrated all its advantages to provide a personalized integration service of logistics, information flow and capital flow for the supply chain, maximizing the resources advantage of third-party logistics and creating a greater supply chain value. Since that, financial logistics value-added services will become the profit growth point of the third party logistics enterprises in the future. It is also a major breakthrough in the optimization of supply chain management, the best way to get win-win results [8].

Acknowledgments The research was supported by the key project of logistics management and technology lab.

References

1. Shi YY (2010) Third-party logistics value-added services. Railw Purch Logist 3:51–52
2. Chen XF (2008) The innovation of supply chain finance services. The Press of Fudan University, Shanghai
3. Chen XF, Zhu DL, Ying WJ (2008) The search of capital constraints and the operational strategic decision of supply chain finance. J Manag Sci 4(2):84–86
4. Zhu CF (2007) Facing up to "value-added services" in business opportunities. Market Weekly 10:16–17
5. He T, Qu L (2007) Analysis of SME financing model based on supply chain. Logist Technol 9 (9):76–77
6. Ramsay J (2007) Purchasing theory and practice: an agenda for change. Eur Bus Rev 10:16–17
7. Liu ZH (2007) China's forerunner of logistics finance – a interview of Chengtong China CEO Hong S.K. Chin Logist Purch 15:6–7
8. Wang ST (2008) Study on the third party logistics innovation based on logistics finance. China Logist Purch Fed 9(9):3–4

Evaluation System of the Supply Chain Stability

Hanlin Liu and Yue Cao

Abstract The stability of the supply chain relates to the supply chain management efficiency and effectiveness, and its concept has been extended to all aspects of business operations. This article combines researches that have been made both at home and abroad, and refers to the idea of the value chain. Based on the stability of the resource input, the stability of the value output and the stability of supply chain collaboration, we establish the three-lever index system to evaluate the performance of the whole supply chain stability, and attempts to evaluate it using analytic hierarchy process (AHP).

Keywords Stability • Evaluation system • Analytic hierarchy process (AHP) • Supply chain management

1 Introduction

With the global marketing competition intensifying, more and more enterprises attach importance to and begin implementing the supply chain management. There is a wide range of supply chain research and application. However, the research of the overall supply chain system stability is still in the exploratory stage. The studies are focused on the supplier partnerships and uncertainties. And research of stability of the supply chain structure as a whole is still very rare. The stability of the supply chain is paid more and more attention. How to define the moderate stability of the supply chain is the important issue of supply chain success and growth.

This article starts with the reason of fluctuations in the supply chain, leading to the meaning of the supply chain stability. From dealing with the generally unexpected

H. Liu (✉) • Y. Cao
School of Economics an Management, Beijing Jiaotong University,
100044 Beijing, People's Republic of China
e-mail: liuhanlingo@126.com; 09245060@bjtu.edu.cn

Z. Zhang et al. (eds.), *LISS 2012: Proceedings of 2nd International Conference on Logistics, Informatics and Service Science*, DOI 10.1007/978-3-642-32054-5_31,

events, it expands to the whole process of operational planning decisions, and further explains the significance of the stability of the supply chain. The second part of the article combines the researches of domestic and foreign supply chain stability, and establishes the evaluation index system based on the linear supply chain model. Finally, on the foundation of the second part of the general measure index of supply chain stability, attempts are made to introduce the analytic hierarchy process (AHP) in order to fulfill the measurement and evaluation of the stability of the whole supply chain and provide ideas to enterprises.

2 The Reasons of Supply Chain Stability and Volatility

The stability of the supply chain is: the member companies' supply chain operation of their own and the convergence status between companies have remained within a certain range, and the input and output of a whole chain of open systems maintain balanced.

According to its nature, the risks in the supply chain are divided into emergencies and operational risks [5]. Emergency usually means a material adverse effect on system events caused by natural disasters or man-made factors. For example, after the 2001 "911" terrorist attacks, companies in accordance with global supply chains and JIT operation of Ford, Toyota and other large manufacturing companies suffered supply shortages and had various degrees of distress.

Operational risk usually refers to the abnormal fluctuations in the supply chain when the system deviates from the plan due to market demand, supply and production of raw materials. It is generally caused by changes in the industry, as well as in decision making, particularly significant in the competitive and updating industry or expansion industry. For example, over the past decade, Chinese enterprises have set off another round of merger boom. But unfortunately, the cases of failure are more than success ones [3].

3 The Establishment of Impact Factors and Indicators of the Supply Chain Stability

3.1 Domestic and Foreign Research

At present, the supply chain partnership has caused research and concern by domestic and foreign scholars. Bimbirg studies the control issues between the alliance partners from five aspects: (1) the absolute and relative input level; (2) the symmetry of return; (3) the degree of mutual trust; (4) the size of the uncertainty; (5) the length of the formation time of the relationship. The importance of these factors varies with the changes in the environment [7], during which different combinations affect the

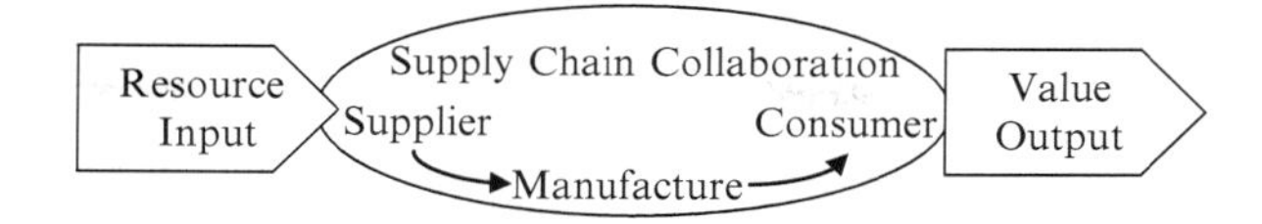

Fig. 1 The model of the supply chain analysis

stability of the union. Badaracco and some other people think, only when the alliance participants have their own resources in products, technologies, capabilities, financial strength [1], as well as talented personnel can the alliance be a solid. And Barney further divides resources into physical capital, human capital and organizational capital and into the finance, entity, management, human resource, organization and technology. Shan Miyuan and some other people think profit structure is an important factor of the strategic alliance stability, and introduces four multi-organizational game model to describe it. R·Axelrod's research results show that if the alliance companies' profits structure changes [8], it can make the whole union cooperation from unstable into stable. Jian Zhaoquan thinks in comparison with the immediate profit, cooperative enterprises pay more attention to future earnings, if the future is sufficiently important relative to the current, cooperation is stable, and the enterprises which pay more attention to future benefits would choose cooperation. Liao Chenglin and some other people propose that building the framework of the balance power in the supply chain can reduce the opportunistic behavior of partners [6], control the switching costs, and is conducive to the stability of the supply chain, thus preventing supply chain destruction due to changes of internal and external structure of the supply chain. Wang lihu and some other people from the customer satisfaction' perspective research the satisfaction with the special performance in the supply chain structure, put forward the concept of corporate satisfaction and think enterprise satisfaction is an important parameter of the supply chain structure stability, and it is closely related to the flexibility of the supply chain structure.

3.2 Supply Chain Stability Index System

According to these scholars who study the stability of the supply chain and business alliances, we conclude that the impact factors of the supply chain stability include three aspects: the level of resources node companies have and input, earnings structure and benefit-sharing, supply chain structure and node companies' relations. Therefore, we establish the chain structure of the supply chain, and the evaluation system based on impact factors. In this article, specific analysis of the influencing factors on the supply chain stability uses the chain structure model. Factors affecting the stability of the supply chain include the stability of the resource input, the stability of the value output and the stability of supply chain collaboration (Fig. 1).

3.2.1 The Stability of the Resource Input

The stability of the resources input refers to the stability of the capital, human resource, technology and material supply. The supply chain is an open system, which need the necessary resource input in order to maintain stability.

Capital factor: The smooth flow of funds to each node in the supply chain is crucial. Capital flow problems during operation of a link or a node will result in the operation of the entire supply chain disruption. It can be measured by the return on investment, cost of capital, asset turnover, and other values.
Human Resource factor: Human resources are not only with the pilot but also strategic for future changes. And it is essential to the development of the entire supply chain. It can be measured by overall labor productivity, human cycle, labor cost and other indicators.
Technology factor: New techniques and methods leading to changes in the structure of the industry chain will enable the supply chain business to increase instability. It can be measured mainly by external technical dependence and the cost of technical use.
The Materials Supply Factors: In addition to the necessary raw materials, the related facilities for the supply of materials are also very important. It is mainly measured by the supply risk and supply costs.

3.2.2 The Stability of the Value Output

The stability of the value output is: Supply chain must be able to output the desired value to consumers and society so as to maintain stability. If the value output of the entire supply chain was instable, it would affect the valuation of the entire value chain. It is mainly measured by tangible products, intangible assets, customer demand satisfaction and other parts.

Tangible product factor: The tangible product is a concentrated expression of the supply chain efficiency. It is also a prerequisite for sustainable development of supply chain and the main aspect of the output value. It is mainly measured by the profit on sales, product life cycle, product turnover, product cost performance and other indicators.
Intangible assets factor: Although it does not reflect the benefits of supply chain, it is of great significance to the long-term development. It is mainly measured by marketing share, goodwill and brand, industry influence and social benefits of corporate culture.
Customer demand satisfaction factor: Meeting specific customers and specific needs for specialized supply chain strategy is essential for the market segments. It is mainly measured by the proportion of customized products, service quality after sales, on-time delivery of orders, product qualification and other indicators.
Added value factor: It refers to the increase in the value of the merger and foreign investment of enterprises and the entire supply chain, as well as research and

development input and output. It is mainly measured by added value to products, new product development, capital gains and other indicators.

3.2.3 The Stability of Supply Chain Collaboration

The stability of supply chain collaboration refers to planning, coordinating and controlling logistics, information and funds between the participating organizations and departments in the supply chain. It is mainly reflected by how to enhance cooperation, strengthen the co-ordination of resources and improve the management level. It includes inter-firm logistics efficiency, responsiveness among enterprises, the dependence between enterprises, supply chain costs and other factors.

Logistics efficiency among enterprises factor: The supply chain is mainly composed of three aspects – business flow, logistics and information flow. Logistics is the foundation and the primary issue. It is measured by the total inventory, turnover rate and other indicators.
Responsiveness among enterprises factor: Caused by information and other reasons, the node enterprise from another gets the products or income different from expected, leading to fluctuations in the supply chain. It is measured by on-time delivery rate, production and demand rate between node enterprises and other indicators.
The dependence between enterprises factor: Profit is the driving force of the supply chain to gather together. The position of the main chain throughout the supply chain determines whether the supply chain is loose or close. It is measured by core enterprise business proportion of suppliers and distributors.
Supply chain costs among enterprises factor: The entire supply chain operating cost is very important compared to competitive enterprise and industry level. If the cost is relatively high, the node enterprises may consider switching costs, leading to the instability of the supply chain. It is mainly measured by the rate of supplier's qualified products, delivery, price and other indicators (Fig. 2).

4 Using Analytic Hierarchy Process (AHP) to Assess the Stability of the Supply Chain

The Analytic Hierarchy Process is a qualitative and quantitative combined, systematic and hierarchical analysis method. So we take the analytic hierarchy process for example and fulfill the measurement and evaluation on the stability of the whole supply chain [4].

4.1 Evaluation of Dimensionless

The indicators in the evaluation system used to analyze the system have relative indicators and absolute indicators, positive indicator, reverse indicator and

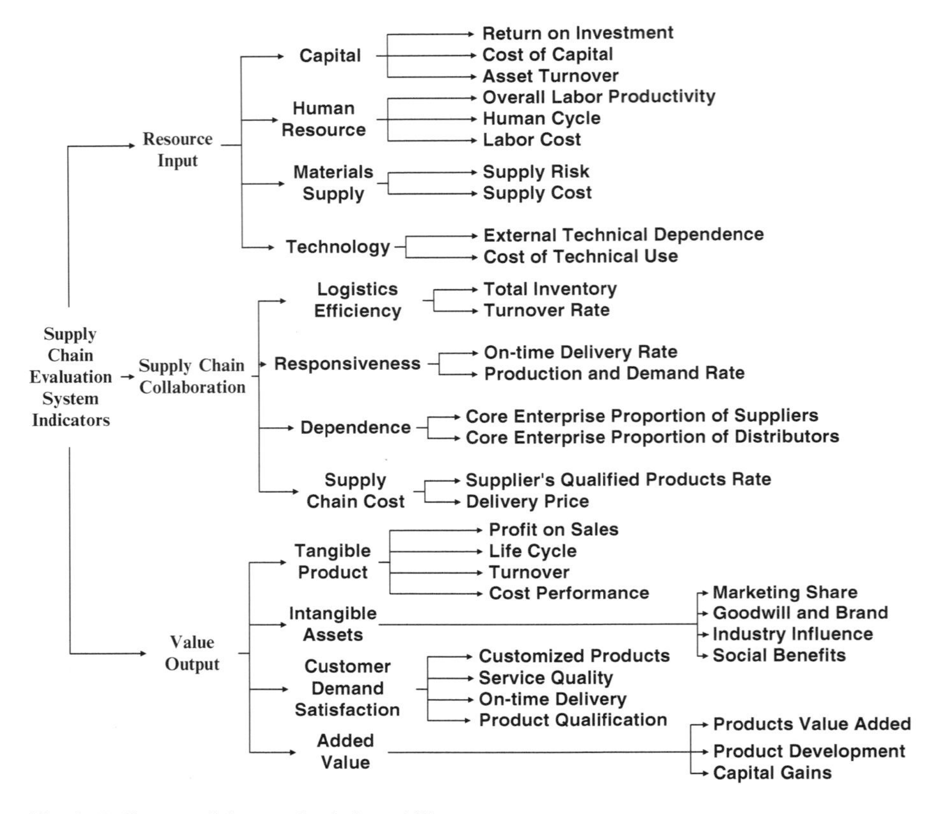

Fig. 2 Indicators of the supply chain stability

moderate indicators. They have different influences on result. Because of different dimension, indicators of the initial value are not comparable. If there was no dimensional process, we could not make a comprehensive evaluation of the stability of the supply chain.

4.2 Build the Judgment Matrix

Survey of experts can be used in the judgment matrix building. By comparing the relative importance between indicators, the judgment matrix can be built. Suppose there are n indicators, their weights are ω_1, $\omega_2 \ldots \ldots \omega_n$. Make pair-wise comparisons for those indicators.

$$A = \begin{bmatrix} \frac{\omega_1}{\omega_1} & \frac{\omega_1}{\omega_2} & \cdots & \frac{\omega_1}{\omega_n} \\ \frac{\omega_2}{\omega_1} & \frac{\omega_2}{\omega_2} & \cdots & \frac{\omega_2}{\omega_n} \\ \cdots & \cdots & \cdots & \cdots \\ \frac{\omega_n}{\omega_1} & \frac{\omega_n}{\omega_2} & \cdots & \frac{\omega_n}{\omega_n} \end{bmatrix} = \begin{bmatrix} a_{11} & a_{12} & \cdots & a_{1n} \\ a_{21} & a_{21} & \cdots & a_{23} \\ \cdots & \cdots & \cdots & \cdots \\ a_{n1} & a_{n2} & \cdots & a_{nn} \end{bmatrix} \tag{1}$$

4.3 Level Single Ranking and the Consistency Checking

The index level single ranking weights of the program layer are B_j (j = 1, 2...n) by Analytic Hierarchy Process, after the dimensionless process, the original value of the evaluation index is transformed to the evaluation value R_j (j = 1, 2... n). And the final score of the comprehensive evaluation of supply chain performance evaluation is

$$V = \sum_{j=1}^{n} R_j B_j \ ,0 \leq R_j \leq 1 \ (j = 1,2\ldots n) \sum_{j=1}^{n} B_j = 1 \quad (j = 1,2\ldots n) \qquad (2)$$

The comprehensive evaluation criteria of supply chain performance evaluation is: if $V \in [0.9, 1.0]$, then it is in the first level, the supply chain is very stable and the risk coefficient is very low; if $V \in [0.7,0.9]$, then it is in the second level, the supply chain is stable and the risk coefficient is low; if $V \in [0.5,0.7]$, then it is in the third level, the supply chain is unstable and the risk coefficient is high; if $V \in [0,0.5]$, then it is in the third level, the supply chain is very unstable and the risk coefficient is very high.

5 Conclusion

The current research of supply chain stability evaluation is still in the exploratory stage, on the contents of which foreign and domestic researchers have different views. This article expands the concept of stability to the operation links in addition to emergencies. By analyzing the structure of the supply chain, and on the foundation of the chain structure of the supply chain, we analyze the impact factors of the supply chain stability and the evaluation indicators. Based on the analytic hierarchy process (AHP), the processing of measured data is proposed to get the evaluation results of the supply chain stability.

But this article selects only the abstract theory of the related indicators of the supply chain stability to study, and the evaluation method is relatively simple with only fuzzy analytic hierarchy process. Specific introduction of a comprehensive evaluation to measure the overall stability of the supply chain and selecting some real enterprises for test will be the next focus of the study.

References

1. Zeng Wenjie, Ma Shihua (2010) The impact of supply chain relationship dynamics on collaboration. Chin J Manag 2:1–8
2. Chen Yao, Sheng Bubin (2009) Empirical study of supply chain alliance stability. J Manag World 11:178–179

3. Xiao Yuming, Wang Xianyu (2008) Early-warning analysis on stability of supply chain based on entropy theory. J Ind Eng Eng Manag 3(22):57–64
4. Song Xiang-nuan (2008) Application of analytic hierarchy process to evaluation of integrated supply chain performance. J Tianjin Univ Commer 3:27–30
5. Meg Green (2005) Executives: supply chain is greatest risk: Best's review. November 102
6. Zhang Rongyao, Wang Kan, Diao Zhaofeng (2004) Analysis on the cooperation among enterprises on supply chain. Logist Technol 2:56–58
7. Li Gui chun, Li Congdong, LI Longzhu (2004) Study on supply chain performance measurement target systems and measurement methods. J Ind Eng Manag 1:104–106
8. Qi Fangzhong, Weng Xiaobin (2004) Research on the system framework of synergic decision-making. Mie China 33(8):82–85

Research on Circulation Efficiency Evaluation of Agricultural Products Based on Supply Chain Management

Zhao Feng

Abstract The author designed a basic agricultural products circulation mode based on supply-chain management, and put forward an evaluation method of agricultural products circulation efficiency. By applying Data Envelopment Analysis, the author measured the agricultural products circulation efficiency of Guangxi's 14 cities. The results showed that: (the total performance of Guangxi agricultural products circulation is well;) more than 50% regions' agricultural products circulation are at or close to the optimal production status.

Keywords Agricultural products • Circulation efficiency • Supply chain • Evolution

1 Introduction

It is shown that the economic development have inevitably transformed from the extensive mode to the intensive mode when we consider the experience of many countries in the world. And the economic efficiency should be studied deeply in the process of economic development changes [1]. However, the studies about this problem are mostly concentrated in the areas of production. And now, the efficiency of circulation should receive more attention with the continuous improvement of its status. With China's rural economic reform, the total shortage of agricultural development has largely been resolved. In fact, the main obstacle that handicaps the further development of agriculture has changed from production to market circulation, and inefficient distribution of agricultural products has become an important constraint on China's agriculture development.

Z. Feng (✉)
College of Business Administration, Guangxi University of Finance & Economics, Nanning, People's Republic of China
e-mail: deane318@163.com

Z. Zhang et al. (eds.), *LISS 2012: Proceedings of 2nd International Conference on Logistics, Informatics and Service Science*, DOI 10.1007/978-3-642-32054-5_32,

As a link between the production and the consumption of agricultural products, the agricultural market is also an important platform between small-scale production and large-scale market [2]. According to statistics, in 2009, China's turnover of agricultural products wholesale market transactions is more than two trillion Yuan while more than 100 million Yuan of single deal is calculated [3]. The integration and construction of a flexible, fast response, high efficiency and high satisfaction agricultural products supply chain is the basis of agricultural modernization, and also is the inevitable choice for agricultural products circulation enterprises to meet the social needs. Therefore, improving agricultural products circulation efficiency is very important in the agricultural industrialization and rural economic development. In particular, it is significant to solve the issues of agriculture, farmer and rural area and promote urban and rural integration process.

2 Main Modes of China's Agricultural Products Circulation

Under the action of market mechanism, China's agricultural products circulation mode is the farm-industry-trade combination system which is made of agricultural suppliers, intermediaries, processing distribution businesses and consumers at the present stage. According to the different agricultural products business entities, there are diverse circulation modes as choices.

2.1 Large-Scale Agricultural Products Wholesale Market-Oriented Agricultural Products Circulation Mode

In this mode, large-scale agricultural products wholesale operators are served as the core enterprise of agricultural products supplies, and carry out the organization, control and management of entire supply chain operations. On this basis, an integrative agricultural products circulation mode which includes agricultural production, acquisition, processing, sale, storage preservation, distribution and market information provision has formed. That is, farmers/production base, wholesale carriers, retailers and consumers together constitute the agricultural products supply chain.

2.2 Chain Supermarket-Oriented Agricultural Products Circulation Mode

Focusing on the market demand, a chain supermarket operator construct the production base or unites with agricultural products distributors, processing enterprises, or establish a long-term cooperative relationship with agricultural commodities producers. And then provides stores with more variety of agricultural products through self-built products logistics distribution center. Due to high added value, fast market response and high standardization, chain supermarket-oriented agricultural products circulation mode will prove to be mainstream in the coming period.

2.3 Large-Scale Processing Enterprise-Oriented Agricultural Products Circulation Mode

In this mode, large-scale agricultural products processing enterprises are served as the core enterprise of agricultural products supplies. They establish community exclusive stores themselves or together with others or access to supermarket directly, and control the sales terminal. Thus organize and operate the whole agricultural products circulation supply chain. Due to fewer circulation links, prompt information feedbacks, high market sensitivity, fast distribution and high consumer satisfaction, the added value and technological content of agricultural products have enhanced greatly.

3 Evaluation Model of Agricultural Products Circulation Efficiency Based on Supply Chain Management

3.1 Abstract Basic Model of Agricultural Products Circulation

The main participation entities include producers (farmers), origin wholesalers, transportation/distribution companies, processing enterprises, farmers cooperatives, sale areas wholesalers, retail terminals and consumers. Different types of agricultural products circulation modes are different from each other [4]. In view of the diversity and multi-level nature of agricultural products circulation channels, the author abstracts the agricultural products circulation process, and comes up with a universal agricultural products circulation efficiency evaluation model based on abstract basic mode of agricultural products circulation, as shown in Fig. 1.

3.2 Evaluation Index System of Agricultural Products Circulation

As seen in Fig. 1, agricultural products circulation process is made of five links such as processing, transportation, warehousing, distribution and sales. It reflects the all links that certain agricultural products experienced after they are outputted. The agricultural products circulation programs are worked out according to the compound

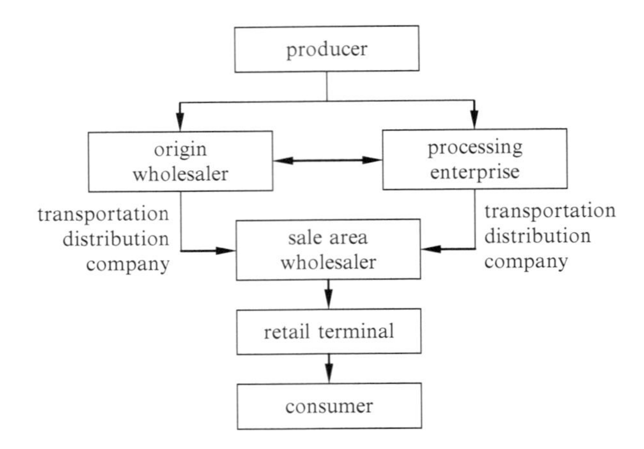

Fig. 1 Basic agricultural products circulation model

mode of five links. In this sense, the efficiency evaluation of agricultural products circulation is essentially the efficiency evaluation of each circulation program. From the aspects of circulation velocity, circulation cost, circulation revenue, circulation quantity and circulation safety, the efficiency of agricultural products circulation program can be evaluated accurately and comprehensively. Accordingly, the evaluation index can be decomposed into velocity efficiency index, economic efficiency index and qualitative efficiency index. In order to acquisition data accessibly, each type of index will be decomposed into a number of detailed indicators.

4 Evaluation Method of Agricultural Products Circulation Efficiency

In this paper, author evaluates agricultural products circulation efficiency by applying data envelopment analysis (DEA). With regard to data envelopment analysis, it's a well-known Non-parametric method developed [5]. The method skillfully solves the comparison and evaluation problems between different dimensions, multiple-input, multiple-output similar decision-making units (DMU). The idea of CCR model is: supposing there are N decision-making units and each decision-making unit is provided with M kinds of "input" and S types of "output". Let $x_j = \left(x_{1j}, x_{2j} \ldots x_{mj}\right)\mathrm{T} \geq 0$, $x_j \in \mathrm{R} + m$; $\mathrm{Y_j} = \left(\mathrm{y_{1j}}, \mathrm{y_{2j}} \cdots \mathrm{y_{sj}}\right)\mathrm{T} \geq 0$, $\mathrm{Y_j} \in \mathrm{R}^+$ were respectively the input variable and output variable of $\mathrm{DMU_j}$ ($j = 1, 2 \ldots n$). On this basis, the author establishes a relatively effective model with non-Archimedean infinitesimal for the j-th decision-making unit. For convenience, the author works out the linear programming model by applying Charnes-Cooper Transformation, and transforms it based on Linear Programming Duality Theory, and then adds slack variables s^+ and s^- [6]. On this basis, the paper gets the dual programming model:

$$\min\left[\theta - \omega\left(e^{\wedge T}s^- + e^T s^+\right)\right]$$

$$s.t\sum_{j=1}^{n}\lambda_j x_j + s^- = \theta x_0, \sum_{j=1}^{n}\lambda_j x_j - s^+ = y_0.$$

$$\hat{e} = (1,\ldots,1)^T \in R^m, e = (1,\ldots,1)^T \in R^n.\lambda_j, s^-, s^+ \geq 0.j = 1,\ldots,n$$

Based on input, the CCR model is mainly used to evaluate the DMU_0 overall efficiency, that is a comprehensive technical and scale efficiency. The Meaning of θ^* is the optimal ratio of required investment and actual investment under the conditions of keeping output unchanged. Suppose the optimal solution for the model is $\lambda^*, s^{*-}, s^{*+}, \theta^*$, then it can be concluded: ① if $\theta^* = 1$, then DMU_0 is DEA effective; ② if $\theta^* = 1$, and $s^{*-} = 1, s^{*+} = 1$, then DMU_0 is DEA effective; ③ if $\theta^* < 1$, then DMU_0 is DEA ineffective, $1 - \theta^*$ refers to the input proportion which can be compressed before achieves efficiency frontier.

DEA not only provides the method which appraises each kind of agricultural products circulation efficiency whether has achieved the maximization, but also indicates the improvement direction and extent of each variable to achieve maximization.

The key of using DEA method is the selection of appropriate input/output indicators. In Western Economics, the productive factors are generally divided into four types of labor, land, capital and entrepreneurial talent. However, considering the difficulty of collecting data, the researchers normally adope fixed assets and employees as input indicators, and adope annual turnover or overall retail sales of social consumption products as output indicators.

5 Conclusion

With the agricultural products circulation input–output data, the author measures the agricultural products circulation efficiency of Guangxi's 14 cities by applying DEA. The results show that: with the deepening of the rural circulation system reform and the continuous improvement of rural infrastructure, the aggregate performance of Guangxi agricultural products circulation is well; among 14 cities, more than 50% regions' agricultural products circulation are at or close to the optimal production status. But during the circulation operation, there are still problems such as low labor rate and lack of asset utilization. Therefore, the agricultural products circulation managers and operators should optimize employment structure and improve asset utilization to improve circulation efficiency.

References

1. Banker RD, Charnes A, Cooper WW (1984) Some models for estimating technical and scale inefficiencies in data envelopment analysis. Manag Sci 30:1078–1092
2. Stevens OC (1989) Integrating the supply chain. Int J Phys Distrib Mater Manag 18:3–8
3. Kumar R, Husain N (2002) Marketing efficiency and prince spread in marketing of grain: a study of Hamirpur District. India J Agric Econ 7:390–397
4. Anrooy RV (2003) Vertical cooperation and marketing efficiency in the aquaculture products marketing chain: a national perspective from Vietnam. In: Paper presented to the Aquamarkets 2003 Conference. Manila, Philippines, 2–4 June
5. Kou R (2008) Research on the analytical framework of agricultural products circulation efficiency. J China Bus Mark 5:20–23
6. Charnes A, Cooper WW, Rhodes E (1978) Measuring the efficiency of decision making units. Eur J Oper Res 2:429–444

A Study of Eco-Performane of Logistics Services in Food Supply Chains

Dong Li and Zurina Hanafi

Abstract Transportation is one of the main contributors of greenhouse gases which give direct negatives impact on environment. Management of logistics services plays an important role in maintaining business competitiveness as well as social responsibility. Optimising logistics service with integrated economic and ecological objectives can help to reduce negative impact on the environment by reducing the amount of carbon emissions and improving operations efficiency. This study focuses on multimodal transportation planning and optimal strategies with a UK food supply chain case under carbon emissions control. The research investigates and identifies impact of the policies on logistics performance.

Keywords Multimodal transportation planning • Carbon emission policy • Fresh produce

1 Introduction

In the last decade, over 50% of fresh produce in the UK market were imported [1]. It has been a great challenge to achieve both economic and ecological objectives in the international transportation services. Research on transportation planning has been extensively reported in the literature [2–6]. Some research on logistics planning considering environmental impact has been reported [7–10].

However, research is still rare on interactions of supply chain economic and ecological performance with carbon control policies [11]. This study focuses on fresh produce supply chain case in the UK and investigates impact of different carbon emission control policies on operations of food logistics industry. The research aims to identify optimal strategies of multimodal transportation of supply

D. Li (✉) • Z. Hanafi
Management School, University of Liverpool, Liverpool, UK
e-mail: Dongli@liv.ac.uk; zurina@liverpool.ac.uk

Z. Zhang et al. (eds.), *LISS 2012: Proceedings of 2nd International Conference on Logistics, Informatics and Service Science*, DOI 10.1007/978-3-642-32054-5_33,

networks under carbon policies, and provide a policy making reference to facilitate understanding of industrial reaction to government environmental policies on carbon emission. The research outcome is expected to have a generic contribution to multimodal transportation planning and government policy making in carbon emission control.

2 Eco-Logistics Planning

An optimisation model as seen in Eq. 1 is proposed to generate solutions and analyse behaviour of the supply network under different carbon control policies. Optimisation models are widely used in solving multimodal freight transportation problem [12–14].

In this paper, a mixed-integer programming is developed with four main elements: cost, time, distance, and mode of transportation, to analyse the economic and ecological performance of the logistics network, in particular the carbon emission policy impact on strategic options of supply chain design. The objective function of the model is to minimise the total cost (see Eq. 1), with consideration of policies of carbon emissions trading and carbon tax [15]. The modes of transportation in this study are road, rail, ship and their combinations.

Objective function:

$$\text{Min } CT = \sum_{k=1}^{p} \sum_{j=1}^{n} \sum_{i=1}^{m} \left(\left(TC_{i,j} * YT_{i,j,k} \right) + \left(CC_{i,j,k} * YT_{i,j,k} * YP_{i,j,k} \right) \right) * X_{i,j,k} \quad (1)$$

Subject to:

$$\sum_{k=1}^{p} \sum_{j=1}^{n} \sum_{i=1}^{m} CC_{i,j,k} * YT_{i,j,k} * YP_{i,j,k} * X_{i,j,k} = CL; \sum_{j=1}^{n} X_{i,j,k} = D_i;$$
$$\sum_{j=1}^{n} TT_{i,j} * YT_{i,j,k} \leq RT_t; X_{i,j,k} \geq 0; YT_{i,j,k} \in \{0,1\}; \qquad YP_{i,j,k} \in \{0,1\}.$$

Notations: i – centre index; j – transportation mode; k – carbon emissions policies;

CL = carbon limit; TCi,j – transportation cost to centre (maritime port, rail freight terminal or a regional distribution centre) i with transportation mode j;

RTi – Required time for trip to a port or regional distribution centre i; TTi,j – time taken to centre i by transport mode j; CCi,j,k – carbon emissions cost to a centre i with transport mode j and carbon policy k; YTi,j,k – 1 if transportation mode j is used, 0 otherwise; YPi,j,k – 1 if policy k is chosen, 0 otherwise; Di – demand at centre i.

Interviews for data and business process mapping have been conducted with the case company which is a fresh produce logistics service provider. Some data such as carbon emission factor for transportation and carbon price are obtained from public sources [16]. At present, logistics companies have been mainly using road transportation for distribution of fresh produce in the UK. Road transportation has an advantage of door to door delivery with faster services. 40-ft refrigerated containers are normally used with heavy goods vehicles (HGVs) for the service. However, HGVs consumes enormous amount of fuels and creates environmental impact. Through the optimisation analysis, solutions of the logistics service network with different carbon policies can be identified with an insight into impacts of policies on shift between transportation modes and best strategies of logistics services.

3 Analysis and Finding

Firstly the model is analysed without considering carbon emission and associated costs. The model suggests distribution of fresh produce from all ports to all RDCs using road transportation. When carbon emission is considered, there is a significant impact on the present transportation practice. With carbon tax, multimodal options are selected (77% for road only and 23% for multimodal with rail plus road). On the other hand, with carbon emission trading, the best solution is suggesting road only and multimodal at 87 and 13% respectively. Carbon tax has a greater impact due to higher direct cost to the operations.

3.1 Carbon Tax vs. Carbon Emission Trading

Two common carbon control policies, Cap-and-trade (CT) for carbon emission trading scheme and carbon tax scheme, are involved in this research. CT scheme has an annual allowances allocated to the participants as a cap. Participants who face high abatement costs can continue emission by buying additional allowances, while those who face low abatement costs can take abatement action and sell their surplus allowances for a profit [15]. Carbon tax is based on consumption of fossil fuels.

To identify potential business behavior with government carbon control policies, the analysis is performed with different carbon charge rates as sensitivity analysis. Optimal carbon charges by carbon tax and carbon emission trading is investigated in the research as seen in Fig. 1. As carbon charges are highly dependent on government policies, the government enforcement plays a significant role in managing carbon emissions. The total cost includes transportation cost and carbon cost with consideration of transportation time limit. The analysis of total cost demonstrated optimum carbon charges with different carbon emission limit (CET in the Fig. 1) and carbon tax. It can be seen that, as cap increases, the optimal carbon charge for minimum total cost also increases. The carbon tax scheme in this case has the lowest optimal carbon charge.

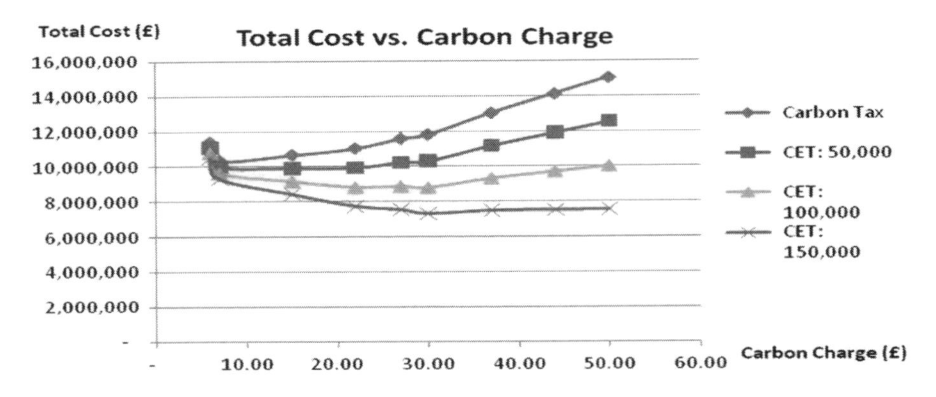

Fig. 1 Total cost fluctuations with different carbon charge

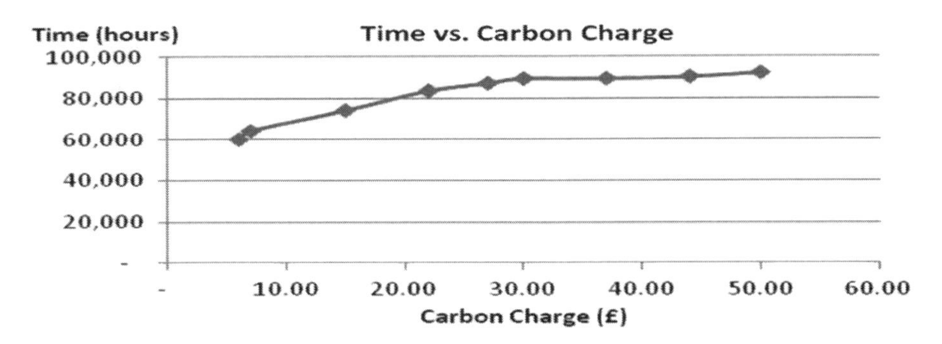

Fig. 2 Travel time with different carbon charge

3.2 *Time Performance and Transportation Mode Selection*

For fresh produce supply chains, time is an important factor. The performance in overall delivery time is analysed. The time spent in transportation processes with different carbon charging rates is shown in Fig. 2. In the analysis, time is a constraint for a trip to ensure food shelf-life requirements being met. As seen in Fig. 2, the travel time for each journey increases as the price of carbon mission increases. Therefore, the carbon charge is positively related to the logistics network performance in time.

To investigate impact of carbon control policy on transportation network configuration, the percentage of selected multimodal transportation routes with different carbon charges is analysed (see Fig. 3). Result shows that the higher the carbon price, the higher the percentage of multimodal transportation is chosen.

The optimal carbon charge is observed when carbon tax is chosen at the rate of £7 per ton of carbon emission, with a given travel time limit. Multimodal transportation accounted at 8 and 92% is for road only. If the CT scheme is chosen, with 50k

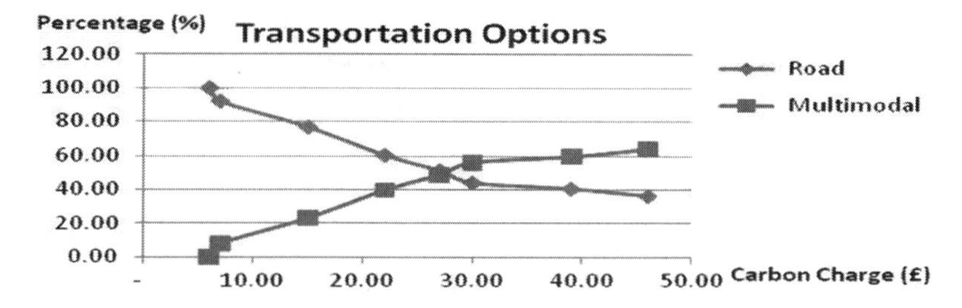

Fig. 3 Transportation options between road only and multimodal

ton of carbon limit, the optimum cost is at the price of £15 per ton of carbon emission. When the carbon limit is set at 100k ton and above, total cost is decreasing until the carbon price reaches £30. Higher than this charge, multimodal transportation becomes the favorable option.

4 Conclusion

This research has investigated impacts of the carbon emission policies on transportation operations for fresh produce industry. A network design approach for fresh produce logistics services under carbon emission control is proposed. With introduced carbon control policies, optimal decisions on transportation planning in fresh produce logistics will be affected by policies to be applied, carbon emission limits and carbon prices to be involved. The higher the charge on carbon emission, the more the allocation would be made to multimodal transportation routes. But the time spent may be increased in such cases, due to time spent in transportation mode transfer. The performance in costs can be optimised with given carbon charges and carbon policy through transportation mode selection. On the other hand, optimal carbon charges can be set to obtain lowest overall costs in the logistics operations. It can be seen that the policy to be applied by governments can play an effective role to shape the logistics network and affect economic and ecologic performance of businesses. This research outcome can be generalised to other industries for development of strategies with given carbon control policies, and for government to set up policies to encourage best business practice.

References

1. EFFP (2010) Driving change in the fresh produce sector, european food and farming partnerships, final report, April 2012, http://archive.defra.gov.uk/foodfarm/food/policy/partnership/fvtf/documents/effp-report.pdf. Accessed on 30 Mar 2012
2. Rondinelli D, Berry M (2000) Multimodal transportation, logistics and the environment: managing interactions in a global economy. Eur Manag J 18(4):398–410

3. Macharis C, Bontekoning YM (2004) Opportunities for OR in intermodal freight transport research: a review. EurJ Operat Res 153:400–416
4. Southworth F, Peterson BE (2000) Intermodal and international freight network modeling. Transp Res Part C 8:147–166
5. Hasan MK (2008) Multimodal, multicommodity international freight simultaneous transportation network equilibrium model. Springer Telecommun Syst 40:39–54
6. Yang X, Low JMW, Tang LC (2011) Analysis of intermodal freight from China to Indian Ocean: a goal programming approach. J Transp Geogr 19:515–527
7. Facanha C, Horwath A (2006) Environmental assessment of freight transportation in the U.S. Int J Life Cycle Assess 11(4):229–239
8. Lopez I, Rodriguez J, Buron JM, Garcia A (2009) A methodology for evaluating environmental impacts of railway freight transportation policies. Energy Policy 37(12):5393–5398
9. Sadegheih A, Li D, Sribenjachot S, Drake PR (2010) Applying mixed integer programming for green supply chain management. South Afr J Ind Eng 21(2):13–24
10. Hoen KMR, Tan T, Fransoo JC, van Houtum GJ (2011) Effect of carbon emission regulations on transport mode selection under stochastic demand. Flex Serv Manuf J, doi:10.1007/s10696-012-9151-6
11. Li D, Hanafi Z, Drake P (2010) Logistics network optimization under carbon emission control for the paper recycling industry. In: Proceedings of the logistics research network annual conference (LRN 2010), Harrogate, UK, 8th–10th September 2010
12. Yamada T, Russ BF, Castro J, Taniguchi E (2009) Designing multimodal freight transport networks: a heuristic approach and applications. Transp Sci 43(2):129–143
13. Caramia M, Guerrioro F (2009) A heuristic approach to long-haul freight transportation with multiple objective functions. Omega 37:600–614
14. Banomyong R, Beresford AKC (2001) Multimodal transport: the case of Laotian garment exporters. Int J Phys Distrib Logist 31(9):663–685
15. Sorrell S, Sijm J (2003) Carbon trading in the policy mix. Oxf Rev Econ Policy 19(3):420–437
16. WRAP (2008) CO_2 impacts of transporting the UK's recovered paper and plastic bottles to China. Waste and resources action programme, final report, August 2008, http://www.envirocentre.ie/includes/documents/CO2_Impact_of_Export_Report_v8_1Aug08.1bd19928.pdf. Accessed on 30 Mar 2012

Research on a Combined Port Cargo-Throughput-Forecast Model

Chi Zhang, Lei Huang, and Zhichao Zhao

Abstract This paper study and analyze the present port cargo throughput forecast model, then the authors develop a combined port cargo-throughput-forecast model. The combined model is verified by a real port in China to obtain relatively higher forecast accuracy when it is not easy to find more information.

Keywords Cargo throughput • Combined forecast model • Logistic growth curve model • Gray forecast model

1 Introduction

Cargo throughput is very important for a port, it is not only the most basic production index for measuring the port development, but also a significant reference to organize its production, make its development plans and construction. In the mean time, the amount of the cargo throughput may reflect the economic situation and the development level of the port city. The correctness and rationality of the forecast means much to various aspects in the development of ports including the scientific port layout, the scale of investment in infrastructure, business strategy, development strategy and the collection and distribution of integrated transport plan. This paper studies the forecast model of cargo throughput of the port, and then takes one port in China as an example to verify the model accuracy, in order to provide a strong reference to the port cargo throughput forecasts.

C. Zhang (✉) • L. Huang • Z. Zhao
School of Economics and Management, Beijing Jiaotong University, Beijing, People's Republic of China
e-mail: crazyzhc98@gmail.com

Z. Zhang et al. (eds.), *LISS 2012: Proceedings of 2nd International Conference on Logistics, Informatics and Service Science*, DOI 10.1007/978-3-642-32054-5_34,

2 The Selection of Combination Forecast Model

Combined forecast method doesn't directly use of the historical and current data modeling, but establishes an appropriate combination of various kinds of individual forecast models to obtain the optimal value to forecast the port cargo throughput [4].

We can easily find the combination forecast model can reduce the error of a single prediction model to a larger extent. This paper will make an organic combination of gray forecast model and Logistic population projection model, to achieve more accurate port forecast model.

2.1 *Grey Forecast Model*

Gray Forecast Model, abbreviated as GM(1, 1), is as follows [1]:

Firstly, there is an original sequence denoted by $x^{(0)}$, it is made up of $x^{(0)}(1)$, $x^{(0)}(2), \ldots, x^{(0)}(n)$.

The corresponding differential equation is that:

$$\frac{dx^{(1)}}{dt} + ax^{(1)} = u \tag{1}$$

2.2 *Logistic Growth Curve Model*

Logistic curve equation is [2]:

$$y = \frac{k}{1 + me^{-at}} \tag{2}$$

In the Eq. (2), "t" stands for the point of the time sequence, generally natural numbers; "a", "m" is the undetermined coefficient; "k" is a given saturation value by the forecasters according to the actual situation and the growing trend of "y".

2.3 *Combined Forecast Model*

The greatest concern in the combined forecast is how to calculate the weighted mean coefficient to make the combined forecast model to improve the prediction accuracy [5]. Suppose "m", "n" is the coefficient, "m" stands for how many ways to forecast the cargo throughput, while "n" stands for how many time periods we use. yij (i = 1,2,. . .,n, j = 1,2,. . .,m) means the predictive value in the "j"th way and the "i"th time period. "rj" stands for the weight of the different forecast methods, it must meet the constraints.

Table 1 The port cargo throughput of nearly 11 years

Year	2002	03	04	05	06	07	08	09	10	11
CT (Mt)	1,532	1,719	2,152	2,504	3,028	3,432	3,470	3,639	4,110	4,510

CT represents Cargo Throughput

Suppose yi0 as the combined predictive value in the "i"th time period, so that

$$y_{i0} = \sum_{j=1}^{m} r_j y_{ij} \tag{3}$$

3 A Port Cargo Throughput Forecast

A Port in Southern China is an omnibus advocate hub port. In 2011, the port cargo throughput reached 451 million tons, among the global port 4. The Port has 50 million-ton berths; 14 million-ton loading and unloading buoy; 23 million-ton loading and unloading of anchorage. Considered from any factors like the historical, economic and cultural, the Port is a representative in the ports of China and the world. With the Port cargo throughput of nearly 11 years as the sample, this paper studies the port cargo throughput prediction on combination forecasting model (Table 1).

3.1 Establish GM (1,1) Model

According to the Port historical data, we can calculate through formula (1) for gray forecast model parameters: a = −0.1, u = 1,567.7, while we can also get the small error probability is 1, variance is 0.1906. These two values indicate that the model accuracy for A, so that we can use this result as forecast model [3]:

$$\frac{dx^{(1)}}{dt} - 0.1X^{(1)} = 1567.7 \tag{4}$$

4 Establish Logistic Growth Curve Model

Setting the initial value[x,y] = [1,18] in formula (2), we can get the undetermined coefficient "m" equals 4.225, "a" equals 0.2574. Finally, the Logistic growth curve equation is:

$$y = \frac{5418}{1 + 4.225e^{-0.2574t}} \tag{5}$$

Table 2 Comparison of three models

		Logistic		GM(1,1)		Combination	
Year	Original (Mt)	Fitted (Mt)	Error/%	Fitted (Mt)	Error/%	Fitted (Mt)	Error/%
2002	1,532	1,550.942	1.2	1,797.0	17.2	1,673.356	9.2
2003	1,719	1,850.679	7.6	1,997.7	16.2	1,923.822	11.9
2004	2,152	2,175.748	1.1	2,220.9	3.2	2,198.212	2.1
2005	2,504	2,517.617	0.5	2,469.0	−1.3	2,493.43	−0.4
2006	3,028	2,865.72	−5.3	2,744.8	−9.3	2,805.562	−7.3
2007	3,432	3,208.701	−6.5	3,051.4	−11.1	3,130.444	−8.8
2008	3,470	3,535.858	1.9	3,392.2	−2.2	3,464.388	−0.2
2009	3,639	3,838.411	5.4	3,771.1	3.6	3,804.924	4.5
2010	4,110	4,110.309	0	4,192.4	2.0	4,151.149	1.0
2011	4,510	4,348.437	−3.5	4,660.7	3.3	4,503.788	−0.1

5 Establish Combined Forecast Model

Setting the weight of GM (1,1) model, Logistic population model as y_1, y_2, we can calculate them from formula (3): $r_1 = 0.4975$, $r_2 = 0.5025$. Finally, combined forecast model is:

$$y = 0.4975y_1 + 0.5025y_2 \quad (6)$$

5.1 *Three Forecast Model Comparison*

This paper compares the error of three forecast methods to verify the superiority of the combined model (Table 2).

We can see that the combined forecast model is better than the others, and it is closer to the true value. This indicates that the combined forecast model is valid. It can be seen from the above forecast: The combined forecast method forecast that the port cargo throughput in 2015 will be 644 million tons. This will provide a basis for the Port to make the port development strategy.

6 Conclusion

The port facility capacity in China is so fast that the port construction may be developed in advance. However, if the capacity is serious surplus, not only its economic and social benefits can not work, but will cause the waste of idle facilities and resources. Predicting the cargo throughput of the port is of great importance for state and local to formulate the port development strategy.

Acknowledgement This research was supported by "Research of Logistics Resource Integration And Scheduling Optimization" under the National Natural Science Foundation 71132008. The authors also wish to thank the port in the paper for support in data collection.

References

1. Jiang F, Lei K (2009) Grey prediction of port cargo throughput based on GM(1,1,α) model. Logist Technol 9:68–70
2. Xu CX, Yan YX, Zhang P (2006) Prediction model of port cargo throughput based on system dynamics. Port Waterw Eng 5:691
3. Chen XY, Gu H (2010) Gray linear regression model in port cargo throughput prediction. Port Waterw Eng 5:90–92
4. Huang J, Cai QD, Yi WX (2010) LM-BP neural network-based coastal port cargo throughput simulation and prediction. Port Waterw Eng 7:63–65, 103
5. Chen TT, Chen YY (2009) Port cargo throughput forecast based on BP neural network. Comput Mod 10:183

Research on Dynamic Berth Assignment of Bulk Cargo Port Based on Ant Colony Algorithm

Chunfang Guo, Zhongliang Guan, and Yan Song

Abstract Berth assignment is the basic problem of the port ship-scheduling. Considering the different importance of different ships, this paper put forward a model for the use of ship resources, through the improvement of traditional ant colony algorithm, the ship-berth matching relation constraint matrix forms by ontology reasoning. Finally, the model based on improved ant colony algorithm and ontology reasoning is applied to the combination of the dynamic berth assignment problem solving process.

Keywords Ship-scheduling system • Improved ant colony algorithm • Ontology reasoning • Dynamic berth assignment

1 Introduction

According to statistics, bulk cargo transportation accounts for one-third of the total world seaborne coal, ore and grain-based three bulks accounted for about 60% of the total bulk cargo transportation. Traditional ship scheduling and production deployment of resources has been unable to meet the frequent ship out of port and cargo handling needs; improving the operating efficiency of cargo transportation, thereby increasing the overall level of production scheduling/Port Holdings, is the key to realize of bulk groceries Port Holdings ship scheduling information and the intelligence.

C. Guo (✉) • Z. Guan • Y. Song
School of Economics and Management, Beijing Jiaotong University,
100044 Beijing, People's Republic of China
e-mail: Chfguo@bjtu.edu.cn

Z. Zhang et al. (eds.), *LISS 2012: Proceedings of 2nd International Conference on Logistics, Informatics and Service Science*, DOI 10.1007/978-3-642-32054-5_35,

Table 1 Comparison of multi-development and traditional ant colony optimization

	Optimal value		Mean value		Worst value		Standard deviation	
Oliver 30	*	#	*	#	*	#	*	#
Calculation results	426.78	427.50	427.00	436.23	427.50	444.83	0.35	5.22

* Presents improved ant colony algorithm.
# Presents traditional ant colony algorithm.

2 Improved Ant Colony Algorithm Based on Mixing Behavior

The improved algorithm made better in convergence speed and the target value minimizing certain, however, it is prevalence exist that the algorithm become stagnation and iterative means does not significantly reduced. In this paper, with the first iteration, use variance of the solution set to represent the dispersion of solution, also known as the "concentration". Greater concentration means greater dispersion of solution, and slower Convergence rate, this need to increase the distribution of pheromone amount in optimal path. And lower concentration means lower dispersion of solution, and speeder Convergence rate, need to decrease the distribution of pheromone amount in optimal path.

$$s_i = \frac{\sum_{j=1}^{m} \left(l_{ij} - \bar{l}_i\right)^2}{m} \quad 0 < i \leq n \tag{1}$$

$$s_i^{\max} = \left(l_i^{\max} - \bar{l}_i\right)^2 \quad 0 < i \leq n \tag{2}$$

$$\varepsilon_i = \frac{s_i}{s_i^{\max}} = \frac{\sum_{j=1}^{m} \left(l_{ij} - \bar{l}_i\right)^2}{m \cdot \left(l_i^{\max} - \bar{l}_i\right)^2} \quad 0 < i \leq n \tag{3}$$

$$\Delta\tau_i = \varepsilon \cdot Q / l_i^{best} \quad 0 < i \leq n \tag{4}$$

Formula (1) said in the first i iterations of the ant colony, the discrete dispersion degree of passed path l_{ij} and Path average $\bar{l}_i$ of all the m ants. Formula (2) said the difference of after the longest path and the average path that the ants passed in the first i iterations. Formula (3) is the Standard deviation coefficient, it said that in the first i iterations. Formula (4) guide the total pheromone update by the value generated by standard deviation coefficient, adjust the allocation of the pheromone automatically, make the standard deviation coefficient produce positive feedback to the distribution of pheromone. We raised the improved hybrid ant colony algorithm based on the standard deviation on the basis of the maximum and minimum ant colony algorithm.

As is shown in Table 1, Improved ant colony algorithm with the traditional ant colony algorithm, the results of 10 times running.

Combined with (1) and (2) results, we presents improved ant colony algorithm based on hybrid behavior, respectively combine the maximum and minimum and standard deviation coefficient optimization organic, Improve the convergence efficiency of traditional ant colony algorithm, reduce the precocious extent, and get better target.

Table 2 Factors of priority about ship using resources

Serial	Factor	Description
1	The importance of goods	The timeliness of the goods, whether the key material, whether it is dangerous goods, etc.
2	Importance of customer	Customers' cargo throughput the port, costs and the payment
3	Trade types	Ocean or coastal areas, domestic or foreign trade
4	Collection and distribution method	Ship loading and unloading of goods shipped that way to train, automobile, barge or pipeline or local acts out port
5	Arrival time	Time of ship's arrival in Anchorage

3 Sort of Priority Based on the Ship Resources

The object of this survey includes the ports corporation scheduling manager, line dispatchers of a port, we conducted a survey of the factors that affect the sort of ship job priority in their professional and work areas, and we used statistical methods like computing and testing dealing with the questionnaire. In view of our cargo transportation is strong regional, and the geographical situations are different, The articles selected for the Guangzhou Port Group as a whole object, respectively involve its subordinate four Ports Corporation: Company A, Company B, C and D Ports Corporation, acquired factors as is shown in Table 2.

In addition, we use number '1' to number '5' to represents five important degree respectively.

This study distributed 60 questionnaires and recovered 51, excluding individual invalid questionnaires, 45 valid questionnaires left. According to the actual work of the cargo transportation ship scheduling, initially identified the priority of the ship operating factors, stripping the cross or related higher impact factors through in-depth analysis of the questionnaire to determine the main influencing factors. Explore factor analysis (EFA) was used in this study. Then we can extract the first three factors that the importance of the goods, the importance of customers, type of trade as the main factors of the ship operating priority.

4 Berth Assignment Problem Implementation Based on Improved Ant Colony Algorithm

4.1 Ship – Berth Constraints

When the port make ship-berth selection, there is docked constraint set between specialized berths and different types of cargo according to the specialized adapt requirements of the port cargo types to the berths. Constraints of the collection contains two aspects:

1. ships – berth physical condition (boat length, type width and maximum draft and other property and the length of the berth, draft, etc.) must be met;

2. ship – berth job match degree. The corresponding relations of the various types of berths and the ship can be shown as following model:

$$\sum_{i=1}^{n} v_i = V, \sum_{j=1}^{m} b_j = B \quad \sum V_{c\arg o} = \sum B_{type}$$
$$V_{c\arg o} \in B_{type} = \left(V^1_{c\arg o}, V^2_{c\arg o} \cdots V^n_{c\arg o} \right) \Rightarrow berth(v_i) = b_j \quad (5)$$

v_i is Ship waiting for berthing operations, *berth*(v_i) is the berth ship v_i docked, $c\arg o(v_i)$ is the information of the goods ship v_i carried, B_{type} said the services type that the berths provide, formula (5) said that only when the goods information on ship corresponding to the requirements of loading and unloading cargo information of the berths provided, it can choose the berth.

Based on the two types of constraints above, corresponding to ships – berth constraints concept , merge the ant colony algorithm, the constraint conditions turns into (5), Ship – berth constraint matrix.

$$A = \begin{bmatrix} y_{11} & y_{12} & y_{13} & \cdots & y_{1n} \\ y_{21} & \ddots & & & \vdots \\ y_{31} & \ddots & \ddots & & \vdots \\ \vdots & & y_{ij} & \ddots & \vdots \\ \vdots & & & \ddots & \vdots \\ y_{m1} & \cdots & \cdots & \cdots & y_{mn} \end{bmatrix} \quad y_{ij} = \begin{cases} 1 & \text{ship } v_i \; can\, berth\, at\, b_j & 0 < j \le m \\ 0 & \text{ship } v_i \; can\, not\, berth\, at\, b_j & 0 < i \le m \end{cases} \quad (6)$$

4.2 Prioritization of the Vessel Operation

As said before, since the scarcity of port productive resources, it is inevitable happening the phenomenon of Inbound vessels operating waiting in line, In this paper, the ships resource using priority model based on the Analytic Hierarchy Process was raised to measure the different resources using priority of the ship.

$$P_i = \alpha \cdot v_i^{customer} + \beta \cdot v_i^{c\arg o} + \gamma \cdot v_i^{trade} \quad (7)$$

Calculate marine resources using priority sequence *Vodr*, Sort the ship in *Vodr* in accordance with the resources using priority.

When all the ships complete one operation, it said that the berth assignment problem obtained a solution.

$$WorkT(t) = \begin{bmatrix} \frac{Ws_1 - Wl_1(t)}{Lvb_1} & \cdots & \cdots & \cdots & \frac{Ws_1 - Wl_1(t)}{Lvb_n} \\ & \ddots & & & \\ \vdots & & \frac{Ws_i - Wl_i(t)}{Lvb_j} & & \vdots \\ \vdots & & & \ddots & \vdots \\ \frac{Ws_m - Wl_m(t)}{Lvb_1} & \cdots & \cdots & \cdots & \frac{Ws_1 - Wl_1(t)}{Lvb_n} \end{bmatrix} \tag{8}$$

$$\frac{Ws_i - Wl_i(t)}{Lvb_j} = \begin{cases} \frac{Ws_i - Wl_i(t)}{Lvb_j} & y_{ij} = 1 \\ +\infty & y_{ij} = 0 \end{cases}$$

$$ShipAllowed[i] = \begin{cases} 1 & v_i \text{ Allowed} \\ 0 & v_i \text{ Not Allowed} \end{cases} \text{—Ship visit table}$$

berth Finish Time[j]—Ship j Job completion time

Transition probability at time t

$$p[i] = \begin{cases} \frac{\tau_{ij}^{\alpha}(t)\eta_{ij}^{\beta}(t)}{\sum_{r \in allowed_k} \tau_{ij}^{\alpha}(t)\eta_{ij}^{\beta}(t)} & ShipAllowed[i] = 1 \ And \ y_{ij} = 1 \\ 0 & else \end{cases} \tag{9}$$

5 Calculation Example and Conclusion

In this section, we use actual production data of a bulk cargo port to calculate the solution methods and models above. The result shows ships all reflect tend to choice higher operating efficiency berths; View on berth type, all ships – berth match the conditions of the ship – berth match constraints, indicating that the ship – berth reasoning constraints played a role. In addition, the ship's resource using prioritization and waiting time sort are consistent; it also indicated the important role of marine resource use priority that played a prominent role in improving ship scheduling quality.

Through the questionnaire analysis of ship resource usage priority affecting factor, raised the ship resources priority model, and combined with knowledge-based reasoning, put forward the ship -berth match collection, and using the improved ant colony algorithm model for solving.

Acknowledgement This research was supported by "the Fundamental Research Funds for the Central Universities" under Grant 2011JBM042. The authors also wish to thank the port in the paper for support in data collection.

References

1. YuMeng, JiYu-duo (2007) Based on multi-agent container terminal production scheduling. Wuhan University of Technology
2. LiuJian (2008) Improved virtual enterprise partner selection and integration of genetic algorithm and ant colony algorithm. Beijing University of Aeronautics and Astronautics

Three-Level and Dynamic Optimization Model for Allocating Medical Resources Based on Epidemic Diffusion Model

Ming Liu and Peiyong Zhang

Abstract In this paper, a three-level and dynamic linear programming model for allocating medical resources based on epidemic diffusion model is proposed. The epidemic diffusion model is used to construct the forecasting mechanism for dynamic demand of medical resources. Heuristic algorithm coupled with MTLAB mathematical programming solver is adopted to solve the model. A numerical example is presented for testing the model's practical applicability. The result may provide some guidelines for decision-makers who are in charge of medical resources allocation in an epidemics control effort.

Keywords Epidemic diffusion • Time-varying demand • Time-space network • Dynamic optimization

1 Introduction

Over the past few years, disastrous epidemic events such as SARS and H1N1 significantly impact people's life. The outbreak of infections in Europe is another recent example. Actually, many recent research efforts have been devoted to understanding the prevention and control of epidemics, such as [1, 2]. The major purpose of these articles is to compare the performance of the following two strategies, the traced vaccination (TV) strategy and the mass vaccination (MV) strategy. Another stream of research is on the development of epidemic diffusion models by applying complex network theory to traditional compartment models, such as [3–5]. For example, Ref. [4] presented some suggestions for the epidemic prevention and infection control in the Wenchuan earthquake areas. These above

M. Liu (✉) • P. Zhang
Department of Management Science and Engineering, Nanjing University of Science and Technology, Nanjing, Jiangsu 210094, P.R. China
e-mail: liumingseu@126.com

Z. Zhang et al. (eds.), *LISS 2012: Proceedings of 2nd International Conference on Logistics, Informatics and Service Science*, DOI 10.1007/978-3-642-32054-5_36,

mentioned works represent some of the research on various differential equation models for epidemic diffusion and control. However, after an epidemic outbreak, public officials are faced with many critical issues, one of the most important of which being how to ensure the availability and supply of medical resources. Refs. [6–8] presented some optimization approaches to the operation of medical resources allocation in response to the time-varying demand during the crucial rescue period. However, most research on medical resources allocation studies a static problem taking no consideration of the time evolution and dynamic nature of the demand.

The remainder of the paper is organized as follows: Sect. 2 is the problem description and model formulation. A numerical example is presented in Sect. 3. Limitations of the proposed model and future research directions are proposed in Sect. 4.

2 Problem Description and Model Formulation

As work in [9], this article focuses on the recovered stage of epidemic rescue. Optimization goal in such stage is to construct an integrated, dynamic and multi-level emergency logistics network, which includes the national strategic storages (NSS), the urban health departments (UHD), the area disease prevention and control center (ADPC), and the emergency designated hospitals (EDH). The entire recovered stage of epidemic rescue process is decomposed into several mutually correlated sub-problems (i.e. *n* decision-making cycles). The research idea of such rescue stage is shown as Fig. 1.

In this paper, people in disaster area are divided into four classes: the susceptible people (S), the exposed people (E), the infected people (I), and the recovered people (R). The SEIR epidemic diffusion model, the time-varying forecasting model, and the dynamic demand and inventory model for the UHD in [10] are adopted to depict the epidemic diffusion rule. Notations used in the following optimization model are specified as follows.

nc_{ij}: Unit replenishment cost of medical resources from NSS i to UHD j.

ce_{jk}: Unit distribution cost of medical resources from UHD j to ADPC k.

eh_{kl}: Unit distribution cost of medical resources from ADPC k to EDH l.

ns_i: Amount of medical resources supplied by NSS i in each rescue cycle.

V_{cap}: Capacity of UHD.

d_{lt}: Demand for medical resources in EDH l in rescue cycle t.

d^{v}_{jt}: Demand for medical resources in UHD j in rescue cycle t.

P_{jt}: Total output of medical resources in UHD j in rescue cycle t.

V_{jt}: Inventory of medical resources in UHD j in rescue cycle t.

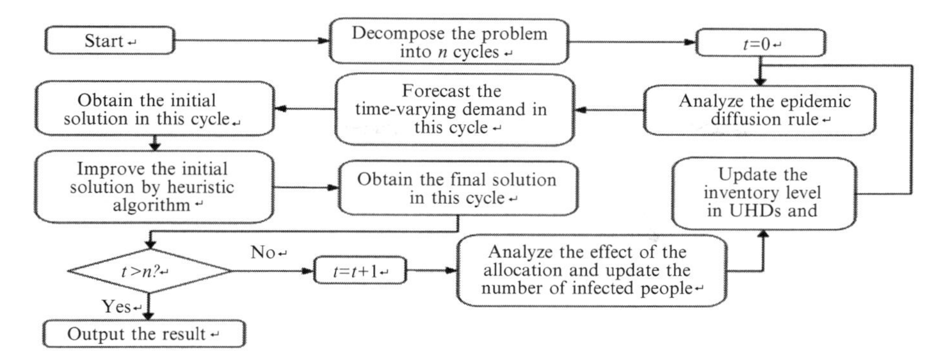

Fig. 1 Operational procedure of the dynamic medicine logistics network

x_{ijt}: Amount of medical resources that will be transported from NSS i to UHD j in rescue cycle t.

y_{jkt}: Amount of medical resources that will be transported from UHD j to ADPC k in rescue cycle t.

z_{klt}: Amount of medical resources that will be transported from ADPC k to EDH l in rescue cycle t.

TC: Total rescue cost of the three-level medical logistics network.

N: Set of NSSs.

C: Set of UHDs.

E: Set of ADPCs.

H: Set of EDHs.

T: Set of decision-making cycles.

Thus, the three-level and dynamic optimization model can be formulated as follows:

$$Min\,TC = \sum_{t\in T}\sum_{i\in N}\sum_{j\in C} x_{ijt}nc_{ij} + \sum_{t\in T}\sum_{j\in C}\sum_{k\in E} y_{jkt}ce_{jk} + \sum_{t\in T}\sum_{k\in E}\sum_{l\in H} z_{klt}eh_{kl} \tag{1}$$

$$\text{s.t.} \sum_{j\in C} x_{ijt} \le ns_i, \forall i \in N, t \in T \tag{2}$$

$$\sum_{i\in N} x_{ijt} = d^v_{jt}, \forall j \in C, t \in T \tag{3}$$

$$d^v_{jt} = V_{cap}, \forall j \in C, t = 0 \tag{4}$$

$$d^v_{jt} = P_{jt-1}, \forall j \in C, t = 1, 2, \ldots, T \tag{5}$$

$$P_{jt} = \sum_{k\in E} y_{jkt}, \forall j \in C, t \in T \tag{6}$$

$$\sum_{k \in E} y_{jkt} \leq V_{cap}, \forall j \in C, t \in T \tag{7}$$

$$\sum_{j \in C} \sum_{k \in E} y_{jkt} = \sum_{k \in E} \sum_{l \in H} z_{klt}, \forall t \in T \tag{8}$$

$$\sum_{k \in E} z_{klt} = d_{lt}, \forall l \in H, t \in T \tag{9}$$

$$d_{lt} = aI_l(t), \forall l \in H, t = 0 \tag{10}$$

$$d_{lt} = \prod_{i=0}^{t-1} (1 + \eta_{li}) \left(1 - \frac{\theta}{\Gamma}\right)^t d_{l0}, \forall l \in H, t = 1, 2, \ldots, T \tag{11}$$

$$\prod_{i=0}^{t-1} (1 + \eta_{li}) = (1 + \eta_{l0})(1 + \eta_{l1}) \cdots (1 + \eta_{lt-1}), \forall l \in H, t = 1, 2, \ldots, T \tag{12}$$

$$x_{ijt} \geq 0, \forall i \in N, j \in C, t \in T \tag{13}$$

$$y_{jkt} \geq 0, \forall j \in C, k \in E, t \in T \tag{14}$$

$$z_{klt} \geq 0, \forall k \in E, l \in H, t \in T \tag{15}$$

Herein, Eq. (1) is to minimize the total rescue cost of the three-level medical distribution network. Equations (2), (3) and (7), (8), (9) are constraints for flow conservation. Equations (4), (5) and (6) are the dynamic demand models in the upper level sub-problem. Equations (10), (11) and (12) are the time-varying demand models in the lower level sub-problem. At last, Eqs. (13), (14) and (15) ensure all the arc flows in the time-space network within their bounds. As Fig. 2 shows, the solution procedure for the proposed optimization model is presented in [10, 11].

3 Numerical Example

To test how well the proposed model may be applied in an actual event, a numerical example is presented to illustrate its efficiency. Assume there is a smallpox outbreak in a city, which has 3 NSSs, 4 UHDs, 6 ADPCs and 8 EDHs. The three NSS can supply 400, 420 and 450 unit of medical resources in each rescue cycle, and

Table 1 Transferred arcs in each rescue cycle[a]

Cycle	Arcs need to be transferred Before	After	Cycle	Arcs need to be transferred Before	After
$t = 0$	N2 → C1	N2 → N1 → C1	$t = 1$	N2 → C1	N2 → N1 → C1
	N3 → C1	N3 → N1 → C1		N3 → C1	N3 → N1 → C1
$t = 2$	N2 → C1	N2 → N1 → C1	$t = 3$	N2 → C1	N2 → N1 → C1

[a] There is no adjustment when $t = 4, 5, 6, 7, 8, 9, 10$

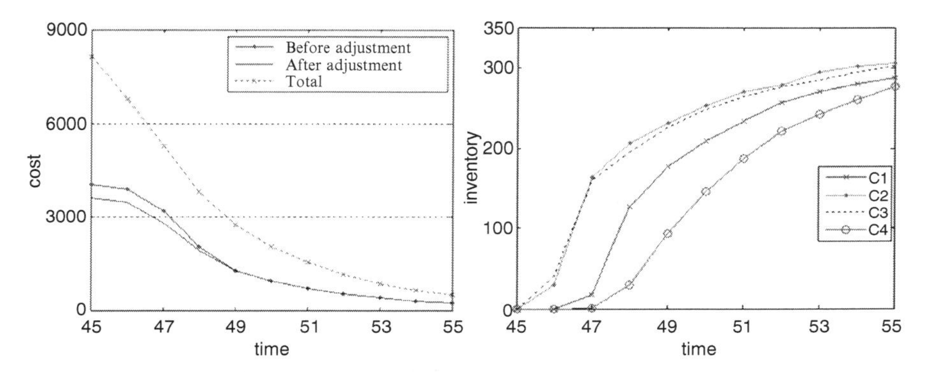

Fig. 2 Total rescue cost at each rescue cycle and inventory level in different UHD over time

suppose that capacity of each UHD is 320. Other values of parameters in SEIR epidemic diffusion model, and the unit cost from the supply points to the demand points are given in [9, 10]. Numerical simulation shows some interesting conclusions, which are given as follows:

1. The time-varying demand is way below traditional demand, suggesting that the allocation of medical resources in the early periods will significantly reduce the demand in the following periods.
2. Some replenishment arcs need to be transferred from the direct shipment delivery system to the hub-and-spoke delivery system, which will decrease the total rescue cost. The transferred arcs in each rescue cycle are shown in Table 1.
3. With the application of the three-level and dynamic optimization model, the total rescue cost can be controlled effectively, and meanwhile, inventory level in each UHD can be restored and raised gradually. Thus, such optimization model achieves a win-win rescue effect. These changes are shown in Fig. 2.

4 Conclusions

In this article, we develop a discrete time-space network model to study the medical resources allocation problem in the recovered stage when an epidemic outbreak. The novelty of our model against the existing works in literature is characterized by the following three aspects: (1) The model couples a multi-stage linear programming for optimal allocation of medical resources with a proactive forecasting mechanism cultivated from the epidemic diffusion dynamics. (2) A win-win emergency rescue effect is achieved by the integrated and dynamic optimization model. (3) The medicine logistics operation problem has been decomposed into several mutually correlated sub-problems, and then be solved systematically. Thus, the result will be much more suitable for real operations. As the limitation of the model, it is developed for the medical resources allocation in a geographic area where an epidemic disease has been spreading and it does not consider possible cross area diffusion between two or more geographic areas.

Acknowledgments This work has been partially supported by the MOE (Ministry of Education in China) Project of Humanities and Social Sciences (No. 11YJCZH109), and the project in Nanjing University of Science and Technology (2011YBXM96), (2011XQTR10), (AE88072) and (JGQN1102). The authors wish to thank Prof. Ding Zhang from State University of New York (USA), who read earlier versions of this paper. His helpful suggestions improved the presentation significantly.

References

1. Wein LM, Craft DL, Kaplan EH (2003) Emergency response to an anthrax attack. Proc Natl Acad Sci 7:4346–4351
2. Kaplan EH, Craft DL, Wein LM (2003) Analyzing bioterror response logistics: the case of smallpox. Math Biosci 1:33–72
3. Jung E, Iwami S, Takeuchi Y (2009) Optimal control strategy for prevention of avian influenza pandemic. J Theor Biol 2:220–229
4. Wang HY, Wang XP, Zeng AZ (2009) Optimal material distribution decisions based on epidemic diffusion rule and stochastic latent period for emergency rescue. Int J Math Oper Res 1:76–96
5. Zhu YF, Gu L, Yu JX (2009) Analysis on the epidemiological characteristics of *Escherichia coli* O157: H7 infection in Xuzhou, Jiangsu, China, 1999. J Nanjing Med Univ 1:20–24
6. Sheu JB (2007) An emergency logistics distribution approach for quick response to urgent relief demand in disasters. Transp Res Part E 6:687–709
7. Yan SY, Shih YL (2009) Optimal scheduling of emergency roadway repair and subsequent relief distribution. Comput Oper Res 6:2049–2065
8. Liu M, Zhao LD (2009) Optimization of the emergency materials distribution network with time windows in anti-bioterrorism system. Int J Innov Comput Info Control 5(11):3615–3624
9. Liu M, Zhao LD (2011) Analysis for epidemic diffusion and emergency demand in an anti-bioterrorism system. Int J Math Model Numer Optim 1:51–68
10. Liu M, Zhao LD (2011) An integrated and dynamic optimization model for the multi-level emergency logistics network in anti-bioterrorism system. Int J Syst Sci. doi:10.1080/00207721.2010.547629
11. Liu JY, Li CL, Chan CY (2003) Mixed truck delivery systems with both hub-and-spoke and direct shipment. Transp Res Part E 4:325–339

Optimal Selling Strategy in Dual-Channel Supply Chains

Yong He, Houfei Song, and Peng Zhang

Abstract This paper focuses on the rate of consumer's preference for direct channel and the sales cost of the retail channel. By assuming that the supply chain is centralized or decentralized, we construct the mathematical models and analyze the impacts of sales cost and the rate of consumer's preference on the decisions of manufacturer and retailer. Then we offer the optimal strategies for manufacturer and retailer when they are under different situations.

Keywords Selling strategy • Dual-channel supply chains • Rate of consumer's preference

1 Introduction

Nowadays, online channel has become a critical sales channel. Hence, manufacturers and retailers also transform the sales model to appeal to the different consumption habits of consumers, also to ensure a greater advantage in the market competition.

Many literatures in dual channel supply chain address the issue how manufacturers and retailers could benefit from their own advantages, mainly concentrated on the sales price and service levels [1–9]. Unlike other studies, in this paper, we start by deriving the optimal quantities for both channels by employing the customer's utility theory, and then analyze how the prices affect the manufacturer's decisions. Besides, since the sales cost of retail channel plays an important role in dual channel supply chain, we derive the optimal strategies when the sales cost is considered.

Y. He (✉) • H. Song • P. Zhang
School of Economics and Management, Southeast University, Sipailou 2, 210096 Nanjing, P.R. China
e-mail: heyong@126.com; songhoufei@126.com; aaazpzp@sina.com

Z. Zhang et al. (eds.), *LISS 2012: Proceedings of 2nd International Conference on Logistics, Informatics and Service Science*, DOI 10.1007/978-3-642-32054-5_37,

2 Model Assumptions and Notations

Q is the size of potential market; θ is the rate of consumers' preference for direct channel, $\theta \in [0, 1]$; c is the manufacturer's production costs; Δc is the sales cost of retail channel; p_r and p_d are the sales prices of retail channel and direct channel, respectively, $p_r > p_d$; q_r and q_d are the demands in retail channel and direct channel, respectively.

According to the literature of Ferrer et al. [10], here we assume that consumers are willing to pay the price U for retail channel, where $U \in [0, Q]$ uniformly distributed in this domain. The customer with product valuation U would derive the utility $U_r = U - p_r$ from retail channel or $U_d = \theta U - p_d$ from direct channel. If $U_d \geq 0$ consumers choose to purchase products from direct channel; similarly, if $U_r \geq 0$ consumers choose to purchase products from retail channel. In this paper, we consider the competition exists between the dual-channel. So, if $U_d > U_r$ consumers choose to purchase products from the direct channel, on the contrary, if $U_d < U_r$ consumers choose to purchase products from the retail channel. Then, we can get the following Proposition:

Proposition 1. *If* $\theta \in \left[\frac{p_d}{p_r}, 1 - \frac{p_r - p_d}{Q}\right]$, *the prices are:*

$$p_r = Q - q_r - \theta q_d \tag{1}$$

$$p_d = \theta(Q - q_r - q_d) \tag{2}$$

If $\theta < \frac{p_d}{p_r}$, *all products are sold through the retail channel and the price of retail channel is* $p_r = Q - q_r$;

If $1 - \frac{p_r - p_d}{Q} \leq \theta \leq 1$, *all products are sold through the direct channel and the price of direct channel is* $p_d = \theta(Q - q_d)$.

All proofs in this paper can be offered if being requested.

3 The Centralized Model

For centralized model, we consider the manufacturer owns direct channel and retail channel, so we can get the centralized model with profit maximization:

$$\begin{aligned} \max_{q_r, q_d} \; & (p_r - c - \Delta c)q_r + (p_d - c)q_d \\ & s.t. \quad q_r \geq 0, \; q_d \geq 0 \end{aligned} \tag{3}$$

Table 1 Optimal strategies under centralized model

	$\Delta c \leq \frac{c}{\theta} - c$	$\frac{c}{\theta} - c < \Delta c < Q(1-\theta)$	$\Delta c \geq Q(1-\theta)$
q_r^{z*}	$\frac{Q - c - \Delta c}{2}$	$\frac{Q(1-\theta) - \Delta c}{2(1-\theta)}$	0
q_d^{z*}	0	$\frac{c\theta - c + \Delta c\theta}{2\theta(1-\theta)}$	$\frac{\theta Q - c}{2\theta}$
$q = q_r^{z*} + q_d^{z*}$	$\frac{Q - c - \Delta c}{2}$	$\frac{\theta Q - c}{2\theta}$	$\frac{\theta Q - c}{2\theta}$
p_r	$\frac{Q + c + \Delta c}{2}$	$\frac{Q}{2} + \frac{c + \Delta c}{2}$	0
p_d	0	$\frac{\theta Q + c}{2}$	$\frac{\theta Q + c}{2\theta}$
Π_r^{z*}	$\left(\frac{Q - c - \Delta c}{2}\right)^2$	$\frac{(Q - c - \Delta c)(Q - Q\theta - \Delta c)}{4(1-\theta)}$	0
Π_d^{z*}	0	$\frac{(c\theta - c + \theta\Delta c)(Q\theta - c)}{4\theta(1-\theta)}$	$\frac{(Q\theta - c)(Q\theta + c - 2c\theta)}{4\theta^2}$

Substituting (1) and (2) into (3), we get the optimal solutions to prices (p_r, p_d), sales quantities at two channels $\left(q_r^{z*}, q_d^{z*}\right)$, profits of the retail channel $\left(\Pi_r^{z*}\right)$ and the direct channel $\left(\Pi_d^{z*}\right)$ for a centralized model. The results are shown in Table 1.

4 The Decentralized Model

4.1 Retailer's Problem

The retailer's best response to the rate of consumer's preference for direct channel θ, wholesale price w, and direct channel quantities q_d. So making retailer's profit maximization model:

$$\begin{aligned} &\max_{q_r} \ (p_r - w - \Delta c)q_r \\ &s.t. \quad p_r = Q - q_r - \theta q_d \end{aligned} \tag{4}$$

we obtain:

$$q_r^* = \frac{1}{2}(Q - w - \Delta c - \theta q_d) \tag{5}$$

4.2 Manufacturer's Problem

We use Stackelberg game to analyze the manufacturer's optimal decision-making when the manufacturer is dominant. From the above response of retailer, we can derive the manufacturer's decisions. The manufacturer's profit maximization model is:

$$\begin{aligned} &\max_{q_d, w}\ (w-c)\,q_r^* + (p_d - c)q_d \\ &s.t.\ \ p_d = \theta(Q - q_r - q_d),\ \ p_d \ge w \end{aligned} \tag{6}$$

We employ the Lagrangian methods to solve it and the optimal solutions are:

$$\begin{cases} q_d^* = \dfrac{\theta(c + Q\theta) + 2\Delta c\theta - 2c}{2\theta(2-\theta)} \\ w^* = \dfrac{c + Q\theta}{2} \end{cases} \tag{7}$$

From (7), we know that w is increasing in θ, Q and c, respectively; q_d^* is increasing in θ, Q and Δc, respectively, but decreasing in c. Hence, the cost leadership strategy is feasible in this case.

We can get the optimal quantities of manufacturer and retailer as follows:

$$\begin{cases} q_r^* = \dfrac{Q(1-\theta) - \Delta c}{2-\theta} \\ q_d^* = \dfrac{c\theta + Q\theta^2 + 2\Delta c\theta - 2c}{2\theta(2-\theta)} \end{cases} \tag{8}$$

From (8), we can get following Proposition:

Proposition 2. (I) $q_d^* \ge q_d^{z*}$; *(II) If* $\frac{c}{\theta} - c < \Delta c < Q(1-\theta)$, $q_r^{z*} > {q_r}^*$.

Proposition 3. (I) *If* $\Delta c < Q(1-\theta)$, $p_r > w^* + \Delta c$ *and* $p_d = w^*$; *(II) If* $\Delta c \ge Q(1-\theta)$ *the retail channel will be closed.*

Proposition 3 shows that $p_d = w^*$, the manufacturer has two options: First, all of its product wholesale to retailer; second, continue to conduct direct sale. But by analyzing, we can get the manufacturer's maximum profits in first case and second case are $\frac{(Q-c-\Delta c)^2}{8}$ and $\frac{(Q\theta-c)^2}{4\theta}$, respectively. Since $\frac{(Q\theta-c)^2}{4\theta} - \frac{(Q-c-\Delta c)^2}{8} > 0$. We know that manufacturer will continue to use the direct channel.

Let Π_r^* and Π_d^* be the profits of retail channel and direct channel under decentralized model, then we can summarize the optimal strategies under decentralized model in Table 2.

Comparing Table 2 with Table 1, we know that if $\Delta c < Q(1-\theta)$, profit of retail channel under centralized model is more than decentralized mode; if $\Delta c \ge Q(1-\theta)$, the retailer's profit is zero for the centralized model and decentralized model; if

Table 2 Optimal strategies under decentralized model

	$\Delta c<Q(1-\theta)$	$\Delta c \geq Q(1-\theta)$
q_r^*	$\frac{Q(1-\theta)-\Delta c}{2-\theta}$	0
q_d^*	$\frac{c\theta+Q\theta^2+2\Delta c\theta-2c}{2\theta(2-\theta)}$	$\frac{\theta Q-c}{2\theta}$
$q=q_r^*+q_d^*$	$\frac{\theta Q-c}{2\theta}$	$\frac{\theta Q-c}{2\theta}$
p_r	$\frac{c(2-\theta)+2\Delta c(1-\theta)+Q(2-\theta^2)}{2(2-\theta)}$	0
p_d	$\frac{\theta Q+c}{2}$	$\frac{\theta Q+c}{2}$
Π_r^*	$\frac{[Q(1-\theta)-\Delta c]^2}{(2-\theta)^2}$	$\frac{(Q\theta-c)(Q\theta+c-2c\theta)}{4\theta^2}$
Π_d^*	$\frac{(Q\theta-c)^2}{4\theta}$	0

$\Delta c\leq \frac{c}{\theta}-c$, profit of direct channel for the centralized model is lower than the decentralized model; if $\Delta c \geq Q(1-\theta)$ the direct sale profit for the centralized model is equal to the decentralized mode; if $\frac{c}{\theta}-c<\Delta c<Q(1-\theta)$, profit of direct channel for centralized model lower than the decentralized model.

5 Conclusions

In this paper, we analyze optimal strategies of the centralized model and decentralized model respectively. We find the condition of the existence of direct channel and retail channel, and prove that the sales cost of the retail channel and consumer's preference rate for direct channel strongly influence the manufacturer's and retailer's decisions and profits.

In this paper, just one retailer has been discussed. In the future work, we will take more retailers into account.

Acknowledgments This work is supported by the National Natural Science Foundation of China (No. 71001025). Also, this research is partly supported by the Program for New Century Excellent Talents in University (No. NCET-10-0327) and the Ministry of Education of China: Grant-in-aid for Humanity and Social Science Research (No. 11YJCZH139).

References

1. Chen KY, Kaya M, Özer Ö (2008) Dual sales channel management with service competition. Manuf Serv Oper Manag 10:654–675
2. Dumrongsiri A, Fan M, Jain A, Moinzadeh K (2008) A supply chain model with direct and retail channels. Eur J Oper Res 187:691–718

3. Hua G, Wang S, Cheng TCE (2010) Price and lead time decisions in dual-channel supply chains. Eur J Oper Res 205:113–126
4. Cai G, Zhang ZG, Zhang M (2009) Game theoretical perspectives on dual-channel supply chain competition with price discounts and pricing schemes. Int J Prod Econ 117:80–96
5. Chun SH, Rhee BD, Park SY, Kim JC (2011) Emerging dual channel system and manufacturer's direct retail channel strategy. Int Rev Econ Finance 20:812–825
6. Liu B, Zhang R, Xiao M (2010) Joint decision on production and pricing for online dual channel supply chain system. Appl Math Model 34:4208–4218
7. Chen J, Bell PC (2012) Implementing market segmentation using full-refund and no-refund customer returns policies in a dual-channel supply chain structure. Int J Prod Econ 136:56–66
8. Chen J, Zhang H, Sun Y (2012) Implementing coordination contracts in a manufacturer Stackelberg dual-channel supply chain. Omega 40:571–583
9. Huang S, Yang C, Zhang X (2012) Pricing and production decisions in dual-channel supply chains with demand disruptions. Comput Ind Eng 62:70–83
10. Ferrer G, Swaminathan JM (2010) Managing new and differentiated remanufactured products. Eur J Oper Res 203:370–379

The Optimization Model and Empirical Analysis for Vehicle Routing Problems with Time Windows Based on C-W Algorithm

Lijuan Fan and Qiuli Qin

Abstract Vehicle routing problem is a kind of optimization scheduling problems which researches how to realize transportation cost optimization though programming driving route reasonably. Considering the vehicle routing problems with time window restriction, we establish the proper model related to the typical practical situation based on c-w saving algorithm. The case study indicates that saving algorithm is reasonable, and it also shows that this calculation might be simple and convenient and this method could be easy to realize by computer. However, the precision of the solution will be reduced if the number of customers increases with growing solution space.

Keywords Vehicle routing problem • Time windows restriction • C-W saving algorithm • Logistics distribution

1 Introduction

The Vehicle Routing Problems (VRP) was first proposed by Dantzig and Ramser in 1959 [1]. And the constraint makes Vehicle Routing Problem with Time Window closer to the reality of logistics distribution than Vehicle Routing Problem [2–5]. The goal of Vehicle Routing Problem is to achieve the minimum transportation cost and the minimum time, the problem can be transformed into the problem of multi-objective optimal operation of fully-loaded vehicle of multi-type vehicles multi-cargo kinds [6]. The paper raised Vehicle Routing Problem with Time Windows that meets the constraints of the multi-objective, considering customer's time requirements, establishes the proper model, and used it to solve the practical problem at last [7, 8].

L. Fan (✉) • Q. Qin
School of Economics and Management, Beijing Jiaotong University, Beijing, P.R. China
e-mail: 11125194@bjtu.edu.cn; qlqiu@bjtu.edu.cn

Z. Zhang et al. (eds.), *LISS 2012: Proceedings of 2nd International Conference on Logistics, Informatics and Service Science*, DOI 10.1007/978-3-642-32054-5_38,

2 The Description of the Problem and the Model

To simply the model, the following assumptions and explanations is essential: distribution center has some same vehicles; each vehicle starts from the distribution center to deliver goods to the designated customer along the a specific route, and then returns to the distribution center; the goods that every customer needs can only be transported by a vehicle; different customers have different time arrival requirements; all expense is related to transportation distance. Parameter symbolic and variables of the model are defined as follows:

M: M vehicles, number 1,2,. . .,M; β: distribution costs of the average unit distance; g_i: distribution requirements of each distribution terminal *i*; Q: average load of per vehicle; d_{ij}: the distance from i to j; c_2: opportunity cost of unit time that causes by transportation vehicle waiting at the joint. If there is flexible time window constrain, its value is an identified a certain value; if not, its value is a large number; c_3: value of unit time in which to transportation vehicles arrive at the distribution terminal after the specified time. t_{ijk}: the time that the k vehicle takes from i to j, and ignoring the loading time of each point and traffic conditions and other factors; x_{ijk}: indicates the k vehicle drive from i to j, if it is from i to j, the value of x_{ijk} is 1; otherwise it is 0; y_{ijk}: decision variable, which indicates the task of distribution site is completed by the vehicle K, if it's K, the value of y_{ijk} is 1; otherwise it is 0; ET_i is the starting point of the arrival time that customer requires, LT_i is the end point of the arrival time that customer requires, T_i is the time that the goods is transported by estimating.

The mathematical modeling of vehicle routing problem is as follows:

$$\mathrm{Min}\left\{\sum_{i=0}^{N}\sum_{j=0}^{N}\sum_{k=1}^{M}\beta d_{ij}x_{ijk}+\sum_{i=1}^{N}\mathrm{Max}[ET_i-T_i,0]+c_3\sum_{i=1}^{N}\mathrm{Max}[T_i-LT_i,0]\right\} \tag{1}$$

$$\left\{\sum_{i=1}^{N} g_i y_{ik} \le Q, \quad k=1,2\ldots M\right. \tag{2}$$

$$\left\{\sum_{k=1}^{M} y_{ik} = 1, \quad i=1,2\ldots N\right. \tag{3}$$

$$\left\{\sum_{i=0}^{N} x_{ijk} = y_{ik}, \quad j=1,2\ldots N,\ k=1,2\ldots M\right. \tag{4}$$

$$\left\{\sum_{k=1}^{M} y_{ok} = M\right. \tag{5}$$

$$\left\{\sum_{j=0}^{N} x_{ijk} = y_{ik},\quad i = 1,2\ldots N,\ k = 1,2\ldots M\right. \tag{6}$$

$$\left\{T_i = \sum_{i=0}^{N} x_{ijk} t_{ijk}\right. \tag{7}$$

$$\left\{x_{ijk}\left(T_j - T_i\right) \geq 0,\quad i,\ j = 1,2\ldots N,\ k = 1,2\ldots M\right. \tag{8}$$

3 Algorithm Principle and Procedures

We consider to use C-W saving algorithm of heuristic algorithm. Its basic idea is: firstly the distribution terminal is connected with distribution center, and we calculate the transporting cost; then, we obtain the saving distance when any two distribution terminals are connected in a route, value represents total saved distance when two distribution terminals are connected in a route.

$$s(i,j) = c_{io} + c_{0j} - c_{ij} \tag{9}$$

$c_{ij}\left(\beta d_{ij}\right)$ represents the cost from point i to j for each vehicle, by C-W saving algorithm, we obtain the saved cost when point i and point j are connected together: $s(i,j) = c_{io} + c_{0j} - c_{ij}$

We use EF_j to represent when connecting between point i and j, the delaying time that the vehicle reaches the point j later than point j in the original route can be obtained by the following formula:

$$EF_j = T_i + t_{ij} - T_j \tag{10}$$

Obviously, $EF_j < 0$, vehicle reaches the point j earlier; $EF_j = 0$, vehicle reaches the point j on time; $EF_j > 0$ vehicle reaches the point j later.

In order to illustrate the problem, now the parameters are defined as follows:

$$\Delta_j^- = \min_{i \geq j}\{T_\gamma - ET_\gamma\} \tag{11}$$

$$\Delta_j^+ = \min_{i \geq j}\{LT_\gamma - T_\gamma\} \tag{12}$$

Δ_j^- – maximum allowable advanced arriving time in that vehicle doesn't need to wait for point j when each task behind point j on the route is performed.

Δ_j^+ – maximum allowable arriving time that each task behind the point j on the route does not violate time window constraint points.

It's necessary to check if the transport time violates time window constraints.

When $EF_j < 0$, if $|EF_j| \le \Delta_j^-$, it doesn't need to wait when vehicle performs the task behind j, or it will need to wait for some time;

When $EF_j > 0$, if $EF_j \le \Delta_j^+$, it doesn't delay when vehicle performs the task behind j, or it will delay.

4 Empirical Analysis

We choose S-company as the example to analyse. It is assumed there are twelve distribution tasks assigned to the distribution centre of Zhengzhou City, distribution demand of the distribution sites is g (unit: 10,000 pieces) and the distribution site is i. The tasks are performed by the vehicles sent from distribution centre whose total capacity Q is 30,000 pieces. Characteristics of distribution sites in Table 1.

It is assumed that travel time of vehicle is proportional to distance, and speed of the vehicle is set as 50 km/h, the travel time from i to j is $t_{ij} = d_{ij}/50$. Initially, when the vehicle drives from the distribution center in Zhengzhou city to distribution sites, if $ET_i \le t_{0i} \le LT_i$, and $T_i = t_{0i}$, if $t_{0i} \le ET_i$, and $T_i = ET_i$, then we calculate it according to C-W saving algorithm, and make frame table of route. Here we choose the route assignment of point 1, 7, 9 as the example to illustrate application process:

Firstly, calculate the saved cost among the points 1, 7 and 9. Arrange the result in the ascending order. The dates are given in Table 2. And then structure lines and the calculation process are shown in Table 3 followed:

Table 1 Characteristics and requirements of tasks on different distribution sites

i	0	1	2	3	4	5
City	Zhenzhou	Luoyang	Kaifeng	Xuchang	Zhoukou	Luohe
g	0.5	1.0	1.4	0.8	0.5	0.3
[ET,LT]	[5,10]	[5,15]	[7,17]	[6,12]	[5,15]	[9,17]

i	6	7	8	9	10	11
City	Pingdingsha	Xinxiang	Shangqiu	Zhumadian	Xinyang	Sanmenxia
g	1.2	1.5	0.2	0.6	0.3	0.5
[ET,LT]	[12,27]	[10,20]	[7,18]	[9,18]	[8,20]	[10,20]

Table 2 Lines structure process table

(i,j)	(7,9)	(1,7)	(1,9)
S(i,j)	351	44	38

Table 3 Lines structure process table

i–j	Position of two points	$q = \sum g_i$	$EF_{ij} = T_i + t_{ij} - T_j$	Δ_i^- or Δ_i^+	Connection type	$T_k = T_k + EF_j (k \geq j)$
7–9	Not on line	$q = 2.1 < Q$	$EF_9 = 8.98$	$\Delta_9^+ = 9$	$7 \to 9$	$T_9 = 17.98$
1–7	Exterior point	$q = 3.1 > Q$			No	
9–1	Exterior point	$q = 3.1 > Q$			No	

We get the result of the arrangement: $0 \to 1 \to 0\,(60\ \text{km})$ and $0 \to 7 \to 9 \to 0$ $(1{,}149\ \text{km})$. Similarly the other two lines arrangements are: $0 \to 4 \to 2 \to 3 \to 5$ $\to 0\,(932\ \text{km})$ and $0 \to 8 \to 11 \to 10 \to 6 \to 0\,(1{,}500\ \text{km})$.

By calculating the total distribution mileage is 3,632 km, it can save nearly half of the mileage compared to send goods for each user individually (total 6,520 km).

5 Conclusions

It is shown from the analysis: the realization of the lowest cost of distribution is due to the premise that the distribution tasks are completed successfully, namely the distribution cost and distribution services reach a balance. But in the algorithm there are some disadvantages that some edge points are difficult to combine, to affect optimization rate. It is also necessary in the future to determine the research to modify the method, in order to make it more suitable for solving Vehicle Routing Problem of under capacity logistics distribution vehicle with time window constraints.

References

1. Dantzig GB, Ramser JH (1959) The truck dispatching problem. Manag Sci 6(1):80–91
2. Li F, Zheng Q, Qiu JR, Ye CM (2010) Optimal algorithms for vehicle routing with time windows. Pract Recognit Math 40:176–181
3. Wang XP, Zhang K, Hu XP (2011) Vehicle routing problem based on fuzzy time windows. J Ind Eng Eng Manag 25:148–153
4. Chen YY, Han J (2005) On vehicle routing problems with time windows. Logist Technol A03:48–50
5. Li J, Guo YH (2009) Research on optimal scheduling method of logistics distribution vehicle optimal. Logist Technol Appl 4:25–40
6. Xiao JH (2008) Transportation route optimization of H Company. Beijing Jiaotong University, Beijing
7. Zhang J, Liu L, Gong YC (2009) Modeling and algorithm of location-routing problems. J Beihang Univ 7:55–56
8. Melita K, Steve K (2011) Understanding map projections. GIS by ESRI TM, Environmental System Research Institute, Inc. Redland, USA, pp 90–92

Study on the Pricing Model of China's Parallel Rail Lines Under the Diversified Property Rights

Shaoni Zhou, Qiusheng Zhang, and Xiaowei Wu

Abstract In this paper, integrated with China railway management practice, we put forward the market-oriented pricing mechanism which railway network in China should adopt, and make a detailed analysis of the influencing factors. Based on these, this paper constructs a market pricing model for the parallel rail lines considering the price offering mechanism and the subsidy mechanism. It also defines the specific application conditions of the model.

Keywords Diversified property rights • Parallel rail lines • Pricing model • Market pricing mechanism

1 Introduction

According to the railway developing project in China, the diversified property rights will become an inevitable trend. In the period when the railways were entirely state-owned, the cost of the rail network acted as the internal costs [1]. While under the diversified ownership background, the pricing of the rail network affects lots of aspects. Most of all, because the parallel lines belonging to different owners provide similar service and can be easily substituted, competitions come up [2]. It is urgent to solve the problem of how to compete fairly for the national railway and joint venture railway, especially when there are both state-owned rail company and joint-ventures which each has one rail line between the network nodes.

S. Zhou (✉) • Q. Zhang • X. Wu
School of Economics and Management, Beijing Jiaotong University, Beijing 100044, P.R. China
e-mail: snzhou@bjtu.edu.cn

Z. Zhang et al. (eds.), *LISS 2012: Proceedings of 2nd International Conference on Logistics, Informatics and Service Science*, DOI 10.1007/978-3-642-32054-5_39,

2 Basic Relationships and Analysis of Influential Factors in the Rail Network Pricing

2.1 Basic Relationships in the Rail Network Pricing

The great scale of the joint ventures needs a market pricing mechanism which can adapt to the overall requirements of railway diversified ownership and properly handle the relationship among different stakeholders.

With the diversified property rights of China rail network, pricing stakeholders can be divided into three levels, their relationships are shown in Fig. 1.

Thus, the pricing mechanism of the rail network under the diversified ownership really would impact these stakeholders and must emphasize their interests [3]. That is, the state-owned railway and joint-venture railway should develop a competitive offering system; transportation companies should have the ability to seek the lowest bid seller of rail network resources; the Ministry of Railways will strive to reduce the level of subsidy which may be only given to the railway infrastructure company for it is more cost sank [4]; and being a rational investment entity, the external capital will investigate the income and the government subsidies in the investment decision [5].

2.2 Analysis of the Influential Factors for the Rail Network Pricing

There are three main influential factors for the rail network pricing. The first one is cost, that is, the price of any services can be regarded as compensation of its cost. The cost of the rail network can be classified into the following five levels: the variable operating cost of the rail network, the variable and fixed operating cost of

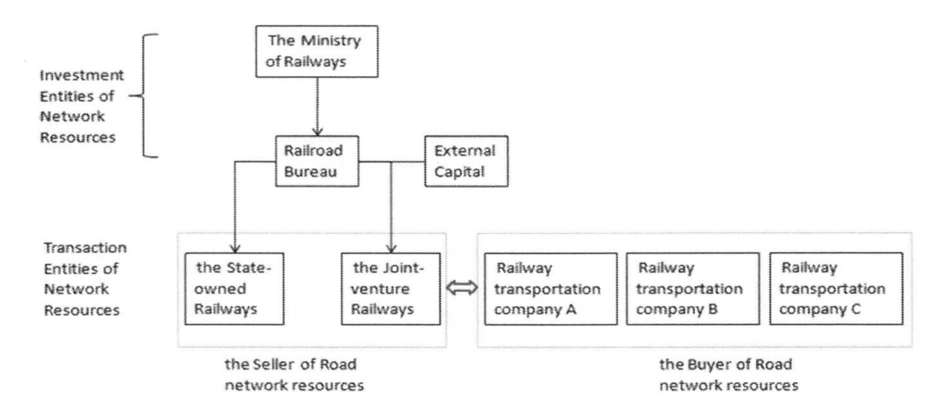

Fig. 1 Stakeholders in rail network pricing

the rail network, the full operating cost of the rail network, the full operating and construction costs of the rail network and the full operating and construction costs of the rail network and the equity cost of it [6]. The second is the supply–demand relationship which is a reflection of the relationship between production and consumption. The supply sides of rail network resource are the railway infrastructure companies and the consumer sides are the rail passengers and freight transport companies. The third lies at the government's actions. In real economic environment, the government's actions will largely affect the price of services and investor returns for the government can set the price of transport services and implement the price subsidies.

3 The Pricing Model of the Parallel Rail Lines

3.1 Specific Conditions of the Pricing Model

The model should possess the following basic conditions when applied to the parallel rail line [7]: the rail network service providers are state-owned railways and joint-venture railways, the number of the rail network service buyer is large; there are two forms to purchase the rail network resources, which are the negotiation in a fixed time and the temporary negotiations according to the provisional transport plan; the fixed period purchase price includes two parts, one is in the form of rail network card, which is a fixed cost and provides the buyer the right of using the network in the range of a certain passage, and another part is in the form of user fees, which is a kind of variable cost by paying the network usage fees according to the actual mileage, speed, load and the rail grade.

3.2 Construction of the Pricing Model

1. The pricing mechanism
   The state-owned railway and the joint venture railway both proposed their rail network card price, they are W_1 and W_2, and rates of fee for per unit, they are ω_1 and ω_2. The card price of the state-owned rail network is lower than that of the joint venture rail network, that is $W_1 < W_2$, but the rates of fee for per unit of the state-owned rail is higher than that of the joint venture rail, that is $\omega_1 > \omega_2$. These will produce the equilibrium point α^* of the train traffic volume in the transaction of the railway resources. That is, when the train traffic volume is on this point, the transaction prices by the state-owned rail and the joint venture rail are the same; when the actual train traffic volume is $\alpha > \alpha^*$, it is cheaper to choose the joint venture rail, vice versa. Those are shown in Fig. 2, and the equilibrium point $\alpha^* = \frac{W_2 - W_1}{\omega_1 - \omega_2}$.

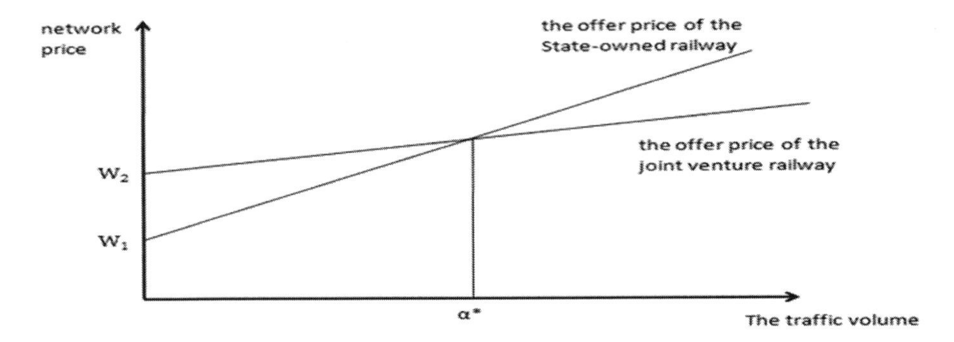

Fig. 2 Offering mechanisms of network resources

This quotation mechanism has three notable characteristics shown in the following. First, the trading price is divided into two parts, namely the fixed card price and the variable use fee part, which can be understood as the compensation of the fixed and the variable costs. Second, this quotation system encourages more rail transportation companies to choose the joint-venture railways, which can in turn increase its profitability and attract more external capital to invest in the railways. Third, it makes the state-owned railway have more remaining rail network resources and can bear the temporary transport demand. The social responsibility of the state-owned railway can be completed without sacrificing the interests of the joint ventures.

2. Subsidy mechanism
As the quotation system described above, the rail transport companies are divided into two categories, one is those choosing the state-owned rail network resources, with a total number of m, each traffic amount is $\alpha_{1,i}$ (i = 1...m), the other is those choosing the joint-venture rail network resources, with a total number of n, each traffic amount is $\alpha_{2,j}$ (j = 1...n).

The annual income of the state-owned railway is:

$$R_1 = mW_1 + \sum_{i=1}^{m} \alpha_{1,i}\omega_1 \tag{1}$$

The annual income of the joint-venture railway is:

$$R_2 = nW_2 + \sum_{j=1}^{n} \alpha_{2,j}\omega_2 \tag{2}$$

Supposing the service life of the state-owned railway and the joint-venture railway is M and N, the total income of the state-owned railway is:

$$TR_1^M = \sum_{I=1}^{M} R_1^I \ (I = 1 \ldots M) \tag{3}$$

The total income of the joint-venture railway is:

$$TR_2^N = \sum_{J=1}^{N} R_2^J \; (J = 1 \ldots N) \tag{4}$$

The variable operating cost of the state-owned railway is:

$$VC_1 = \sum_{i=1}^{m} \alpha_{1,i} c_1 \tag{5}$$

c_1 is the cost of per transport unit;

The variable operating cost of the joint-venture railway is:

$$VC_2 = \sum_{j=1}^{n} \alpha_{2,j} c_2 \tag{6}$$

c_2 is the cost of per transport unit;

For the state-owned railway, the fixed operating cost (excluding the interest cost) is OC_1, the annual interest cost is FI_1, and the fixed investment on the rail network construction is FC_1; for the joint-venture railway, the fixed operating cost (excluding the interest cost) is OC_2, the annual interest cost is FI_2, and the fixed investment on the rail network construction is FC_2.

The total cost of the state-owned railway is:

$$TC_1 = \sum_{I=0}^{M} \left(VC_1^I + OC_1^I + FI_1^I\right) + FC_1 \tag{7}$$

The total cost of the joint-venture railways is:

$$TC_2 = \sum_{J=0}^{N} \left(VC_2^J + OC_2^J + FI_2^J\right) + FC_2 \tag{8}$$

Under the circumstance of no expansion, the government subsidy to the state-owned railway and the joint-venture railway are S_1 and S_2, the average social profit margins is r, the external equity investment of the joint-venture railway is γ, then the goal of the governmental subsidy is:

$$S_1 + R_1 \geq VC_1 + OC_1 + FI_1 \tag{9}$$

$$S_2 + R_2 \geq VC_2 + OC_2 + FI_2 + FC_2 * \gamma * r \tag{10}$$

substitute (1) and (2):

$$S_1 \geq \sum_{i=1}^{m} \alpha_{1,i}(c_1 - \omega_1) + OC_1 + FI_1 - mW_1 \tag{11}$$

$$S_2 \geq \sum_{j=1}^{n} \alpha_{2,j}(c_2 - \omega_2) + OC_2 + FI_2 + FC_2 * \gamma * r - nW_2 \tag{12}$$

Under the circumstance of government subsidy is the least and taking the equals sign, deformation of Eq. (11) is:

$$\begin{aligned} S_1 &= VS_1 + FS_1 \\ &= \sum\nolimits_{i=1}^{m} \alpha_{1,i} c_1 (1 - \omega_1 / c_1) + (OC_1 + FI_1)\left(1 - \frac{mW_1}{OC_2 + FI_1}\right) \end{aligned} \tag{13}$$

For the state-owned railway, VS_1, FS_1 are subsidy to the variable cost and the fixed cost, the compensation rate of the variable cost is ω_1/c_1, the government subsidy rate is $(1 - \omega_1/c_1)$; the compensation rate of the fixed cost is $(mW\_1)/(OC\_1 + FI\_1)$, the government subsidy rate is $(1 - (mW\_1)/(OC\_1 + FI\_1))$.

$$\begin{aligned} S_2 &= VS_2 + FS_2 \\ &= \sum\nolimits_{j-1}^{n} \alpha_{2,j} c_2 (1 - \omega_2 / c_2) + (OC_2 + FI_2 + FC_2 * \gamma * r) \\ &\quad \times \left(1 - \frac{nW_2}{OC_2 - FI_2 - FC_2 * \gamma * r}\right) \end{aligned} \tag{14}$$

Just like the state-owned railway, for the joint-venture railway, VS_2, FS_2 are subsidy to the variable cost, the fixed cost and the economic profit, the compensation rate of the variable cost is ω_2/c_2, the government subsidy rate is $(1 - \omega_2/c_2)$; the compensation rate of the fixed cost and the economic profit is $(nW\_2)/(OC\_2 + FI\_2 + FC\_2 * \gamma * r)$, the government subsidy rate is $(1 - (nW\_2)/(OC\_2 + FI\_2 + FC\_2 * \gamma * r))$.

4 Conclusions

In order to attract the outside capital, one must establish the market-oriented pricing mechanism of the rail network.

This paper mainly analyzes the stakeholders and their relationships under the rail network pricing of the diversified property, designs a clear mechanism of competitive quotation among the sellers, and the buyers should have an independent bargaining right and capacity. On the above basis, the Ministry of Railways can reduce subsidy to the rail network company.

This model still has some problems which can be perfected gradually, such as the government should find the equilibrium point α^* which depends on series of predictions, the average social profit margin is an important exogenous variable for it is an opportunity cost for external capital and so on.

References

1. Coase RH (1960) The problem of social cost [J]. J Law Econ 3:1–44
2. Willianmson OE (1979) Transaction-cost economics: the governance of contractual relations [J]. J Law Econ 22:233–261
3. Andersson H, Ögren M (2007) Noise charges in railway infrastructure: a pricing schedule based on the marginal cost principle. Transp Policy 14(3):204–213
4. Li Minghui, Zhao Jian (2008) Synergy value: crucial reason for difficulty of railway investing and financing in china [J]. Compr Transp 5:9–14
5. Li Daian (2006) Research on the problems of costing and pricing of the infrastructure of our railway [D]. Beijing Jiaotong University, Beijing
6. Nikolova C (2008) User charges for the railway infrastructure in Bulgaria. Transp Res Part A Policy Pract 42(3):487–502
7. Rong Chaohe (2006) Boosting railway investment institutional reform in china from property relations [J]. Compr Transp 1:32–36

Analysis and Prediction of Logistics Enterprise Competitiveness by Using a Real GA-Based Support Vector Machine

Ning Ding, Hanqing Li, and Hongqi Wang

Abstract This research is aimed at establishing the forecast and analysis diagnosis models for competitiveness of logistics enterprise through integrating a real-valued genetic algorithm to determine the optimum parameters and SVM to perform learning and classification on data. The result of the proposed GA-SVM can satisfy a predicted accuracy of up to 95.56% for all the tested logistics enterprise competitive data. Notably, there are only 12 influential feature included in the proposed model, while the six features are ordinary and easily accessible from National Bureau of Statistics. The proposed GA-SVM is available for objective description forecast and evaluation of a logistics enterprise competitiveness and stability of steady development.

Keywords SVM • Logistics enterprise competitiveness • GA • Forecast

1 Introduction

Chinese logistics enterprises have just started to develop in recent years. The regulations about logistics enterprise began to implement until 2005 May 1st in China [1]. Therefore, it is necessary to establish a feasible evaluation index system in order to evaluate the development level of Chinese logistics enterprises.

Xie [2] pointed out that, in order to survive in an increasingly competitive marketplace, many companies are turning to data mining techniques for churn

N. Ding (✉)
School of Human and Development, China Agriculture University,
Haidian District, Beijing 100083, P.R. China
e-mail: 517276872@qq.com

H. Li • H. Wang
School of Economics and Management, Beijing Jiaotong University,
100044 Beijing, P.R. China

Z. Zhang et al. (eds.), *LISS 2012: Proceedings of 2nd International Conference on Logistics, Informatics and Service Science*, DOI 10.1007/978-3-642-32054-5_40,

analysis. Mahesh Pal [3] has analyzed the SVM were state of art classification algorithms and perform well in terms of classification accuracy in comparison to multinomial logistic regression based classification algorithm as well as other classifier for land cover classifications. Muniz [4] compared logistic regression (LR), probabilistic neural network (PNN) and support vector machine (SVM) classifiers for discriminating between normal and PD subjects in assessing the effects of DBS-STN on ground reaction force (GRF) with and without medication.

2 Brief Description of the Research Method

2.1 *Support Vector Machine*

SVM, proposed by Vapnik [5], was mainly used to find out a separating hyperplane to separate two classes of data from the given data set [2]. Let each entry of data be (x_i, y_i), $(i = 1, 2, 3, \ldots, n)$, $x \in R^d$, $y \in \{-1, +1\}$, and x_i is input data, y_i represents category, demotes the sample quantity, and demeans the input dimension. For any x_i on the separating hyperplane, the condition (1) Should be satisfied. As usual, (2) Denotes the decision functions, where w is the normal vector of the hyperplane and means the bias value. For any given entry of test data, if $f(x) \geq 0$, the entry of data could be classified as "+1"; if $f(x) < 0$, it could be classified as "-1". SVM could be classified as liner or non-linear based on the problem types [6]. When data could be categorized as two types, the linear SVM could find a hyperplane with the maximum margin width $\frac{1}{2}\|w\|$ to separate the data into different types by finding the minimum of $\frac{1}{2}\|w\|^2$ subject to (3) For solving the above problem, the Lagrange optimization approach could be adopted for carrying out the resolution process easily. Constraint Eq. (3) could be replaced by Lagrange multipliers. The Lagrange function is expressed as (4) where the Lagrange multiplier $a_i \geq 0,\ i = 1, 2, \ldots, \mathrm{n}$, corresponds to each inequality with the constraint Eq. (3). As such, the original problem to find the minimum $\frac{1}{2}\|w\|^2$ has been converted into finding the minimum L_p with the constraint equation $a_i \geq 0$. However, it is still difficult for the non-linear SVM to find the optimal solutions. For dealing with this situation, the Lagrange Dual Optimization Problem is used to make the solution process easier, which is formulated (5).

$$w \cdot x + b = 0 \tag{1}$$

$$f(x) = w \cdot x + b \tag{2}$$

$$y_i[(w \cdot x) + b - 1] \geq 0, \quad i = 1, 2, \ldots, \mathrm{n} \tag{3}$$

$$L(w,b,a)=\frac{1}{2}(w\cdot w)-\sum_{i=1}^{n}a_i\{y_i[(w\cdot x_i)+b]\}+\sum_{i=1}^{n}\alpha i \tag{4}$$

$$Max\,L_D(\alpha)=\sum_{i=1}^{n}a_i-\frac{1}{2}\sum_{i,j=1}^{n}\alpha_i\,\alpha_j\,y_i\,y_j\left(x_i\cdot x_j\right) \tag{5}$$

$$\sum_{i=1}^{n}\alpha_i y_i=0\quad \alpha_i\geq 0,\quad i=1,2,\ldots,\mathrm{n} \tag{6}$$

$$y_i[(w\cdot x_i)+b-1]+\xi_i\geq 0,\ \xi_i\geq 0,\quad i=1,2,\ldots,\mathrm{n} \tag{7}$$

$$Min\frac{1}{2}\|w\|^2+C\left|\sum_{i=1}^{n}\xi_i\right|,c>0 \tag{8}$$

$$Max\,L_D(\alpha)=\sum_{i=1}^{n}a_i-\frac{1}{2}\sum_{i,j=1}^{n}\alpha_i\alpha_j y_i y_j\left(x_i\cdot x_j\right) \tag{9}$$

$$\sum_{i=1}^{n}\alpha_i y_i=0 \tag{10}$$

$$\int\!\!\int k(xi,\ xj)g\,(x_i)\,g\,\left(x_j\right)dx_i dx_j>0,\ g\ \in\ L_2 \tag{11}$$

Once α_i is found, the optimum w and b can be obtained and therefore the decision function $f(x,a,b)$ can be determined through Eq. (4). In addition, for the linear SVM to process non-separable data, Vapnik [5] indicated that slack variables ξ_i could be added into the constraints as (4) and (7).

When errors happen to the classification of training data, ξ_i should be larger than 0. Therefore, a lower $\sum\xi$ should be preferred when determining the separating hyperplane. For this purpose, a cost parameter $c>0$ is added to control the allowable error ξ_i. The objective function should be changed from the solution for the minimum $\frac{1}{2}\|w\|^2$ into the solution for (8). For simplification, Eq. (8) could be transformed into the dual problem as (9). Based on above descriptions, it is simple to use linear SVM to separate the two different categories of data, if data could be fully separated by a linear function; otherwise, a parameter C is required to control the allowable errors. However, in the real world, not all data could be separated by linear hyperplane. Boser et al. [7] made the comparison between the linear and non-linear problems, and found if the original data are transferred to another feature space of high dimension (ϕ: Rd $\rightarrow$ F) through the mapping function ϕ, thereafter, the linear classification was conducted within the space and the process could find better effect. If data $\left(x_i,x_j\right)$ are transferred to the feature space of a high dimension, i.e., $\phi(x_i) \cdots \phi(x_j)$, the corresponding term in the dual problem (6) should be changed. The dot product of $\phi(x_i)$ and $\phi(x_j)$ was defined by the Kernel function, $K(x_i,\ x_j)$, thus, the optimization by the linear or non-linear SVM finally becomes $Max\, L_D(\alpha) = \sum_{i=1}^{n} a_i - \frac{1}{2}\sum_{i,j=1}^{n} \alpha_i\alpha_j y_i y_j (x_i \cdot x_j)$ Subject to (10).

2.2 Genetic Algorithm

GA coding strategies mainly include two sectors; one sector recommends the least digits for coding usage, such as binary codes; another one recommends using the real-valued coding based on calculation convenience and accuracy [8]. Binary codes are adopted for the decision variables in solving the discrete problems; however, it probably causes conflict between accuracy and efficiency when the problems are featured with continuity in that the calculation burden is quickly increased. The adoption of real-valued coding does not only improve the accuracy, but also significantly increases the efficiency in the larger search space, so that more practical applications adopt the real-value coding for solution.

2.3 GA-SVM Model Discriminant Analysis

As mentioned before, a kernel function is required in SVM for transforming the training data. This study adopts RBF as the kernel function to establish support vector classifiers, since the classification performance is significant when the knowledge concerning the data set is lacking. Therefore, there are two parameters, C and δ2, required within the SVM algorithm for accurate settings, since they are closely related to the learning and predicting performance. However, determining the values exactly is difficult for SVM. Tay and Cao [9] Suggested that C should range from 10 to 1,000 and δ^2 from 1 to 100, so that the established models can achieve much better results. Generally, to find the best C and δ^2 a given parameter is first fixed, and then within the value ranges another parameter is changed and cross-comparison is made using the grid-search algorithm. This method was conducted with a series of selections and comparisons, and it will face the problems of lower efficiency and inferior accuracy when conducting a wider search. However, GA for reproduction could provide the solution for this study. The scheme of an integration of real GA and SVM is shown in Fig. 1 to establish a classification model that could be used to determine whether a business crisis is approaching.

The final converged solutions should be affected by the possibility of genetic operations or parameters For example, the possibility for mutation and population size was determined after several experimentation cycles to make the training data sets have the maximum accuracy.

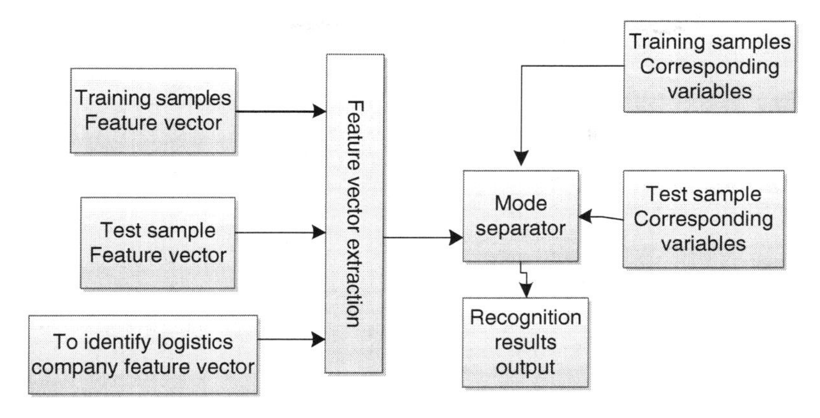

Fig. 1 Competitiveness evaluation model of listed logistics companies

3 Design of the Diagnostic Model

This article builds the appraisal model of the finance of listed companies through empirical method. The model includes two processes: the training of the model and the testing of the model. If the model which has undergone the two processes has a good recognition capability, it will be open to specific finance appraisals to further prove its validity. The structure of the model shows in graph 1.

In the graph, black arrows represent the direction of data flow of training and testing samples; white arrows represent the direction of data flow in the actual finance appraisal process. From the graph 1, it can be concluded that the first step of building the model is to build the training sample and testing sample, and train the pattern classifier after the input of training sample. When the training is complete, the result can be confirmed by the input of training sample. Pattern classifiers which only pass the test can be used to appraise the finance condition of the listed companies. If the finance condition of the listed companies complies well with the results from the pattern classifiers, the model is proved to be successful.

4 Conclusion

This paper object of study is the finance data of logistics enterprises. Through the empirical method attracted several conclusions as follows:

1. The finance data of logistics enterprises in our country is effective. It has strong ability to predict.
2. Through an empirical test of sample and out sample, the model show the logistics enterprises competitive power.

References

1. Wang Yi, Zeng Libin (2008) Design on the evaluation index system of logistics enterprises' competitiveness in China. Econ Manag 22(11):54–57
2. Xie Y, Li X (2009) Customer churn prediction using improved balanced random forests. Expert Syst Appl 36:5445–5449
3. Mahesh Pal (2012) Multinomial logistic regression-based feature selection for hyperspectral data. Int J Appl Earth Obs Geoinformation 14:214–220
4. Liu Sheng, Li Yanyan (2007) Parameter selection algorithm for support vector machines based on adaptive genetic algorithm. J Harbin Eng Univ 28(4):398–402
5. Vapnik V (1995) The nature of statistical learning theory, Springer, New York
6. Burges CJC (1998) A tutorial on support vector machines for pattern recognition. Data Min Knowl Discov 2(2):121–167
7. Boser B, Guyon I, Vapnik V (1992) A training algorithm for optimal margin classifiers. Fifth annual workshop on computational learning theory. ACM Press, New York
8. Haupt RL, Haupt SE (1998) Practical genetic algorithms, Wiley, New York
9. Tay FEH, Cao L (2001) Application of support vector machines in financial time series forecasting. Omega 29(4):309–317

The Research of 3G Development Capability of Chinese Telecom Operators

Sha Jing and Ding Huiping

Abstract With the arrival of 3G, there is a comprehensive restructuring for China communications industry. The six original operators have been integrated into three operators. They are China Mobile, China Telecom and China Unicom, all of which cover both mobile and fixed business service. For Chinese telecom operators, it is very important to provide a good competitive environment in the promotion of the industry. This article starts the domestic three operators' introduction. Then it states the competition ability for every operator in the 3G market. It indicates that the operators should seize the opportunities for the purpose of the long-term development.

Keywords Operators • Strategy • SWOT • The 3rd generation

1 Introduction

At present, the state has provided a loose policy environment for the development of 3G, the whole 3G industry chain development can stimulate the domestic demand and pull our country economy certainly. The race for the 3G market is the strategic target for the major operators, and the effective operation needs to have a few basic conditions in the 3G market [1]. Firstly, the mature technology base is essential. Secondly, the high quality of the 3G communication network coverage is needed. Thirdly, the good competition environment between operators is very important. Finally, it's the right business model or business. In China, 3G license had been distributed in January 2009. With the coming of 3G, there are high-speed data network, low price and the instant access methods.

S. Jing • D. Huiping (✉)
School of Economics and Management, Beijing Jiaotong University, Beijing, P.R. China
e-mail: zishuijing_123@126.com

Z. Zhang et al. (eds.), *LISS 2012: Proceedings of 2nd International Conference on Logistics, Informatics and Service Science*, DOI 10.1007/978-3-642-32054-5_41,

2 The SWOT Analysis on Domestic Telecommunication Operators

2.1 *The SWOT Analysis on China Telecom*

As the main Chinese telecom enterprise and the largest subject based network operators, China Telecom has the world's biggest network in the fixed telephone field. The employees of the company can be found all over the Chinese 31 provincial level enterprises in the nationwide operation of telecommunications services. Creating a world class telecom enterprise is the goal of China Telecom. It would promote "focusing on customer information innovation" strategy in operation [2]. At the same time it has strong fixed network resources and the business operation advantages (Table 1).

2.2 *The SWOT Analysis of China Mobile*

China Mobile is the only one that focuses on the development of mobile communication operating for many years. In the process of development, China Mobile always plays the leading role. At present it has the biggest network size and customer scale all around the world. And China Mobile has a strong capital and the advantage in R&D [3]. The arrival of 3G will provide more challenges than opportunities for China Mobile. Now China Mobile is facing with the shortage of the solid line resource obviously and the huge challenge coming from the business competition (Table 2).

2.3 *The SWOT Analysis of China Unicom*

China Unicom has set up branches in Chinese 31 provinces and many foreign countries. By the end of 2010, there are more than 311 million subscribers. As the

Table 1 China telecom SWOT analysis

Items	Evaluation
Strengths	It has the Chinese largest fixed communication network and strongest channel; It has the closest relations with the local government and customers; It has low cost but high speed in CDMA2000 network upgrades
Weaknesses	It is insufficient in the northern network coverage; The historical burden which is from the former China Telecom is heavy; Cash flow is short relatively; No mobile operation experience; It need to be improved in the service quality
Opportunities	It has strong competitive strength in the optical broadband field; The demand in 3G business market is growing rapidly; Asymmetric control policy
Threats	It is difficult for the wireless local call to turn nets; The scale of Mobile is smaller

Table 2 China mobile SWOT analysis

Items	Evaluation
Strengths	China's largest mobile service provider; The most global users and the world's largest mobile communication network; The abundant capital and technology strength; The advantages of low cost and excellent brand
Weaknesses	The shortage of commercial operation experience; The weak TD industrial chain; lacking of the fixed network resources, as well as the network and terminal
Opportunities	There is space for China Mobile to improve in the Mobile communications market; It may be enhanced on the wireless Internet drive function; The fast development of the mobile Internet and the Internet of Things; China encourages the large enterprise to develop home and abroad with a good prospect
Threats	The penetration rate of Mobile phone increases; Great changes have taken place in the communication market; There are the threats from anti-monopoly and technical progress

Table 3 China Unicom SWOT analysis

Items	Evaluation
Strengths	It has strong fixed network foundation in 10 provinces of north; A lot of experience in mobile operation; It has the industrial chain and mature technology of 3G international network resources (WCDMA); The largest number of terminal
Weaknesses	The weakest financial strength; The lack of fixed network access resources in northern; The small rural network coverage; The high operation cost; The small scale of the users; The low brand awareness; Low customer satisfaction; The high losing rate; GSM network covers insufficiently
Opportunities	Asymmetric regulatory policy leans to it; Fewer mobile users in the northern market
Threats	China Unicom is facing with China Telecom's powerful impact on 3G; There is a bad influence on China Unicom's fixed-line broadband business

late comer of domestic mobile operators, China Unicom lagged far behind rivals in brand construction [4]. Because of the brand appeal power gap, it leads to a low satisfaction of the China Unicom's users. Therefore the consumer losing rate is very high (Table 3).

3 The Comparison of Competition Ability of Telecommunication Operators in 3G

After the separation and reorganization in 2008, the three basic operators have done market positioning respectively. First, China Telecom is the main domestic broadband and the biggest basic network operator, which has the largest fixed telephone network, covering both urban and rural. Then, China Mobile is based on GSM and TD-SCDMA mode of mobile communication network. It only focuses on the development of mobile communication operating. Finally, China Unicom is the only comprehensive operator which can provide comprehensive telecommunication basic business. It makes use of the WCDMA network forming a large size in the national scope with the extensive 3G advanced technology.

3.1 Network

It is the key point that realizing the 3G network seamless coverage to optimize and the upgrading the Internet. China Unicom makes use of WCDMA. China Telecom utilizes CDMA2000. China Mobile uses TD-SCDMA. WCDMA and CDMA2000 network technology has been relatively mature. The speed of China Telecom network upgrades fastest, because CDMA2000 is original by CDMA evolution. China Unicom is dominant in network maturity with WCDMA which is mature than others.

3.2 Terminal Promotion

In 2010, China Telecom can already provide more than 600 cell phone terminal for the consumers, 3G mobile phone accounts for only more than 100, although from the fourth quarter of 2009, China Telecom terminal manufacturers had released 68 kinds of 3G phones, which price is about 1,000 RMB in 2010. However the CDMA intelligent terminals are still less, and the number of relatively low end terminals are more. This kind of situation will drag down China Telecom 3G user's quality overall. The ARPU value will also decline inevitably. In 2010 China Unicom WCDMA terminal types of more than 100, the brand and benefit of the effect is not ideal. While China Mobile phone TD has about more than 60 types, comparing with WCDMA and CDMA, optional scope is still limited in the respects of brand and the design. In order to break through the bottleneck problem of the terminal, operators treat the promotion of the development of the 3G terminal and sales as one of the key work of the past 2 years.

3.3 Comparison of the Standard Treatment

The operators have been pushing unified mobile phone charging standards. China Unicom national 3G charges used the "long city flood integration" structure, canceling the long distance and roaming charges. China Mobile and China Telecom also introduced similar packages and money standards gradually. China Telecom "e surfing" 3G package type is more. It is divided into travel packages, wireless broadband package, long distance chat, mass combo and fashion set meal. Different area the package price also has difference more or less. China Mobile wants to maintain existing users better, also reduce fees to some extent and carry out family planning. At present, the telecom market, especially in the traditional voice market gradually saturated, 3G market is not based on the 2G in the market of the transverse propulsion. It is about the application and business abilities. China Unicom 3G Nation has no long-distance phone fees, no past and no answer fee, all phone in nation change into A local calls, which is very suitable for the people who connects with travel, mobile office and market trends to use, according to the user group, launched A, B and C three kinds of different packages, each with different levels in the package.

4 Conclusion and Suggestions

At the beginning of 2009, the operators take different 3G format with the new industry. China Unicom gets the license of WCDMA, in mature technology level with a strong construction industry chain. China Mobile can attract industry chain of other units TD-SCDWA construction with strong capital capability and market share. And China Telecom's fixed-line is mainly in the southern market with high speed development of economy on industries. The three operators respectively take different products and service strategies under the guidance of combinative themselves. It has made progress in charges, channel, brand and customer foster. According to the development experience of Japan and South Korea in 3G, the next 2 or 3 years the 3G market share will get big. It's necessary for operators to pay close attention to the dynamic change.

References

1. Wang Zhenying (2010) 3G market "three wars"-2009 3G operators to strategy and tactics to summarize and comment on. China's telecom industry 140(8):51–53
2. Chang Baoqian (2008) China Unicom company 3G communications market marketing strategy research. Xiamen University, Xiamen
3. Chang Jiandong (2007) China telecom 3G mobile business development strategy. Suzhou University, Suzhou
4. Li Jianjia (2009) China Unicom company competition ability analysis and business strategy research. Jilin University, Changchun

Pricing and Coordination Research for TPL Based on Different Logistics Service Level

Xuehui He, Wei Li, and Kai Nie

Abstract In this paper, we consider a situation that different logistics service level can influence the market demand, TPL service and the pricing decision models are constructed by using game theory. The equilibrium prices, service levels under different systematic states of two TPL enterprises are given. And the conclusion of this paper shows that the strong ability of logistics service does not necessarily have a competitive advantage when under the separate decision, pricing equilibrium under joint decisions not only make both sides get more income, but is also advantageous to raise the level of service. The conclusion also shows that revenue sharing is a good coordination mechanism for logistics service union, and its revenue sharing percentage depends on the negotiation skills of both sides.

Keywords TPL • Service level • Service price • Revenue sharing • Game theory

1 Introduction

The concept of Third Part Logistics (TPL) originated from 1980s and has developed into a certain size of industry in developed countries. Due to the emerging of TPL, the competition among TPL becomes fiercer. Intangible services usually have to face up to more uncertain market environment compared to tangible productions, and their market demand is not only influenced by price standard but also by service level. Boyer K K's study [1] showed that logistics service level can directly affect the demand of guests. Therefore, this paper assumes that the actual demand of the logistics service in this thesis is about the linear function of service price and the provided service level.

Some achievements of TPL research haven been made since 1980s. In summary, it can be divided into two aspects: first are the studies out the similarity of TPL

X. He (✉) • W. Li • K. Nie
College of Economics and Trade, Hunan University, Changsha 410079, P.R. China
e-mail: hexuehui2009@yahoo.cn; liweihncs@126.com

Z. Zhang et al. (eds.), *LISS 2012: Proceedings of 2nd International Conference on Logistics, Informatics and Service Science*, DOI 10.1007/978-3-642-32054-5_42,

services, which is to elaborate TPL's concept, its importance and merits and flaws. For instance, Boyson [2] believes that enterprises can reduce costs rapidly by delegating logistics business to the professional TPL service companies; Berglund [3] believes that storage, process, transportation and other TPL services can achieve economy of scale, while TPL service can provide various types of service according to the diversified demands made by guests. Recently, while TPL enterprises become increasingly important, Chinese scholars begin to pay attention to TPL, and most of their studies are carried out from angle of TPL similarity. Second are studies on pricing TPL services. Lambert.DM [4] found that reasonable logistics service price is a key factor in a successful cooperation relationship; Albert Y Ha [5] analyzed how to make decisions on price and delivery to win the favor of the guests under the circumstances of competition and non-cooperation between two delivery service suppliers; Alberto De Marco [6] proved that maintenance cost of warehouses is correlated to the performance of the logistic service; Weijers S [7] showed that sustainability is handled by the logistics company as an integral part of the corporate strategy. Some Chinese scholars also studied TPL from that aspect: Qi Ershi [8] studied the pricing policies of competition and cooperation between TPL companies and the subordinate logistics company of port enterprises, and built the advanced Bertrand competition model in competitive relationship, equivalent in consultation and cooperation model and remedies cooperation model in cooperative relationship.

Most studies on the influences of service levels on the variation of market demands are carried out from the view of supply chain. For example, Xu Minghui [9] used game model to study the supply chain formed by one supplier and one retailer, and respectively explored the decisions made by suppliers and retailers on the base of different service providers; Wu Qing [10] used dynamic game model to study coordinated design of contracts between customer companies and TPL service providers in the situation that logistics service level can influence product demands of the consumer enterprises. The research perspective of most scholars on the cooperation of market demands influenced by service levels, is from the upstream and downstream companies of the supply chain, but there is few studies on horizontal cooperation management of the similar enterprises. Taking into account the intangibility and non-repurchase of logistics service, this paper studies the pricing strategy and the definition of the service level in independent decision-makings and joint decision-makings of two logistics enterprises in different service levels. On this basis, we explored coordinated pricing policy based on revenue sharing contract.

2 Model Introduction

Suppose there are two logistics enterprises supplying service in the market, namely logistics enterprise 1 and logistics enterprise 2. Suppose the actual demand of logistics service is related to its price and service level. The demand functions of two enterprises are as follow respectively:

$$D_1 = a - p_1 + \alpha p_2 + H_1 \tag{1}$$

$$D_2 = a - p_2 + \alpha p_1 + H_2 \tag{2}$$

α is a alternative coefficient of price, $\alpha \in [0, 1]$; Assume that the potential market demand scale of two logistics enterprises (the price and service level are considered) is equal to a.

Assume that the relationship between the cost of logistics service and logistics service level is a strictly increasing convex function, the relationships between service costs C_i and service level H_i are as follow: firstly, the cost of services will raise as the service level increase, and the marginal costs will increase too; secondly, the stronger service ability provided, the smaller the marginal cost is. Set the relationship of service level and the service cost is: $C_i = k_i H_i^2/2$, $\forall i \in \{1, 2\}$ k_i is a related coefficient of logistics service provider, which reflect the ability of logistics enterprise, $k_i > \frac{1}{2}$. This second form of cost function is commonly used in the research [9, 11]. π_1 as the profit of Logistics enterprise 1 and π_2 as the profit of Logistics enterprise 2, then, profit functions of two enterprises are as follow respectively:

$$\pi_1 = p_1(a - p_1 + \alpha p_2 + H_1) - \frac{1}{2}k_1 H_1^2 \tag{3}$$

$$\pi_2 = p_2(a - p_2 + \alpha p_1 + H_2) - \frac{1}{2}k_2 H_2^2 \tag{4}$$

3 Solution of Nash Game Under Independent Decision

Assume that two logistics enterprise have equal status in the market, they conduct Nash game. In competitive relationship, two logistics enterprises make independent decision and determine its service price and service level respectively.

$$\frac{\partial \pi_1}{\partial p_1} = a - 2p_1 + \alpha p_2 + H_1 = 0 \qquad \frac{\partial \pi_1}{\partial H_1} = p_1 - k_1 H_1 = 0$$

$$\frac{\partial \pi_2}{\partial p_2} = a - 2p_2 + \alpha p_1 + H_2 = 0 \qquad \frac{\partial \pi_2}{\partial H_2} = p_2 - k_2 H_2 = 0$$

Solve the Nash equilibrium by above four formulas, and conclude that two logistics enterprises should adopt the pricing strategy and service level are as follow in order to pursue their maximum profit when under the completion relationship.

Conclusion 1: Conduct Nash game under independent decision, pricing and service level of two enterprises are such as below:

$$p_1^1 = \frac{ak_1[k_2(2+\alpha) - 1]}{k_1k_2(4-\alpha^2) - 2(k_1+k_2) + 1} \qquad H_1^1 = \frac{a[k_2(2+\alpha) - 1]}{k_1k_2(4-\alpha^2) - 2(k_1+k_2) + 1}$$

$$p_2^1 = \frac{ak_2[k_1(2+\alpha) - 1]}{k_1k_2(4-\alpha^2) - 2(k_1+k_2) + 1} \qquad H_2^1 = \frac{a[k_1(2+\alpha) - 1]}{k_1k_2(4-\alpha^2) - 2(k_1+k_2) + 1}$$

Then, their profits are:

$$\pi_1^1 = \frac{a^2k_1(2k_1-1)[k_2(2+\alpha)-1]^2}{2[k_1k_2(4-\alpha^2)-2(k_1+k_2)+1]^2} \quad \pi_2^1 = \frac{a^2k_2(2k_2-1)[k_1(2+\alpha)-1]^2}{2[k_1k_2(4-\alpha^2)-2(k_1+k_2)+1]^2}$$

From the above conclusion 1 we can see: when conduct Nash game, their profit difference only caused by the different of the cost. As the two logistics enterprises are in the same authority. The relationship of optimal service level and potential market demand scale is a positive linear correlation, that is, when the potential market demand scale increases, two logistics enterprises' optimal service level will increase.

Deduction 1: p_1 and p_2, π_1 and π_2 are increase function with alternative coefficient α if k_1, k_2 are fixed. It shows that the greater the degree of substitution, the higher price level under Nash equilibrium when two logistics enterprises are in monopolistic competition. As k_1, k_2 are fixed, the competition between two enterprises is price, according to the Bertrand price competition model, the enterprises will compete to reduce their prices in order to have a bigger share of the market, so when the competition is fierce, that α is great, the two enterprises' original price will be higher. Only in this way, the enterprises will have a bigger price cut space under the equilibrium level.

Deduction 2: $\Delta H = H_1 - H_2 = \frac{a(2+\alpha)(k_2-k_1)}{k_1k_2(4-\alpha^2)-2(k_1+k_2)+1}$ as the difference degree of service level that provided by two logistics enterprises, if α is fixed, when ΔH is bigger, then, the difference of the equilibrium price $\Delta p = p_1 - p_2 = \frac{a(k_2-k_1)}{k_1k_2(4-\alpha^2)-2(k_1+k_2)+1}$ is bigger too. It shows that the service price and the service level are related, the higher service level is, the higher price is.

Deduction 3: (1) when α is fixed, if $k_1 = k_2$, then, $p_1 = p_2$, $H_1 = H_2$ and $\pi_1 = \pi_2$. It shows that the service price and service level will be the same if two enterprises have the same ability to provide service when conducting Nash game. In this case, profit level of two logistics enterprises are the same. (2) when α is fixed, if $k_1 > k_2$, then, $p_1 < p_2$, $H_1 < H_2$, but profit level of two enterprises can not be determined. $k_1 > k_2$ reflects that the ability to provide service of enterprise 2 is stronger than that of enterprise 1, so the service level and the service price are higher than that of enterprise 1 too. When two enterprise conduct Nash game under independent condition, although logistics enterprise 2 has advantages in service capability, but the higher service price makes the profit level not necessarily higher than logistics enterprise 1 that's the service ability is weaker, which is accord with the reality condition.

From the above we can see that: the enterprise which has a strong ability to provide service may not have the absolute competitive advantage when it is in a competition condition. The reason is that the enterprise which has a strong ability is not willing to focus its energy on enhancing the service level, but to get more market share by cutting the price to achieve its maximize profit purposes. Next, relation of two logistics enterprises are discussed when they are in cooperation condition and compose service union.

4 Equilibrium Solution Under Joint Decision

Suppose that two logistics enterprises cooperate and together form a service union, in order to achieve win-win objective, they do the behavior such as information sharing, cost sharing etc. Under this condition, the pricing decisions and service level are to maximize the total profits. The profit function of the whole system is:

$$\pi = p_1(a - p_1 + \alpha p_2 + H_1) + p_2(a - p_2 + \alpha p_1 + H_2) - \frac{1}{2}k_1H_1^2 - \frac{1}{2}k_2H_2^2 \quad (5)$$

Consider a first-order we can get:

$$\frac{\partial \pi}{\partial p_1} = a - 2p_1 + 2\alpha p_2 + H_1 = 0 \quad \frac{\partial \pi}{\partial H_1} = p_1 - k_1H_1 = 0$$

$$\frac{\partial \pi}{\partial p_2} = a - 2p_2 + 2\alpha p_1 + H_2 = 0 \quad \frac{\partial \pi}{\partial H_2} = p_2 - k_2H_2 = 0$$

By solving the above four formulas, conclude that two logistics enterprises should adopt the pricing strategy and service level below in order to maximize the total profit when under the cooperation relationship.

Conclusion 2: when doing joint decision, pricing and service level of two enterprises are below:

$$p_1^2 = \frac{ak_1[2k_2(1+\alpha) - 1]}{4k_1k_2(1-\alpha^2) - 2(k_1+k_2) + 1} \quad H_1^2 = \frac{a[2k_2(1+\alpha) - 1]}{4k_1k_2(1-\alpha^2) - 2(k_1+k_2) + 1}$$

$$p_2^2 = \frac{ak_2[2k_1(1+\alpha) - 1]}{4k_1k_2(1-\alpha^2) - 2(k_1+k_2) + 1} \quad H_2^2 = \frac{a[2k_1(1+\alpha) - 1]}{4k_1k_2(1-\alpha^2) - 2(k_1+k_2) + 1}$$

Then, their total profits are: $\pi = \frac{a^2[4k_1k_2(1+\alpha) - (k_1+k_2)]}{2[4k_1k_2(1-\alpha^2) - 2(k_1+k_2) + 1]}$

Comparing with conclusion 1, we can get: $H_1^1 < H_1^2, H_2^1 < H_2^2, p_1^1 < p_1^2, p_2^1 < p_2^2$ and $\pi_1^1 + \pi_2^1 < \pi$.then can get the following deduction.

Deduction 4: (1) Comparing with independent decision, the total profits of joint decision are higher than that of independent decision. The following can be obtained: the efficiency of competition is lower than the efficiency of cooperating, both sides have the potential incentive to cooperate. (2) Comparing with independent decision, the service level and service price of joint decision are higher than that of independent decision. It shows that their goal is to maximize the total profits when both are under the cooperation relationship, there is no need to lower the price to get a bigger share of the market, instead by improving the service level. So the service level and the service price in the cooperation condition are higher than that in the competition condition.

5 Pricing Coordinated Strategy Based on Revenue Sharing

By deduction 4, it is known that the total profits under joint decision are higher than that under independent decision. So there are potential incentive for the enterprises to cooperate. In order to make two logistics enterprises cooperate, it has been made sure that each enterprise's profit under cooperation is not less than that under competition, which is involved pricing and the profits coordination problems.

In order to set up reasonable prices and make the profit distribution equitable, this paper discusses one kind of coordination mechanism which was commonly used in the literature–revenue sharing contract [12]. Assume that the profit of enterprise 1 accounts for the proportion λ of total profits, which is the parameters of the revenue sharing contract, the profit of enterprise 2 accounts for the proportion $(1-\lambda)$ of total profits. Assume that the retained profits of logistics enterprise i are $\pi_i^{\min}(i=1,2)$. Then the models of revenue sharing contract are:

Logistics enterprise 1:

$$\begin{cases} \max\ \pi_1 = \lambda\left[p_1(a-p_1+\alpha p_2+H_1)+p_2(a-p_2+\alpha p_1+H_2)-\frac{1}{2}k_1H_1^2-\frac{1}{2}k_2H_2^2\right] \\ s.t.(IR)\pi_1 \geq \pi_1^{\min} \\ \lambda \in (0,1) \end{cases}$$

Logistics enterprise 2:

$$\begin{cases} \max\ \pi_2 = (1-\lambda)\left[p_1(a-p_1+\alpha p_2+H_1)+p_2(a-p_2+\alpha p_1+H_2)-\frac{1}{2}k_1H_1^2-\frac{1}{2}k_2H_2^2\right] \\ s.t.(IR)\pi_2 \geq \pi_2^{\min} \\ \lambda \in (0,1) \end{cases}$$

IR is called the participation constraint of logistics enterprise $i\,(i=1,2)$, to guarantee that logistics enterprise will accept the contract. The proportion λ of revenue sharing depends on the bargaining ability of each side. If $\lambda=1$, it reflects that logistics enterprise 1 is in absolute advantage position in the process of coordinating, and get all the profits, $\lambda=0$ is said that logistics enterprise 2 get all the profits. Apparently this is difficult to achieve effective agreement, therefore, reasonable revenue sharing proportion should meet $0<\lambda<1$.

6 Example Analysis

Assume the market demand function of logistics enterprise 1 is $D_1=10-p_1+0.5\,p_2+H_1$, the market demand function of logistics enterprise 1 is $D_2=10-p_2+0.5\,p_1+H_2$, command $k_1=2$, $k_2=1$. The related results in Table 1 are as below.

From the Table 1 we can see that the service level has improved when logistics enterprises are in joint decision, and the total profits are increased too. So a union between logistics enterprises, making a horizontal cooperative, not only can make

Table 1 Related results under independent decision and joint decision

	Independent decision	Comparison	Joint decision
Service price of enterprise 1	12	<	40
Service level of enterprise 1	6	<	20
Service price of enterprise 2	16	<	50
Service level of enterprise 2	16	<	50
Profit of enterprise 1	108	<	200
Profit of enterprise 2	128	<	250
Total profits	236	<	450

the enterprise win more profits, but also can avoid the monopolistic competition among the industry, which make enterprises intend to improve the logistics service level, and promote the healthy development of whole logistics market.

7 Conclusion and Outlook

Based on the service system consisted of two logistics enterprises, considering the situation that market demand is affected by the level of logistics service, this paper discusses the pricing strategies, the determination of the service level as well as coordination and cooperation strategy of two logistics enterprises in the independent decision and joint decision, and reach the following conclusions by constructing a game model.

(1) For monopolistic competition, two logistics enterprises in different logistics service level are inclining to maximize their own profit by price competition, rather than increasing the service level. (2) Enterprises with logistics services higher ability, do not have an absolute competitive advantage in their monopolistic competition, so there is no potential incentive for the enterprises striving to improve their service level which is not conducive to the development of the whole logistics market level. (3) The total profits of two logistics enterprises in joint decision are higher than the total profits earned by their own independent decision. The more competitive the logistics enterprises are, the more they should cooperate to determine the reasonable service level and reasonable pricing. This will not only increase the system's total profits, but also can make the choice of logistics services more flexible. (4) Revenue sharing contract can help enterprises solve the coordination problems of interest in the joint decision. The proportion λ of revenue sharing depends on the negotiating ability of each side.

With the development of TPL enterprises, the scale of the logistics services market is becoming larger. The fact is there are much more than two logistics enterprises in one region, so extend the model by considering the reality can be an important direction for future research; at the same time, this paper aims to discuss the pricing and coordinating of logistics enterprises under complete information, but the logistics services market has become increasingly complex, asymmetric information is very often, therefore the pricing and coordinating problem of logistics companies under incomplete information can be an important work for future research too.

Acknowledgments This study was supported by philosophy social science fund of Hunan province (2010YBA048), general project of the ministry of education on humanities and social science research (11YJC790084) and Research fund for the Doctoral Program of higher education (20110161120032).

References

1. Boyer KK, Hult GTM (2005) Extending the supply chain: integrating operations and marketing in the online grocery industry. J Oper Manag 23:642–661
2. Boyson S, Corisi T, Dresner M, Rahinovich E (1999) Managing effective third party logistics relationships: what does it take. J Bus Logist 20:73–100
3. Berglund, M (1997) Third-party logistics providers towards a conceptual strategic model. Tek Lic thesis no. 642
4. Lambert DM, Emmerlhainz MA, Gardner JT (1999) Building successful logistics partnerships. J Bus Logist 20:165–181
5. Ha AY, Li L, Ng SM (2003) Price and delivery logistics competition in a supply chain. Manage Sci 49(9):1139–1153
6. de Marco A, Mangano G (2011) Relationship between logistic service and maintenance costs of warehouses. Facilities 29:411–421
7. Weijers S, Glöckner H-H, Pieters R (2012) Logistic service providers and sustainable physical distribution. Log Forum 8:157–165
8. Qi Ershi, Jiang Hong, Huo Yanfang (2010) Pricing analysis of logistics service between TPL and port enterprises. J Tianjin Univ 43(9):385–389
9. Xu Minghui, Yu Gang, Zhang Hanqin (2006) Game analysis on supply chain with providing service. J Manag Sci China 18–26
10. Wu Qing, Dan Bin (2008) TPL coordination contract based on logistics service level influencing the market demand. J Manag Sci China 64–74
11. Xie T, Li J (2008) Pricing game analysis for third-party logistics services. J Syst Eng 23:751–757
12. Chen J, Zhou L, Shao X-F (2008) Capacity investment decision of service system with exterior flexible capacity cooperating. Syst Eng Theory Pract 28:59–64

ERP Implementation Risk Assessment Based on Analytic Hierarchy Process

Li Huang, Bing Zhu, and Bing Han

Abstract Many companies have launched Enterprise Resource Planning systems nowadays, however only a few of which have achieved their business goals. The ignorance of ERP implementation risks would be the major reason. This paper uses the analytic hierarchy process to examine major risk factors that affect the successfulness of ERP implementation so as to help companies minimize the related risks so as to achieve their competitive advantages.

Keywords ERP implementation • Risk • Risk assessment • Analytic hierarchy process

1 Introduction

The successful implementation of ERP needs to consider a variety of risk factors. Cheng [1] referred that macro-risks should be considered during implementation, including risks from the following aspects, such as politics, economics, laws and regulations etc. Zhang [2] thought that internal risks should be considered firstly in the case of stability of environment, risks associated with business process reengineering, software and human resource should be taken into account according to progress of ERP implementation.

L. Huang (✉) • B. Zhu • B. Han
School of Economics and Management, Beijing Jiaotong University, No. 3 Shang Yuan Cun, Hai Dian District, Beijing 100044, P.R. China
e-mail: 11120688@bjtu.edu.cn

Z. Zhang et al. (eds.), *LISS 2012: Proceedings of 2nd International Conference on Logistics, Informatics and Service Science*, DOI 10.1007/978-3-642-32054-5_43,

2 Evaluating ERP Implementation Risks

Risk assessment [3] aims to evaluate the consequence of each risk events and determine the severity, providing a basis for companies' decisions. This paper mainly uses the analytic hierarchy process [4], which is suitable for complex and obscure problems. Four steps will be used in AHP modeling:

1. Establish hierarchical system of risk index;
2. Establish all judgment matrix of each level. The above hierarchy of elements will decide the standard of the next hierarchy of elements of $A_1, A_2 \ldots A_{ii}$ and then give them corresponding weights according to the relative importance of A_1, $A_2 \ldots A_{ii}$ under this standard, which will be given certain value according to the proportional scale of 1–9. For example,1 shows that factor i is as important as factor j; 3 shows that factor i is slight important than factor j; 5 shows that factor i is less important than factor j; 7 shows that factor i is more important than factor j; 9 shows that factor i is much more important than factor j; others show that the comparison results of factor i and factor j are in the middle of the above results; reciprocal shows that comparison results of factor j and factor i are the reciprocal of comparison results of i and j.
3. Calculate the relative weight in the criterion and exam their consistency. The characteristic root of matrix A is $Aw = \lambda_{\max} w$, and get the w that has been regularized as the weightiness of elements of A_1, $A_2 \ldots A_{ii}$ in the standard of C_k. (1) multiply the elements of A according to line; (2) get products and open n power separately; (3) make rad vectors normalization and get sort weight vectors w; (4) calculate $\lambda_{\max}$ using $\lambda_{\max} = \sum_{i=1}^{n} (Aw)_i / nw_i$, and the $(Aw)_i$ stands for the i-th element of Aw [5].
4. Do some consistency test, as the following: calculate the consistency index of CI that can get from $CI = (\max - n)/(n-1)$ in which n is the order of the judgment matrix and then calculate the consistency ratio of CR, $CR = CI/RI$
5. Calculate combination weight of each element and CI step by step. If calculation results of CI, RI and CR in layer $k-1$ can be got respectively, and then corresponding index of layer k is: $CI_k = \left(CI_k^1 \ldots CI_k^m\right) a^{k-1}$, $RI_k = \left(RI_k^1 \ldots RI_k^m\right)$ a^{k-1}, $CR_k = CR_{k-1} + CI_k / RI_k$, where CI_k^1 and RI_k^1 are respectively stand for consistency index and average random consistency index in layer $k-1$ in the standard. Consistency of layer k will be accepted if $CR_k < 0.1$, or, the comparison matrix must be adjusted.
6. Do some risk control analysis according to the important degree among risks, and then control the focus of risk factors and keep the risk level under a certain level.

3 Case Analysis of TISCO

TISCO (Taiyuan Iron & Steel Co) is an oversize steel joint enterprise that can do iron mine mining steel production, processing, distribution and trade. Devote themselves to the research and development and the production and processing in stainless steel, high-grade fine carbon steel and special steel. At present, TISCO is equipped with the information-based foundation in finance and business, but each system stays in relatively independent state and can't display value that should be found in enterprise development strategy, logistics management, human resources, production management control and financial management. Now risk assessment [6] will be analyzed using AHP from five parts of TISCO: software risk, business process reengineering risk, management changes risk, human resources risk, as the following (Fig. 1):

Use the method of experts' interviews and list all levels of judgment matrix according to expert, as the following (Tables 1, 2, 3, 4, and 5):

Second step is to calculate according to formula:

1. The matrix calculation of layer A-B

   (a) Calculate geometric average of judgment matrix

$$w_1' = (1*3*1/4*1/2) \wedge (1/4) = 0.783 \quad w_2' = (1/3*1/6*1/5*1) \wedge (1/4) = 0.325$$
$$w_3' = (4*6*3*1) \wedge (1/4) = 2.913 \quad w_4' = (2*5*1*1/3) \wedge (1/4) = 1.351$$

   (b) Make $w_1\prime$ standard into w_i, $W_i' = w_1' + w_2' + w_3' + w_4' = 5.372$

$$w_i = w_1'/W_i' \; w_1 = 0.146, \; w_2 = 0.0605, \; w_3 = 0.542, \; \mathrm{w}_4 = 0.251,$$
$$w = (0.146, 0.0605, 0.542, 0.251)$$

Calculate characteristic root of $\lambda_{\max}$ and make judgment matrix of layer A-B for D_w

$$D_w = \begin{pmatrix} 1 & 3 & 1/4 & 1/2 \\ 1/3 & 1 & 1/6 & 1/5 \\ 4 & 6 & 1 & 3 \\ 2 & 5 & 1/3 & 1 \end{pmatrix} \begin{pmatrix} 0.146 \\ 0.0605 \\ 0.542 \\ 0.251 \end{pmatrix} = \begin{pmatrix} 0.589 \\ 0.250 \\ 2.242 \\ 1.026 \end{pmatrix}$$

$$\lambda_{\max} = 0.589/4*0.146 + 0.250/4*0.0605 + 2.242/4*0.542 + 1.026/4*0.251 = 4.098$$
$$CI = (4.098 - 4)/(4 - 1) = 0.033$$

Look-up table and then get $CR<0.1$, so the matrix of comparative judgment in layer A-B meets the requirements of consistency

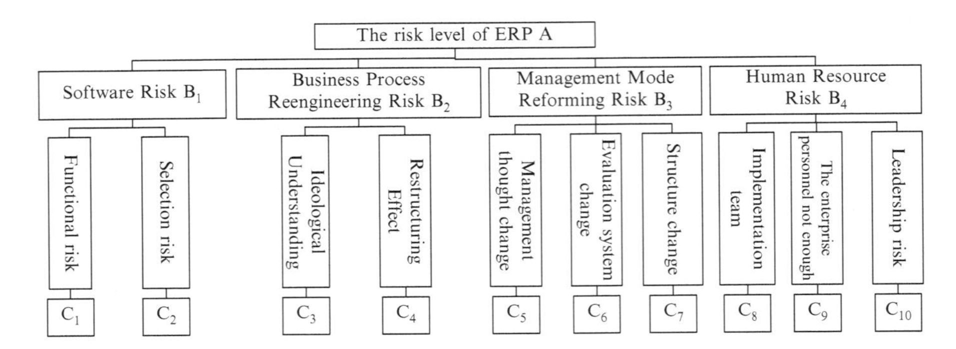

Fig. 1 The risk structure of TISCO

Table 1 The judgment matrix of layer *A*-*B*

A	B_1	B_2	B_3	B_4
B_1	1	3	1/4	1/2
B_2	1/3	1	1/6	1/5
B_3	4	6	1	3
B_4	2	5	1/3	1

Table 2 The judgment matrix of layer B_1-C

B_1	C_1	C_2
C_1	1	3
C_2	1/3	1

Table 3 The judgment matrix of layer B_2-C

B_2	C_3	C_4
C_3	1	2
C_4	1/2	1

Table 4 The judgment matrix of layer B_3-C

B_3	C_5	C_6	C_7
C_5	1	2	1/3
C_6	1/2	1	1/5
C_7	3	5	1

Table 5 The judgment matrix of layer B_4-C

B_4	C_8	C_9	C_{10}
C_8	1	3	2
C_9	1/3	1	1/2
C_{10}	1/2	2	1

Table 6 The matrix ranking of *B*-*C* layer

B / *C*	B_1 0.146	B_2 0.0605	B_3 0.542	B_4 0.251	*w*
C_1	0.750				0.1095
C_2	0.250				0.0365
C_3		0.667			0.0404
C_4		0.333			0.0201
C_5			0.321		0.174
C_6			0.107		0.0580
C_7			0.571		0.309
C_8				0.540	0.1361
C_9				0.163	0.0409
C_{10}				0.297	0.0745

2. The matrix calculation of layer *B*-*C*, similarly available
   The matrix of layer (Table 6):

$$B_1 - C, w = (0.750,\ 0.250); \qquad B_2 - C, w = (0.667,\ 0.333);$$
$$B_3 \text{ - } C, w = (0.321,\ 0.107,\ 0.571); \quad B_4 \text{ - } C, w = (0.540,\ 0.163,\ 0.297)$$

Third step is to rank the results

Examine consistency of layer *C* and then get that all meet requirements.

Rank the results according to the importance and then get three main risk factors: organizational structure adjustment risk, management thought transformation risk and implementation team organization risk. The powerful control around these risk factors can avoid financial losses and resources waste.

4 Conclusions

This paper is committed to using analytic hierarchy process to research risk factors in the process of ERP implementation, so as to identify the major risks and control them, which will help companies minimize risks. The case of TISCO is used to show that how to use analytic hierarchy process, sort these risks according to importance and identify the key control objects, which can help companies avoid loss that is due to the immediate launch of project. ERP implementation requires a lot of manpower, financial and material resources, if risk assessment is ignored, companies will bear huge losses when they fail. A scientific risk assessment combined with practical companies summary and analysis can provide great convenience to the next risk control. What's more, companies can't grasp all risk factors that affect ERP implementation, so the effectiveness of control measures becomes very important, which can help us identify high probability of risk factors and sort them, and then get the key research objects, where is the value of this study.

References

1. Yanfang Cheng (2009) Management research of ERP implementation risk in iron and steel enterprise. China University of Geosciences, Beijing
2. Yuesheng Zhang (2009) Risk management study of enterprise' ERP project implementation. Jiangsu University, Beijing
3. Zhong Liu (2008) Based on the analysis of internal risk in the project of ERP risk factors and countermeasures [J]. Sci Technol Info Dev Econ 18:166–168
4. Mingyang Qi (2009) The study which based in the theory of fuzzy of risk assessment in ERP project of medium-sized enterprise. Wuhan University, Wuhan
5. Juping Liang (2006) The analysis and control of risk in the ERP project management. Master's thesis, Huazhong Normal University, Wuhan
6. Yan Zhang (2006) The risk management research of ERP implementation. Jiangsu University, Zhenjiang

Research on Performance Evaluation of IT Projects Based Value Management

Wang Xindi, Chen Li, and Sheng Fushen

Abstract It is very important to evaluate the performance of IT projects. Based on the value management of IT projects, this paper analyzes critical success factors, designs a prototype according to the overall balance and project life-cycle to evaluate the performance of IT project, establishes procedures of evaluation include preparation, implementation, dispute, documentation, monitoring. The results of this paper will help to establish monitoring mechanisms of IT projects to improve the effectiveness of resource allocation and investment returns.

Keywords IT project • Performance evaluation • Value management

1 Introduction

With the development of information technology, more and more enterprises implement many IT projects in all fields. It is very important to improve performance of the IT projects.

The enterprises take advantage of modern information technology and resources through the development and wider use of deepening continuously to improve business decisions, development, management, property management efficiency, and improve economic efficiency and market competitiveness. Business can be assisted through project management and decision-making and accurate grasp of

W. Xindi
School of Economics and Management, Beijing Jiaotong University,
Beijing 100044, P.R. China
e-mail: xdwang@bjtu.edu.cn

C. Li
School of Electrical Engineering, Beijing Jiaotong University, Beijing 100044, P.R. China

S. Fushen (✉)
Service School, Tianjin Campus of Naval Engineering University, Tianjin 110802, P.R. China

Z. Zhang et al. (eds.), *LISS 2012: Proceedings of 2nd International Conference on Logistics, Informatics and Service Science*, DOI 10.1007/978-3-642-32054-5_44,

market information in a timely manner in order to gain more business opportunities, improved the response capacity of the market, promoted the accuracy and predictability of decision-making in order to enhance greatly the competitiveness of enterprises.

According to the results of evaluation, the enterprises can adjust the investment field and direction in order to improve the efficiency in the use of funds, the results of the evaluation can guide the sector to increase attention to improve the quality of scientific research [1]. It is very important to promote of IT projects management, and improve efficiency in the end.

This paper will discuss the overall framework to evaluate IT project performance from three dimensions such as the critical success factors of perspective, the evaluation of the balance point of view, the process of project performance perspective.

2 Analysis of the Value Management of IT Project

The purpose of evaluation will be different from all kinds of IT projects. Value management is a powerful management tool to ensure the smooth evaluation progress to make sure of the value of specific IT projects.

In particular, the two main lines of specific performance focus on four areas: to promote application of the results, to improve project benefits, to strengthen project management, to meet the expectations of all parties. To promote the application of these results will be shown to expand market share and to promote social progress, to improve project benefits mainly refers to direct benefits, indirect benefits, social benefits etc. To strengthen project management covers the life cycle of projects included the opening of project, the implementation of projects, the end rollout of the project etc. To meet performance expectations for the projects means not only to enhance the strength of the parties, but also meet the expectations of all parties. For details, Please see Value Management Map of IT projects (Fig. 1).

3 A KPI Prototype Design for Performance Evaluation of IT Project

It can achieve resources input and configuration effectively through the IT project performance evaluation, then the appropriate policies and incentives system will be established to ensure the information technology to promote vigorously. However there must be a set of KPI system to support.

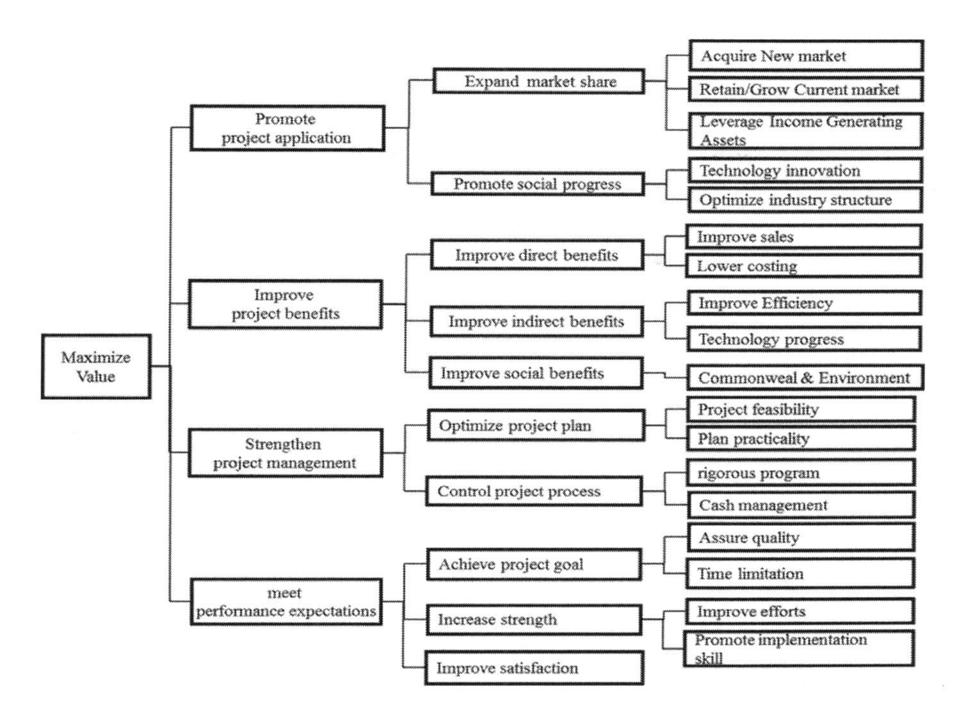

Fig. 1 Value management of IT projects

3.1 Based the Overall Balance of IT Value

Recently, assessment of the value of IT, or information technology project performance evaluation, demonstrated the effectiveness of the Balanced Scorecard implementation. Because lack of the overall balance is always a problem for the comprehensive performance evaluation of IT investment projects, we can use the Balanced Scorecard thinking to the project performance evaluation. In the process of implementation, more new elements should be added on the basis of carrying forward the tradition of the Balanced Scorecard.

Performance evaluation of IT investment projects not only emphasize one indicator, but stressed that the balance of various indicators, such "balance" embodied in several areas as short-term and long-term, financial and non-financial, process and result, internal and external [2].

3.2 Focus on Life-Cycle of IT Projects

There are four phases in the life cycle of projects such as project build, project implementation, project rollout, project tracking. The purpose of evaluation and testing standards should be differed from the every phases [3].

In the phase of project build, the key objective of project evaluation is to test technological innovation and practicality, feasibility of the research program, technical strength and research base is expected to prospect, for meaningful project, technical program, projects such as the implementation of conditions Feasibility study related factors; mid-term evaluation mainly focused on the project control, the results of an evaluation and recommended follow-up work, from the implementation of projects in the state to identify the main Problems, the continued implementation of the project and possible conditions, and the original target of bias, and make further improvements, to find a solution; evaluation of the latter part of the application of the performance evaluation, usually after the end of the project, the main evaluation criteria Consideration of intellectual property rights, patents and technology applications such areas as well as potential social and economic benefits, the main purpose is for the quality of projects have been completed, the results of output, whether the targets are met, the potential social impact assessment; tracking evaluation generally refers to The completion of the project after a period of time (sometimes for years) the evaluation, with emphasis on technological innovation and integration, key technology breakthroughs and master, the output of independent intellectual property rights, technical standards development, economic and social benefits, and so on.

3.3 A KPI Prototype for Performance Evaluation of IT Project

It is a comprehensive work to build the prototype of performance evaluation of IT projects. The prototype of performance evaluation should be based on the value analysis of such projects, drawing in the Balanced Scorecard performance management concepts and advanced process management thinking, cross-cutting projects throughout the life cycle analysis of the key success factors (CSF) and to achieve the interaction between the various elements of balance, then ensure multi-dimensional and all-round performance evaluation [4].

A KPI prototype of performance evaluation brings out as Table 1.

4 Process Design of Performance Evaluation

Evaluation process is a detailed activities description from preparation to evaluation report submitted throughout the various stages of performance evaluation life cycle, is also the basic way to carry out evaluation of the specific description. There are not specific evaluation procedures expressly provided in China, so random changes will be taken place in project evaluation activities.

For IT project evaluation, it is the most important to develop evaluation procedures as soon as possible in order to ensure evaluation activities have a legal framework for all participants. Any parties, including project evaluation

Table 1 A prototype of KPI of IT project

Aspects	CSF	KPI	Quan./qual.	Phase
Marketing & customer	Marketing	Market share of the project results	Quantitative	Project tracking
		Product return rate	Quantitative	Project tracking
	Customer	Customer retention	Quantitative	Project tracking
		Customer rating	Quantitative	Project tracking
	Satisfaction of shareholders	Satisfaction of investors	Qualitative	Project implementation
		Leadership satisfaction index	Qualitative	Project implementation
		Authorities satisfaction index	Qualitative	Project implementation
Finacial	Direct economic benefit	Revenue increasing per million expenditure	Qualitative	Project rollout
		Cost reduction	Quantitative	Project tracking
	Indirect economic benefit	The degree of automation technology and equipment	Quantitative	Project rollout
		Flexibility in new product sales	Quantitative	Project tracking
Process	Pre-project conditions	Reasonability of project objective	Qualitative	Project build
		Possibility of delivery	Qualitative	Project build
	Project management	Integrity of management system	Qualitative	Project implementation
		Level of quality management	Qualitative	Project implementation
	Fund management	The implementation rate of use of funds	Qualitative	Project implementation
	Target completion	Target completion rate	Qualitative	Project rollout
		The timeliness of project completion	Qualitative	Project rollout
Innovation & development	Technical innovation	Income ratio of new products	Quantitative	Project tracking
	Patent	Number of granted patent per million expenditure	Quantitative	Project rollout
	Research	Number of publication per million expenditure	Quantitative	Project rollout
	Awards	Number of major achievement per million expenditure	Quantitative	Project rollout
	Staff development	Number of employments per million expenditure	Quantitative	Project tracking
	Resource utilization	Reduction of energy consumption	Quantitative	Project tracking
	Environmental impact	Environmental quality index	Quantitative	Project tracking

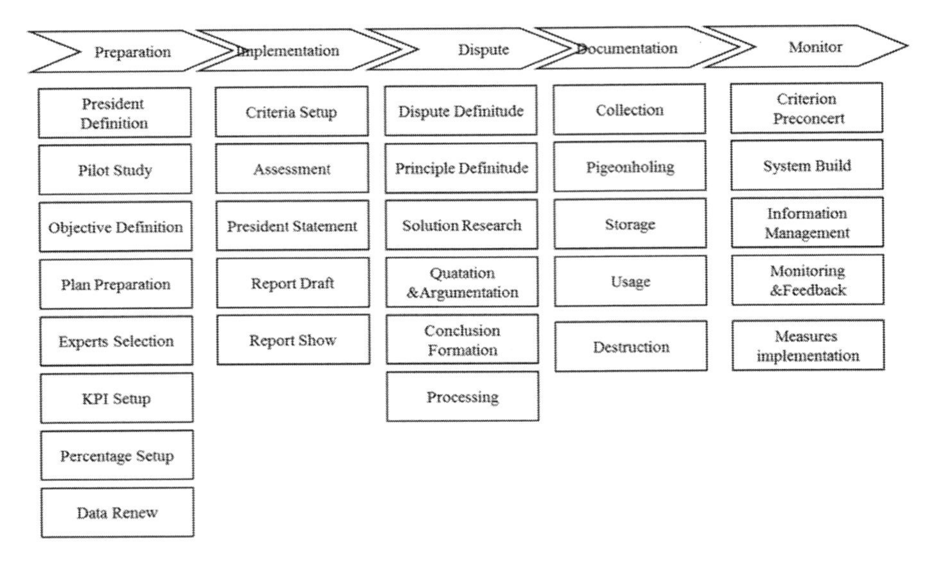

Fig. 2 The performance evaluation of IT projects

commissioned party cannot breach. It should be punished for those under the guise of evaluation in the name self-interest, or corrupt persons in the evaluation activities. Evaluation activities must follow the standard procedures, it is necessary to reflect the impartiality to ensure the standard evaluation process [5].

Performance evaluation of IT project can be designed into five basic processes such as preparation, implementation, dispute, documentation, and monitor. Each basic process includes multiple operating activities [6]. In details, please see the performance evaluation process (Fig. 2).

Preparation: It is the most important process during the procedure of performance evaluation. The suitable president will be defined firstly, then president organize to define the objectives and plan of this evaluation case, experts will be selected from resource pools according to requirements of evaluation. It is another important task to set up KPI and percentage for this case. Finally, data collection will be taken up to match the evaluation framework built before [7].

Implementation: It is in the priority to setup criteria in this process, assessment report will be draft after experts group evaluate this case in terms of evaluation framework and criteria.

Dispute: It is a mechanism to deal with arguments happened in the procedure of evaluation.

Documentation: It is necessary to record the evaluation procedure, documentation focuses on collection, pigeonholing, storage, usage, destruction, which covers life cycle of documents management.

Monitor: By monitoring the evaluation, organization can affirm the achievements, sum up experiences, identify problems and improve measures to improve and develop a sound evaluation system.

5 Conclusions

The above analysis provides a performance evaluation method of IT projects. Although the framework has a certain degree of universality, according to the difference of project's investor, features of technology, project's objectives and content, there will be inevitably the specific elements in practical application in the future, however it is very important to analyzer performance of IT projects based value management, the specific KPI can be selected form balanced aspects, the procedures also can be processed through evaluation cycle.

Acknowledgments The research was supported by the key project of logistics management and technology lab.

References

1. Ke Jian (2007) Research on IT performance evaluation based on BSC [J]. Mod Manuf Eng 4:23–24, J
2. Sun Jie (2010) Research on the PPP project performance appraisal [M]. Economic Science Press, pp 66–67
3. HUA Bin, HE Li (2007) Research on management and decision model for S&T project assessment & acceptance [J]. Sci Sci Sci Technol Manag 2:42 (Ch)
4. Yu Benhai (2009) Study on IT project performance evaluation and process improvement model [M]. Publishing House of Electronics Industry, pp 121–123 (Ch)
5. Xiao Li (2004) Research on the classification, objectives and procedures of the national technological project evaluation [J]. Technol Manag 3:12–15 (Ch)
6. JohnWard (2007) Strategic planning for information systems [J]. China Machine Press 2:37–38
7. Paul Sanghera (2011) Fundamentals of effective program management [M]. J Ross Publishing, pp 137–139

A Declarative Approach for Modeling Logistics Service Processes

Ying Wang, Lei Huang, and Yi Guo

Abstract A methodology integrating the declarative and imperative process modeling approach is proposed for modeling the logistics service processes. The characteristics of the logistics service process are discussed in detail. Through investigation of the service departments and services involved, the process is decomposed into loosely-structured parts modeled by a declarative approach and highly structured parts modeled by an imperative approach. The loosely-structured and highly-structured parts are then integrated to form the model for the logistics service process. Finally, the model is verified through process execution. A case study is elaborated for a Chinese bulk port.

Keywords Logistics service process • Process modeling • Declarative approach • Imperative approach

1 Introduction

The logistics industry is considered as a classic example of the service-based industries, which hold an increasingly dynamic and pivotal role in today's knowledge-based economies [1]. As a process-oriented business [2], logistics service is defined as the process of delivering products and/or services to customers, in a way that creates added value to customers [3]. There is a widespread recognition of the importance of correctly modeling service processes as an effective way to provide a comprehensive understanding of the process and to enhance the service quality to clients [4]. However, surprisingly few articles published in this respect for logistics service process modeling.

Y. Wang (✉) • L. Huang • Y. Guo
School of Economics and Management, Beijing Jiaotong University,
100044 Beijing, P.R. China
e-mail: yingw2002@sina.com; lhuang@bjtu.edu.cn; 11113164@bjtu.edu.cn

Z. Zhang et al. (eds.), *LISS 2012: Proceedings of 2nd International Conference on Logistics, Informatics and Service Science*, DOI 10.1007/978-3-642-32054-5_45,

This paper aims to construct the logistics service process model using the integration of the declarative and imperative approach. The remainder of this paper is structured as follows: Sect. 2 provides the state of the art of service process modeling in logistics. Followed by, the methodology is discussed in Sect. 3. Section 4 elaborates the case study for a port. Finally, Sect. 5 concludes the paper and discusses the future work.

2 Service Process Modeling in Logistics: State of the Art

2.1 Port Logistics Service Process

We identify the basic characteristics of logistics service as: Firstly, better service performance is achieved through offering a wide variety of logistics services [5]; Secondly, logistics is an integrated process that involves a network of organizations; Thirdly, logistics service is a process that creates value [4, 6].

Port logistics service process can be divided into primary processes which satisfy external customers of the port and support processes which satisfy only internal customers [7]. Primary processes are the core processes that create the core competencies of the port. The primary processes for port logistics service are often loosely structured and cross administrative boundaries, requiring high degree of flexibility. On the other hand, the processes inside a specific administrative organization are often highly-structured, requiring strict specification of the service procedure.

2.2 Imperative and Declarative Process Modeling Approaches

Imperative process modeling approach, such as Petri net and YAWL (Yet Another Workflow Language), strictly specifies how the process will be executed and is considered not flexible enough to support loosely-structured processes. In contrast, declarative approach specifies a set of constraints as business rules that should be followed during the execution so as to support loosely-structured process models [8].

3 Methodology

In this section, we present a methodology for modeling the logistics service processes by decomposing the processes into highly-structured and loosely-structured parts (see Fig. 1).

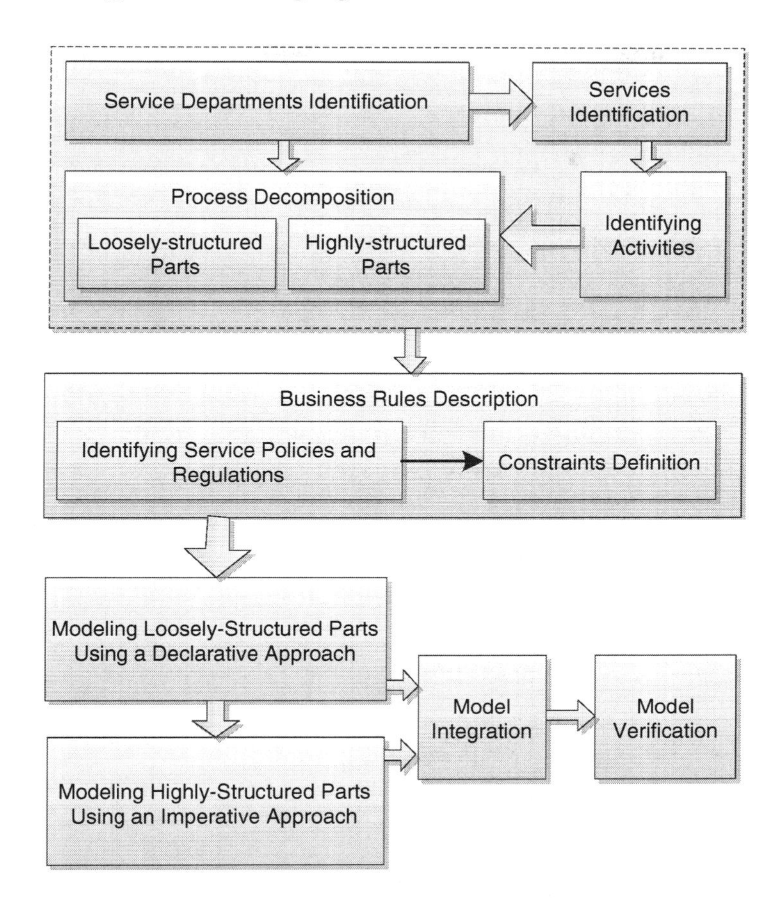

Fig. 1 The methodology roadmap for modeling the logistics service processes by decomposing the processes into highly-structured and loosely-structured parts

First of all, *the service departments and services conducted by each department are investigated*. The service process can be decomposed into the loosely-structured and highly-structured parts according to the service organization structure.

Next, *service policies and regulations and the corresponding business rules are investigated and summarized*. Integration of complex process logic into process models as business rules using a declarative approach [9] would support dynamic changes and provide a set of construction rules that enforce business constraints and policies [10] for the service processes.

Thirdly, we *construct the model* of the loosely-structured parts using a declarative approach while the highly-structured parts using an imperative approach. These parts are then mixed together to form an integrated process model for the complex logistics service process.

Lastly, *the model is verified* against dead activities and conflicting constraints by analyses of the states during process execution.

4 A Case for a Chinese Port

4.1 *Scenario*

Services for inward and outgoing cargo handling constitute the core logistics services for the port. The service departments involved in this process are identified as Port Entrance, Ship Scheduling Department, Human Resource Department, Business Managers, Commercial Document Administration, Wagon Balance Department, Warehouse Department, Cargo Inspector, and Accounting Office. The process is then decomposed into loosely-structured and highly-structured parts according to the service organization structure.

4.2 *Service Policies and Regulations*

Some of the service policies and regulations and the corresponding business rules in the inward cargo handling service process are investigated and summarized in Table 1.

4.3 *The Model*

Firstly, we construct the model for the loosely-structured parts of the process using a declarative approach (see Fig. 2).

Next, the highly structured parts of the process are modeled by YAWL (see Fig. 3).

Table 1 Except of constraints defined in the service process model

Rule ID	Service policies and regulations	Constraints
C. 1	A contract has to be signed	*Exactly1* ("*Contract Signing*")
C. 2	Billing terms must be appointed after the contract signed	*Succession* ("*Contract Signing*", "*Appoint Billing Terms to the Tally Book*")
C. 3	Loading certificate has to be created after the contract signed	*Succession* ("*Contract Signing*", "*Loading Certificate Creation and Checking*")
C. 4	Billing terms must be appointed to each tally book.	*Succession* ("*Tally Book Input*", "*Appoint Billing Terms to the Tally Book*")
C. 5	Ship arrivals must be reported before the ship scheduling	*Succession* ("*Ship Arrival Report*", "*Ship Scheduling*")

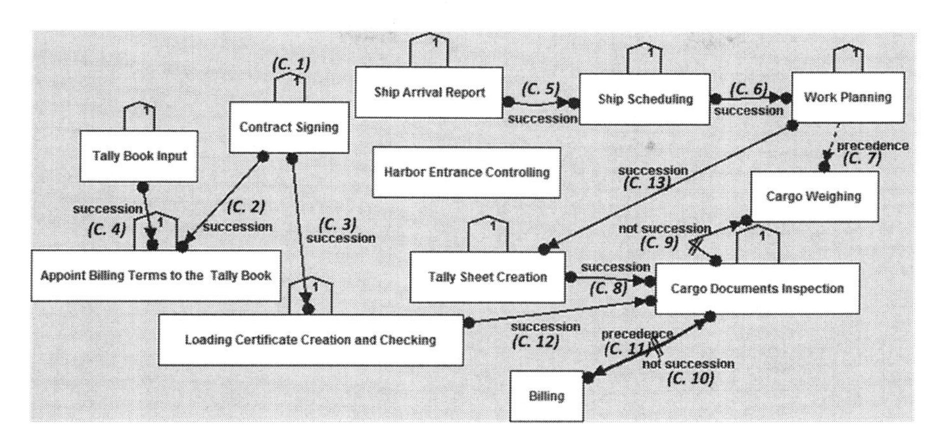

Fig. 2 Modeling loosely-structured parts of the process using DECLARE

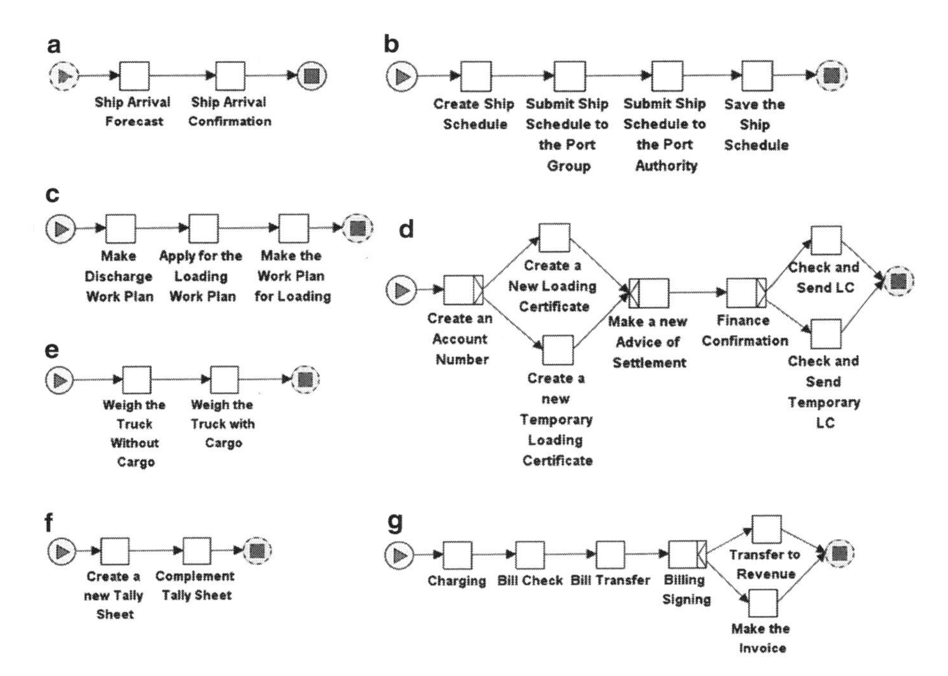

Fig. 3 Modeling highly-structured parts of the process using YAWL. (**a**) Ship arrival report. (**b**) Ship scheduling. (**c**) Work planning. (**d**) Loading certificate creation and checking. (**e**) Cargo weighing. (**f**) Tally sheet creation. (**g**) Billing

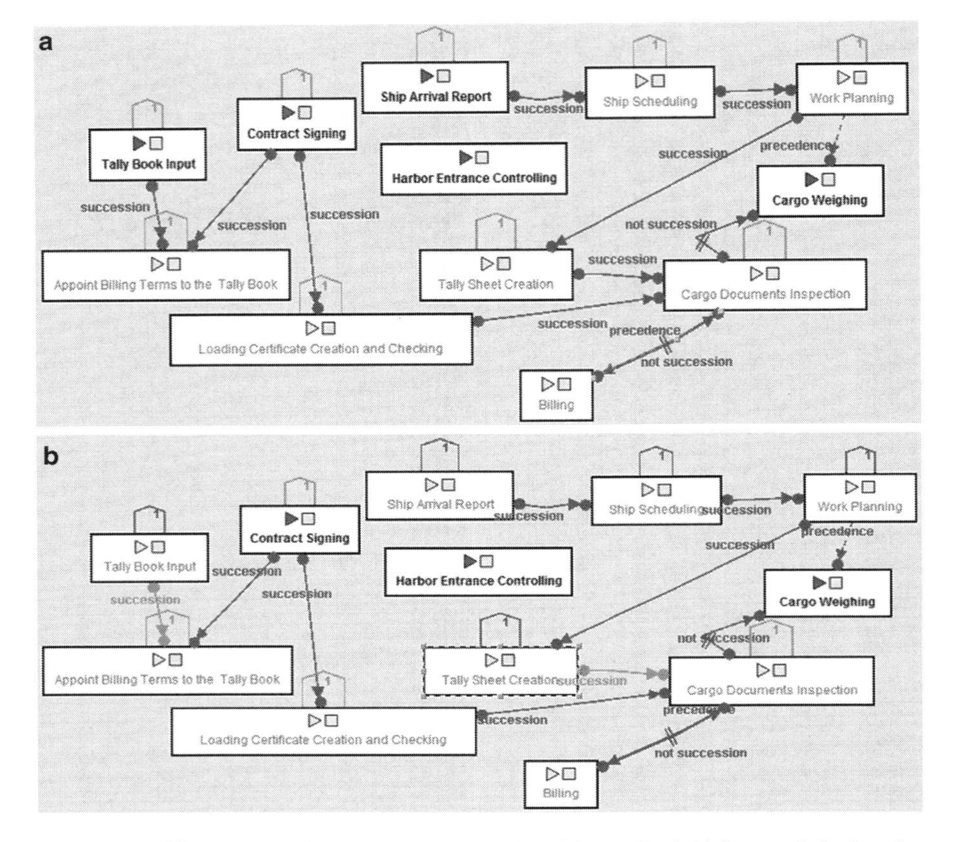

Fig. 4 The different state of the loosely-structured model. (**a**) The initial state of the loosely-structured model. (**b**) "Loading certificate creation and checking" cannot be enabled without the contract

4.4 Model Verification

Fig. 4a shows the initial state of the loosely-structured model. Further analysis shows that this model provides good support for both compliance and flexibility for the logistics service process. For example, serious violation of the business rules, such as Loading Certificate creation without a contract cannot be enabled during the process execution (see Fig. 4b).

5 Conclusion and Future Work

In this paper, we present a methodology for modeling the logistics service processes by integrating the declarative and imperative process modeling approach. A case study is elaborated for a Chinese bulk port. Through investigation of the service departments and services involved, the process is decomposed into loosely-structured parts modeled by a declarative approach and highly structured parts modeled by an imperative approach. Future work could be conducted for conformance checking of the logistics service process using the model as a reference model.

Acknowledgements This research is supported by the Natural Science Foundation of China under Grant Nos. KBA212004533. In addition, the authors would also like to thank the anonymous reviewers for their valuable comments.

References

1. Chapman RL, Soosay C, Kandampully J (2003) Innovation in logistic services and the new business model: a conceptual framework. Int J Phys Distrib Logist Manag 33(7):630–650
2. Rutner SM, Langley CJ Jr (2000) Logistics value: definition, process and measurement. Int J Logist Manag 11(2):73–82
3. Van der Veeken DJM, Rutten WGMM (1998) Logistics service management: opportunities for differentiation. Int J Logist Manag 9(2):91–98
4. Aguilar-Saven RS (2004) Business process modelling: review and framework. Int J Prod Econ 90(2):129–149
5. Lai K (2004) Service capability and performance of logistics service providers. Transp Res Part E Logist Transp Rev 40(5):385–399
6. Mentzer JT, Rutner SM, Matsuno K (1997) Application of the means-end value hierarchy model to understanding logistics service value. Int J Phys Distrib Logist Manag 27(9/10):630–643
7. Hill AV et al (2002) Research opportunities in service process design. J Oper Manag 20 (2):189–202
8. Pesic M, Schonenberg H, van der Aalst WMP (2007) DECLARE: full support for loosely-structured processes, enterprise distributed object computing conference, 2007. EDOC 2007. 11th IEEE International. p 287, 15–19 Oct 2007
9. Lu R, Sadiq S (2007) A survey of comparative business process modeling approaches. Springer, Berlin/New York
10. Zur Muehlen M, Indulska M (2010) Modeling languages for business processes and business rules: a representational analysis. Info Syst 35(4):379–390

A Model for the Door-to-Airport Passenger Transportation Problem

Yuan Kong, Jianbing Liu, and Ji Yang

Abstract This paper addresses Door-to-Airport Passenger Transportation Problem (DAPTP) in flight ticket sales companies in China, where passengers call free-get-to-airport-ride service a few days in advance to schedule a ride. The problem is formulated under the framework of the fleet size and mix vehicle routing problem with time windows (FSMVRPTW). A mixed integer programming model is presented, and service quality is factored in constraints, by introducing passenger satisfaction degree functions that consider time window and ride time.

Keywords Passenger transportation • Door-to-airport • Service quality

1 Introduction

Chinese made more than 621 million trips by air in 2011, up 10%, and 53 airports had more than ten million trips [1]. It is forecast that airline trips will rise to one billion in 2020. The dramatic growth in the market provides a more competition in air transportation than ever. A large number of Flight Ticket Corp. (FTC) engaging

Y. Kong (✉)
Key Laboratory of Urban Operation and Management,
Beijing Academy of Science and Technology, Beijing, P.R. China

Wireless Technology Innovation Institute, Beijing University of Posts
and Telecommunications, Beijing, P.R. China
e-mail: yokery@126.com

J. Liu
Key Laboratory of Urban Operation and Management,
Beijing Academy of Science and Technology, Beijing, P.R. China

J. Yang
Wireless Technology Innovation Institute, Beijing University of Posts
and Telecommunications, Beijing, P.R. China

Z. Zhang et al. (eds.), *LISS 2012: Proceedings of 2nd International Conference on Logistics, Informatics and Service Science*, DOI 10.1007/978-3-642-32054-5_46,

in sailing tickets through outlet, telephone and internet was activated. The agency not only sails ticket, but also provides services that satisfy certain customer needs, such as in-house print-on-demand feature, convenience in terms of seat availability and aircraft type. Despite pricing as the major competitive variable, it is not always as the responsive to demand changes. Therefore, to increase FTC's market shares, adding a sense of uniqueness in satisfying customer needs after high quality market research has become more important as the strengthened competitive environment.

To get to airport, travelers depart from downtown where could be 1,000 miles away from airport. In China, only Beijing and Shanghai provide a high-speed railway linked service. The large portion of travelers has to drive many miles from and to airport by bus links. Passengers will spend more time for transportation. Add to the fact of complicated security at airports and customary flight delays, and air travel does not look very appealing.

As a consequence of these impediments, the trend seems that some of the target markets easily lose and more Chinese prefer traveling by train rather than by air.

These characteristics in China have promoted FTCs to intensify their efforts in offering better service to passengers. At this point, several FTCs provide "free-get-to-airport-ride" service either independently or cooperating with a third-party logics company. Zhong Shan FTC in Shenyang is the first.

The availability of the new free ride service suggests a new passenger transportation business: door-to-airport, an on-demand service in which travelers call 1 day or a few days in advance to route and schedule transportation. The advantages of such a system are obvious. The service, which land closer to where they live, gives travelers the option of using the convenient rides, without cost, much time consuming, early or late arrival, and heavy baggage transit. By sharing the car with other passengers, aggregation can greatly increase profits while still ensuring a very convenient service.

For business to be profitable, each revenue ticket should pay for a load of a passenger ride. The key to reduce such load factors is approximately optimized routing systems.

The free-get-to-airport-ride has generated a lot of media interest, such as NanFang Daily, Yang cheng Evening News, sina.com, sohu.com, much, much, more.

To effectively manage operations on-demand, to-airport ride service. An offline routing system to construct minimum operation cost for a period of time is a key component.

In this paper, we introduce a mixed-integer linear program (MILP) with service quality constraints for offline DAPTP.

2 Problem Description and Model Formulation

The door-to-airport problem is concerned with routing and scheduling a set of requests to airport for airline passengers during a period (typically, a day). A request specifies an origin address, a desired time window at the origin, a latest time-limit at the airport, a desired excess ride time between origin and airport, and the number of passengers. In this schedule, waiting is allowed if vehicles arrive early. The desired time windows and the desired excess ride times are hard, but these constraints, as behaviors of quality of service, can be violated with an endurable duration. Vehicles of a given fleet depart from a common depot and end at an airport. Each vehicle has a seat capacity limitation. A dispatcher has to decide which vehicles to use to satisfy all demands and what routes and schedules will be, i.e., the sequencing of stops and associated departure times.

The DAP is an example of pickup and delivery problem dealing with passenger transportation, which is known in the name of dial-a-ride problems. Dial-a-ride problems have received a fair amount of attention in the vehicle routing literature. Cordeau and Laporte [2] provide overviews of dial-a-ride problems. Even though door-to-airport and dial-a-ride have many common characteristics in controlling user inconvenience, there are some notable differences. The dial-a-ride problem often arises in social services contexts, e.g., transportation of the elderly, whereas the door-to-airport problem is encountered exclusively in business settings. As a result, there tends to be less flexibility in specification of requests and takes more on cost saving factors. Secondly, since air passengers always travel together with his families or partners, the demand-weight more than one in a request is available. Furthermore, in door-to-airport environments, for safety or scheduling reasons, a vehicle is supposed to terminate at the destination, airport, with one trip, whereas in dial-a-ride environment a vehicle can scheduled until the shift time met. Other relevant dial-a-ride papers include [3–5].

Following the traditional Vehicle Routing Problem (VRP), this study defined a graph $G = \{N, A\}$, where $N = \{0, 1, \ldots, n + 1\}$. Nodes 0 and $n + 1$ represent the depot and the airport, respectively. $C = \{1, \ldots, n\}$is the set of requests. $A \subseteq N \times N$ represents the travel possibilities between nodes. Normally $N \times N$ is complete; except that no other arcs are incident to node 0 and no arcs are incident from node $n + 1$. $V = \{1, \ldots, K\}$ is the set of available vehicles. There is a travel cost c^k_{ij}, $\forall(i, j) \in A$, $\forall k \in V$ that depend on the vehicle. There are vehicle-specific acquisition costs F^k and capacities Q^k. t_{ij} is the vehicle independent travel time between nodes. Each request i is associated with a nonnegative load q_i ($q_0 = 0$, $q_i > 0$, $i \in C$ and $q_{n+1} = 0$), a desired service start time window at the origin $[e_i, l_i]$, and a deadline arrival at the airport d_i. Finally, let γ ($\gamma \geq 1$) denotes the weight of excess ride time over the Direct Airport Time (DAT, $t_{i,n+1}$) for all requests.

Service quality scenarios constrained by time window and excess ride time are considered. A linear delay of the latest arrival time and excess ride time equation are used as follows:

$$l_{\beta i} = l_i + U_t * (1 - \beta),\ i \in \mathrm{C} \text{ and } \gamma_\beta = \gamma + U\gamma * (1 - \beta) \tag{1}$$

Where U_t and U_γ are the relaxation widths of time windows and excess ride time, and β ($0 \leq \beta \leq 1$) denotes service quality.

We formulate the problem base on the well-known vehicle flow formulation for the VRP [6]. Hence, there are two types of decision variables in the formulation. For each triplet (i, j, k) with $(i, j) \in A$, $k \in V$, x^k_{ij} is a binary decision variable that expressed whether vehicle k travels directly from node i to node j. For each pair (i, k) with $i \in N$, $k \in V$, y_i^k is a real variable that determines the exact departure time at this node if it is served by the vehicle.

Using these variables and a very large positive constant M (see for Bazaraa et al. [7]), door-to-airport problem under service level β can be formulated as follows:

$$Min. \sum_{k\in V}\sum_{j\in C} F^k x^k_{oj} + \sum_{k\in V}\sum_{(i,j)\in A} c^k_{ij} x^k_{ij} \tag{2}$$

$$s.t \sum_{k\in V}\sum_{j\in N} x^k_{ij} = 1, \quad \forall i \in C \tag{3}$$

$$\sum_{i\in C} q_i \sum_{j\in N} x^k_{ij} \le Q^k, \quad \forall k \in V \tag{4}$$

$$\sum_{i\in N} x^k_{iu} - \sum_{j\in N} x^k_{uj} = 0, \quad \forall u \in C, k \in V \tag{5}$$

$$y^k_i + t_{ij} - y^k_j \le \left(1 - x^k_{ij}\right)M, \quad \forall (i,j) \in A,\ k \in V \tag{6}$$

$$e_i \sum_{j\in N} x^k_{ij} \le y^k_i \le (l_i + (1-\beta)U_t) \sum_{j\in N} x^k_{ij}, \quad \forall i \in C, k \in V \tag{7}$$

$$y^k_{n+1} - y^k_i \le (\gamma + (1-\beta)U_\gamma) t_{i,n+1} \sum_{j\in N} x^k_{ij}, \quad \forall i \in C, k \in V \tag{8}$$

$$y^k_{n+1} \le d_i \sum_{j\in N} x^k_{ij}, \quad \forall i \in C, k \in V \tag{9}$$

$$x^k_{i0} = x^k_{n+1,i} = x^k_{i,i} = 0, \quad \forall i \in N, k \in V \tag{10}$$

$$x^k_{ij} \in \{0, 1\}, \quad \forall (i,j) \in A, k \in V \tag{11}$$

In the objective (2), the first linear term represents the sum of used vehicle costs, whereas the second nonlinear term is the arc travel cost.

Constraint (3) states that all customers must be visited by a vehicle; inequality (4) gives capacity constraint for each vehicle; constraint (5) is the flow balance. Constraint (6) guarantees that the arrival times at two consecutive requests allow for service and travel time. Constraints (7 and 8) make sure the service level is no less than β and (9) is the deadline constraint at the airport. Constraint (10 and 11) is a binary constraint.

3 Conclusions

This paper proposes a mixed-integer programming model for the offline mix fleets DAPTP. The proposed model can be used by flight service organizations providing Door-to-airport passengers transportation. They incorporated the important features of the service quality constraints with vehicle routing and scheduling.

Acknowledgments This research is financially supported by the Scientific Research Program of Beijing Academy of science and Technology (PXM2011_178215_000007), and International S&T Cooperation Program of China (ISTCP 2010DFA12780).

References

1. The civil aviation industry development statistical bulletin (2011) http://www.caac.gov.cn/I1/K3/201205/P020120507306080305446.pdf
2. Cordeau J-F, Laporte G (2007) The dial-a-ride problem: models and algorithms. Ann Oper Res 153(1):29–46
3. Calvo RW, Colorni A (2007) An effective and fast heuristic for the dial-a-ride problem. 4OR Q J Oper Res 5(1):61–73
4. Häme L (2011) An adaptive insertion algorithm for the single-vehicle dial-a-ride problem with narrow time windows. Eur J Oper Res 209(1):11–22
5. Paquette J, Cordeau J, Laporte G (2009) Quality of service in dial-a-ride operations. Comput Ind Eng 56(4):1721–1734
6. Cordeau J-F, Desaulniers G, Desrosiers J, Solomon MM, Soumis F (2001) VRP with time windows. In: Toth P, Vigo D (eds) The vehicle routing problem. SIAM, Philadelphia
7. Bazaraa MS, Jarvis JJ, Sherali HD (2005) Linear programming and network flows, 3rd edn. Wiley, New York

Analysis of the Influence Factor in Urban Residents Travel Based on Rough Set Theory

Liping Shao and Xiaodan Shi

Abstract To study the characteristics of urban residents travel is essential and of importance in designing urban transport system as well as to the city system reliability. Hence, in this paper, the major purpose is to propose a mathematical approach based on quantitative analysis of the combined weights in rough set method, which is to find the priority of influence factor in urban residents travel mode choice. As the result, we can not only get the rank of influence factor in urban residents travel mode choice, but also provide a new idea for extension the applications of rough set method.

Keywords Urban residents • Residents travel mode selection • Rough set • Combined weights

1 Introduction

Urban transport is an important part of complex and giant urban system. Traditional computational modeling approach is difficult to describe the selection process of the traffic way accurately [1, 2]. Therefore, this paper use rough set theory methods to establish the model in calculate of weights of the single factors and combined factors in urban residents travel mode choice, and provide scientific decision-making support for the urban transport construction and the actual operation and management [4–6].

L. Shao (✉) • X. Shi
School of Economics and Management, Beijing Jiaotong University,
Beijing 100044, P.R. China
e-mail: lpshao@bjtu.edu.cn; 11120700@bjtu.edu.cn

Z. Zhang et al. (eds.), *LISS 2012: Proceedings of 2nd International Conference on Logistics, Informatics and Service Science*, DOI 10.1007/978-3-642-32054-5_47,

2 To Obtain and Choose Influencing Factors of Urban Residents Travel Based on Rough Set Theory

2.1 Selection of the Influencing Factors

The project determined the travel choice index system of urban residents, as shown in Table 1, in which five factors are the travel time, travel price, trip distance, travel safety, travel comfort, a factor evaluation results is the chosen way to travel, through data analysis and pre-survey [3]. Travel time is influenced by accuracy of time, whether it is the peak travel, and waiting time. Trip distance consists of driving distance and transfer distance. Travel security includes driving safety and property security. Travel comfort is affected by whether there are seat, is crowding, the convenience of the station facilities and whether or not to carry bulky items.

2.2 Knowledge Representation of Influence Factors

Based on rough set theory, we use 40 resident records in the research survey as a research object, and delete data that can not be applied. Then assume the remaining data set is set $U = \{x_1, x_2, x_3, x_4, x_5 \ldots x_{27}\}$ and a, b, c, d, e represents the importance of time accuracy, the importance of travel costs, the importance of total travel distance, the importance of to travel security and the importance of travel comfort respectively, which constitute the condition attribute set of the factor analysis $C = \{a, b, c, d, e\}$, and constitute a set of decision attributes D.

Then, in the conditional attribute with the number '1' on behalf of the questionnaire options 'very unimportant', with the number '2', '3', '4' and '5' represents 'more unimportant', 'general', 'more important' and 'very important' respectively; Similarly, in the decision-making property, with a digital '0'on behalf of the travel way for the 'bicycle tube', the number '1', '2', '3' and '4' represents 'public transport vehicles', 'subway', 'Taxi' and 'self-driving' respectively. It can be constructed a two-dimensional information for decision-making table to describe the urban trips factors selecting knowledge representation system, as shown in Table 1:

3 Rough Set-Based Urban Residents Travel Factor Calculation and Analysis

3.1 Weights of Single Factors

According to the definition of rough set theory:

Assume condition attribute set $C = (a, b, c, d, e)$, the decision attribute set is D, then, $U/D = \{(x_7, x_{16}, x_{21}), \ldots, (x_9, x_{12})\}$ $Pos_C(D) = \{x_1, x_2, x_3, \ldots, x_{27}\}$

Therefore, $\gamma_c = Card(Pos_C(D))/Card(U) = 27/27 = 1$

Table 1 Knowledge representation system of influence factors in urban travel mode choice

	Condition attributes					Decision attribute
Serial number	a. Time accuracy	b. Travel costs	c. Travel distance	d. Travel security	e. Travel comfort	D. Travel mode
x_1	1	4	4	5	5	2
x_2	1	5	4	5	3	1
x_3	2	3	2	3	3	2
x_4	2	4	3	3	3	1
x_5	3	1	3	5	4	1
x_6	3	2	3	5	4	1
x_7	3	2	5	5	5	0
x_8	3	3	2	5	5	3
x_9	3	3	2	4	2	4
x_{10}	3	3	3	4	2	1
x_{11}	3	3	3	5	4	3
x_{12}	3	3	3	4	5	4
x_{13}	3	3	4	5	2	3
x_{14}	3	4	2	5	3	1
x_{15}	3	4	3	4	3	1
x_{16}	3	4	5	4	5	0
x_{17}	4	1	3	3	4	1
x_{18}	4	2	4	5	5	1
x_{19}	4	3	2	5	4	2
x_{20}	4	3	3	5	4	3
x_{21}	4	3	4	5	5	0
x_{22}	4	3	4	5	4	1
x_{23}	4	4	4	5	4	1
x_{24}	4	4	4	5	4	3
x_{25}	4	4	5	5	3	1
x_{26}	5	3	3	4	5	3
x_{27}	5	3	4	5	4	1

When we remove the property a, $U/(C-\mathrm{a}) = U/ind(b,c,d,e)$

Importance of attribute a is: $Sig_{C-a}(D) = 1 - \frac{Card(Pos_{C-a}(D))}{Card(Pos_C(D))} = 1 - \frac{23}{27} = \frac{4}{27}$

Similarly, the importance of attribute a, b, c, d, e are: 8/27, 7/27, 2/27, 6/27.

These are the calculated weight of individual factors according to the basic principles of rough sets. Taking into accounted the combination of factors similar to (a, b) and (a, c), decision-making results changes a lot.

3.2 *Weights of Combined Factors*

According to the definitions above, we can combine two properties even also be three or more, to calculate the combined weights. Hence, we can see that single influencing factors on the decision-making results impact is not great, but when combined with other attributes, decision-making results may change a lot.

When we remove the property (a, b),

$$U/(C-(a,b)) = U/ind(c,d,e)$$

Therefore, the importance of a combination of a and b is as follows:

$$Sig_{C-(a,b)}(D) = 1 - \frac{Card\left(Pos_{C-(a,c)}(D)\right)}{Card(Pos_C(D))} = 1 - \frac{12}{27} = \frac{15}{27}$$

Similarly, we have the importance of (a, c), (a, d), (a, e), (b, c), (b, d), (b, e), (c, d), (c, e), (d, e) are 17/27, 4/27, 14/27, 18/27, 10/27, 15/27, 10/27, 14/27, 9/27.

We can get the weight $W_{(a_i,a_j)}$ of binary combination through normalized:

$$W_{(a_i,a_j)} = \frac{Sig_{C-(a_i-a_j)}(D)}{\sum_{i=1}^{n}\sum_{j=i}^{n} Sig_{C-(a_i-a_j)}(D)} (i \geq j, i = 1, 2, \ldots, n, j = i, \ldots, n)$$

We can calculate weights of the binary combination as follows:

$$W_{(a,b)} = \frac{15}{126}, \quad W_{(a,c)} = \frac{17}{126}, \quad W_{(a,d)} = \frac{4}{126}, \quad W_{(a,e)} = \frac{14}{126}, \quad W_{(b,c)} = \frac{18}{126},$$

$$W_{(b,d)} = \frac{10}{126}, \quad W_{(b,e)} = \frac{15}{126}, \quad W_{(c,d)} = \frac{10}{126}, \quad W_{(c,e)} = \frac{14}{126}, \quad W_{(d,e)} = \frac{9}{126}.$$

It can be found through the above calculation; the binary combinations of attributes weights have the following properties:

1. $$W_{(a_i,a_j)} = W_{(a_j,a_i)} (i,j = 1, 2, \ldots, n)$$

2. $$W_{(a_i,a_i)} = Min\left\{W_{(a_i,a_j)}\right\} (i,j = 1, 2, \ldots, n)$$

The mathematical sense of the nature of 2 is that a single attribute weight is less than or equal to any combination of weights. This also shows that the above the rough sets approach to weight of binary combination is consistent with the objective reality.

3.3 Analysis of Calculation Results

Actual calculation results of single urban trips factor show that influencing factors weights descending order for the travel costs, the total travel distance, travel comfort, time accuracy and safety.

Actual calculation results of combination factors weights show that the weight of combination of travel costs and the total travel distance is the highest weight; And time accuracy and the total travel distance; lowest weight is the accuracy of time and travel security. Comparing q with the single factor, the combination weight of travel costs and the total travel distance is always the most important factors to affect the travel mode choice of urban residents, while the time accuracy as a single factor, but its combined total distance weights is larger.

4 Conclusion

In this paper, we applied rough set theory to quantitative analysis of combined weighting problems. And an example of main influence factor problem in urban residents travel mode choice was used to illustrate the proposed approach. We have two interesting results, one is in the main factors that influence the travel mode choice of residents, the most important influence factor is travel cost; this is the first contribution in our study. In addition, two properties are combined for our needs, three or more properties will be combined for further research, and this one is the second contribution in our study.

To sum up, we propose this new approach, which presented some contributions in traffic designing and management field. This study presents both theoretical and practical significance. However, the weak point still exists, such as the different weighting analysis method and c the extension of influence factors in the system.

Therefore, we suggest that we can combine relative methods, such as fuzzy or GA with our study, and get more experiment data. Then, the research can be developed in future research.

References

1. Bogers EAI, Viti F, Hoogendoorn SP (2005) Joint modeling of ATIS, habit and learning impacts on route choice by laboratory simulator experiments. Transportation research board annual meeting, Washington, D.C., 2005
2. Bowman John L, Ben-Akiva Moshe (2001) Activity-based disaggregate travel demand model system with activity schedules. Transp Res 35(1):1–28
3. Kalmanje S, Kockelman KM (2004) Credit-based congestion pricing: travel, land value and welfare impacts. Transportation research board annual meeting, Washington, D.C., 2004
4. Peers S, Jeong Whon Yu (2005) A hybrid model for driver route choice incorporating en-route attributes and real-time information effects. Netw Spat Econ 5(1):21–40
5. Liubing En, Juanzhi Cai, Liyan Ling (2008) Development of a multinomial logit model for trave mode choice of residents. J Highw Transp Res Dev, 1(5):120–124
6. Dell'Orco M, Circella G, Sassanelli D (2008) A hybrid approach to combine fuzziness and randomness in travel choice prediction. Eur J Oper Res 185(2):648–658

Research on Management Strategies of Reverse Logistics in E-Commerce Environments

Wenming Wang, Yan Liu, and Yingjie Wei

Abstract With the rapid development of e-commerce, online transactions are becoming more and more popular, which causes the flourish of logistics industry in turn. But the high rate of returning has become a great challenge to companies, thus leaving floor to considerations of the effects of reverse logistics. Starting from the connotation and characteristics of the reverse logistics in e-commerce, the article first probes into its causes, then analyzes the its current management problems in China, and finally proposes suggestive strategies for every management flaw respectively.

Keywords Electronic commerce • Reverse logistics • Management strategies

1 Introduction

The reverse logistics in e-commerce environment mainly refers to the reverse logistics about returns or exchange, namely the entity transfer process in which the downstream customers return the goods which don't comply with requirements of the order to upstream suppliers [1]. When products purchased online turn out to be are of inferior quality or not satisfying in other aspects, they will be returned to suppliers or manufacturers under their claims for returns or exchange from merchants. The specific flow of reverse logistics should include logistics flow, information flow, capital flow, and all or part of business flow [2].

Due to most of domestic merchants' ignorance of the reverse logistics currently, the problem about returns or exchange is always one of the largest contradictions

W. Wang (✉) • Y. Liu • Y. Wei
School of Economics and Management, China University of Petroleum,
Qingdao 266555, P.R. China
e-mail: wenmingw@upc.edu.cn; liuyanlacey@126.com; flyjie1111@gmail.com

Z. Zhang et al. (eds.), *LISS 2012: Proceedings of 2nd International Conference on Logistics, Informatics and Service Science*, DOI 10.1007/978-3-642-32054-5_48,

between the merchants and customers. Therefore, the development of reverse logistics is imperative in e-commerce.

2 Characteristics of Reverse Logistics

1. Uncertainty.
   Due to the uncertainty of the type, quantity, generation time and distribution of the reverse logistics to some extent, it's quite hard to predict and calculate a routine [3].
2. High Treatment Costs
   Scale benefits of storage and transport can't be used due to the random and sporadic product transfer, resulted from the uncertain source of refund and exchange.
3. Complexity
   The process of returns or exchange is flooded with complex treatments, variable processing forms and complicated structures, while each stage of the process is affected by the result of the reverse logistics planning.
4. Low Speed
   The tardiness usually represents itself in aspects of slow accumulative speed of logistics volume, the complexity of treatment process, etc. The whole process involves the artificial testing, judgment, classification, processing, and so on, which are laborious and time-consuming.

3 Cause Analysis of Reverse Logistics

1. Information asymmetry caused by the network shopping
   In the electronic commerce mode, it's difficult for customers to get a full view of the characteristics of the purchased commodities, for they are based on the visual perception of pictures or instructions got from Internet. The gap between the marketing information online and consumer expectations give rise to refund or exchange
2. Driven by competition among online merchants
   As the market competition becomes increasingly fierce, in order to attract more customers some online merchants tend to beautify goods photos or make overstatements of goods, while others vie in promoting preferential returning conditions such as "accepted return in case of dissatisfaction". These do work efficiently in touting customers, but result in kinds of problems in returning on the other hand.
3. Problems of Commodity
   Many conditions can cause the returns problems, such as the discrepancy between the actual products and description, flaws or quality problems, wrong

size, quantity and even goods, merchandise being close to or exceed the shelf life, damaged or wrong goods caused by negligence in logistics distribution, etc.

4. Instability of consumer preferences
   Blind shopping is apt to occur because of the characteristics of network shopping, for consumers cannot fully understand quality, function and other related information about products even before using. The purchase decision made out of curiosity or impulse, is prone to lead to requirements of returns.
5. Imperfections of the forward logistics
   A certain degree of damage to goods may occur in the process of transportation, loading and unloading, handling and distribution, such as the loss of components, expired goods, metamorphism caused by poor commodity preservation, and extension, damage or mismatch in the distribution process [4].

4 Problems Lying in the Management of Reverse Logistics

1. Online vendors do not attach enough importance to reverse logistics
   In the traditional operation mode of returns management, much attention of most online merchants is paid to attracting customers instead of highlighting returns management [5]. In case of reverse logistics, most businesses are in a daze of the process, while the big differences in terms of number, species and location increase the difficulty of returns management.
2. Lack of efficient returns working mechanism
   Costumers are concerned with the way in which returns problems are handled, so to a large extent the returns efficiency will influence customers' purchasing decisions. However, in China, the returns procedures are complicated and slow, costly and tricky.
3. Low processing efficiency and high cost
   At the present stage because of the lack of specialized third-party reverse logistics services, many of our enterprises have to build their own costly reverse logistics management systems. Excessive reverse logistics costs hinder enterprises' enthusiasm for reverse logistics management, and the lagging reverse logistics will inhibit consumers' purchasing desires in turn, thus reducing corporate profits at the end.
4. Poor service quality of returns logistics
   Returns logistics service is very poor at present. On the one hand it's hard for customers to get in touch with the customer-service staff, on the other hand even if the touch is successful it's still difficult to get a satisfactory exchanged processing result.

5 Management Strategies for Reverse Logistics

5.1 To Optimize the Online Trading Links to Reduce Avoidable Reverse Logistics

1. Try to provide real and detailed information to reduce the possibilities of wrong purchasing as well as returns; other measures such as the message board and virtual community should be established in website, in order to strengthen communication with consumers and overcome information asymmetry by providing evaluations, recommendations and questions.
2. The businessman should try to minimize the amount of returns derived from impulsive purchasing. When the order is created online, consumers still have the opportunity to reconsider or even to cancel their orders in a given period.

5.2 To Develop Reasonable Returns Policies

Businesses should implement positive returns policies for the overall interests of both parties. Firstly, rational return price, whether at a full refund or discount refund according to the original wholesale price, should be formulated to maximize the overall interests. Secondly, we must determine the best return ratio by allowing appropriate amount discount or price discount before the delivery, to reduce the uncertainty of returns, to better balance the costs and benefits. Thirdly, returns responsibilities should be made clear. Clear the ownership of the returned goods in the contracts signed by the manufacturers, suppliers and retailers, to avoid the disputes over powers and responsibilities among different parties. Finally, return policy should be listed in a prominent position on consumer shopping pages, making them clear to customers.

5.3 To Establish an Efficient Reverse Logistics Information System

It's necessary to establish a rational reverse logistics information system for the well running of reverse logistics. By establishing a perfect information system in use of network and information technology, we can store collected information in database and codes, systematize the commodity tracking, realize the electronic management of returns information, thus improving the reverse logistics management by realizing the bidirectional, timely and complete information exchange in the whole supplying chain from suppliers to customers. We can compile the code for every return reason, store information of return status and classification, track recovery process and conduct inspection and control of the returned items, and finally classify the disposition results, so as to make the information system an effective source for companies to get customer and market information.

5.4 Choose Appropriate Reverse Logistics Mode

Reverse logistics systems can be generally divided into self-supporting ones and third-party ones. Considering the characteristics of reverse logistics in e-commerce environments and the increasing development status of third-party reverse logistics providers, merchants can benefit a lot by taking advantage of the third-party reverse logistics.

Firstly, by outsourcing part of their non-core reverse logistics business to the TPL providers, merchants can effectively reduce costs, save money, and reduce risks. Meanwhile, companies can concentrate their focus on the key processes, which helps to improve their operational capacity, profits and transaction rates. Secondly, by means of the powerful information platform and advanced hardware and software systems of third-party reverse logistics providers, businesses can achieve real-time tracking, and reduce the online return rates by improving sales and after-sales services according to feedback. In this system, once a customer use the information networks to issue a return logistics request, based on that the e-business applications and third-party logistics companies can quickly grasp the amount and distributions of starting point in reverse logistics, thus eliminating the predictable differences. Finally, by use of the third party reverse logistics, industry scale advantage can be better achieved; in addition, by organizing customer groups to carry out recovery and transport activities, the logistics will get more intensive and efficient. Its operation flow model is shown in Fig. 1 below.

Third-party logistics companies can take advantage of its extensive distribution network to consider the consumers' requests for returns and then make decisions according to the accessible return standards of the enterprise. Select goods that can be returned, and then get in touch with vendors so as to take next step in Return Logistics. By using of third-party logistics for professional management, vendors can benefit not only from significant social effects but also good economic effects.

6 Conclusion

With the further development of e-commerce, for enterprises, the achievement of greater benefits will be greatly influenced by the way in which the existing resources are utilized for the reduction of reverse logistics possibilities, as well as the management efficiency in case of reverse logistics in e-commerce environment. Reverse logistics is a very weak link in the logistics system. Based on the further understanding of reverse logistics connotation in e-commerce environment,

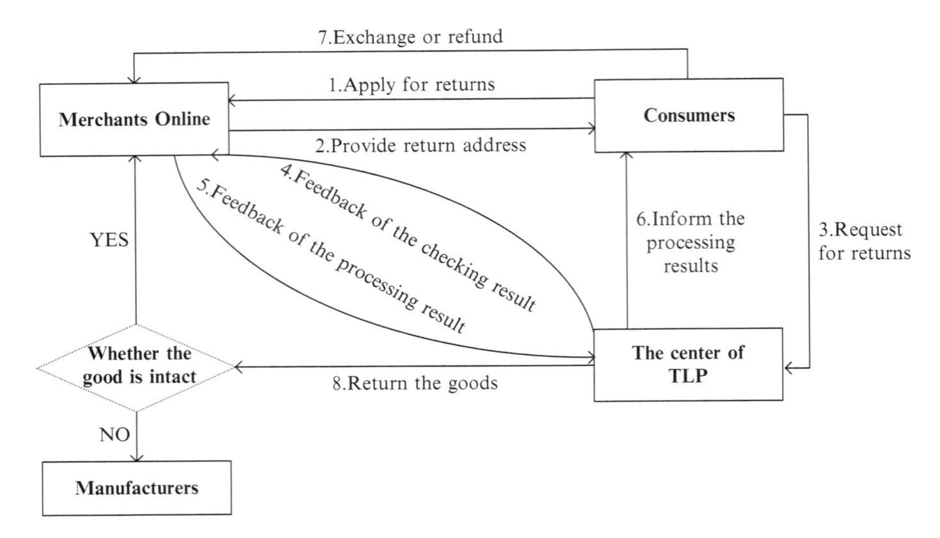

Fig. 1 Reverse logistics system model based on TPL

together with analysis of practical problems, such as enterprises' lack of management awareness, vacancy of effective returns mechanism, processing inefficiencies and poor service attitude, we have designed a set of more rational management strategies. By applying strategies such as optimizing the trading process, developing reasonable return policies, and selecting appropriate reverse logistics modes, hopefully the healthy development of reverse logistics could be realized.

References

1. Han Fang (2006) Establishment and operation of the reverse logistics system in the electronic commerce environment. Logist Technol 25(3):69–71
2. Li Ran (2011) Analysis of the returns management in reverse logistics for garment enterprises. Logist Sci Tech (9):86–88
3. Du Hong (2010) Study on return reverse logistics of B2C e-commerce transactions. Logist Sci Tech (4):12–14
4. Ni Ming, Liao Ruihui (2009) Four types of reverse logistics model based on e-commerce environment. Libr Info Serv 53(18):137–139
5. Zhou Min, Ni Ming (2010) How to make decision about outsourcing reverse logistics under e-commerce. Libr Info Serv 54(6):132–135

Constructing Butting Elements System Between Agro-products Market and Production Base Based on Coordination Game Theory

Fen Peng and Anguo Xu

Abstract Based on the analysis of the butting system from the perspective of the coordination game, the paper puts forward that there is comprehensive policy compensation and external pay-out among the parties involved. Therefore, among all the companies in the butting system, there are certain key factors which will have a great impact on the efficiency of the system. The paper constructs an integrated model composed by such elements as "strategic butt, interests butt, organization butt, information butt, standards butt, process butt, and culture butt" and proposes the path for butting elements at all levels respectively.

Keywords Agricultural products circulation • Butt • Coordination game

1 Introduction

Along with the growth of e-commerce and professional logistics, the current circulation system heads towards "*the shortening of channel, the equivalence of node scale, the balance of interests and the sharing of organizational resource*", thus pushing the growth of the circulation system of agricultural products [1]. This paper delves into the butting of agricultural product market and the base, that is, the circulation system that links the consumers of agricultural products to the production base.

The butting element refers to the basic factor that lies in organizations and their closely related logistics environment and coordinates the circulation system. It generally includes organizational factor, environmental factor and their binding factor. Only the realization of effective connection of each organization, can guarantee the coordination of the overall butting process and the efficiency of the joint operation.

F. Peng (✉)
School of Management, Minzu University of China, 100081 Beijing, P.R. China
e-mail: Pengfen2009@sina.com

A. Xu
School of Economics and Management, Beijing Jiaotong University, 100044 Beijing, P.R. China

Z. Zhang et al. (eds.), *LISS 2012: Proceedings of 2nd International Conference on Logistics, Informatics and Service Science*, DOI 10.1007/978-3-642-32054-5_49,

2 Literature Review

Currently, research on coordination game is relatively limited, but some scholars have given a descriptive statement about it. Vincent P. Crawford [2] held that the coordination game was involved in the same reference of participants for different combinations of strategies. If other people could anticipate correctly, they would have a unique solution in multiple Nash equilibrium, namely the similar belief to choose the same behavior among all the related participants [3]. Coordination game, in essence, refers to the game in which all the behavior subjects complement to each other. Yet the related theory hasn't been applied to the circulation of agricultural products.

As the supply chain management theory continuously becomes mature, recent years have witnessed some related researches on the coordination and cooperation of organizations of agricultural product. Hui Wang (2011) studied the coordination issue of distribution-retail supply chain and used the compensation policy to work out an optimal coordination mechanism [4]. Zhao Jiang and Kai Wang (2006) concentrated on five main interests connecting mechanisms between farmers and leading organizations and analyzed the advantages and disadvantages [5].

The existing data reflects that the current research in the field of logistics in our country is developing in depth. Since the particularity of agricultural product logistics makes the research of agricultural products very complex, the associated elements among logistics organizations of agricultural products lack integral and comprehensive understanding. In practice, the elements include not only such "visible" elements as interest, organization and information technology, but also "invisible" elements such as strategy, culture and so on. And in the meantime, the research on the organization mode, constraints, cooperation and coordination mechanisms and operating mechanism of agricultural products circulation has not yet been launched in a systemic way.

Based on the existing research and combined with the latest development, this paper will construct a butting System between agricultural products market and production base under the guidance of coordination game theory with the purpose to supplement and improve the logistic theory of agricultural products and with the expectation to guide the agricultural products logistics in practice.

3 Coordination Game and Butting System

3.1 Coordination Game

Coordination game is a kind of game on the basis of the complementarities of the action subject. Its equilibrium selection is not only related to the payment in game, but also dependent on the complicated anticipation process of all the involved parties. Some properties of coordination game can be used to study butting system, such as strategic complementarities and payment spillover.

In Cooper-John model, the response curve of each action subject is used to describe the characteristics of the coordination game [6]. Action subject i, i = 1, 2, . . ., i, selects a strategy in the interval [0,1]. In other words, the action level or effort of action subject i is the optimal action or the best response function, which serves as a scalar for the parameterization of action subject's revenue.

When the other participants increase their action, the best response for the subject is to increase e_i as a result of strategic complementarities. So, the response curve is upward sloping: $\Phi(e,\theta)$ increase in e. The chart above is an evidence of strategic complementarities, there exists multiple equilibrium e_l, e_m and e_h.

3.2 Coordination Game in Butting System

Butting system of agricultural products, related to multiple subjects, is the circulation process of agricultural products from production to consumption, in which each subject plays its own role on the basis of technical and functional division. It is characterized with strong interdependency or complementarities. It can exert an influence of coordination, but may fall into coordination imbalance which does not attain the Pareto Optimality.

In other words, an action subject in the game selecting a higher level action increases the marginal revenue of the other choosing a higher level action. However, experimental studies of such game theory conducted by Cooper [6] have shown that the outcome of the game may be inclined to the Nash Equilibrium at a lower level of effort on account of risk and other factors, even though both firms understand the results of non-efficiency. Such sub-optimal Pareto equilibrium outcome is known as the coordination failure.

It can be drawn from the above examples that the equilibrium choice of the coordination game in the butting system, different from the common game, depends on how fully similar beliefs the game participants would have. Coordination failure that the butting system does not remain in the optimal Pareto equilibrium is due to policy uncertainty among participants and resulting differences of behavior expectation. The mechanism should be designed to decrease the uncertainty of strategies between people.

3.3 Inspirations on Elements System Construction

There are many examples of discoordination in the supply chain of agricultural products, resulting in the butting obstacles as shown in Fig. 1, these are the concrete manifests of coordination failure.

According to the studies of Chinese and foreign experts, there are many factors that influence the equilibrium choice of coordination game, such as polity-significance factor, risk factor and interactive structure. But polity uncertainty

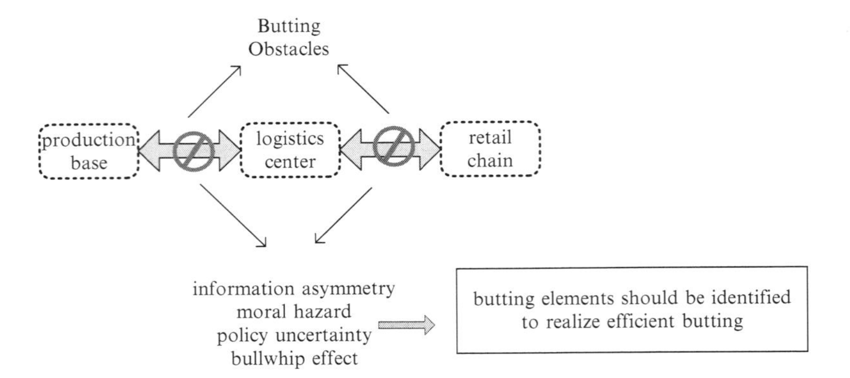

Fig. 1 Butting obstacles

involved among the participants is prominent. As participants can not make sure what kinds of strategies other members will choose, the participant will adopt a strategy of the lower level effort from the safety point, leading to a low efficiency of the overall operation in the non-Pareto optimal equilibrium state. Therefore, the design of relevant mechanisms to increase mutual trust, understanding and information sharing among members in the butting system is very significant in order to reduce policy uncertainty.

From the above analysis, it can be seen that there are some key elements in the enterprises cooperation, which may exist in the corporate policy within the macro-structure level, or in process management within the microscopic view. It should be addressed by the design of the mechanism to achieve the butting of those key elements in order to increase trust and understanding between the cooperative enterprises to reducing policy uncertainty.

4 Butting Elements System Construction

From the above analysis, there will be coordination failure in the butting process between butting subjects and agricultural products from market to base. The reason is that some elements fail to effectively butting among subjects during the butting process of agricultural products from market to base, which affects the efficiency of the overall circulation of agricultural products as well as individual interests. Thereby, it should be addressed by the design of the mechanism to achieve the butting of those key elements in order to increase trust and understanding between the cooperative enterprises to reducing policy uncertainty. The key elements are called butting elements.

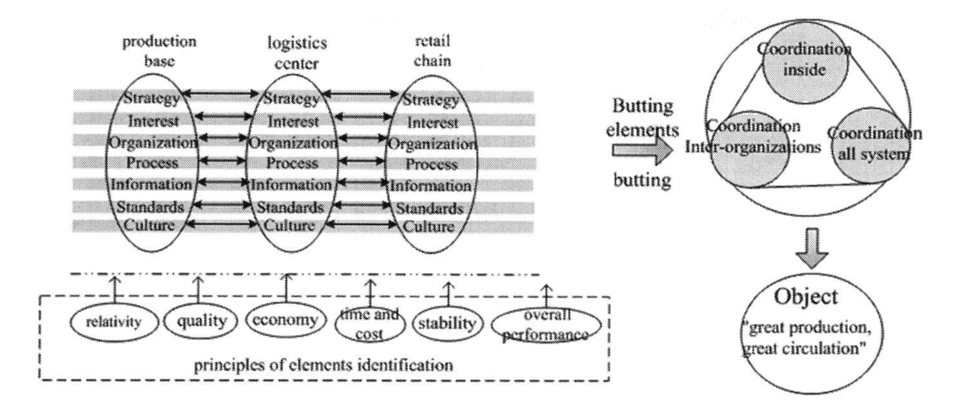

Fig. 2 Butt elements system

4.1 Analysis of Butting Elements

The analysis of butting elements is to identify important factors affect agro-products markets and base running. Butting elements can be recognized in line with the five elements in evaluation of the logistics system, which are quality (quality constancy), quantity (economy and efficiency of transportation), time (time on schedule), location (reasonable transportation scheme) and price (cost control). Based on those elements and combination with the characteristics of the butting system of agricultural products and the system stability requirements, the general principles of elements identification should be as follows: whether the elements affect the quality of agricultural products. Whether the elements affect the economy of the transportation of agricultural products. Whether the elements affect the time and cost of the circulation of agricultural products. Whether the elements affect the overall butting performance of markets and bases. Whether the elements affect the stability of subjects cooperation in the butting.

According to the above principles, the butting elements should be: Strategy: the subjects strategic butting to ensure the coordination of overall objectives. Interest: to balance the overall and individual benefits distribution. Organization: coordination for organization design to facilitate the information delivery and strategic integration. Process: coordination between inventory, order management and other key processes. Information: to achieve information sharing and consistency with reduce information distortion and information asymmetries. Standards: consistency with product quality testing standards and the understanding of quality. Culture: understanding of supply chain management, reducing costs and improving service quality.

Butt elements system is shown in Fig. 2.

After the analysis of elements in the butting system, the realization of elements butting through the appropriate mechanism design should also be in accordance with the corresponding principle:

Firstly, the achievement of overall goals in the system is the most important part. The butting value depends on the extent of the overall system efficiency increased by elements butting; Secondly, it does not require elements to achieve the best or optimal design, but with the emphasis on integrated relations between the various elements of the system; Thirdly, the expected ultimate outcome will be greater than that of individual performance as various elements are combined as a whole.

From the above, the agro-products butting system between bases and markets is not the simple connection of many subjects. At the request of "great production, great circulation" manifested in modern agro-logistics, the subjects of agro-production deeply joint via new forms of organization. In a way of solving interests conflicts between each subject through certain measures, methods and mechanisms, the needs of mutual interests or common goals involving with organizations is meat. As a result, coordinative operation state is achieved.

4.2 Butting Path

The butting of elements in the butting system is not a simultaneous and easy development, but a gradual process along with the deepening of cooperative relations and continuous development among subjects. According to the analysis of the butting mechanism elements, such as strategy, interests, organization, process, information, standards, culture and other elements, the above butting elements can be divided into different implementation levels.

Firstly, the achievement for butting in strategy and interests is the basis of butting system construction. Each subject will unify development goals to achieve the rationalization of resource allocation and maximization of resources usage through strategic butting. Conducting benefits distribution among subjects in accordance with a fair and reasonable proportion is the foundation for cooperation, which should be achieved initially at the beginning of butting system construction.

Secondly, it is the organization and process butting that also can be understood as the tactical butting. With the deepening of cooperative relations between the butting system, subjects can conduct more in-depth elements butting in the medium and micro level. The effective convergence and butting of organizational structure, organizational function and organizational strategy as well as production process, distribution process, customer service process and inventory management processes will play a positive role in promoting to lower the operating costs of the individual enterprise and the butting system as a whole, to improve the response for market demand and to maximize the individual and overall profits.

The third level is the information and standard butting. The information butting only can be achieved on the basis of the organization and process butting. Standard butting needs to be constructed on the basis of standard butting.

Finally, the highest level is culture butting. As different companies have different cultures, it will take quite a few times to the integration. It will possibly achieve a real culture butting after long-term cooperation.

When the four levels butting are achieved, the entire butting system will reach on a higher level. Thus, it will face the continuous optimization of the strategy and organization to a higher stage of development. So, the entire butting process in a butting system is in a continuous and spiral process.

5 Conclusion

Nowadays, with the rapid development of the market, Chinese agriculture has been integrated into the international competitive market economy. The seriously lagging of logistics development of agricultural products becomes the bottleneck of the entire circulation system in China. Therefore, among each key players in the butting system, there are certain key factors which will directly impact the output efficiency of such system. The paper construct the integration model based on the elements such as "strategic butt, interests butt, organization butt, information butt, standards butt, process butt, and culture butt" and puts forward butt path for elements of all levels. The efficient butting of agricultural products from the market to production base is of great significance to promoting the modern logistics development of agricultural products in China.

References

1. Fen Peng, Mingyu Zhang, Weibin Cao (2008) Research on the historical evolution of logistics organization of China agricultural products. Logist Technol 27(11):5–7
2. Crawford VP (1995) Adaptive dynamics in coordination games. Econometrica, Econometric Society 63(1):40–43
3. Liang-qiao Zhang (2007) Coordination game theory and equilibrium selection. Quest (7):50–52
4. Hui Wang (2011) Research on the incentive mechanism of cooperation type supply chain coordination. J Changsha Univ 25(9):102–104
5. Zhao Jiang, Kai Wang (2006) Analyses of advantages and disadvantages about benefit connection mechanism between farmers and leading enterprises in China. Mod Econ Res (1):71–75
6. Cooper RW (2001) Coordination games: complementarities and macroeconomics. Press of Renmin University of China, Beijing

Research on Logistics Distribution Networks of Online Retailers

Wei Wang

Abstract With the rapid development of the online retailer, the logistics more and more become its bottleneck. Because of lack of specialized third-party logistics entrepreneurs, the logistical cost of online retailers is high and the service quality is low. Which method of logistics the online retailer should select becomes the most important issue. Based on theories and practices, we aim to study the networks model of logistics distribution in our country.

Keywords Online retailer • Networks of logistics distribution • Self-service pack station

1 Introduction

As data show, the quantity of Chinese citizen will reach 4.57 Billions, the quantity of the users who shopping online will reach 160,510 thousands [1].[1] Shopping online is more and more famous to the network users. In the year of 2010, the quantity of Chinese online retail reached 513.1 billion Yuan, which has doubled than the year of 2009, and which is approximately 3% of the total of social retail.[2] Online retail is becoming an important power of economic development. However, a important complication of the users disgruntled to shopping online, is the existing mode of logistics distribution trail to the network retail trade development [2]. In

[1] The 27th statistical report on the China internet development, 2011.1, China Internet Network Information Center http://www.cnnic.net.cn/

[2] The report about data monitoring of Chinese electronic commerce market in 2010, 2011.1, China Electronic Commerce Research Center http://b2b.toocle.com/

W. Wang
School of Economics and Management, Beijing Jiaotong University, Beijing, China
e-mail: wawe@263.net

Z. Zhang et al. (eds.), *LISS 2012: Proceedings of 2nd International Conference on Logistics, Informatics and Service Science*, DOI 10.1007/978-3-642-32054-5_50,

the research about disgruntled, 21.2% of users consider that the delivery time is too long, 15.7% of users consider that the couriers have bad service, and 10.8% of users consider that the charge of freight is too expensive.[3] Online retailer urgent needs suitable mode of logistics distribution.

2 Trait of Logistics Distribution Networks

Whether based on the shopping mind of users or based on the variety of retail commodity, online retailer which have trait of its own was different from traditional retailer. It decided that the trait of logistics distribution of online retailers is different from the trait of logistics distribution of traditional retailers.

1. Because of a great variety and small batch of the commodity, distribution produced economies of scale is difficult. It was harder for distribution and higher cost of distribution, that the commodity of online retailer is mostly composed of Consumer goods and daily supplies, and a great variety of the commodity and small batch of the commodity [3].
2. It was difficult to dope out that the variety and amount of the commodity ordered, and it was harder for Inventory management. Because of the convenience and random of commodity ordered online, and separate users who shopping online, the requirement online was difficult to dope out accurately, debasing the cost of logistics was impacted.
3. Because of pressed for time, it was difficult to advance the quality of distribution. It was a important key for winning users that distribute online commodity on time. It need saving, taking and setting commodity as soon as possible, and may take commodity laying waste and losing easily.
4. It made the process of logistics distribution getting complex and beyond control, that the place of distributing commodity is separate and indeterminacy.

3 Summarize of Logistics Distribution Networks

Summarize of logistics distribution networks means: The main objects of distribution are distribution network sites, when the commodity arrive at distribution network site, commodity should consign to the users as the mode of terminal distribution which users choice. The basic distribution process is Fig. 1.

[3] The research newspaper of china online retailer market in 2009, 2009.11, China Internet Network Information Center http://www.cnnic.net.cn/

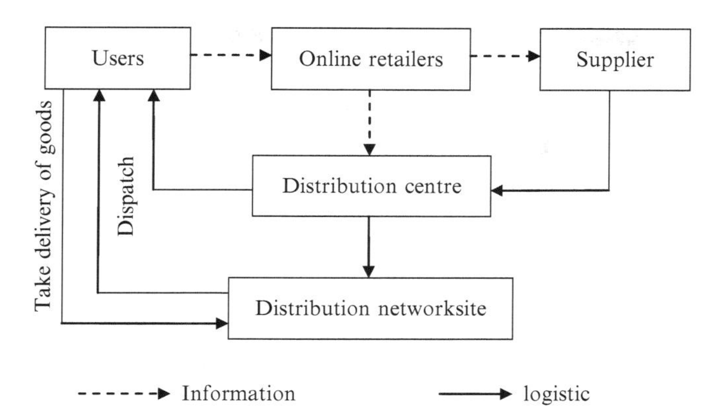

Fig. 1 Networks of logistic distribution flow chart

3.1 Trait of Logistics Distribution Networks

1. The master of distribution was either online retailers or the third party of logistics company. It was chosen by distribution capacity of online retailers and demand of scale by users.
2. There were two mode of terminal distribution which users can choice: The one was sending, means distribution network sites send the commodity to the place where the users appoint; The other was taking, means users take the commodity from distribution network sites on time.
3. There were three mode of terminal distribution by online retailers: The first was sending the commodity to the place where the users appoint by distribution centre which don't need distribution network sites; The second was sending the commodity to distribution network sites, then sending the commodity to the place where the users appoint by distribution network sites; The third was sending the commodity to distribution network sites, then users take the commodity from distribution network sites on time
4. There were three function of distribution network sites. The first was information platform. The information from distribution network sites, distribution centre and suppliers can be shared, then actualized self-motion ordering commodity, Inventory and accounts settlement. The second was transferring by logistics. After users ordered commodity online, the commodity was sent from distribution centre to distribution network sites, then to the users, distribution was scale. The third was more service by logistics. The users can enjoy service like commodity show, cash on delivery, return and replacement commodity from distribution network sites.
5. Distribution network sites can utilize the existing society resources like shop, florist, kiosk in the community, also can Set up a special company which preside over distribution network sites. There were there condition that distribution network sites choice the address: The first was near the place where users live

and work in order to easy to sent and take. The second was near the traffic place in order to convenience by distribution vehicles. The third was doing business as long as possible in order to take users more time.

3.2 Advantage of Logistics Distribution Networks

In contrast to existing mode of logistics distribution, the mode of logistics distribution networks have its own advantage.

1. Reduced the cost of logistics distribution. Distribution network sites decreased terminal distribution, the distribution by vehicles only aimed at distribution network sites, advanced the efficiency of distribution vehicles, reduced time and distance of transit deeply, reduced cost of transit. The mostly objects of distribution was distribution network site, averted sending time after time, economized the time of distribution. Unified the process of taking, return and replacement commodity, without returned commodity recycling specially, reduced the cost of commodity reclaimed [4].
2. Advance quality of logistics distribution. It vehicleried out centralized distribution with the distribution network point as the distribution object, in this way, it will propitious to intensify the controlling and intendancy of the logistics distribution, reduce the mistake rate of distribution and enhance veracity of distribution. It is unnecessary to deliver the commodity in doors, can engendered miniature distribution by light-vans and microbus, abatement using non professional distribution vehicles such as bicycle and motorcycle, reduced the mangle rate of the commodity which delivered in transit, assure the commodity safely.
3. Advance the users' satisfaction rate. Because of the mode which users taking commodity themselves, users was not critical of the time and place of distribution. When users taking commodity, they can choose return and replacement, the flow which was predigested advanced the trust rate by users. On account of using special transportation vehicles, the transportation vehicles information can be control more easily, convenient for users looking for the information of commodity on the way. Because of decreasing terminal distribution, the cost of distribution was reduced deeply, then, the commodity price was debased, and benefitted customers.

4 The Analysis of the Operational Approach of Distribution Network Point

There are two ways of the operation of distribution network point: the one is run business, which use of the existing social resources. For example, building a distribution network which close to the residents of life through supermarkets or newspaper stalls, etc. The successful case is 7-Elevenstore chain group from Japan through this way. Another one is the speciality stores, which set up the third party company or organization will be responsible for the operation of special distribution outlets, and getting profit with the professional operation. Under this way, the successful case is Germany DHL self-help package stand [5].

4.1 Sideline Ways: Set up the Retailers Like Convenience Stores as the Distribution Network Sites

Japanese 7-Elevenstore chain group carry out online retail business joint with many well-known enterprise. These stores open in the neighborhoods and operate the whole day that the customers are able to buy goods at any moment. There configured high speed computer workstation in the stores that the customers can choose or order the goods which have no ready stock, and they can also took delivery of the goods in any 7-Elevenstores.Hence,the stores become the media of online shopping and the logistics distribution terminal of online shopping goods.

There are several advantages as the stores to be distribution network point:

1. Abundant network stores' resource. It's easy to find stores, supermarkets, flower shop and newspaper stands in the area of people's living and working. Whether in the cities or the in the rural areas, whether in the community or out of the community, it can find the small stores which is engaged in terminal retailing.
2. Close to the residents living. Retail shops usually open in the scope of the consumers' daily life and working area. Furthermore, most of them have rather long business time, especially, some stores open 24 h a day. Therefore, with it based, can shorten the radius of distribution, reduce distribution costs, and facilitate clients to access goods by themselves.
3. Save the investment cost of the fixed asset. Use existing social resources as the distribution network point, compared to construct the specialized distribution network point, will reduce the investment of the fixed asset and save the cost.
4. Cut down the promo fees. Because the retail shops have approved by the consumers, They just need add new functionality to the existing, thereby, they can play the role of the distribution network point. And these shops are relatively easy to develop new business on account of owning fixed target customers.
5. Supply the value-added services. The real shops as convenience stores can help the online retailers to sell commodities, show samples, deliver the commodities, collect the payment, change or return products and deliver goods to customers' door, which are value-added services.

4.2 Specialized Ways: Set up the Specialized Organizations like Self-Service Parcel Stations as the Distribution Network Point

The model is an innovative approach taken by the German DHL company in order to cope with the rapid growth of online shopping. Its basic meaning is: They set up self-service parcel stations in the densely populated and transportation convenience areas. Then install electronic locker with sufficient quantity and enough capacity in the parcels stations, and configure the computers and other devices that can access to internet. The express companies deliver the goods that the consumers purchased to the lockers, and the parcel stations give transfer the lockers' number and opening code to the customers. With this code, the customers are able to accomplish the process of picking up the goods without any helps. It saves the cost of human resource on account of picking up the goods by customers. However, it requires higher technology because of complete automatic management. The distribution process of DHL self-service parcel stations as follow:

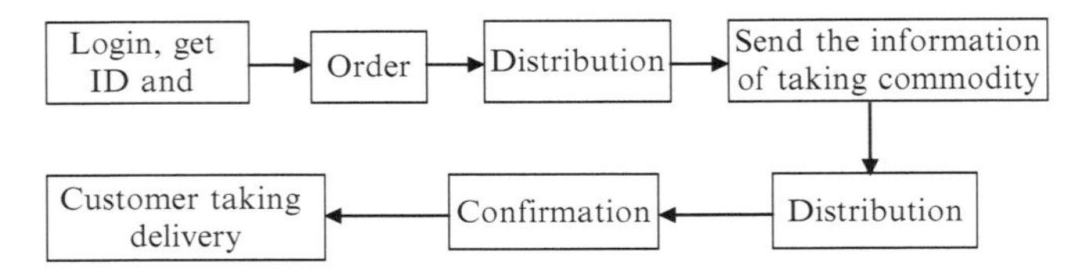

DHL built 2,500 self-service parcel stations in the railway station, shopping mall and supermarket of Germany, which providing all-weather service. Specifically, the parcel stations located in well-planned sections, for instance, on the way work or home, and it save the time of customers taking parcels. Now there are 1.5 millions customers register to become affiliates. With the rising of use rate of parcel stations, 5–10% parcels are delivered through the parcel stations in DHL, and each parcel station delivered 10 parcels everyday.

Compare to the way of the retailers like convenience stores as the distribution network sites, it had some advantages that set up the specialized organizations like self-service parcel stations as the distribution network point.

1. Intensify the control of the logistics distribution. They adopt information-based management, reducing the human element, cutting down the damage rate and enhance the accuracy of delivering the goods in the distribution.
2. Save the cost of management. It reduce the cost of human resource on account of deliver the parcels by customers. At the same time, it cut down the communicational cost between the network retailing companies and the shops. Thereby, they can concentrate energy on the technology of deposit equipment in order to improve the security and automation degree.

5 The Application in China of the Logistics Distribution Networks

5.1 The Application Prospect of the Logistics Distribution Networks in China

There are huge quantity group of online shopping, according to the data of CECRC, the number of using the online retail is nearly 106 millions, still go upward. See Fig. 2 (The value in the year 2011 and 2012 is expected). The quantity of the users who shopping online is growing every year, which is more than 5% of the quantity of Chinese online retail. See Fig. 3.

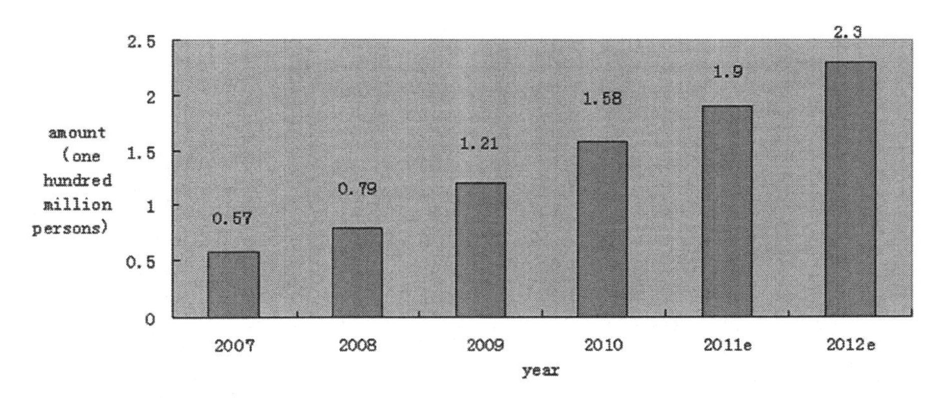

Fig. 2 Customer share of online retailers from 2007 to 2012

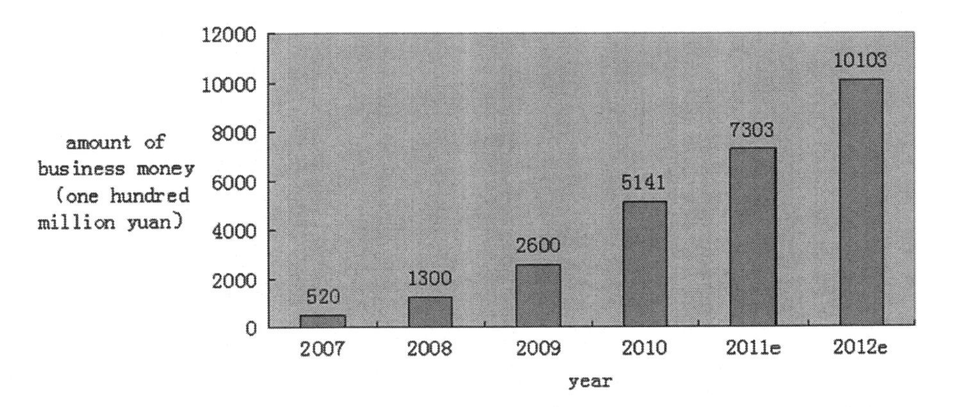

Fig. 3 Exchange scale of online retailers from 2007 to 2010 (Data from: China Internet Network Information Center http://www.cnnic.net.cn/)

5.2 The Problem of the Logistics Distribution Networks Based on the Convenience Stores

1. It is difficult to carry out centralized management. It must become a multi-function service center integrated sell, show, logistics transfer and distribution service in order to play the role of the distribution net point, which need redesign the market strategy and Service concept.
2. The scale of the shops is relatively small, haven't developed a network. Most of the chain enterprises that focused on regional markets mainly open convenience stores in large cities like Beijing, Shanghai, however, develop little cross-regional market. Therefore, there are little nationwide enterprises. And it is hard to carry out logistics distribution networks as there is no support of chain stores in the area.
3. The cooperation between the stores and online retailers exist venture. Online retail companies and stores share the information of order processing, payment settlement, change and return goods, to become an information hub and a logistics transfer area, which requires a large investment.

5.3 The Problem of The Logistics Distribution Networks Based on the Self-Service Parcel Station

1. Large investment on research and development, and long term on payback period. Self-service parcel station strictly required locker safely and expediently, it need higher standard to prevent any burglaries, perfect information systems and autocontrol systems, and achieved by the technical progress.
2. It was hard to construct a distribution network site. This mode need construct many distribution network sites in the cities, so the address was chosen at the place which have more people and convenient traffic to unloading. Under these conditions, it was difficult to choose enough sites which according with the equal condition.
3. It need definite time that users accept the self-service. It need change the conventional service mode which deliver commodity to the customers, and asked users taking commodity themselves by information systems. It need some time to adopt.

6 Conclusion

The rapid development of online retail, provided a good market environment and development space for network type logistics model. As innovative approaches to solve the distribution problem of the terminal, it should be appropriate to try and gradually introduced.

1. Network type logistics model can be piloted in the economically developed areas. The online retail companies can try to select convenience stores as a strategic partner to conduct a joint distribution terminal service. That strength of the logistics enterprises, and the online retail business, or storage cabinet equipment manufacturer can try to establish a self-service parcel stations in densely populated areas, and gradually expand the distribution network.
2. To attract government involvement in network building. The construction of distribution network is not only the logistics cost savings for the online retail business, which have opened up new economic growth point for the convenience stores and other retail outlets, but also bring convenience to people's lives, to guide people to the formation of healthy life. Therefore it has a positive social. What' more, self-service parcel stations with the nature of the universal service. On the other hand, the effective operation of the distribution network need a good infrastructure as a guarantee, the logistics infrastructure has strong nature of public goods and external economies, Therefore the Government should strengthen and attention to the construction and management of the logistics infrastructure.
3. As new phenomena, the self-service parcel stations is an innovative way to solve the distribution problem of the terminal, this method has not caused widespread concern in the industry. But the rapid development of network retail made a large market potential for self-service parcel stations. Recommends that the Government to encourage enterprises to invest in the construction of specialized distribution network through the guiding role of the policy.

References

1. China Internet Network Information Center (2011) The 27th statistical report on the China internet development[R], http://www.apira.org/news.php?id=358
2. Wen Longguan, Yu Bo (2009) Research on logistics distribution of electronic commerce. China Logist Purchas [J] (21):74–75
3. Milner JM, Kouvelis P (2007) Inventory, speculation and sourcing strategies in the presence of online exchanges. Manuf Serv Oper Manage Logist Sci Techcnol 9(3):312–331
4. Kasturi Rangan, Ramchandran Jaikumar (1991) Integrating distribution strategy and tactics: a model and an application. Manage Sci 37(11):1377–1389
5. Williamson OE (2002) The theory of the firm as governance structure: from choice to contract. J Econ Perspect 16(3):171–195

Research on Two-Stage Supply Chain Ordering Strategy Optimization Based on System Dynamics

Lian Qi and Lingjia Su

Abstract This paper constructs a two-stage supply chain ordering model based on system dynamics, and proposes a second exponential smoothing ordering strategy for both supermarket and distribution center. With the goal of minimizing the cost of whole supply chain and the fluctuation of inventory, genetic algorithm is adopted to help improve the model simulation, which can optimize the controllable parameters in the ordering model. Simulation results show that system dynamics can better resolve the optimization problem of complex system in combination with genetic algorithm, and the ordering strategy this paper proposed has a relatively good adaptation for markets with different kinds of traits.

Keywords Ordering strategy • Parameter control • System dynamics • Genetic algorithm

1 Introduction

Traditional researches of ordering strategies in supply chain are mostly based on operation, and solve problems by inventory model. This kind of research method usually has many disadvantages, like overemphasizing mathematical form and demanding large amount of computation, unable to track the logical and numerical relationship among variables. Nowadays, many scholars at home and abroad have introduced system dynamics into the study of ordering strategy in supply chain. Most of them constructed simulation model by system dynamics and analyzed the influences of key parameters have on ordering strategy through "try-and error" method [1]. It is not only inefficient but also inaccurate. So in order to avoid these disadvantages, this paper combines genetic algorithm with system dynamics to

L. Qi (✉) • L. Su
College of Economy and Management, BEIHANG University, Beijing 100191, P.R. China
e-mail: smilenancylian@gmail.com

Z. Zhang et al. (eds.), *LISS 2012: Proceedings of 2nd International Conference on Logistics, Informatics and Service Science*, DOI 10.1007/978-3-642-32054-5_51,

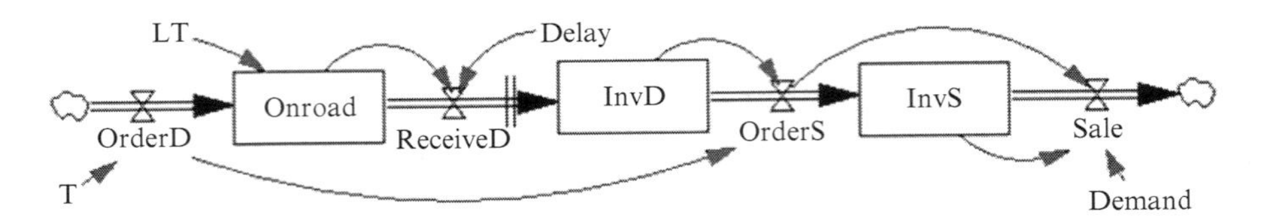

Fig. 1 Basic two-stage supply chain ordering system SD model

construct a two-stage supply chain ordering model, and studies what influences of forecasting techniques and information sharing level have on ordering strategies under different market demand characteristics, and then optimized the model by genetic algorithm.

2 SD Model of Two-Stage Supply Chain Ordering System

The basic flow chart of the two-stage supply chain ordering system is constructed as Fig. 1:

In Fig. 1, T represents ordering period of distribution center, *OrderD* is interpreted as ordering rate of distribution center, LT is interpreted as the ordering lead time of distribution center, *OnRoad* is interpreted as the inventory on road, Re*ceiveD* is interpreted as the goods receive rate of distribution center, Delay is interpreted as the goods receive delay time of distribution center, *InvD* is interpreted as the inventory of distribution center, *OrderS* is interpreted as the shipment rate of distribution center, *InvS* is interpreted as the inventory of supermarket, *SaleS* is interpreted as the sale rate of supermarket, Demand is interpreted as the real market demand.

This paper define the minimization of the whole supply chain's funds cost and inventory cost as the objectives of the SD model. Inventory cost consists of two parts, which are over storage cost and lack storage cost.

(1) Over storage cost
Over storage cost refers to the expense used for maintaining the inventory [2]. It is the sum of inventories in both supermarket and distribution center.

$$\min TCost = CostD + CostS \tag{1}$$

In Fig. 2, CostD represents the inventory cost caused by distribution center, including inventories in warehouse and on road; CostS represents the inventory cost caused by supermarket. Since distribution center and supermarket is usually colse to each other, in order to simplify the model, onroad inventory between distribution center and supermarket will not be taken into consideration. Besides, HO, HD, HS relatively represents the unit inventory cost(yuan/unit/day) of onroad inventory, distribution inventory and supermarket inventory; AvO, AvD, AvS represents the relative average inventory.

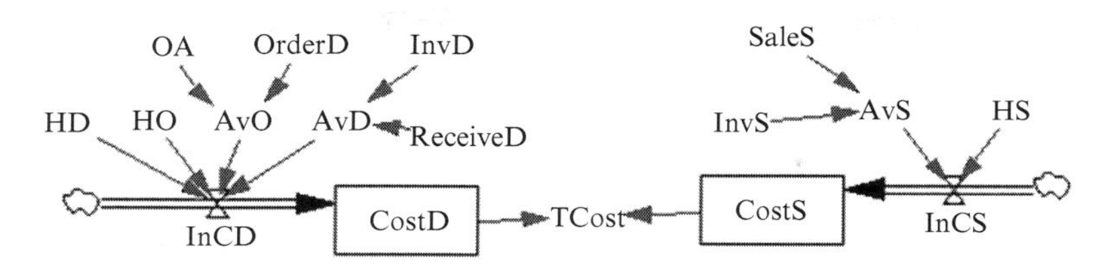

Fig. 2 Inventory cost of system flow chart

(2) Lack storage cost
Lack storage cost refers to the lost when real market demand *Demand* exceeds sale volume *Sale* [3], which equals the product of the difference between market demand and sale volume with the good's market price *P*, namely:

$$\min A = \sum_t (Demand - Sale_t) \times P \tag{2}$$

3 Analysis of Ordering Strategy

3.1 Ordering Strategy of Supermarket

The study of this paper is based on the prerequisite of periodical ordering strategy for supermarket and distribution center. Two main factors should be taken into consideration when managers determine the ordering quantity, the real market demand, and the inventory. Considering all factors mentioned above, the expected inventory of supermarket is defined as formula 3:

$$DInvS = ForcastS + sdS + SS \tag{3}$$

$$SS = f(ForcastS,\ CSL) \tag{4}$$

In the formula, *DInvS* is supermarket expected inventory, *ForcastS* is the supermarket's forecast of market demand, *sdS* is the forecast error. This paper adapt exponential smoothing forecast method to forcast market demand; *SS* is the supermarket's safe inventory which whose calculation method can be referred to reference [4]; *CSL* is goods' obtainable level of supermarket. In order to make the forecast results approach real demand, second exponential smoothing is used to treat with *InvS* [5].

Forecast error *sdS* of *ForcastS* can be acquired by formula 5, and L is the simulation interval length.

$$sdS = \sqrt{\sum_{t=1}^{L} \frac{Demand - ForcS}{L - 1}} \tag{5}$$

In addition, supermarket still need to consider its inventory adjust rate MRS, its formula is (TS is supermarket's inventory adjust time)

$$MRS = (DInvS - InvS)/TS \tag{6}$$

So, the supermarket's ordering strategy can be concluded as:

$$OrderS = ForcastS + sdS + SS + MRS \tag{7}$$

3.2 Ordering Strategy of Distribution Center

The ordering strategy of distribution center should be discussed under two conditions, one is information shared through the supply chain, and the other is the opposite. When information is shared in the supply chain, supermarket and distribution center's forecast base is the same, the real market demand. But when the information isn't shared, the forecast base of distribution center is the order of supermarket $OrderS$ [6]. Second exponential smoothing is applied for the forecast of next period, the formula is as 8.

$$\begin{aligned} InvD &= ForcastD + sdD + SD \\ &= ForcastD(T + LT) + sdD(T + LT) + SD \end{aligned} \tag{8}$$

In the formula, $InvD$ represents the distribution center's inventory of next period, $ForcastD(T + LT)$ and $sdD(T + LT)$ relatively represents the forecast of supermarket's order and forecast error. Because the supermarket adopts every day ordering strategy, distribution center could treat the supermarket's order history with a second exponential smoothing method to obtain $ForcastD(T + LT)$. Calculated as below:

$$\begin{aligned} ForcD(T + LT) &= m(T + TL) + n(1 + 2 + \cdots + (T + LT)) \\ &= \left| m + n\frac{1 + T + LT}{2} \right| \times (T + LT) \\ sdD(T + LT) &= sdD \times \sqrt{T + LT} \\ SD &= f(ForcD(T + LT), CSL) \\ DInvD &= ForcD + SD \end{aligned} \tag{9}$$

The same to supermarket, distribution center also need to consider the inventory adjusts rate MRD, the corresponding formula is (TD is distribution center's inventory adjust time):

$$MRD = (DInvD - InvD)/TD \tag{10}$$

In conclude, the ordering strategy of distribution center is as formula 11:

$$\begin{aligned} OrderD &= ForcastD + sdD + SD + MRD \\ &= ForcastD(T + LT) + sdD(T + LT) + SD + MRD \end{aligned} \tag{11}$$

4 Simulation and Optimization of the Two-Stage Supply Chain Ordering System

4.1 Simulation of the System

Synthesize the analysis above, the system flow chart (SF charts) of distribution center and supermarket ordering SD model under information share and non-share condition are as Figs. 3 and 4. The only differences between two figures are the information shared level and forecast bases. So the two figures can be analyzed together as follows.

The SD model's objective function is:

$$\min TotalCost = \int_{t=0}^{t=L} [(CostD + CostS + Absent \times P_1) + (OrderD \times P_0 - Sale \times P_1) \times R]dt \tag{12}$$

Since different market demand characteristic should be matched with different policy parameters, this paper considers three kinds of demand characteristic, namely, random fluctuation, trend fluctuation and periodical fluctuation.

$$\text{Random fluctuation}: \quad Demand_t = F + \phi_t(t = 1, 2, \ldots, L) \tag{13}$$

$$\text{Trend fluctuation}: \quad Demand_t = at + \phi_t(t = 1, 2, \ldots, L) \tag{14}$$

$$\text{Period fluctuation}: \quad Demand_t = b\sin(t\pi/L)(t = 1, 2, \ldots, L) \tag{15}$$

a, b, F is constant, Φ is chaos function, L is simulation length. Set a = 0.8, b = 10, F = 10, Φ is a normal distribution function with average value 10, variant 3.

The simulation is finished with Vensim_PLE Version5.6b.

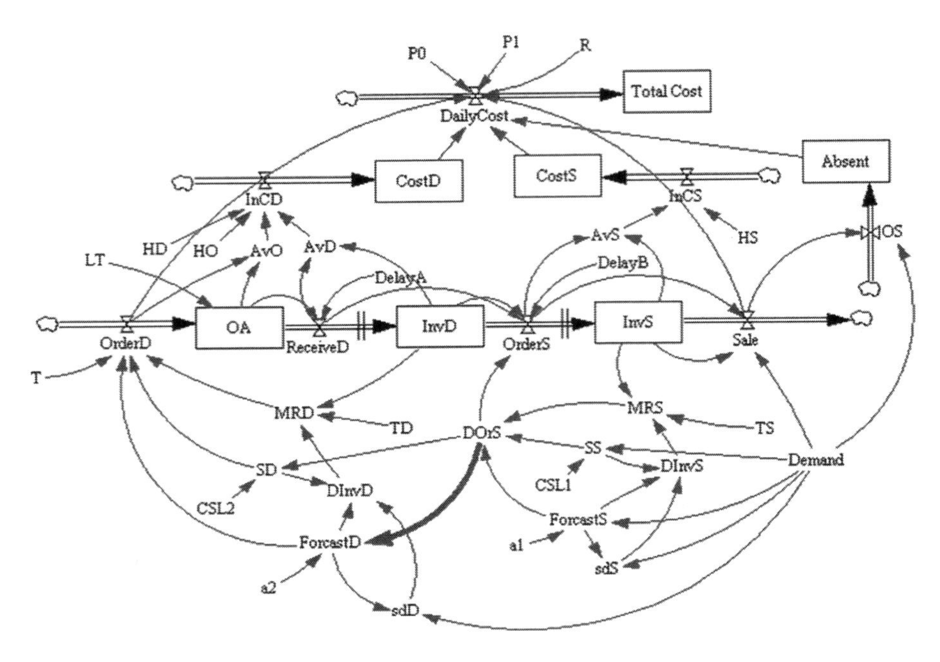

Fig. 3 SF chart when information is shared

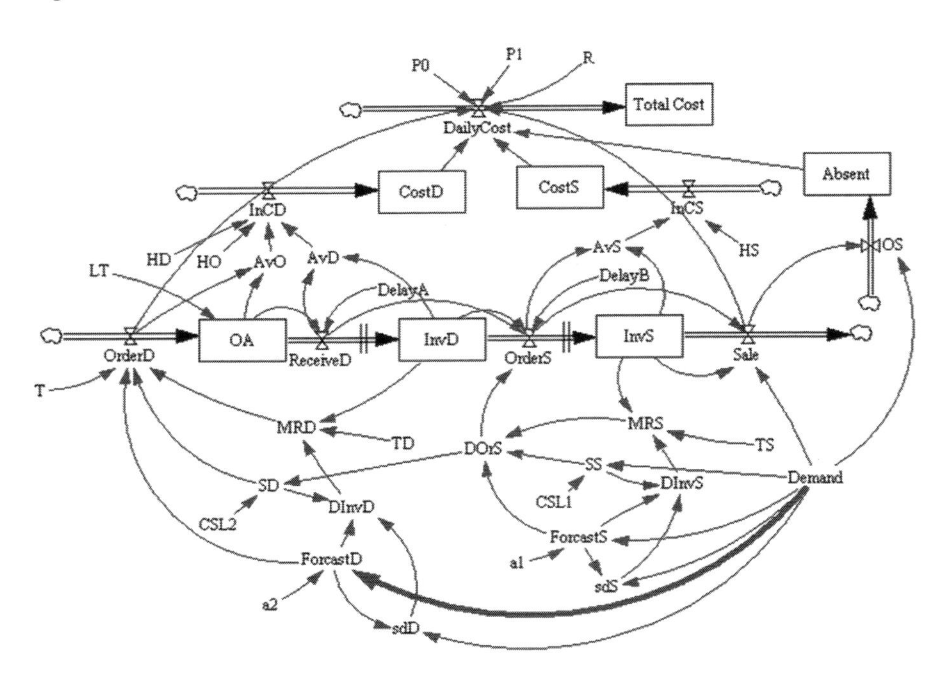

Fig. 4 SF chart when information isn't shared

Table 1 Results of four parameters combine optimization

Information share level		Total share			Non-share		
Demand characteristic		Random fluctuation	Trend fluctuation	Periodical fluctuation	Random fluctuation	Trend fluctuation	Periodical fluctuation
Controllable parameters	α1	0.67	0.74	0.64	0.67	0.74	0.64
	α2	0.08	0.14	0.1	0.08	0.14	0.1
	CSL1	0.95	0.96	0.97	0.95	0.96	0.97
	CSL2	0.94	0.95	0.96	0.96	0.96	0.97
	Total cost (yuan)	744.43	4,570	659.65	849.69	4,523	667.25
Assessment indicators	S inventory variance	1.65	1.88	1.40	2.69	1.32	1.34
	D inventory variance	33.33	301.79	28.18	37.02	300.70	28.03

Note: S represent supermarket; d represents distribution center

4.2 Simulation Results and Optimization of the System

There are four controllable parameters, which are obtainable level *CSL*1 and relative *CSL*2, forecast exponential smoothing coefficient α1 and α2 in the SD model. Adopting "try-and-error" method, this paper studies the combination and optimization of the controllable parameters for the three demand characteristic under information share and non-share condition. Set $DelayA = 3$, $DelayB = 0.5$, the ordering lead-time of distribution center LT = 3, ordering period T = 5, the inventory maintain cost of distribution center HS = 0.05 yuan/unit/day. The initial inventory of distribution center 20 kg, inventory of supermarket 120 kg, and no on-road inventory. Besides, set simulation length L = 365 days, and simulation step DT = 1 day. The results of simulation are showed in Table 1.

The simulation shows model's total cost will change with no regulations when four parameters change at the same time. And if controllable parameters CSL1 and CSL2 are limited to 0.95 or above, assessment indicators are not sensitive to the two parameters. Considering the collaboration of supply chain's up-and-down stream, too many controllable parameters will result in a very complex "try-and-error" process and cause an unnecessary lost. So it's reasonable to set some parameters through counseling for the whole supply chain. Here, set CSL1 = 0.96 and CSL2 = 0.98.

Since "try-and-error" is not an ideal method to resolve complex system, this paper adopt genetic algorithm to optimize the two-stage supply chain ordering system. Define the system goal is the minimum cost of the supply chain, the state transition regulation and the occupied funds could not exceed the whole funds. Adopt binary code, set word length n = 64, group capacity m = 150, apply zoom and league selection based on the objective function value of the order of the arithmetic sequence and consistent cross, take the cross-coefficient Pc = 0.80, the gap coefficient G = 0.64 [7]. In addition, since the function form of control

Table 2 Results of two parameters combine optimization

Information share level		Total share			Non-share		
Demand characteristic		Random fluctuation	Trend fluctuation	Periodical fluctuation	Random fluctuation	Trend fluctuation	Periodical fluctuation
Controllable parameters	α1	0.60	0.68	0.63	0.60	0.68	0.63
	α2	0.24	0.31	0.28	0.24	0.31	0.28
	Total cost (yuan)	713.34	4,236	609.55	759.56	3,122	597.23
Assessment indicators	S inventory variance	1.65	1.88	1.40	2.69	1.32	1.34
	D inventory variance	33.33	301.79	28.18	37.02	300.70	28.03

Note: S represents supermarket; d represents distribution center

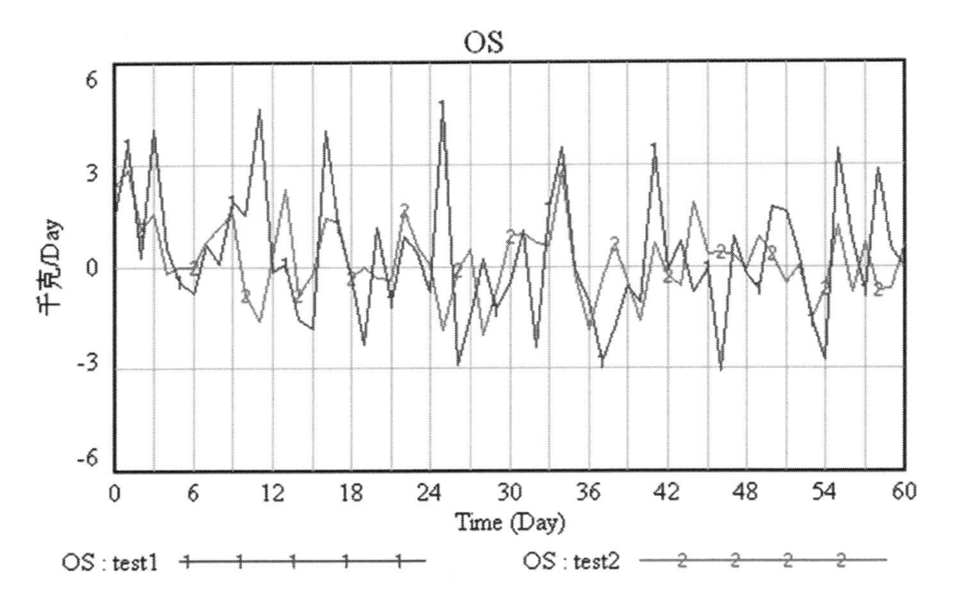

Fig. 5 Goods absent rate of supermarket

strategy in group is different, cross operation is operated in the same control strategy form. In order to reduce the computational load, set Pm = 0.01, when 30% average mutation probabilities of the function value, in inferior sort, in the group is less than 1%, change the mutation probability Pm = 0.005 [8]. The genetic algorithm optimization results of two controllable parameters are shown in Table 2.

The genetic optimization results show that most indicators improve under whatever conditions. It is not only the cost of supply chain decreases, but also both of fluctuation of inventory and absent rate have relieved. Take the supermarket as example, the optimization results are shown as Figs. 5 and 6.

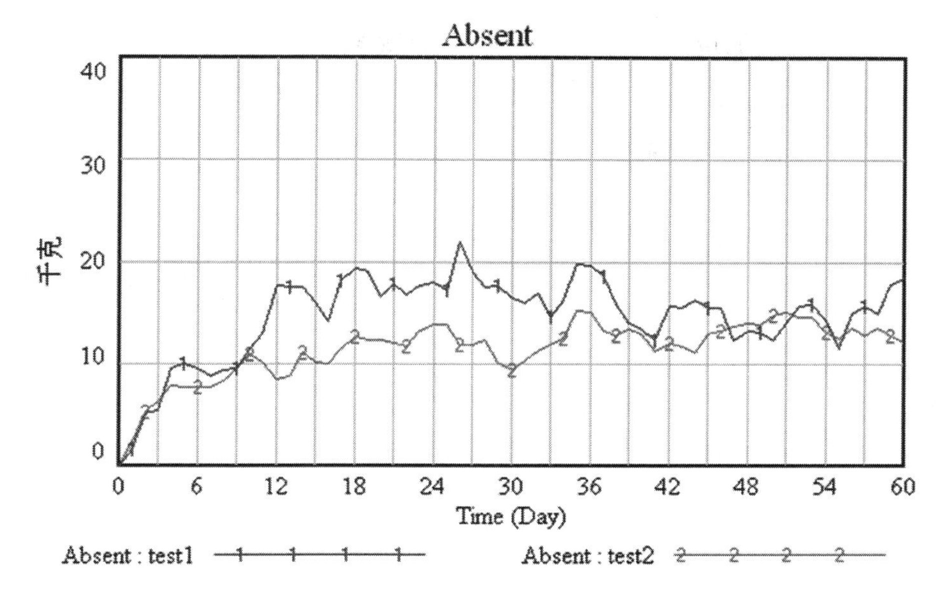

Fig. 6 Total goods absent quantity

5 Conclusion

This paper first analyzes the boundaries and causal relationship of two-stage supply chain ordering system, and constructs the SD simulation model based on the principles and methods of system dynamics. With the goal of minimum cost of the whole supply chain, this paper then studies the ordering strategies for distribution center and supermarket, and determines the optimization of the four controllable parameters combined with genetic algorithm. The simulation results reflected that ordering strategy and cost controlling in the supply chain is a complex system project that relates to the organizational model, control program and policy parameters. Even though the results of this paper is concluded from an individual case, not group ones, since the study stands on the point of practical application and operability, the ordering model possesses the common characteristics, and at the same time, large sums of sensitive analysis of the parameters have been done, the results and conclusion of this paper have some kind of general and guidance meaning.

References

1. Li Xu, Chen Yusheng, Liu Zhengzheng (2007) Application of systems dynamics in the optimization of ordering policy in a distribution center of supermarket. Sci Technol Rev 25:68–72
2. Lin Wen hao (2002) The application of genetic algorithms in estimation of important parameters in system dynamics model. J Fujian Agric Univ(Nat Sci) 3(31):404–407

3. Chen Wen-jia, Mu Dong (2008) Research on warehousing system at distribution center by system dynamics. J Beijing Jiaotong Univ(Soc Sci Ed) 1(7):27–32
4. Meindl P (2008) Supply chain management strategy, planning, and operation. Renmin University of China Press, Beijing
5. Erma Suryani, Shuo-Yan Chou (2010) Demand scenario analysis and planned capacity expansion: a system dynamics framework. Simul Model Pract Theory 18:732–751
6. Wan Jie, Li Min qiang (2003) Analysis and control on the bullwhip effect in the course of forecasting and processing of demand information. J Ind Eng Eng Manag 3(36):369–373
7. Cheng Jin, Wang Hua-wei, He Zu-yu (2002) Research on system dynamics models based-on genetic algorithm. Syst Eng 5(111):77–80
8. Ali Arkan, Seyed Reza Hejazi (2012) Coordinating orders in a two echelon supply chain with controllable lead time and ordering cost using the credit period. Comput Ind Eng 2(62):56–69

The Optimal Taxation of Logistics Industry in China

Dongmei Wang, Cairong Zhou, and Hejie Sun

Abstract The 12th Five-Year Program of People's Republic of China's National Economic and Social Development stressed that "vigorously develop modern logistics industry", tax issue is very important to the development of the logistics industry. At present, the study of the tax burden on China's logistics industry is almost blank. By using Cobb-Douglas production function expansion mode, this paper estimates the optimal tax burden level of China's logistics industry, and puts forward recommendations to improve the tax policy of the logistics industry.

Keywords Logistics industry • Optimal taxation • Production function • China

1 Introduction

The logistics industry in China is still in the early stage, is a new industry, but it links the social production and circulation, it has an important function on the smooth operation of social economy, it will become an accelerator to promote economic development. The logistics industry has a set of characteristics, such as low profit margins, low returns and long cycle on investment, this decides that it can't bear heavy tax burden.

The problem of tax burden is always the most acute, the most sensitive part that relates to the people's livelihood. Because it doesn't only constitute a state tax policy and the core content of the tax system, but also closely related with economic growth and social stability. Dongmei and Songdong [3], Dongmei and Cairong [2] have analyzed tax burden of logistics industry in China, shows that the tax burden of logistics industry above the social average level. High level of tax burden has

D. Wang • C. Zhou (✉) • H. Sun
School of Economics and Management, Beijing Jiaotong University,
Beijing 100044, P.R. China
e-mail: dmw12345@263.net; 10125439@bjtu.edu.cn; hjsun@bjtu.edu.cn

Z. Zhang et al. (eds.), *LISS 2012: Proceedings of 2nd International Conference on Logistics, Informatics and Service Science*, DOI 10.1007/978-3-642-32054-5_52,

restricted the development of logistics industry; the existing tax system has been not adapted to the need of the logistics industry. The problems existing in the tax policy system of logistics industry become the bottleneck of influence the development, the tax system of the logistics needs to be optimized.

Study on the optimal taxation of logistics industry provides the theory basis for tax policy-making, it is extremely important to the improvement of the tax policy in logistics industry. January 1, 2012, business tax changes VAT pilots in transportation industry and modern service industry in Shanghai [4]. After the full implementation of the VAT reform in 2009, it is a symbolic event in the process of the tax system reform. Business tax change VAT is a major reform of promoting the economic structure adjustment and accelerating the transformation of the economic development, is also the important content of the structural tax cuts. How to use the opportunity of this reform, solve the tax problems that inhibit the development of logistics industry, it is an urgent need to study the issue.

2 The Research Theory and Methods

The optimal taxation is determined by the maximum of gross domestic product. This paper uses Cobb-Douglas production function expansion mode as the foundation, which appeared in Barro [1], analysis the relationship between gross domestic product, government spending and capital investment, and deduces the optimal taxation. We consider the economic production and utility function of household, enterprise and government department, and deduce the economy increasing function.

2.1 The Relationship Between Tax Burden and Economic Growth

From the macroscopic, the tax is rooted in economic, only economic developed well, the tax will have plenty of sources. If other conditions remain unchanged situation, the economic growth will have a great influence on the tax. As in a slowing economic growth period, the proportion of taxable GDP will drop accordingly, tax growth will slow accordingly. If the tax grew more rapidly than economic, it may seriously weaken economic growth in the future.

About tax burden and economic growth, the more representative is the view of Supply school that the high tax has a negative impact on economic growth. This view comes from the analysis that tax has an impact on labor and capital: the high taxes for labor, real income of labor will drop, reducing the supply; At the same time improve labor costs, demand also reduce, causing the labor supply and demand gap, and to reduce the possibilities of the realization of the economic efficiency. Also, to capital demanders, higher taxes make investment cost increase and reduce capital needs; And to capital providers, high tax reduced capital gains, capital

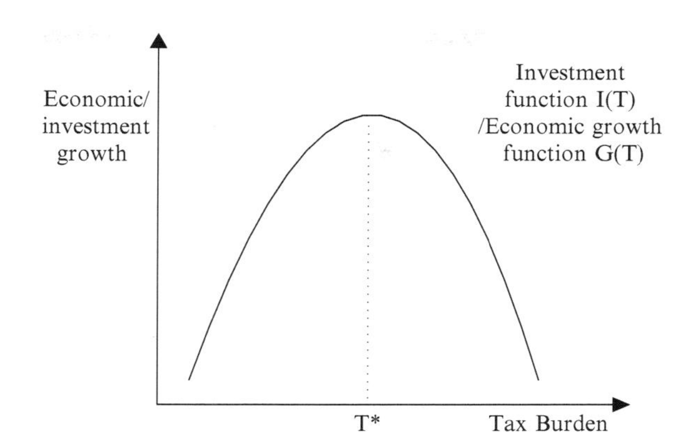

Fig. 1 Economic (investment) growth and macro tax burden

supply also reduced. Due to the high taxes that rising costs of production factors, yields down, and then to produce unfavorable effects on economic growth.

In China, as the Labor supply is rich, technological progress on economic growth contribution is low, so the economic growth is the main impetus for investment. We believe, the government's income scale and public spending level is low, public facilities and public service supply shortages, increases the marginal positive effect of public products provided by taxation, greater than the marginal cost of tax increases. Appropriate raising the proportion of taxes in GDP is good to expand investment and economic growth. As the tax amount has been bigger, tax increase again, because of the law of diminishing marginal returns, positive effect increases of tax provide public products will not make up marginal cost for the marginal tax, continue to improve the proportion of tax in GDP, go against stimulate social investment and economic growth. The net effect of tax depends on the relative size of these two forces. The tax and investment, economic growth is a nonlinear relation, investment. Economic growth is the size of the tax concave function. As shown in Fig. 1.

2.2 *Cobb-Douglas Production Function*

Cobb-Douglas production function is used to predict the countries and regions of industrial system or big enterprise's production and analysis of the development of production a means of economic mathematics model. It is the most widely used in economics a production function form, it in mathematical economics and economic metrology of research and application of important position. Cobb-Douglas production function is that

$$Y = A(t)L^{\alpha}K^{\beta}\mu$$

where Y is industrial output, A (t) is a comprehensive technical level, L is the labor force in number, K is put into capital, generally refers to net value of fixed assets, and α is the elasticity coefficient of labor, β is the capital of the output elasticity coefficient, μ said the influence of random disturbance, $\mu \leq 1$.

From this model, the main factors determining the level of development of industrial systems are the number of inputs of labor, fixed assets, and comprehensive technical level (including management level, the quality of the labor force, to introduce advanced technology, etc.). According to the combination of α and β, it has three types:

① $\alpha + \beta > 1$, it is called an incremental reward type, showed that the technique used to expand the scale of production to increase output is favorable.
② $\alpha + \beta < 1$, it is called a diminishing reward type, showed that the technique used to expand the scale of production to increase output is pyrrhic.
③ $\alpha + \beta = 1$, it is called a same reward type, showed that the production efficiency will not with the enlarging of the production scale and improve, only to improve the technical level, it will improve the economic benefits [6].

2.3 *Economic Growth and the Optimal Taxation Model*

Because of Cobb-Douglas production function is not suitable for use in the logistics industry, so I will find the right model.

2.3.1 The Household Department

I begin with endogenous growth models that build on constant returns to a broad concept of capital. The household departments use Ramsey [5] put forward an infinite-lived household in a closed economy utility function, as given by

$$U = \int_0^\infty \frac{C^{1-\theta} - 1}{1 - \theta} e^{-pt} dt \tag{1}$$

Household budget constraints,

$$\Delta a = ra + w - c \tag{2}$$

where $\theta > 0$, so that marginal utility has the constant elasticity $-\theta$. c is consumption per person, $p > 0$ is the constant rate of time preference, a is the household real assets, r is actual yields, w is wage.

Every person works a given amount of time; that is, there is no labor-leisure choice. As is well known, the maximization of the representative household's overall utility in Eq. (1) implies that the growth rate of consumption at each point in time is given by

$$\gamma = \frac{\Delta c}{c} = \frac{r - p}{\theta} \tag{3}$$

The symbol γ denotes a per capita growth rate. I assume that the technology is sufficiently productive to ensure positive steady-state growth, but not so productive as to yield unbounded utility.

2.3.2 The Enterprise Department

The Enterprise Sector use Cobb-Douglas function expansion mode:

$$\begin{aligned} &Y = AG^{\alpha}L^{\alpha}K^{1-\alpha} \text{ (Enterprise production function)} \\ \text{Or} \quad &y = Ag^{\alpha}k^{1-\alpha} \text{ (Per capita production function)} \end{aligned} \tag{4}$$

2.3.3 The Government Department

The Government departments carry out balance the budget, according to macro tax burden, tax is given by

$$t = g = \tau * y \quad \text{or} \quad \tau = \frac{g}{y} \tag{5}$$

Where t is government revenue and τ is the tax rate. I have normalized the number of households to unity so that g corresponds to aggregate expenditures and t to aggregate revenues. Note that Eq. (5) constrains the government to run a balanced budget.

Substituting Eq. (4) into Eq. (5) yields

$$g = (\tau A)^{\frac{1}{1-\alpha}} k \tag{6}$$

2.3.4 The Determination of the Optimal Taxation

Now, through the above three departments to push the basic formula, determined the optimal taxation.

The production function in Eq. (4) implies that the marginal product of capital is

$$\frac{\partial y}{\partial k} = A(1-\alpha)k^{-\alpha}g^{\alpha} = A(1-\alpha)\left(\frac{g}{k}\right)^{\alpha} \tag{7}$$

Under the tax rate τ, the return on capital is now

$$r = (1-\tau)\frac{\partial y}{\partial k} = A(1-\alpha)K^{-\alpha}g^{\alpha} = A(1-\alpha)(1-\tau)\left(\frac{g}{k}\right)^{\alpha} \tag{8}$$

With the presence of a flat-rate income tax at rate τ, this return is (1 − τ) * (∂y/∂k), where ∂y/∂k is given from Eq. (7).

Therefore, the growth rate of consumption is now

$$\gamma = \frac{\Delta c}{c} = \frac{r-\rho}{\theta} = \frac{1}{\theta}\left[A(1-\alpha)(1-\tau)\left(\frac{g}{k}\right)^{\alpha} - \rho\right] \tag{9}$$

Substituting Eq. (6) into Eq. (9) yields

$$\gamma = \frac{\Delta c}{c} = \frac{r-\rho}{\theta} = \frac{1}{\theta}\left[A^{\frac{1}{(1-\alpha)}}(1-\alpha)(1-\tau)(\tau)^{\frac{\alpha}{(1-\alpha)}} - \rho\right] \tag{10}$$

From Eq. (10), taxes on the influence of the growth rate is reflected in two aspects, 1 − τ representing macro tax burden on capital on behalf of after-tax marginal product negative effects, $\tau^{\frac{\alpha}{(1-\alpha)}}$ representing the government service to capital marginal product is effect after tax. If the economy no transformation dynamic, the government sets the capital k and production y to grow at the same rate τ.

From Eq. (10), derivation of the growth rate y on the tax rate τ, and according to the optimum conditions $\partial\gamma/\partial\tau = 0$, draw a conclusion that τ* = α. In a word, when the economic growth rate reached the maximum, the optimal taxation τ* is the output elasticity of the financial scale α.

The macro tax burden and economic steady growth rate is the relationship between the fall U. When τ < τ*, macro tax burden on the capital marginal product is effect after more than negative effects, economic growth with τ increase with the increase of the macro tax burden; When τ = τ*, negative effects and macro tax burden is equal to effect, the economic growth rate reaches maximum; When τ > τ*, macro tax burden of its economic impact of negative utility more than is utility, make the steady-state growth rate and began with the increase of the macro tax burden, and continue to drop. So in terms of the influence of on macro economy, the optimal macro taxation value should be τ*.

Therefore, the optimal tax rate in calculation, this article uses the Cobb-Douglas production function expansion mode:

$$LnY = C + \alpha_1 \ln G + \alpha_2 \ln K + \gamma \ln L + \varepsilon \tag{11}$$

Where Y is GDP, K is Gross Capital Formation, G is total government spending = budgetary expenditures + non-budgetary expenditures, and L is the total amount of the labor force in the employment. The optimal tax rate is r* = α1.

Study on the optimal taxation of logistics industry, I will use industry data to replace the variables of Cobb-Douglas production function expansion mode. Use the logistics industry business income as Y, because business income is the value of products and services that created by a industry; With the fixed assets and inventory of the logistics industry replace K, because the gross capital formation including the total fixed assets formation and inventory; Use the owner's equity of the logistics industry replaces G government spending, because it is equivalent to government spending to run an enterprise and must invest money, namely enterprise cost of investment–the owners' equity, so this paper used the owner's equity to replace government spending.

So in Eq. (11), Y is business income of the Logistics industry, K is the fixed assets and inventory of the logistics industry, G is the owner's equity of the Logistics industry, and L is the total amount of the labor force in the employment.

3 The Calculation of the Optimal Taxation of Logistics Industry

As a new industry, the statistics of logistics industry is not perfect. In our existing government statistics, it does not have independent statistic index system of logistics. Transportation and Storage industry is the most basic and important part of modern logistics, this paper analyzed the statistical data of transportation and Storage industry. According to the collected statistical data, this paper analyzed the state-owned transportation and storage industry, and listed companies of transportation and storage industry.

3.1 Calculate the Optimal Taxation of the State-Owned Transportation and Storage Industry

The calculation in this part used the state-owned enterprise of transportation and storage industry as the sample. Data mainly comes from 2002 to 2010 *Chinese Fiscal Yearbook, China Statistical Yearbook.*

3.1.1 Unit Root Stationary Test

Because of the existence of the time trend, the economical time series data usually is non-stationary, and used a non-stationary time series to regression another non-stationary time series can lead to a false return. To test whether the regression with meaning, it needs time series variables stationary test, this paper used Eviews software to test the steadiness of the data. Test results as is shown in Table 1.

Table 1 shows that all of the variable value in the statistical sense is located in the critical value, can't refuse to have a unit root and the original hypothesis that all variables is level is not stationary sequence. All one order difference variables are not located in the critical value, refused to have a unit root of the original hypothesis

Table 1 ADF unit root test results

Variables	ADF test statistic	Critical value	Exist unit root or not
lnY	3.955	−2.886	Y
lnG	3.238	−2.886	Y
lnK	5.911	−2.886	Y
lnL	−2.357	−1.996**	N
D(lnY)	0.556	−2.937	Y
D(lnG)	−0.950	−2.937	Y
D(lnK)	0.114	−2.937	Y
DD(lnY)	−1.728	−1.597*	N
DD(lnG)	−4.734	−3.206***	N
DD(lnK)	−3.075	−3.007***	N

*Represents 10% of the level of significance; **represents 5% of the level of significance; ***represents 1% of the level of significance

that is stationary series. Therefore, we can conclude that all of the above variables have a unit root, but one order or second-order difference variables have no unit root, meaning that these sequences or relatively stable sequence. Because the sequence contains sample size is small, so the sequence of second-order difference no unit root the event that the sequence is stable.

3.1.2 Regression Analysis

Through the regression analysis of the statistical software SPSS, we get the maximum likelihood estimation of parameters, standard deviation, coefficient and t-statistics, as shown in Table 2.

So, the regression equation is

$$\mathrm{LnY} = -14.293 + 0.0403 * \mathrm{lnG} + 1.543 * \mathrm{lnK} + 1.610 * \mathrm{lnL}$$
$$\left(\mathrm{Adjust\ R^2} = 0.988,\ \ \mathrm{F} = 221.035\right)$$

Regression showed that, the adjusted R square of the equation is 0.988, namely the adjusted R square is more is better; model can be a good explanation of the relationship between the variables, and through the level of significance test. lnK is in 1% of the level of significance, lnG and lnL are in 5% of the level of significance. Such as equation showed, $\alpha 1$ is 0.040. The optimal taxation of the state-owned transportation and storage industry is 4.0%.

According to our calculations, from year 2001 to 2009, the actual tax burden of the state-owned transportation and storage industry were 4.23, 4.29, 4.54, 4.49, 4.18, 4.91, 5.89, 5.22, 4.96%. They were all greater than the optimal taxation.

Table 2 Estimated coefficient of the state-owned transportation and storage industry

Variables	Coefficient	Standard deviation	t-statistics
Constant	−14.293	4.394	−3.253
lnG	0.040	0.248	0.162
lnK	1.543	0.308	5.016
lnL	1.610	0.512	3.146

Table 3 Estimated coefficient of listed companies in transportation and storage industry

Variables	Coefficient	Standard deviation	t-statistics
Constant	13.953	0.558	24.985
lnG	0.036	0.010	3.744
lnK	0.066	0.013	5.110
lnL	0.729	0.065	11.097

3.2 Calculate the Optimal Taxation of Listed Companies in Transportation and Storage Industry

This part calculated and analyzed the 79 listed companies of the transportation of storage industry that selected in Shanghai stock exchange and Shenzhen stock exchange. Rule out the listed companies that were ST and *ST, 59 listed companies of the transportation of storage industry were selected. The financial data is mainly from CSMAR database, and the related report data sited on Shanghai stock exchange and Shenzhen stock exchange.

Through the regression analysis of the statistical software SPSS, we get the maximum likelihood estimation of parameters, standard deviation, coefficient and t-statistics, as shown in Table 3.

So, the regression equation is

$$\begin{aligned} \mathrm{lnY} &= 13.953 + 0.036 * \mathrm{lnG} + 0.066 * \mathrm{lnK} + 0.724 * \mathrm{lnL} \\ &\left(\mathrm{Adjust}\ \mathrm{R}^2 = 0.721,\ \mathrm{F} = 92.946\right) \end{aligned}$$

Regression showed that, the adjusted R square of the equation is 0.721; model can be a good explanation of the relationship between the variables, and through the level of significance test. lnK, lnG and lnL are in 1% of the level of significance. Such as equation showed, α1 is 0.036. The optimal taxation of listed companies in transportation and storage industry is 3.6%.

According to our calculations, the actual tax burden of listed companies in transportation and storage industry was above 5%, the mean is 6.99%. There are 116 sample data more than the optimal tax, accounting for more than 67% of the total sample.

The calculations show that, the actual tax burden of listed companies in transportation and storage industry is greater than the optimal taxation.

4 Conclusions

Based on the theoretical analysis, I put forward the model used in this paper—Cobb-Douglas production function extension mode. The empirical analysis on the state-owned companies and the listed companies of transportation and storage industry, and come to conclusions that, (1) the optimal taxation of the state-owned transportation and storage industry is 4.0%, lower than the average lever of the actual tax burden from 2001 to 2009. (2) The optimal taxation of the listed companies of transportation and storage industry is 3.6%, lower than the average lever of the actual tax burden.

Affected by lack of Chinese logistics industry statistics, we only estimate the optimal taxation of state-owned logistics enterprises and listed companies of logistics. But, the two samples are the main force of the logistics industry in China, and the optimal tax calculation result is close, so we can believe that the calculation results are representative.

The results show that, the actual tax burden of China's logistics industry is higher than the optimal taxation. This implies that tax problem has restricted the normal operation of the logistics industry in China. Because of the relationship between tax and economic, tax burden achieves the optimal level, it will play the largest role in promoting the economy. Our country should take measures to reduce the tax burden of logistics industry, in order to promote the development of the logistics industry. In March 2009, the state council issued "The logistics industry restructuring and revitalization plan" (Guo Fa [2009] No.8), in August 2011, the general office of the state council issued Guo Fa [2011] No.38 "about promoting the healthy development of the logistics industry policy measures"; the suggestions are clear requirements about "earnestly alleviate the logistics enterprise tax burden". The Government should implement practically the measure to reduce the tax burden of the logistics industry as soon as possible.

Acknowledgment The research was supported by the key project of logistics management and technology lab.

References

1. Barro RJ (1990) Government spending in a simple model of endogenous growth. J Pol Econ 98(5):103–125
2. Dongmei W, Cairong Z (2011) Research on tax burden of transportation and storage industry in China. In: ICEIS2011, vol 4, SciTe Press, Beijing, pp 586–590
3. Dongmei W, Songdong J (2009) Research on taxation burden of logistics industry in china. Circ Econ China (1):25–28
4. Jie X (2012) To continue to push forward the value added tax system reform-improvement of structure adjustment of the taxation system. The Economic Daily, April 1, p 15
5. Ramsey FP (1928) A mathematical theory of saving. Econ J 38(152):543–559
6. Runsheng N (2010) The optimal tax burden and economic growth- empirical study based on the C-D production function and the Barro model. J Hunan Tax Coll 23(110):37–39

New Error Correction Scheme for Multi-level Optical Storage System

Zhang Xiaotian, Pei Jing, and Xu Haizheng

Abstract With the improvement of recording density of multi-level (ML) optical storage system, more serious burst errors will be caused by the same degree of scratches on the surface of disc in ML optical disc than conventional optical recording systems such as DVD. In this paper we present a new error correction code (ECC) scheme on the basis of improving original ECC scheme. The scheme is composed of two blocks. One is used to generate erasure information through the error-only decoding and the other carries out error and erasure decoding according to the erasure information provided from one. The code rate of the proposed scheme is up to 87.9% and the maximum correctable length of burst errors is 7104 bytes. The simulation result shows that the new scheme is qualified and effective for multi-level (ML) optical system.

Keywords ECC • Picket code • Channel coding • Multi-level • Optical disc

1 Introduction

Multi-level recording technology is a key technology to increase significantly the storage capacity without changing the optical parameters and mechanical structure of optical disc systems. In the optical disk system, the process of accessing data can be viewed as a special digital channel. As part of the channel coding, error correction code is generally used to improve the efficiency and reliability of the data sequence transmitted in the optical disk system. The original data are encoded with error correction code (ECC) to generate the data stream which contains a certain amount of redundancy information, and they will be acquired through the

Z. Xiaotian (✉) • P. Jing • X. Haizheng
Department of Precision Instrument and Mechanology, Optical Memory National Engineering Research Center (OMNERC), Tsinghua University, Beijing 100084, P.R. China
e-mail: zxt09@mails.tsinghua.edu.cn

Z. Zhang et al. (eds.), *LISS 2012: Proceedings of 2nd International Conference on Logistics, Informatics and Service Science*, DOI 10.1007/978-3-642-32054-5_53,

decoding process after the data stream is read out from optical disc. Even if some errors are produced by noise of channel, pollutions on the disc, servo system out of synchronization and so on, ECC can also check and correct them. Different methods of ECC are adopted to correct errors in different optical disc systems. The error correction codes of CD, DVD and BD system are designed based on Reed-Solomon code because RS code has a strong ability to correct both burst errors and random errors. The error correction code in DVD system is Reed-Solomon product code (RSPC) and Picket Code is adopted as an ECC in BD system.

Although multi-level technology can greatly increase the capacity of optical disc, it will reduce quality of readout signals and bring more errors in the read channel. Furthermore, the same degree of scratches or pollutions on the surface of disc will cause more serious burst errors because the recording density of ML DVD is substantially higher than that of DVD. RSPC is not qualified for ML DVD and many new ECC schemes are proposed to improve the performance of ECC. Youngki KIM, et al. propose a new ECC with syndrome check code for high-density optical storage [1]. Jun Lee and Seong-Hun Lee present a new error correction strategy for BD systems that performs error and erasure decoding using the only erasure information supplied from a modulation code decoder [2]. Liu Hailong et al. propose a new interleaving scheme for ML DVD based on Reed-Solomon(RS) code [3]. In this paper, a new and efficient ECC scheme is proposed to improve the correction capability for ML DVD.

In §2, we briefly descript the new error correction scheme and simulation results for the code performance are discussed in §3. The conclusion is in §4.

2 The Proposed Error Correction Scheme

2.1 The Structure of Frame

In the proposed error correction scheme the ECC block is composed of LDC block and BIS block. User data and address information unit are encoded respectively in LDC block and BIS block. The user data in the proposed scheme is recorded in 64 KB partitions, called LDC clusters, containing 32 data frame which consists of 2048 bytes of user data and 4 bytes error correction code (EDC). For the purpose of accessing the data on the disc, each data frame is assigned an address unit which contains 4 bytes of identification data (ID) and 2 bytes of ID error detection code (IED). The structure of a data frame is shown in Fig. 1a. Before the encoding procedure, main data of each data frame will be scrambled in order to ensure the differential phase detection and decrease the burden of inhibiting the low frequency components in modulation coding.

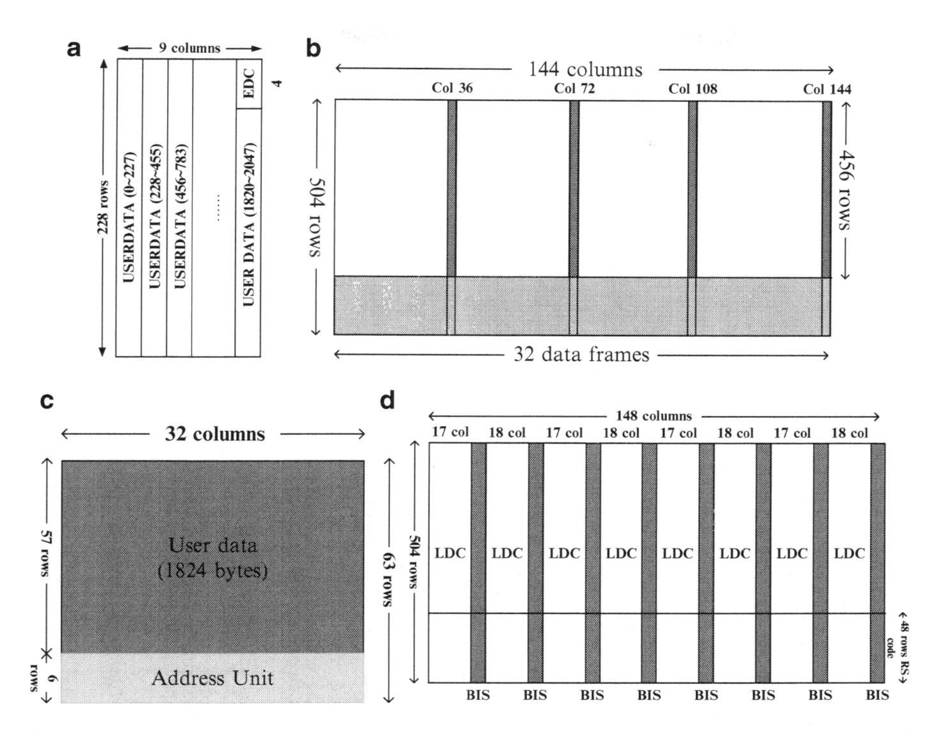

Fig. 1 (**a**) The structure of frame; (**b**) The structure of LDC cluster; (**c**) The structure of BIS cluster; (**d**) The structure of ECC cluster

2.2 The LDC Cluster Encoding Procedure

32 continuous scrambled data frames are combined into one block of data, these data are rearranged into an array of 228 rows × 288 columns by dividing each scrambled data frame into 9 columns. The RS (252, 228, 25) code is calculated in the directions of columns, called Long Distance Code(LDC) and 24 rows of generated LDC code is attached to ECC block every column. The LDC code is calculated by the formula (1):

$$R_j(x) = \sum_{i=228}^{251} D_{i,j}x^{251-i} = I_j(x)x^{24} \bmod G_{ldc}(x) \tag{1}$$

Here, $I_j(x) = \sum_{i=0}^{227} D_{i,j}x^{227-i}$, generated polynomial $G_{ldc}(x) = \prod_{k=1}^{24} (x + a^k)$, a is the primitive element of primitive polynomial based on GF(2^8).

After generating the LDC codeword, the LDC block is interleaved resulting in the LDC cluster. In the interleaving step the 288 columns of height 252 are rearranged into a new array with 144 columns and 504 rows. Each new columns is formed by multiplexing each even column from the LDC block, with the next odd column. As shown in Fig. 1b, after interleaving, User data in the four columns (column number is 36, 72, 108, 144) in the LDC cluster will be taken out as the data in the BIS block discussed below.

2.3 The BIS Cluster Encoding Procedure

BIS block contains 32 address units and 4 columns × 456 rows user data (1824 bytes) which are taken out from LDC block as mentioned above. Each address unit consists of 4 bytes of ID and 2 bytes of IDE. The encoding procedure for BIS block is given as follows:

Step1: 1824 bytes of user data are rearranged into an array of 57 rows × 32 columns and address units are rearranged into an array of 6 rows × 32 columns. The user data and address units are combined into an array of 63 rows × 32 columns as shown in Fig. 1c.

Step2: Calculating RS (64, 32, 32) code in the directions of rows. 32 columns of BIS code is attached to BIS block every row. The BIS code is calculated by formula (2):

$$R_i(x) = \sum_{j=32}^{63} B_{i,j} x^{63-j} = I_i(x)x^{32} \bmod G_{bis}(x) \tag{2}$$

here, $I_i(x) = \sum_{j=0}^{31} B_{i,j} x^{31-j}$, generated polynomial $G_{bis}(x) = \prod_{k=1}^{32} (x + a^k)$, a is the primitive element of primitive polynomial based on GF(2^8).

Step3: After generating the BIS codeword, the BIS block is mapped in an interleaved way into an array of 504 row × 8 columns. In the first step all columns of the BIS block shall be shifted over mod (k × 8, 63) bytes to the bottom, where k is column number, $0 \le k \le 63$. The bytes that shift out at the bottom side are re-entered in the array from the top side. Then the BIS block is split into 8 groups array of 63 rows × 38 columns. These eight groups are combined in turn into an array of 504 rows × 8 columns in the direction of columns. This ultimate formed array is called a BIS cluster.

2.4 The Structure of ECC

64 KB user data and its corresponding address units construct LDC cluster and BIS cluster by encoding. The LDC cluster is split into eight groups of 17 or 18 columns each. In between these eight groups, the eight columns from the BIS cluster are

Table 1 Comparison of the correction capability between RSPC, Picket Code and proposed code

	RSPC	Picket code	Proposed code
Code rate	32 × 1,024/(208 × 182) = 86.6%	64 × 1,024/(496 × 155) = 85.2%	64 × 1,024/(504 × 148) = 87.9%
Information block size	32 KB	64 KB	64 KB
Maximum correctable burst error length	16 × 182 = 2,912 bytes	64 × 155 = 9,920 bytes	48 × 148 = 7,104 bytes
Maximum correctable length of tangential scratches	6.21 mm	8.22 mm	11.90 mm

inserted one by one. After multiplexing the BIS cluster with LDC cluster, an array of 504 rows × 148 columns called ECC cluster is reached finally as shown in Fig. 1d.

3 Simulation Results

The new error correction scheme is realized on a software platform designed by OMNERC. We attain the correctable capability and performance of the proposed scheme by simulation. Table 1 shows the code rate and maximum correctable burst error length about RSPC, Picket Code and the proposed scheme. In this table, The maximum correctable burst error length of the proposed code is 7104 bytes and that of RSPC is 2912 bytes. The maximum correctable capability increases approximately 144.0%. In our signal waveform modulation (SWM) ML DVD system [4] the modulation code is bi-modulation code with code rate of 4/5 and the smallest pit is 2 T with the length of 335 nm. Accordingly, the modulation code is EFM plus with code rate of 8/16 and the smallest pit length is 3 T with length of 400 nm in common DVD system. Thus, the maximum correctable length of tangential scratches on the surface of ML DVD can be calculated with the formula (3). The maximum correctable length of tangential scratches on ML DVD system is improved about 91.6% compared with the common DVD system. The code rate of proposed scheme is up to 87.9% which increases about 3% than that of Picket Code. The track pitch of the SWM ML DVD disc is 0.68 μm and that of DVD is 0.74 μm . According to the storage capacity for DVD 9(8.5 GB), the expected capacity for ML DVD which adopts the proposed ECC scheme is obtained with the formula (4). The large code rate of the proposed scheme will ensure the storage capacity for SWM ML DVD can reach the storage capacity standard of 15 GB.

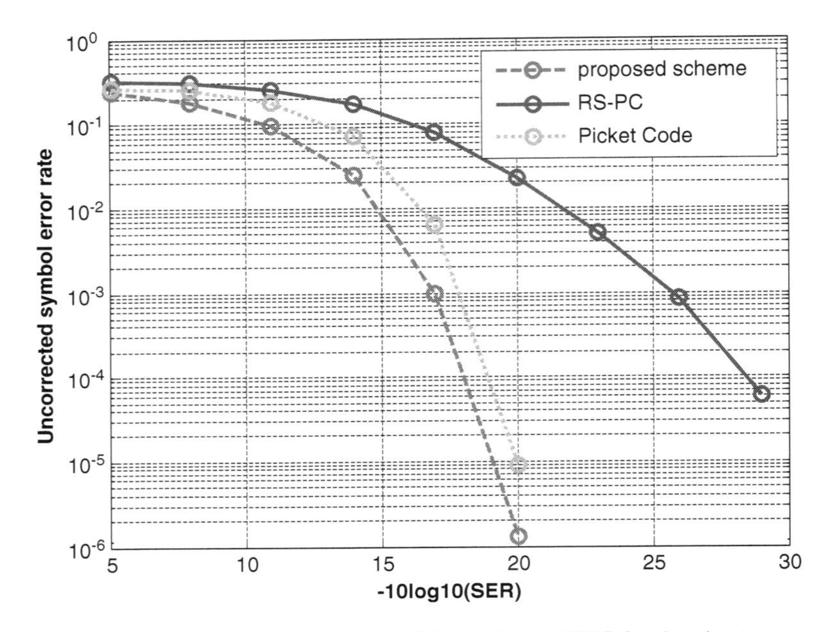

Fig. 2 The performance of proposed scheme, Picket Code and RSPC for short burst errors

$$\frac{7,104 \times 8}{4/5} \times (335/2) \times 10^{-6} = 11.90\ \text{mm} \tag{3}$$

$$8.5 \times \frac{4/5}{8/16} \times \frac{0.74}{0.68} \times \frac{87.9\,\%}{86.6\,\%} = 15.02\,(\text{GB}) \tag{4}$$

In our SWM ML DVD system, the types of errors are mainly consist of random errors and short burst errors in which the length is between 5 and 40 bytes. With the interval between BISs reduced to 17 or 18 columns from 38 columns which the interval is in the Picket Code, the error correction performance of the proposed scheme for short burst errors is more powerful than Picket Code. Figure 2 shows the performance of RSPC, Picket Code and the proposed scheme for short burst errors. In Fig. 2, the X-axis means $-10\log10$ (SER) and the Y-axis means the uncorrected symbol error rate after decoding. From simulation results, we can identify that the proposed scheme has better correction performance for short burst errors than RSPC and Picket Code.

4 Conclusions

To meet the requirement for error correction of multi-level optical disc, we proposed a new ECC scheme for multi-level optical disc. The simulation results confirm that it is a powerful error correction code. The error correction performance of the proposed

ECC scheme is largely better than that of RSPC. The code rate of proposed scheme is 87.9% and the maximum correctable burst error length of the proposed scheme is 7104 bytes. The maximum correctable length of tangential scratches on the SWM ML DVD surface is 11.90 mm. Furthermore, the performance of correct random errors of the proposed scheme is also as well as burst errors. Thus, we conclude that the proposed scheme can be qualified for high-capacity storage systems such as multi-level optical disc.

Acknowledgments This work is supported by the National Natural Science Foundation of China grant 60977005.

References

1. Youngki Kim, Jun Lee, Jaejin Lee (2004) A new error correction algorithm for high-density optical storage systems. Jpn J Appl Phys 43(7B):4867–4869
2. Jun Lee, Seong-Hun Lee (2010) A new error correction technique for BD systems. IEEE Trans Consum Electron 56(2):663–668
3. Liu Hailong, Pan Longfa, Hu Beibei et al (2009) New interleaving scheme for error correction code of multilevel optical storage system. Proc SPIE 7125:71250I-1–71250I-5
4. Tang Yi, Pei Jing et al (2008) Multi-level read-only recording using signal waveform modulation. Opt Express 16(9):6156–6162

A Study on the Performance Evaluation of Third-Party Logistics Enterprises Based on DEA

Xu Zhang, Weixin Luan, and Quande Cai

Abstract As an important part of the transportation industry, the third-party logistics plays an important role in national economy and foreign trade of our country. It is one of the basic industries with strategic meanings in the economic construction of China's socialism market. Based on the research findings of enterprise performance evaluation and combining with concept, characteristics and definition of third-party logistics, this paper discusses the business performance evaluation of third-party logistics enterprises. On this basis, a performance evaluation system of third-party logistics enterprises is established. Through the research on process of business performance evaluation of third-party logistics enterprises, the author introduces DEA theory to the specific evaluation of business performance of third-party logistics enterprises. Problems such as construction of input and output indicator system, choice of DEA model and analysis of evaluation, etc. are solved according to the methods and contents of DEA.

Keywords Third-party Logistics • Performance • DEA • Evaluation

1 An Overview of Performance Evaluation of Third-Party Logistics Enterprises

Third-party logistics is to provide services such as transportation, warehousing, dispatching and packaging, etc. for production enterprises to guarantee their core competitiveness and reduce their production costs [1]. The most distinctive characteristic of logistics enterprise engaging in logistics management is the logistics service customized for production enterprises. Different from the traditional vertical integration based on the control and management of ownership, the

X. Zhang (✉) • W. Luan • Q. Cai
Transportation Management College, Dalian Maritime University, Dalian 116026, P.R. China
e-mail: 86578692@163.com

Z. Zhang et al. (eds.), *LISS 2012: Proceedings of 2nd International Conference on Logistics, Informatics and Service Science*, DOI 10.1007/978-3-642-32054-5_54,

characteristic lays more emphasis on the coordination between organizations as well as management of cooperation and operation. In the circumstance of logistics management [2], the idea and emphasis of management have changed a lot. For example, the manager's attention is transferred from the internal control to coordination between internal and external control, and his target develops from a single enterprise into the enterprises cluster [3, 4]. Therefore, the evaluation on enterprise's operation performance has changed with the emphasis of logistics management. It is difficult for the traditional idea that emphasizes the performance of an independent department to push forward the development of logistics enterprise; hence the traditional evaluation ideas, indicators and methods of operation performance should be changed [5].

2 Process of Business Performance Evaluation of Third-Party Logistics Enterprises

The business performance evaluation of third-party logistics enterprises based on DEA actually means to apply the evaluation methods of DEA to the performance evaluation of third-party logistics. In the application, it is required to complete the preparatory work before evaluation and follow a certain process. The process of business performance evaluation of third-party logistics enterprises based on DEA includes: collection and arrangement of basic data, confirmation of evaluation target, establishment and completion of indicator system, selection and application of DEA model, output of evaluation result and analysis after evaluation, etc. Picture 1 is the flow chart 1 of the study on performance evaluation of third-party logistics enterprises based on DEA (Fig. 1).

3 Establishment of the Model of Business Performance Evaluation of Third-Party Logistics Enterprises

3.1 To Establish the Input/Output Set

The business performance evaluation of third-party logistics enterprises involves numerous indicators, where some correlation may be found. However, the correlative indicators not only make it more difficult to carry out the evaluation [8] but also cause ambiguity of the valuation, thus no unanimous conclusion can be drawn. Therefore, in the practical application, it is necessary to select the rational indicators without correlation to establish an input/output indicator set according to the methods and tools of mathematical theory. For the input/output indicator set established through selection, see Table 1.

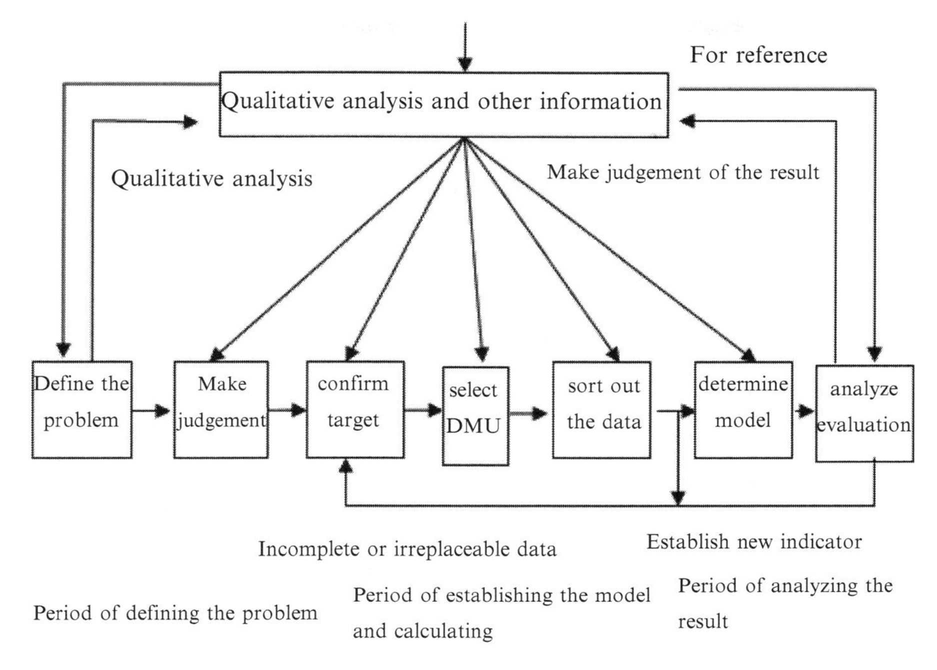

Fig. 1 Performance evaluation of third-party logistics enterprises based on DEA [6, 7]

Table 1 Input indicators and output indicators

Input indicator	Output indicator
Amount of investment in fixed assets	Yield of net assets
Actual load rate	User satisfaction
Capacity utilization	
Operation cost	

3.2 To Select the DEA Model

According to n years' relevant data of a third-party logistics enterprise, we refer to each year requiring evaluation as a decision making unit, which is called DMU_i. In each decision making unit, we choose four indicators, including amount of investment in fixed assets, actual load rate, capacity utilization and operation cost, as the input indicators and two indicators, including yield of net assets and user satisfaction, as the output indicators. Weight vector of input indicators: $v = (D_1, D_2, D_3, D_4)^T$[9]; weight vector of output indicators: $u = (u_1, u_2)^T$. The performance of an enterprise can be calculated by the ratio of the product of output indicators and weight vector of output indicators to the product of input indicators and weight vector of input indicators. Among them, h_j refers to the relative performance of the decision making unit j; x_{ij}

refers to the input i of the decision making unit j; y_{rj} refers to the output r of the decision making unit j; u_r refers to the weight vector of the output indicator r; v_i refers to the weight vector of the input indicator i. Since the output cannot exceed the input, therefore,

$$\sum_{r=1}^{S} u_r y_{rj} / \sum_{i=1}^{m} v_i x_{ij} \leq 1 \tag{1}$$

Among them, x_{ij} refers to the input i of the decision making unit j; y_{rj} refers to the output r of the decision making unit j; u_r refers to the weight vector of the output indicator r; v_i refers to the weight vector of the input indicator i. According to the DEA theory, u_r (refers to the weight vector of the output indicator r) and v_i (refers to the weight vector of the input indicator i) shall be no smaller than ϵ (refers to the non-Archimedean quantity) To sum up, the C^2R model (1) of h_j (refers to the relative performance of the decision making unit j) is as follows:

$$D = \begin{cases} \min \theta \\ s.t. \\ \sum_{j=1}^{n} \lambda_j Y_j + \omega = \theta X_{j_0} \\ \sum_{j=1}^{n} \lambda_j Y_j - \mu = Y_{j_0} \\ \lambda_j \geq 0, j = 1, 2, \ldots n \\ \omega \geq 0, \mu \geq 0 \end{cases} \tag{2}$$

DMU refers to the logistics enterprise requiring performance evaluation; h_j refers to the relative performance of the decision making unit j; u_r refers to the weight vector of the output indicator r; v_i refers to the weight vector of the input indicator i; x_{ij} refers to the input i of the decision making unit j; y_{rj} refers to the output r of the decision making unit j;

4 Case Study

4.1 Evaluation of the Data

In each decision making unit, we choose four indicators, including amount of investment in fixed assets, actual load rate, capacity utilization and operation cost, as the input indicators and two indicators, including yield of net assets and user satisfaction, as the output indicators. The operation data of the company in 2002–2008 is chosen, see Table 2.

Table 2 2000–2008 company-related data

	Input indicator				Output indicator	
Year	Amount of investment in fixed assets (100 million yuan)	Actual load rate	Capacity utilization	Operation cost (100 million yuan)	Yield of net assets	User satisfaction
2002	2.4	0.58	0.67	0.56	0.07	0.84
2003	2.5	0.74	0.65	0.40	0.10	0.93
2004	2.9	0.60	0.74	0.54	0.09	0.82
2005	2.25	0.65	0.86	0.48	0.10	0.95
2006	3	0.82	0.85	0.50	0.13	0.90
2007	2.9	0.59	0.71	0.52	0.08	0.86
2008	2.6	0.79	0.62	0.62	0.15	0.82

Table 3 Evaluation of the results of the efficiency of decision making units

Year	θ	Relative efficiency	Scale efficiency	Technology efficiency
2002	1.000000	DEA efficient	Scale increase	Efficient
2003	0.9372734	DEA inefficient	Scale increase	Inefficient
2004	1.000000	DEA efficient	Scale increase	Efficient
2005	0.9115698	DEA inefficient	Scale increase	Inefficient
2006	0.9855072	DEA inefficient	Scale increase	Inefficient
2007	1.000000	DEA efficient	Scale increase	Efficient
2008	0.9090475	DEA inefficient	Scale increase	Inefficient

Table 4 ω_i, μ_r variable

Year	ω_1	ω_2	ω_3	ω_4	μ_1	μ_2
2002	0.000000	0.000000	0.000000	0.3105E − 02	0.000000	2.500000
2003	0.4377E − 03	0.000000	0.4465E − 02	0.000000	0.000000	2.466509
2004	0.000000	0.000000	0.000000	0.3105E − 02	0.000000	2.500000
2005	0.1566E − 05	0.000000	0.1445E − 01	0.000000	0.000000	2.463702
2006	0.000000	0.000000	0.1449E − 01	0.000000	0.8351756	0.000000
2007	0.000000	0.000000	0.000000	0.3105E − 02	0.000000	2.500000
2008	0.000000	0.000000	0.1408E-01	0.000000	0.8116496	0.000000

4.2 Model Solution

Based on the DEA model, the author entered date into LINGO, for the evaluation of the results of the decision making units, see Table 3.

4.3 Analysis of the Results of Model

Through a comprehensive analysis of Tables 3 and 4, it can be seen that DEA is efficient in 2002, 2004 and 2007, when the input and output reach the optimal state.

At the meantime, the technology and returns to scale are efficient, which shows that the efficiency of input and output in that year reaches the optimum. In other years, DEA is inefficient, namely, the input is not fully used. Basically, the efficiency of third-party logistics enterprises corresponds to the prosperity of the third-party logistics market. In 2007, the market of third-party logistics was booming, hence the profit of third-party logistics enterprise increased relatively. In 2008, due to the influence of financial crisis, the depression occurred to the transportation market, hence it was difficult for the third-party logistics enterprise to survive and its performance went down. In 2003, 2005, 2006 and 2008, the DEA is inefficient, namely, the business performance of the enterprise in these years fails to reach the optimum. However, the increase of business scale tells us that although the performance fails to reach the optimum, yet it still represents growth, which shows that the enterprise is developing towards optimal performance due to the absolute influence of external environment.

5 Conclusion

This paper establishes a relatively rational system for performance evaluation of third-party logistics enterprises, and then gets the final evaluation system through detailed analysis. Starting from the characteristics of third-party logistics industry, the author chooses corresponding indicators such as industrial level, overall competition and future development, etc. and puts forward a simple indicator set composed by several factors for the performance evaluation of third-party logistics enterprises. What's more, this paper adopts DEA model, combines with the actual operation condition of a third-party logistics enterprise and uses LINGO software for calculation to evaluate the performance of the third-party logistics enterprise.

References

1. McAdams JL, Hawk EJ (1999) Organizational performance & rewards: experiences in making the link. American Compensation Association, New York, pp 663–665
2. Cools GK, van Prag M (2000) The value relevance of a single-valued corporate target [J]. An empirical analysis, Working paper, 710–712
3. MeNair CJ, Lynch RL, Cross K (2002) Do financial and nonfinancial performance. Manag Account 13(15):28–36
4. Kaplan R, Norton D (2001) The balanced scorecard measures drive performance. Harv Bus Rev 53(3):71–79
5. Rand GK (2000) Critical chain: the theory of constrains applied to project management. Int J Proj Manag 20(18):263–367
6. Handfield RBG (1996) Supply chain: best practices from the furniture industry. In: Proceedings annual meeting of the decision sciences institute, Boston, America, pp 26–27
7. Lee HL (2000) Information sharing in supply chain. Technol Manag 21(8):77–79
8. PBMSIG (2002) The performance management handbook: establishing integrated performance. Meas Syst 33(6):458–460
9. Chen Lliwen (2003) Theory model of item investment risky returns[J]. J Syst Sci Info 34(2):153–162

Manufacturers' Outsourcing Decision Based on the Quantity Competition

Zhang Chi, Ai XingZheng, and Tang XiaoWo

Abstract With the global competition is becoming increasingly fierce, outsourcing become a common phenomenon. In this paper we study manufacturers' outsourcing decision based on the quantity competition that produces alternative products. For the key parts for products, manufacturers select independent production, or outsourcing. If they select the same outsourcing supplier, the final product's alternative increase, competitions enhance.

Keywords Outsourcing • Alternative products • Quantity competition decision

1 Introduction

With the global competition is becoming increasingly fierce, outsourcing become a common phenomenon. Outsourcing can make the company focus on its product design and marketing [1]. The direct impact of outsourcing makes the customer feel product's difference decreases for the different manufacturers have the same parts, and then the product's alternative enhanced [2].

Outsourcing results in two additional impacts:

1. Double marginalization. When the supply chain composed by upstream suppliers and downstream suppliers in a state of decentralization, the phenomenon of "double marginalization" will appear [3].
2. Outsourcings to famous suppliers likely aggrandize the customs' acceptable value [4].

This paper discusses the outsourcing decision for manufacturers based on the quantity competition.

Z. Chi (✉) • A. XingZheng • T. XiaoWo
Information Technology and Business Management,
Chengdu Neusoft University, Chengdu, China
e-mail: zaipc@163.com

Z. Zhang et al. (eds.), *LISS 2012: Proceedings of 2nd International Conference on Logistics, Informatics and Service Science*, DOI 10.1007/978-3-642-32054-5_55,

2 Basic Model

We study the market with two manufactures produce alternative products. Suppliers can fulfill the manufactures' all parts demand. Each manufacturer select the outsourcing decision or not based on maximize profit. Symbols in this paper are regulated as follows:

The manufacturers and suppliers have the same technology, the unit production costs is c; w: Supplier's wholesale prices; c_0: Supplier's unit production costs; $w > c_0$;

For simplified calculation, assume that a product only need a spare part; and the other parts of the production cost is 0; based on the economics principle, the linear prices function is:

$$p_i = a_i - q_i - bq_j \quad i,j = A,B \quad i \neq j \tag{1}$$

Corresponding inverse demand function is:

$$q_i = [a(1-b) - p_i + bp_i]/(1-b^2) \tag{2}$$

a is the customers' identity value to product. b is the substitution degree between the two product and $0 < b < 1$; the higher the b the higher degree of substitution. When the two manufacturers select the same outsourcing supplier, the substitution degree will enlarge, is b', and $b' > b$; p_i is the retail price of manufacturers i; q_i is the corresponding demand [5].

1. If the two manufacturers all produce their product independent, the profit function of manufacturer i is:

$$\Pi_i = (q_A, q_B) = (a - q_i - bq_j - c)q_i \quad i = A,B \tag{3}$$

Record the two manufacturers all produce their product independent is II, similarly, OI, IO, OO respectively indicate A outsourcing, B production; A production, B outsourcing. Through the profit function (3) we can get the equilibrium quantity is:

$$q_i^{\Pi} = (a-c)/(2+b) \quad i,j = A,B \quad i \neq j \tag{4}$$

Based on (4), if $a \leq c$, then $q_i^{II} \leq 0$, that means manufacturer have no sell, Discuss is meaningless. In this paper, we assume the all equilibrium quantity is positive, $a > c$.

q_i^{II} is increasing function for a, decreasing function for b, decreasing function for. This proved the fact that in order to maintain price manufacturers prefer the lower alternative.

From (1) and (4) can gets the equilibrium price p_i^{II} is:

$$p_i^{\Pi} = \frac{(2-b)(a-c) - cb^2}{4-b^2} \quad i = A,B \tag{5}$$

p_i^{II} is decreasing function for alternative coefficient b. From (3) and (4) can get the manufacturers i's profit function $\pi_i^{\Pi} = \left(q_i^{\Pi}\right)^2$ is also the b's decreasing function. That means the higher the similar degree the lower the equilibrium price, and improve the similar degree will reduce manufacturers' cost, enhanced the competition intension. Meanwhile, it will reduce the manufacturers' profit.

2. If manufacturer i select outsourcing decision and manufacturer j produce their product independent, then i's profit function is:

$$\pi_i(q_A, q_B) = (\tilde{a} - q_i - bq_j - c)q_i \tag{6}$$

   Here a is the customers' identity value for the outsourcing product.

3. If both the two manufacturers select outsourcing decision, then manufacturer i's profit function is:

$$\pi_i(q_A, q_B) = (\tilde{a} - q_i - b'q_j - w)q_i \tag{7}$$

   In the function (6), if there is one manufacturer select outsourcing decision, we assume the outsourcing decision do not affect the products' similarity, which means it will not change the competitive degree, so the product's alternative coefficient is the same. When the two manufacturers all select outsourcing decision, the product's alternative coefficient will increase from b to b', and $b > b'$.

3 The Game Outsourcing Decision Process in the Condition of Single Supplier

We inspect the game behavior in the supply chain with only one supplier and two manufacturers having the same production capacity. The three stages developing game based on the completely information between the Supplier and the manufacturers as follow:

(a) Supplier decides the part's wholesale price: w;
(b) The two manufacturers decide wither to outsourcing or produce their product independent;
(c) The two manufactures decide the production quantity q_A and q_B simultaneously. Applying the reverse induction can get the Nash equilibrium conclusion. Firstly, get the manufacturers' quantity decision; then get the outsourcing decision.

3.1 Manufactures' Quantity Decision

Firstly, calculate the manufactures' equilibrium quantity. For example, when the two manufacturers all select outsourcing decision (O,O), get:

$$\pi_i(q_A, q_B) = (\tilde{a} - q_i - b'q_j - w)q_i$$

$$\text{Command}: \begin{cases} \dfrac{d\Pi_A(q_A)}{dq_A} = -q_A^2 + (\tilde{a} - b'q_B - w)q_A = 0 \\ \dfrac{d\Pi_B(q_B)}{dq_B} = -q_B^2 + (\tilde{a} - b'q_A - w)q_B = 0 \end{cases} \quad \text{Get}: \begin{cases} q_A = \dfrac{\tilde{a} - bq_B - w}{2} \\ q_B = \dfrac{\tilde{a} - bq_A - w}{2} \end{cases}$$

$$\text{Then get: } q_A = q_B = \frac{(\tilde{a} - w)(2 - b')}{4 - b'^2} = \frac{(\tilde{a} - w)}{2 + b'}$$

Through the calculation get $\pi_i^{mn} = (q_i^{mn})^2, m, n = I, O, i = A, B$; that means each manufacture's optimal profit is the square of corresponding order quantity. In the Chart 1 supplied the optimal quantity in the condition of fixed wholesale price of raw material.

From the upper chart get $q_A^{OI} - q_A^{II} = 2[(\tilde{a} - a) - (w - c)]/(4 - b^2)$, the margin decided whether positive or negative of the profit $(\Pi_A^{OI} - \Pi_A^{II})$. In the condition of manufactures B produce the product independent, manufacturer A' outsourcing decision based on the result of $\tilde{a} - a$ and $(w - c)$. When $\tilde{a} - a = w - c$, that means if customs regard the value change equal to the outsourcing cost change, manufacturer A' outsourcing decision will not affect the equilibrium quantity.

When $\tilde{a} - a > w - c$, manufacturer A prefer to outsourcing; when $\tilde{a} - a < w - c$, manufacturer A prefer to produce the product independent.

If the manufacturer's cost of production c_0 less than the manufacturer's cost of production c, that is $c_0 \leq w \leq c$, even if $\tilde{a} < a$, manufacturer A most probably select outsourcing decision. From the Chart 1 get, when there is only one manufacturer select outsourcing decision, if w enlarged, the equilibrium quantity of the outsourcing manufacturer will decrease; for the product independent manufacturer will increase. Some demand will transfer from the outsourcing manufacturer to the product independent manufacturer. When $\tilde{a} > w$, the two manufacturers can select outsourcing decision [6].

B \ A	I	O
I	$q_A^{II} = (a-c)/(2+b)$	$q_A^{IO} = [2(a-c)-(\tilde{a}-w)b]/(4-b^2$
	$q_B^{II} = (a-c)/(2+b)$	$q_B^{IO} = [2(\tilde{a}-w)-(a-c)b]/(4-b^2$
O	$q_A^{OI} = [2(\tilde{a}-w)-(a-c)b]/(4-b^2$	$q_A^{OO} = (\tilde{a}-w)/(2+b')$
	$q_B^{OI} = [2(a-c)-(\tilde{a}-w)b]/(4-b^2$	$q_B^{OO} = (\tilde{a}-w)/(2+b')$

Chart 1 Manufactures' equilibrium quantity

3.2 Manufacturers' Outsourcing Decision

Proposition 3.1. *The sufficient and necessary conditions for (I, I) become a sub game perfect equilibrium is:* $\tilde{a} - a < w - c$

Proving: when A select produce independent, only when $\Pi_B^{II} \geq \Pi_B^{OI}(w)$

$q_B^{II} - q_B^{IO} = \frac{(a-c)(2-b)-[2(\tilde{a}-w)-(a-c)b]}{4-b^2} = \frac{2[(a-c)-(\tilde{a}-w)]}{4-b^2} \geq 0$, that is $\tilde{a} - a \leq w - c$, B will select produce independent. Conversely, when B select produce independent, only when $\tilde{a} - a \leq w - c$, A will select produce independent.

Proposition 3.1 indicate: *when one of the manufacturers select produce independent, only when the customer's value accept degree changed not big than the affection of outsourcing, another manufacturer will select produce independent.*

Proposition 3.2. *The sufficient and necessary conditions for (O,O) become a sub game perfect equilibrium is:* $b < b' \leq b'^*(w)$, *and* $b'^*(w) = (\tilde{a} - w)(4 - b^2)/[2(a - c) - (\tilde{a} - w)b] - 2$

Proving: Because the $q_A^{IO}(w)$ and $q_A^{OO}(w)$ are all positive, and $\pi_i^{mn} = (q_i^{mn})^2$, so whether $\Pi_A^{IO}(w) - \Pi_A^{OO}(w)$ is positive or negative is decided by the $q_A^{IO}(w) - q_A^{OO}(w)$'s positive or negative. Assume:

$$f(b') = q_A^{IO}(w) - q_A^{oo}(w) = \frac{2(a-c) - (\tilde{a} - w)b}{4 - b^2} - \frac{\tilde{a} - w}{2 + b'}$$

Because $\tilde{a} > w$, f (b′) is b''s increasing function. Solve equations f (b′) = 0 get

$$b'^* = \frac{(\tilde{a} - w)(4 - b^2)}{2(a - c) - (\tilde{a} - w)b} - 2$$

Therefore, when manufacturer B select outsourcing decision, only when $b < b' \leq b'^*(w)$, manufacturer A will select outsourcing decision. When $b' > b'^*(w)$, manufacturer A will select produce independent.

On the contrary, according to the same reason, when manufacturer A select outsourcing decision, only when $b<b' \leq b'^*(w)$, manufacturer B will select outsourcing decision; when $b'>b'^*(w)$, manufacturer B will select produce independent.

Proposition 3.2 indicate: *when one of the manufacturers select outsourcing, only when the alternative coefficient of the product become more big, another manufacturer will select outsourcing decision, but the alternative coefficient not big than* $b'^*(w)$.

Proposition 3.3: *The sufficient and necessary conditions for (I, O) and (O, I) become a sub game perfect equilibrium is:* $\tilde{a} - a \geq w - c$ *and* $b' > b'^*(w)$

Proving: The proving process is just similar to the Proposition 3.1 and Proposition 3.2's proving process.

4 Conclusions

1. When one of the manufacturers i selects outsourcing decision, only when the alternative coefficient keeps relatively small, another manufacturer j will select outsourcing decision. If the alternative coefficient keeps relatively larger, competition strength will increase, in this case, manufacturer j's best select is to produce independent. Only when the influence of brand effect greater than the influence of outsourcing, manufacturer j will select outsourcing, that is: $\tilde{a} - a \geq w - c$. Obviously, the higher the manufacturer j's cost of production, the better the supplier's product quality, the manufacturer j would more prefer to outsourcing.
2. When the supplier's price keeps high and quality keeps low, the two manufacturers' best choice is to produce independent. If the change degree of the alternative coefficient keeps small, the two manufacturers' best choice is outsourcing. But when the equilibrium state achieved, may one select outsourcing and another select to produce independent; that means exist Asymmetric state (I, O) and (O, I).

With regard to the supplier's wholesale price w, solving equation $b' = b'^*, b' = \frac{(\tilde{a}-w)(4-b^2)}{2(a-c)-(\tilde{a}-w)b} - 2$, get the wholesale price's borderline $w^* = \tilde{a} - \frac{2(a-c)(b'+2)}{4-b^2+b(2+b')}$, and can get $\frac{dw^*}{db'} = \frac{2(a-c)(b^2-4)}{[b^2-(b'-2)b-4]^2}$, from $(\tilde{a} - c)$ and $0 < b < 1$ can get $\frac{dw^*}{db'} < 0$, $w*$ is b's decreasing function.

Because $b' > b$ get $w^* < \tilde{a} + c - a$.

To sum up, can come to the following conclusion:

1. When $w > \tilde{a} - a + c$, the two manufacturers' best choice is to produce independent;
2. When $w^* < w < \tilde{a} - a + c$, one select outsourcing and another select to produce independent;
3. When $c_0 < w < w^*$, the two manufacturers' best choice is outsourcing.

References

1. Forlani D, Mullins JW (2000) Perceived risks and choices in entrepreneurs new venture decisions. J Bus Ventur 15:305–322
2. Sitkin SB, Pablo AL (1992) Reconceptualizing the determinants of risk behavior. Acad Manage Rev 17:9–38
3. MacCrimmon KR, Wehrung DA (1988) Taking risks the management of uncertainty. Free Press, New York
4. March JG, Shapira Z (1987) Managerial perspectives on risk and risk taking. Manag Sci 33:1404–1418

5. Newall J (2003) Industrial buyer behavior. Eur J Mark 166–211
6. Valla JP (1982) The concept of risk in industrial buyer behavior. Presented at workshop on organizational buyer behavior, European Institute for advanced studies in Management, Brussels, 9–10 Dec 1982

Study on Mobile E-Commerce Business Process Optimization

Rongxiang Li

Abstract With the continuous development of e-commerce, mobile communication technology and wireless internet technology, mobile e-commerce has become a new business mode. Yet in the early development of the mobile e-commerce, it still exists many problems, such as high costs, low efficiency and small consumer group etc. This paper executes optimization on the mobile e-commerce according to characteristics of consumption habits of mobile e-commerce consumption groups and tries to solve existing problems in the development of mobile e-commerce.

Keywords Mobile e-commerce • Business process • Optimization

1 Development Overview of Mobile E-Commerce

The fast development of communication technology and internet technology has become the environmental basis of the mobile e-commerce. According to the IResearch Consulting's survey on China mobile e-commerce industry, in 2009 China's mobile e-commerce users already reached 36.684 million and increased 117.7% comparing to 2008. In 2011 the market scale of China mobile e-commerce reached 39.31 billion and it accounted for 30.5% of the overall market scale. It can be seen from the above data that mobile terminals scale and mobile e-commerce market transaction scale are very huge in our country. Mobile e-commerce will become the most potential and the hottest market in this new historical period.

However, at present the mobile e-commerce lack of innovation and operation mode. And its mobile terminal payment is not convenient enough. The clarity of its merchandise browse is not high and the input of information is slow. All of these

---

R. Li (✉)
Computer Science Department, Northeast Petroleum University,
Qinhuangdao City, Hebei Province, P.R. China
e-mail: lrxteacher@163.com

Z. Zhang et al. (eds.), *LISS 2012: Proceedings of 2nd International Conference on Logistics, Informatics and Service Science*, DOI 10.1007/978-3-642-32054-5_56,

above reasons led to the mobile terminal users' low using frequency of mobile e-commerce. In order to improve the satisfaction of terminal users, we need a more humane mobile e-commerce process.

2 Introduction of Mobile E-Commerce Business Process

2.1 Definition of Mobile E-Commerce Business Process

Mobile e-commerce is a new e-commerce model to transmit data via mobile communications networks and using mobile terminal devices, such as mobile telephone or PDA, to carry out various business activities [1]. It makes it possible for customers to carry on shopping, business trading, online payments, other business activities, finance activities and related service activities at any time and place. Mobile e-commerce overcomes modern business's limitations on time and space. It is a new integrated information services which is at the closest distance to the target consumer groups and has great market demand. Mobile e-commerce business process is mean of the purpose of accomplishing company goals or transaction tasks, using mobile e-commerce information technology to conduct a series of related business activities. Mobile e-commerce business process optimization means to fundamentally rethink the business process of the company, using IT and mobile terminal equipment to seek significant improvement in cost, quality, service and speed [2]. The purpose of the optimization is to reduce business costs, promote the efficiency of business process, and enhance user experience and satisfaction. The key to the optimization is "how to run the present business in a better way".

2.2 General Process of Mobile E-Commerce

Business process is very important for mobile e-commerce. The future mobile e-commerce system is an integration of the actual logistics and capital flow. And it is an information flow that reflects logistics process. The general process of mobile e-commerce mainly includes six steps: preparations before transactions, business negotiations, signing the contracts, contracts implementation, payments and after-sale services.

According to the characteristic analysis of the mobile e-commerce terminal users' consumption behavior, mobile e-commerce platform should has the following functions, such as searching for products and services quickly, browsing products and service information conveniently and simplifying transaction functions so as to make it possible for mobile terminal users to carry out online transactions on mobile devices conveniently. So on the basis of general e-commerce process, mobile e-commerce process should better meet various needs such as user search, browse and payment.

2.3 *Analysis of Mobile E-Commerce Business Process*

Mobile e-commerce process lack of humanity and its function is not perfect. With the development of technologies, mobile devices have been constantly updated. E-commerce is also developing rapidly. When the majority of consumers just get used to the e-commerce transaction model, now they have to adapt to the mobile e-commerce terminals and platforms. Therefore, it requires the mobile e-commerce business process to be humanistic, secure and easy to be operated. But now, the searching function of the mobile e-commerce platform is not perfect enough. Its convenience of operation and learning, searching ability and correcting ability also need to be improved.

The payment of the mobile e-commerce is inconvenient and its security is low. Now mobile e-commerce is still in early stages of development. There are still many problems, such as low security, inconvenience of mobile terminal applications etc. For instance, when using Mobile CTOC E-commerce site, the security of mobile payment has following risks: consuming phone memory, stealing all the information and data in the phone, spreading illegal and harmful information, forced consumption caused by the code to control the phone leading to the owner's sharp increase in communication costs and information costs [3].

The service timeliness of the mobile e-commerce is poor. There are communication delays, SMS delays and even packet loss phenomenon. The existing mobile network still has certain limitations, which need to be improved by the related technical departments. It is the basic service for mobile e-commence to send related information among sellers, payment operators and banks accurately. This information include handling fees, communication costs, information costs, discounted charges, bad debts, refunds and on-time delivery of relevant information. Cooperated trading parties should improve service efficiency in accordance with country's finance and taxation management regulations.

3 Optimization Model of Mobile E-Commerce Business Process

3.1 *Process Optimization Objectives*

The main objectives of mobile e-commerce business process are to embody the humane of process design, improve service efficiency, promote utilization rate and satisfaction of clients, and increase effective consumer groups. It will achieve open and global commodity businesses, decrease the cumbersome links in trading, lead to direct transaction between producers and consumers and change the whole socio-economic running pattern eventually.

The optimized mobile e-commerce business process will change the traditional business process to be electronic, digital and mobile. Firstly, it uses electron flow to replace real logistics which can reduce manpower, material resources and expenses.

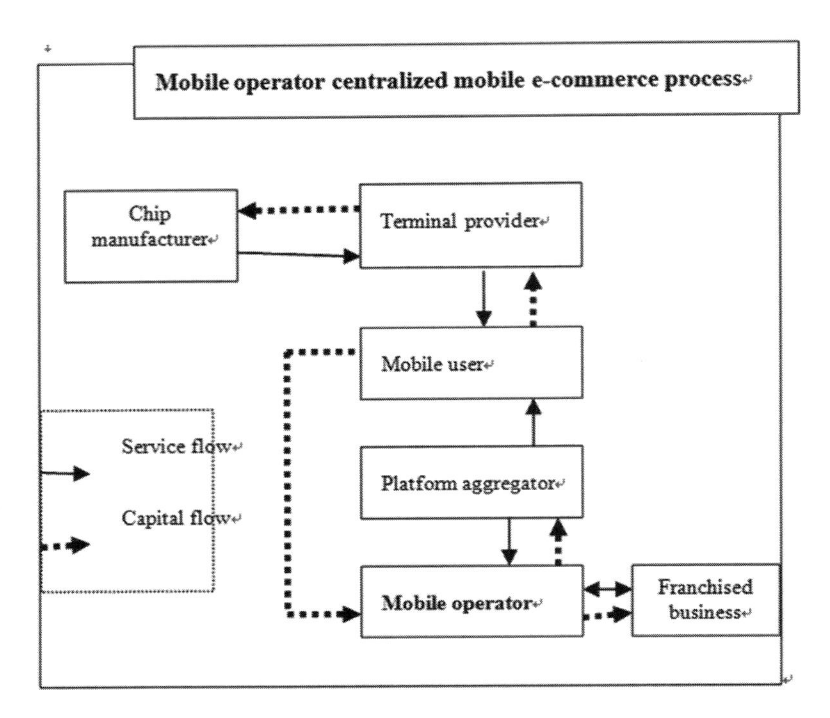

Fig. 1 Mobile e-commerce process optimization model

Secondly, it breaks the restrictions of time and space, making it possible to conduct the transaction at any given time or places, which greatly improves trading efficiency. The whole business process includes consumer, seller, mobile network platform, payment platform, mobile banking, logistics and many other subjects. After the optimization, the business process should be of clear level and division of labor. It will not appear to be disorganized and each link will be a clear division of labor and close cooperation. So that it can realize the common beneficial targets of mobile e-commerce value chain finally.

3.2 *Optimized Mobile E-Commerce Business Process*

Optimized mobile e-commerce process focuses on end-user experience of mobile e-commerce platform, and with the huge amount of customer information of mobile operator, integrates the resources of commercial service providers, to develop a secure third party payment platform [4]. Optimization model is as Fig. 1.

The above process integrating the traditional process, optimizing past complex parts, makes the mobile e-commerce process even faster, more convenient, and easy-to-understand.

By mobile devices, the consumer visit mobile e-commerce website platform to position target product and to search service information. When it find target product and service information, the consumer need to conduct lateral comparison among obtained information through different platforms, and then choose the best purchase program, and confirm product order.

Through the order, the seller confirms with the consumer, informs the way of charging, and makes preparation for delivery – logistics and distribution. After receiving goods and acceptance checking, the consumer delivers the payment by mobile billing platform. Then third-party platform informs buyer and seller the trading result, and then the transaction is completed. After the completion of the transaction, the consumer gives evaluation about the product and service, therefore the seller gets consumer's feedback in time.

3.3 Value Analysis of Optimization Model

The value of the mobile e-commerce process lies in the following aspects:

Firstly, shopping interface should be clear, concise and convenient hence to save consumers' time on adapting user interface. So it will be easy to understand and operate. Information publishing can be combined with mobile electronics characteristics, which can be achieved by adding telephone counseling, SMS, MMS and e-mail to notify consumers.

Secondly, in technical aspect, it should enhance the technical support of related mobile companies and network departments, relying on the production website construction and webpage designer's feeling over preferable model to display the goods to consumers.

Thirdly, Mobile payment is a business method through mobile terminal to pay the purchased goods or complete payment activities. By combing the consumer's bank accounts with phone number, consumers can operate their own bank accounts via SMS, voice, GPRS, etc. to complete payment, consumption and other functions, and through text messages and other ways to inform the result of transaction and account changes [5]. There are different trading methods. When facing complex trading activities, we need to establish a unified price standard and price system, hence to solve mobile price problems in mobile e-commerce activities.

Finally, in logistics aspect, using automatic logo recognition methods, establishing perfect designated goods pick-up locations, and with the support of electronic automatic identification and goods selection devices, it can realize a network system with multi-company and multi-client intent delivery process.

The optimized mobile e-commerce process achieves the integration of mobile e-commerce operators, platform operators and other resources. It chooses the adapted operation mode and provides a safe and humane business environment for terminal users finally.

4 Mobile E-Commerce Process Optimization for Yiwu Small Commodity City at Qinhuangdao

4.1 Mobile E-Commerce Project Overview of Yiwu Small Commodity City

Qinhuangdao Yiwu small commodity city was developed and constructed by Qinhuangdao North Logistics Company. It was put into operation in 2011. Its main functions include: small commodity wholesale trading center, logistics information center, transportation center, e-commerce center and services that support for projects such as financial, insurance, industry and commerce, taxation and catering.

Yiwu small commodity city innovates business ideas, takes entity shops as the basis, and signs an agreement with China Unicom to establish virtual online "Vendor Network" and "mobile phone shop". This is the new pattern for mobile e-commerce management which merges three shops to one.

The main functions of "mobile phone shop" include: first, communication, namely, building call center in the park of management companies in Qinhuangdao Yiwu small commodity to ensure the timely communications. The communications between internal group networks are free of charge. At the same time, provides free 3 G mobile phones for the majority of merchants to achieve better communications. Second, sales, namely, provides "Vendor Network" e-commerce platform shops and phone shops to help merchants to do network marketing. Third, logistics, namely, let the logistics companies which pass through the examinations locate in the small commodity city, and achieve unified pickup and delivery management. Fourth, settlement, namely, Minsheng Bank will provide POS machines, VIP services and payment platform on networks. Minsheng Banks will also provide mobile phone payment tools which achieve capital payment based on 3 G phone and provide more comprehensive and convenient electronic settlement methods for merchants. Five, security, namely, it has network eyes which can not only watch shops and goods through the internet and mobile phones, but also can carry out real-time monitoring. So the shopkeepers are more at ease. Six, management, namely, through network office automation management software which was developed by China Unicom to interact with shopkeepers, release news, download files, and execute remote video conference and training.

The appearance of "mobile phone shop" added a large sales channel for network shops. In fact, it achieves mobile office which means the merchants can process customer consultation, deliveries and settlement business at any time and space. It can not only enhance the work efficiency greatly, but also conduct its own product marketing in the mobile phone customer groups. Meantime, "mobile phone shop" can also use mobile phones to operate "network eye" which can achieve the real-time watch to monitor shops and goods at a full range. It can also achieve a lot more humane functions.

At present, Qinghuangdao Yiwu small commodity city draws lessons form the business model of Zhejiang, and improves the model by establishing online stores and mobile phone shops and merging the three shops to one. Its business process has become more and more perfect, but there are still some deficiencies and defects. The small commodity city invested a great amount of capitals. Though the software and hardware environment of mobile e-commerce are complete at present, its marketing is still insufficient and not yet formed an effective user groups and the loyal customers.

4.2 Mobile E-Commerce Business Process Optimization of Yiwu Small Commodity City

Through the research of Yiwu small commodity city and "mobile phone shop", we summed up a set of appropriate solutions for its mobile e-commerce process. Before the transaction, the buyer searches commodity supply in Qinghunagdao Yiwu small commodity city according to their own needs, makes a shopping plan, carries out market research and analysis and understands the market condition. Though the mobile communication terminals, they can understand the new situations whenever and wherever, and then complete the consulting activities. The businesses will do market positioning, determining the goods supply and do preparation s for shops. Before setting up shops, the businesses should contact everything and sign the contracts to ensure that customers can receive goods in best time. Buyer and businesses can login the vendor network and register through mobile devices. Buyers search in the goods released by businesses, select goods and then determine which they want to purchase. Buyers use mobile phone bank tools or other online tools for payments. Then through information, businesses will inform the buyer that payment is successful completed and they will deliver goods immediately. Buyers will confirm the good receipt after they get the goods and then the money will be paid to the businesses from the bank. When there is money circulating among buyers, sellers and banks, both the buyer and seller will receive the responding mobile phone SMS alerts to enhance the security of payment. At last, buyers and sellers will evaluate each other. The evaluation can improve the trust for other consumers to purchase again.

Along with the advance of the network information integration, Qinghuangdao Yiwu small commodity city will realize all kinds of modern information communication technologies, fully utilize and expand the network advantages, realize the synchronous office exchange at different time and different position and improve work efficiency. It will achieve centralized management, unified control and try to be the first-class logistics information management park.

References

1. Zeng Sheng (2010) Simulation research based on Arena mobile e-commerce process. Beijing Jiaotong University Master's Thesis 07
2. Sun Dajiang (2011) Problems and countermeasures of China's mobile e-commerce. Cooper Econ Technol 20:126–127
3. Cao Shurong, Tian Cui (2010) Research on domestic and foreign mobile e-commerce development. Technol Sq 7:222–223
4. The twenty-sixth statistical report of Internet development of china. Internet Network Information Center of China. [EB/OL]. http://www.cnnic.cn/hlwfzyj/hlwfzzx/qwfb/201101/t20110124_31170.htm. 15 July 2010
5. Wu juebo, Cao hui, Lu zhelu (2012) Foreign mobile e-commerce business model and its enlightenment to China. E-commerce 2:21–23

Ordering Decision-Making Model for a Dual Sourcing Supply Chain with Disruptions

Jingjing Zhu and Shaochuan Fu

Abstract This article mainly captures the trade-off between ordering policies and disruption risks for an unreliable dual sourcing supply network. Stochastic newsvendor models are presented under both the unconstrained and fill rate constraint cases. The models can be applicable for different types of disruptions related among others to the supply of raw materials, the production process, and the distribution system, as well as security breaches and natural disasters. Through the model, we obtain some important managerial insights and evaluate the merit of contingency strategies in managing uncertain supply chains.

Keywords Supply chain disruption • Dual sourcing • Ordering decision • Risk aversion • Stochastic model

1 Introduction

As companies throughout all industries continue to globalize their operations and outsource significant portions of their value chain, they often end up relying heavily on order replenishments from distant suppliers [1]. The use of long-distance sourcing and the reliance on few key suppliers are exposing procurement process to increasing risk and disruption. Such trends have placed enormous pressures on supply chains.

This paper focus mainly on the disruption risks of a dual sourcing supply chain. Specifically, generic newsvendor stochastic ordering models for risk-neutral and risk-averse decision-makers are proposed for a supply chain network of two unreliable competing suppliers and one retailer. The main objective is to capture

---

J. Zhu (✉) • S. Fu
School of Economics and Management, Beijing Jiaotong University,
100044 Beijing, P.R. China
e-mail: 11120683@bjtu.edu.cn; fushaochuan@263.net

Z. Zhang et al. (eds.), *LISS 2012: Proceedings of 2nd International Conference on Logistics, Informatics and Service Science*, DOI 10.1007/978-3-642-32054-5_57,

the trade-off between ordering policies and disruption risks, assuming that both suppliers are susceptible to disruption risks. The consideration of two suppliers with different procurement prices and disruption probabilities, differentiates this work from the existing literature for dual-sourcing supply chains.

2 Relevant Literature

The design and execution of appropriate approaches can play a critical role in handling risks and disruptions. Towards this direction, the literature dealing with the joint tackling of yield/inventory and risk management appears to be growing during the last decade. Xia et al. [2] developed a deterministic EOQ-type inventory model for a two-stage supply chain that is susceptible to production-rate disruptions. Tiaojun Xiao and Xiangtong Qi [3] investigated a one supplier–two competing retailers supply chain that experiences a disruption in cost and demand during a single period. Moreover, Chahar and Taafe [4] formulated a stochastic programming model for the single sourcing case, in which the supplier is susceptible to risks of disruption.

Proceeding to the dual and multiple sourcing research papers, Tomlin [5] developed a Markov chain single-product model by considering capacity constraints for both suppliers and order quantity flexibility for the reliable vendor. Amanda and Lawrence [6] consider one case where a firm's only sourcing option is an unreliable supplier subject to disruptions and yield uncertainty, and a second case where a second, reliable (but more expensive) supplier is available. They develop models for both cases to determine the optimal order and reserve quantities.

Although the literature covers several risk and yield management settings, in this paper, we attempt to bridge explicitly disruption management and risk aversion issues.

3 Model Formulation

3.1 Unconstrained Model

We propose a single period inventory system where a single ordering decision is to be made before the sales period begins (thus, emergency replenishment is not allowed), so as to maximize expected total profit. When a disruption occurs the supplier can provide nothing to the retailer. We denote with p_j the probability of a supply chain disruption. Demand X is assumed to be a positive stochastic random variable with probability density function f(x) and cumulative distribution function F(x). c_j is the unit purchase cost paid to supplier j ($j = 1, 2$). The unit selling price is denoted by s and it is assumed that $s > c_j$ ($j = 1, 2$). The surplus stock that remains unsold at the end of the period can be sold to a secondary market at a unit salvage value g, it is assumed that $g < c_j$ ($j = 1, 2$). In addition, b indicates the lost sales cost.

Initially, when there is no disruption, the expected profit $\pi_0(Q_1, Q_2)$ is obtained by the classical newsvendor problem analysis:

$$\pi_0(Q_1, Q_2) = \int_0^{Q_1+Q_2} [sx - c_1Q_1 - c_2Q_2 + g(Q_1 + Q_2 - x)] f(x)dx + \int_{Q_1+Q_2}^{\infty} [s(Q_1 + Q_2) - c_1Q_1 - c_2Q_2 - b(x - Q_1 - Q_2)] f(x)dx \quad (1)$$

When a disruption occurs to the first supplier's channel (with probability p_1), only a portion of Q_2 initially ordered from supplier 2 can now be employed to satisfy demand. The expected profit is:

$$\pi_1(Q_1, Q_2) = \int_0^{Q_2} [sx - c_2Q_2 + g(Q_2 - x)] f(x)dx + \int_{Q_2}^{\infty} [sQ_2 - c_2Q_2 - b(x - Q_2)] f(x)dx \quad (2)$$

Similarly, when a disruption occurs to the second supplier's channel (with probability p_2), the expected profit is:

$$\pi_2(Q_1, Q_2) = \int_0^{Q_1} [sx - c_1Q_1 + g(Q_1 - x)] f(x)dx + \int_{Q_1}^{\infty} [sQ_1 - c_1Q_1 - b(x - Q_1)] f(x)dx \quad (3)$$

Moreover, when disruptions occur simultaneously to both suppliers (with probability p_1p_2), the expected profit $\pi_{12}(Q_1, Q_2)$ is given by:

$$\pi_{12}(Q_1, Q_2) = -\int_0^{Q_1+Q_2} bxf(x)dx - \int_{Q_1+Q_2}^{\infty} bxf(x)dx \quad (4)$$

Finally, the total weighted expected profit $\pi(Q_1, Q_2)$, considering all possible combinations of disruption events on none, on one or on both supply chains is:

$$\pi(Q_1, Q_2) = (1 - p_1)(1 - p_2)\pi_0(Q_1, Q_2) + p_1(1 - p_2)\pi_1(Q_1, Q_2) + (1 - p_1)p_2\pi_2(Q_1, Q_2) + p_1p_2\pi_{12}(Q_1, Q_2) \quad (5)$$

Therefore, the optimization model which represents the maximization of the total weighted expected profit $\pi(Q_1,Q_2)$, is:

$$(\text{P}) : \max \quad \pi(Q_1, Q_2)$$

Through calculation, Eq. (5) is proved to be negative definite and thus concave to the optimal order lot sizes. Thus, the maximum value of $\pi(Q_1, Q_2)$ is attained for Q_1^* and Q_2^* (optimal order lot sizes) by solving the system of Eqs. (6).

$$(1 - p_2)F(Q_1 + Q_2) + p_2F(Q_1) = (s - c_1 + b)/(s - g + b) \tag{6}$$

$$(1 - p_1)\,F(Q_1 + Q_2) + p_1F(Q_2) = (s - c_2 + b)/(s - g + b) \tag{7}$$

3.2 *Model with Fill Rate Constraint*

The basic model (P) corresponds to risk neutral decision-makers. Model (P) can be extended through the consideration of a fill rate constraint in order to take also into account the risk aversion factor. Fill rate r measures the part of the stochastic demand that is met from the delivered quantity of products. As motivated earlier, risk-averse decision makers would prefer more "conservative" policies that lead to larger order quantities by setting an appropriate service level constraint. The resulting optimization model, which represents the maximization of the total weighted expected profit subject to a fill rate constraint is:

$$(\mathrm{Pr}) : \max \quad \pi(Q_1, Q_2)$$

Subject to:

$$r > r_0$$

With:

$$r = 1 - \frac{\text{Expected number of stockout units}}{\text{Mean demand}} = 1 - \frac{E(n(Q1,\ Q2))}{\mu}$$

$$E(n(Q_1, Q_2)) = (1 - p_1)(1 - p_2)\int_{Q1+Q2}^{\infty}(x - Q_1 - Q_2)f(x)dx + p_1(1 - p_2)$$
$$\int_{Q2}^{\infty}(x - Q_2)f(x)dx + (1 - p_1)p_2\int_{Q1}^{\infty}(x - Q_1)f(x)dx + p_1p_2\int_{0}^{\infty}xf(x)dx$$

The Lagrangian relaxation can be used to obtain the optimal order quantities Q_1^*, Q_2^* and the global maximum value for problem (Pr).

Table 1 Optimal ordering quantity and total weighted expected profit based on p_1 and p_2

p_2 \ p_1 (Q_1^*, Q_2^*, π)	0	0.05	0.1	0.15	0.2
0	(600,0,4200)	(600,0,3615)	(462,138,3092)	(308,292,2862)	(231,369,2746)
0.05	(600,0,4200)	(600,0,3615)	(509,95,3071)	(384,228,2753)	(308,308,2562)
0.1	(600,0,4200)	(600,0,3615)	(534,73,3060)	(432,187,2684)	(363,264,2430)
0.15	(600,0,4200)	(600,0,3615)	(550,59,3053)	(466,158,2636)	(404,231,2331)
0.2	(600,0,4200)	(600,0,3615)	(560,49,3048)	(490,137,2601)	(436,205,2254)

4 Numerical Analysis

We assume that the unit selling price is $s = 45$, the unit purchase cost paid to supplier 1 is $c_1 = 21$, to supplier 2 is $c_2 = 24$, the salvage value of unsold products is $g = 10$, the shortage cost is $b = 15$. Moreover, it has been assumed that the retail firm faces a demand with a uniform distribution pattern; the maximum demand is equal to 1,000 units, while the minimum demand is equal to 0 unit. Table 1 illustrates the effect of various combinations of p_1 and p_2 on the optimal ordering quantity and then total weighted expected profit.

It is observed that when the probabilities of a disruption on the first channel lower than 5%, the retailer utilizes only supplier 1.While as the probabilities increases, the optimal solution moves from a solution that mainly utilizes the first supply chain to a solution that mainly utilizes the second one. When the disruption probabilities of two suppliers are the same, it is obvious that ordering more form supplier 1 brings better result because of his lower purchase cost.

5 Summary and Conclusions

An effective disruption management strategy that enhances supply chain resilience is a necessary component of a firm's overall hedging strategy. In this article, we examined the trade-off between ordering policies and disruption risks for an unreliable dual sourcing supply network. Stochastic models are presented under both the unconstrained and fill rate constraint cases. While this paper considered only the profit of retailer and the objective was retailer's profit maximization. In fact, suppliers also make decisions in consideration of their own benefit. Thus, further research can extend to the whole supply chain, exploring the best solution for both retailers and suppliers, finally the whole supply chain. Future research directions to this work could also include the extension of the proposed models for multiple types of products, for more than two supply sources, and for supply chains of more tiers.

Acknowledgments We thank the reviewers for comments that helped to sharpen the final version of the article.

References

1. Patterson JL (2007) Supply base risk assessment and contingency planning in emerging markets. In Electrical insulation conference and electrical manufacturing expo, Nashville, TN, United states 10:342–347
2. Xia YS, Yang MH, Golany B, Gilbert SM, Yu G (2004) Real-time disruption management in a two-stage production and inventory system. IIE Trans 36:111–125
3. Tiaojun Xiao, Xiangtong Qi (2008) Price competition, cost and demand disruptions and coordination of a supply chain with one manufacturer and two competing retailers. Int J Manag Sci 36:741–753
4. Chahar K, Taafe K (2009) Risk averse demand selection with all-or-nothing orders. Omega 37:996–1006
5. Tomlin B (2006) On the value of mitigation and contingency strategies for managing supply chain disruption risks. Manag Sci 52:639–657
6. Schmitt AJ, Snyder LV (2010) Infinite-horizon models for inventory control under yield uncertainty and disruptions. Comput Oper Res 39:850–862

Establishment and Application of Logistics Enterprises' Low-Carbon Factors Index System

Meng Yang and Yuan Tian

Abstract In the background of the low-carbon economy, this paper establishes the logistics enterprises' low-carbon management factors index system from the perspective of energy consumption. Then we calculate the connection degree of the indicators by grey connection analysis. The connection degree is greater than 0.5, indicating that various factors have a significant impact on the low-carbon management of logistics enterprises, so as to provide ideas for logistics enterprises to implement low-carbon management.

Keywords Logistics enterprises • Low-carbon • Energy consumption • Grey connection analysis

1 Introduction

In recent years, the abnormal climate and other environmental issues are becoming increasingly serious. The low-carbon revolution of the globalization takes human into an era of low-carbon economy based on "low energy consumption, low pollution and low emissions". The basic position of logistics and its characteristics of high energy consumption, high emissions determine its special status in low-carbon economy.

At present, the academic community hasn't given the standard definition of the low carbon logistics. Wu and Dunn [1] proposed an environmentally responsible logistics system. Low-carbon logistics is based on low-carbon economy and green logistics theory [2]. Bai Jing [3] and Luo Wenli [4] discussed the specific relationship of low-carbon economy and logistics industry. In the aspects of logistics

---

M. Yang (✉) • Y. Tian
School of Economics and Management, Beijing Jiaotong University,
Beijing 100044, P.R. China
e-mail: 11120676@bjtu.edu.cn

Z. Zhang et al. (eds.), *LISS 2012: Proceedings of 2nd International Conference on Logistics, Informatics and Service Science*, DOI 10.1007/978-3-642-32054-5_58,

enterprises, Jiang Fa [5] discussed the positive and effective measures which should be taken by the government and logistics industry.

Overall, the low-carbon logistics is just developing, and most of the articles about low-carbon logistics describe the necessity of implementing low-carbon management on the macro level. Until now little has been studied deeply about low-carbon impact factors on the level of logistics companies. This article intends to contribute to the development of this field of research by jointly establishing the logistics enterprises' low-carbon factors index system and taking the demonstration analysis by grey connection theory on the level of logistics companies, so as to find the key low-carbon impact factors and provide ideas for logistics companies to implement low-carbon management.

2 Establishment of Low-Carbon Factors Index System

In the research of the logistics Enterprises' low-carbon impact factors, it is difficult to study the carbon emissions of the logistics enterprises directly. There are significant difficulties in the index quantification and the data acquisition. So this paper analyzes the energy consumption of the logistics enterprises and establishes the low-carbon impact factors index system based on reducing energy consumption.

2.1 Logistics Enterprises' Energy Consumption Analysis

Logistics is a joint operating system of many links with the complex characteristics of a variety of energy consumption and factors affecting energy consumption. The type and quantity of the energy consumption by logistics activities, such as transportation, storage, packing, processing and other business processes are different.

The transport process: Transport is one of the main function elements of logistics, and also the main cause of energy consumption of the logistics system. Its energy consumption consists of two parts: the direct consumption by a variety of transport means or facilities, mainly the burning of gasoline, diesel and other energy sources; the other one is the energy consumption of the services in transportation activities by the transport organization. The transport energy consumption we discuss is the former.

The warehousing and handling process: This part of energy consumption is about the equipment running. Handling is very frequent in the logistics activities. Storage and handling machinery drive and work in different ways, so the form of main energy consumption is different. The energy of the small and medium-sized warehouse is oil, and the energy used by the large and automated warehouse is electricity.

The packaging and processing: There are mainly three types of packaging machinery: fill packaging machinery, wrapping and banding machinery, packaging technology machinery. The types of processing machinery are different. And different forms of packaging and processing mean different forms of energy consumption.

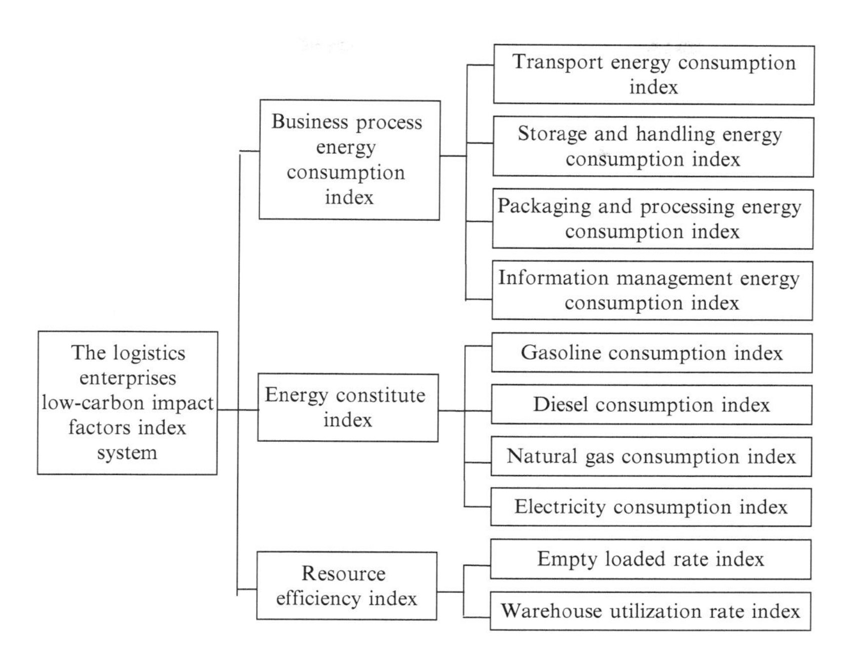

Fig. 1 Low-carbon impact factors index system of logistics enterprises

2.2 Logistics Enterprises' Low-Carbon Factors Index System

According to the analysis of the characteristics of the logistics system functional elements and its energy consumption, considering the general principles of science, rationality, feasibility, ultimately we establish the logistics enterprises low-carbon impact factors index system, shown in Fig. 1.

The Business Process Energy Consumption Index: The establishment of four basic indicators is based on the characteristics of the logistics enterprises' business processes and the energy consumption. According to the analysis above we know that the transport energy consumption is the largest in all business activities of the logistics enterprises. The main consumption is fuel and carbon emission is the most. Facilities and equipment's energy consumption plays an important role in the storage and handling, and also a small amount of fuel consumption. Packaging and processing is also the main facilities and equipment's energy consumption. Information management mainly consumes electricity.

The Energy Constitute Index: According to the analysis above we know that the energy consumed by the logistics companies is gasoline, diesel, natural gas, electricity and so on. And different companies' proportion of energy consumption is different. The energy constitute index analyzes the amount of different energy consumption.

The Resource Efficiency Index: Resource efficiency also plays a certain effect in the low-carbon management of the logistics companies. The efficiency of resource use will directly affect the consumption of corporate resources. This article analyzes the enterprise resource use efficiency in two aspects: Empty loaded rate index and warehouse utilization rate index.

3 Application of Low-Carbon Factors Index System

3.1 Grey Relational Analysis

The basic idea of grey relational analysis is to determine whether closely linked according to the degree of similarity of the sequence of curve geometry. The closer the curve, the greater the degree of connection between the corresponding sequence, on the contrary, the smaller. Grey relational analysis is a method to describe and compare the trend of a system's development and change quantitatively. Its essence is a collection of relations of time-series data, and the purpose is to seek the key relationships between the various elements in the system, to identify the important factors affecting the target, finally to analyze and determine the degree of influence between the elements and the contribution of elements to the system's master behavior.

The calculation analysis steps can be followed:

1. Seek the beginning to value of the sequence: Because the dimension (or unit) of the various factors in the index system is not necessarily the same, the data is difficult to directly compare, and the proportion of geometric curves are also different. Therefore, the dimensions (or units) of the raw data need to be eliminated, in order to be converted to a sequence of comparable data.

$$x_i^{'} = x_i/x_i(1) = \left\{x_i^{'}(1), x_i^{'}(2), \ldots, x_i^{'}(n)\right\}$$
$$i = 1, 2, \ldots, m$$

Among them, "x_i" means the i-th impact factors; "x_0" means the benchmark index; "m" represents the number of the impact factors. In this article m = 10; "n" represents the year, n = 5.

2. Seek the Sequence Difference:

$$\Delta_i(k) = \left|x_0^{'}(k) - x_i^{'}(k)\right|$$
$$\Delta_i = \{\Delta_i(1), \Delta_i(2), \ldots, \Delta_i(n)\}$$
$$i = 1, 2, \ldots, m; \ k = 1, 2, \ldots, n$$

Among them, "$\Delta_i(k)$" means the sequence difference corresponding to the i-th element in the k-th year, and "Δ_i" means the horizontal amount of the sequence difference.

3. Seek the maximum and the minimum differential of the poles:

$$M = \max_i \max_k \Delta_i(k) \quad m = \min_i \min_k \Delta_i(k) \qquad i = 1, 2, \ldots, m;\ k = 1, 2, \ldots, n$$

4. Seek the correlation coefficient:

$$r_i(k) = \frac{m + \epsilon M}{\Delta_i(k) + \epsilon M} \qquad i = 1, 2, \ldots, m;\ k = 1, 2, \ldots, n$$

Among them, "$r_i(k)$" means the correlation coefficient of the i-th element and the benchmark index in the k-th year, and set "ϵ" $= 0.5$.

5. Calculate the correlation degree:

$$r_i = \frac{1}{n} \sum_{k=1}^{n} r_i(k) \quad i = 1, 2, \ldots, m;\ k = 1, 2, \ldots, n$$

Among them, "r_i" means the correlation degree of the i-th element and the benchmark index.

3.2 Demonstration Analysis of the Logistics Enterprise

According to the low-carbon impact factors index system established previously, we set the explanatory variables: "Y_1" means the transport energy consumption, "Y_2" means the storage and handling energy consumption, "Y_3" means packaging and processing energy consumption, "Y_4" means the information management energy consumption, "Y_5" means the gasoline consumption, "Y_6" means the diesel consumption, "Y_7" means the natural gas consumption, "Y_8" means the electricity consumption, "Y_9" means the empty loaded rate, "Y_{10}" means the warehouse utilization rate. Then establish the grey system composed of the explanatory variables above based on the statistical data of one logistics enterprise, as shown in Table 1. Finally calculate the degree of influence of the explanatory variables to the logistics energy consumption, as shown in Table 1.

The calculation result of the grey correlation degree of this logistics enterprise's low-carbon impact factors index system is that: transport energy consumption $Y_1 >$ diesel consumption $Y_6 >$ empty loaded rate $Y_9 >$ storage and handling energy consumption $Y_2 >$ warehouse utilization rate $Y_{10} >$ packaging and processing energy consumption $Y_3 >$ information management energy consumption $Y_4 >$ electricity consumption $Y_8 >$ gasoline consumption $Y_5 >$ gas

Table 1 The original data table of logistics energy consumption and its related factors

Index \ Time	2006	2007	2008	2009	2010	Grey correlation degree	Grey correlation sequence
Logistics energy consumption (million tons standard coal) X	50	61	70	88	94		
Transport energy consumption (million tons) Y_1	26	29	30	37	41	0.987253	1
Storage and handling energy consumption (million tons standard coal) Y_2	10	12	15	20	30	0.863970	4
Packaging and processing energy consumption (million tons standard coal) Y_3	7	10	16	20	25	0.759352	6
Information management energy consumption (million tons standard coal) Y_4	4	5	11	23	25	0.735629	7
Gasoline consumption (million tons) Y_5	20	25	26	35	37	0.583928	9
Diesel consumption (million tons) Y_6	4	3	3	3	3	0.925786	2
Gas consumption (million m^3) Y_7	5	6	5	7	9	0.557462	10
Electricity consumption (million kilowatt-hour) Y_8	150	172	180	188	193	0.713694	8
Empty loaded rate (%) Y_9	50%	50%	43%	41%	37%	0.872591	3
Warehouse utilization rate (%) Y_{10}	70%	75%	80%	77%	79%	0.834276	5

consumption Y_7. And we can find the following characteristics of the logistics enterprises' low-carbon impact factors:

1. The grey correlation degrees of all the indexes of the system are greater than 0.5, which indicates that the index system is scientific and instructive.
2. The grey correlation degree of the transport energy consumption is the highest in the index system, which means that transport plays a key role in the carbon management of the logistics enterprises. In addition, logistics' transport activities consume mainly the diesel, and the grey correlation degree of diesel fuel consumption is also high, which validates the importance of transport energy consumption index.
3. The resource's use efficiency also has a huge impact, such as empty loaded rate and warehouse utilization rate listed in the table. Therefore, the logistics enterprises should improve management efficiency and avoid waste of the resources.

4 Conclusions

By analyzing the characteristics of the energy consumption in the logistics business processes, this paper establishes the logistics enterprises' low-carbon impact factors index system and takes the demonstration analysis by grey connection theory based on one logistics enterprise's data to validate the scientific nature of the index system. From the study in this paper, we know that transport energy consumption and resource efficiency play a key role in the low-carbon management of the logistics enterprises, which are the key indicators of the index system.

Acknowledgement Supported by "the Fundamental Research Funds for the Central Universities", "the National Natural Science Foundation Project with item number: 71132008" and the key project of logistics management and technology lab.

References

1. Wu HJ, Dunn S (1995) Environmentally responsible logistics systems. Int J Phys Distrib Logist Manag 25(2):20
2. Tao Jing (2010) Low carbon logistics under the low-carbon economy. China Econ Trade Guide (12):72–72
3. Bai Jing (2010) The development of logistics in the low-carbon economy era. Logist Technol (2):48–50
4. Luo Wenli (2010) Explore the low-carbon logistics. China Logist Purch (1):48–51
5. Jiang Fa (2010) The behavior of government and logistics industry under the low-carbon economy. Ind Technol Forum (6):147–148

The Information Construction of Third-Party Warehousing in the Cold Chain Logistics

Qin Zhang, Keming Zhang, and Bohui Song

Abstract In recent years, the development of cold chain logistics attracts people's attention, and keeping the quality of goods has become the demand of modern life. Since time is a key factor in cold chain logistics, and goods in cold chain logistics with high requirement to the humidity and temperature, it is vitally important for the third-party warehousing in the cold chain logistics to practice information construction, for it can reducing the risk of business operations, optimize operational performance. In this article, firstly, we will analyze the status of cold chain logistics in third-party warehousing company to find the shortcomings in the actual operation. Then propose how to proceed with the information construction of warehousing, thereby improving their shortcomings, and improve the efficiency of third-party warehousing.

Keywords Cold chain logistics • Third-party warehousing • Information construction

1 Introduction

Combining the advantages of third-party storage and the shortcomings of cold chain logistics, more shippers tend to outsource the storage business to third-party storage enterprises. To meet this challenge, as third-party warehousing companies can take advantage of modern IT to improve the storage management and increase efficiency, thereby can enhancing their competitiveness.

Q. Zhang (✉) • K. Zhang • B. Song
School of Economics and Management, Beijing Jiaotong University,
Beijing 100044, P.R. China
e-mail: 11120679@bjtu.edu.cn

Z. Zhang et al. (eds.), *LISS 2012: Proceedings of 2nd International Conference on Logistics, Informatics and Service Science*, DOI 10.1007/978-3-642-32054-5_59,

1.1 Research Status of Foreign Information Construction in Warehousing

Foreign logistics is relatively more mature. Foreign countries have basically formed a paperless, automated storage and operating system. Daniel E-Oleary [1] introduced that the ERP (Enterprise Resource Planning) system helps to management the resource of the enterprise and improve the company's efficiency. Robert Jendry [2] introduces the technology of DEM (Dynamic Enterprise Module) which can meet the requirement of dynamic management. These systems promote the development of warehousing. At the same time, bar codes, EDI (electronic data interchange), and RFID (Radio Frequency Identification) have been widely used in warehouse. Automated warehouse helps to realize mechanization and automation in warehousing operations.

1.2 Research Status of Domestic Information Construction in Warehousing

China's logistics only can reach a low degree of informatization, but with a rapid speed of development. Through studying about the development of domestic and foreign logistics, Gui Li et al. [3] pointed out that one of the development tendencies of domestic logistics is the information construction. Wan Jianye and Wang Yunpeng [4] studied the information model of warehouse, system structure, operational flow of management in the logistic company. Furthermore, Jiang Chaofeng [5] studied the status and development tendency of warehousing in China, and gives some advices for information construction. Besides, many researchers [6, 7] also pay attention to the information technology applied in the warehousing.

1.3 Summary

The innovation of this paper is proposed to the information construction of third-party warehousing in the cold chain logistics based on the characteristics of cold chain logistics. The goal of this paper is to put forward a scheme to strengthen informatization construction of third-party warehousing in the cold chain logistics.

The organization of the paper is as follows. First we will give a brief overview of the information construction status of domestic and foreign warehousing then we will give a brief representation of the shortage of third-party warehousing in cold chain logistics. Furthermore, we will make suggestion to improve the above shortage. At last we will conclude and give suggestions for further research.

2 The Analysis of Third-Party Warehousing in Cold Chain Logistics

2.1 The Shortages of Third-Party Warehousing in Cold Chain Logistics

The shortages of third-party warehousing can be concluded as below:

1. The low capacity of refrigerator and low space utilization rate: According to statistics, the existing refrigeration capacity has reached more than five million tons. However, it can not meet the actual demand. In addition, space utilization rate is very low, the traditional refrigerator design is generally about 5 m high, but the shelves layer cold storage utilization rate is below 50%.
2. Low utilization rate for each year: Take Lanzhou, China for example, most of the refrigerators store fresh vegetables from May to October, the rest of the time, the refrigerator usually shut down, thus the idle period up to 6 months.
3. Difficult to guarantee quality: The stored goods have a shelf life, if they are not stored well, it is easily to occurring expired goods. when the expired goods can't be disposed timely, they may affect the quality of other items.
4. Low operational efficiency: Among the usage of cold storage customers, because of each shippers' goods are not the same, A lot of information produced during the storage period which have to be record relying on manual entry, so there exist low efficiency, high error rate and the update is not timely.

2.2 The Benefits of the Information Construction of Warehousing

The information construction of third-party warehousing can bring the following advantages:

1. Rationalization of storage: Through information construction, the third-party storage can use information system to record storage time and update storage time. When picking up the goods, the computer will be able to give instructions in time to ensure "first in, first out". Thus effectively raise the turnover rate of the goods.
2. Quickly locate the position: Positioning system will save a lot of seeking time, prevent mistakes by human and will be easy to check. The realization of this function is attributed to the usage of the advantage of information system.
3. Monitor the whole process of the storage: It can easily monitor the storage and flow of goods in cold storage. Besides it can also helps to monitor the temperature and humidity in the warehouse to ensure the quality of the goods.
4. Share the information: Information construction of storage enables enterprises to share the information of each business segment on a unified information

platform. At the same time, third-party warehousing companies can analyze data from information systems in order to effectively support the warehousing business decisions, statistics, information dissemination.

3 The Information Construction of Third-Party Warehousing

Through related study and research, we believed that the main steps of information construction of the third-party warehousing including the following aspects.

3.1 Standardize Operational Process

For third-party warehousing in the cold chain logistics, its main workflow is as below: inbound → cold storage → outbound. Handling, storage allocation, and information processing of customers' storage are included in this main process. After make a clear about this process, third-party warehousing company needs to use some tools to analyze the existing process in order to improve the operational process, pave the way for information construction of warehousing.

3.2 Subdivision of the Functional Module

According to actual demand, the division of the functional modules in third-party warehousing can be shown in Fig. 1:

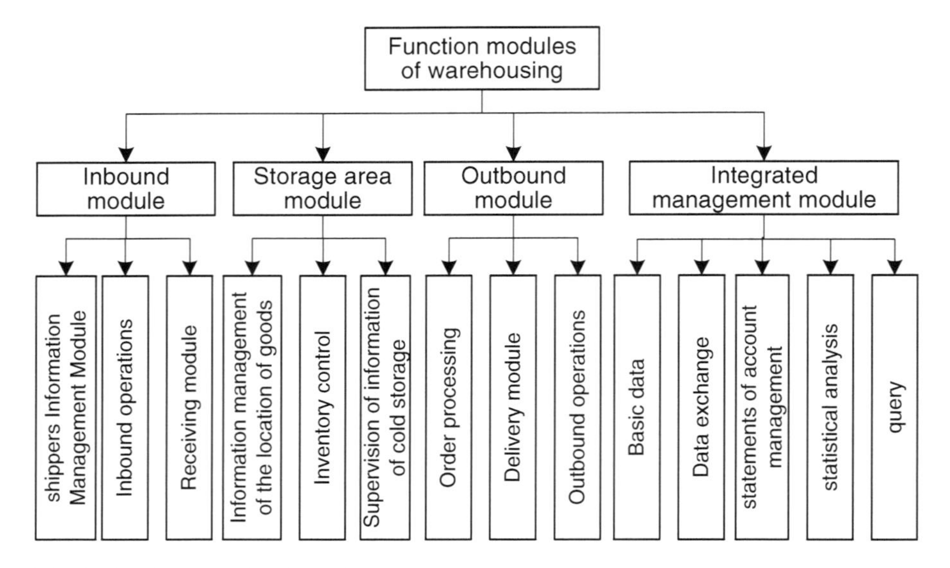

Fig. 1 Function modules of warehousing

1. Inbound module: This module is used to receive the inventory documents and analyze the information. In order to reserve the empty spaces for the shippers, we can query receiving module. The read and write machine can obtain the information of the goods through the label, then the related data will be transmitted to the back office system, and the system will check detail information of the documents.
2. Storage area module: This module is mainly used to plan, use and manage the district of warehouse, and they can help the store personnel to carry out the classification, cargo distribution and furnishings, goods usage and operations. Recording the storage time and updating the information of goods can contribute to the discovery of expired goods. The supervision module can achieve to the purpose of real-time monitoring of the temperature and humidity in the refrigerator.
3. Outbound module: This module is mainly used to receive the delivery bills and query the quality and quantity of the required goods. Then register the outbound information to ensure that the actual quantity and quality of outbound goods matches the shippers' delivery bills and give the vacated storage areas vacancy signs.
4. Integrated management module: Basic data module is to manage the equipment and personnel information with data coding and manage the data , as well as manage the bar code, radio frequency standard information. Data exchange module provide a data exchange platform. Account management module can form the financial statements. Statistical analysis module and query module form different forms which can improve management.

3.3 Hardware and Software Support for the Information Construction

In the process of information construction, via GPS (Global Positioning System), GIS (geographic information system), bar code technology and RFID (radio frequency identification) can accurately locate the location and transmit various documents, images and data, thus can improve logistics efficiency.

Software supporting includes the three aspects: (1) computer software. Third-party cold chain logistics providers and software provider should collaborate to develop the core module of cold chain logistics, for other non-core modules third-party enterprise can use commercial available software, such as financial software. (2) information. By using them, companies can improve enterprise management and support decision-making. (3) people. It is also crucial to train the staffs. The training not only includes the business training and operations training, but also includes transmit the management ideas and concepts in the construction of logistics information for high-level personnel.

4 Conclusions

The role of information construction is to improve efficiency and standardize the management. In this article, we give a brief introduction of the information construction of third-party warehousing in cold chain logistics.

Because of the lack of practical experience, this study remains in the theoretical stage and has a lot of limitations. Future research directions can be combined with the actual operation of the third-party storage in the cold chain logistics, and apply specific research on the information construction process of storage, researchers also can do some research from the technical aspects of the information construction of warehousing.

Acknowledgements Supported by "The research on development model of agricultural products logistics which based on supply chain collaboration" with item number "2011JBM054" and the key project of logistics management and technology lab.

References

1. Oleary DE (2000) Enterprise resource planning system. Cambridge University Press, Cambridge
2. Robert J (2000) BaanERP business solutions. Pruna Publishing, Monarch
3. Gui Li, Xiong Chan, Zhang Yingjiang (2009) Overview of the development and trendency of domestic and foreign logistics. Mod Bus Trade 6:101–102
4. Wan Jianye, Wang Yunpeng (2009) The research on the warehousing information system which based on manufacturing companies. J Liaoning Univ Technol (Nat Sci) 29:315–321
5. Jiang Chaofeng (2011) The status and development trendency of warehousing in China. China Logist Times 1:58–59
6. Guo Bao Ding Hu Company (2009) The planning and design of cold storage in modern cold chain logistics. Logist Technol 1:56–57
7. Yuan De (2011) The profit model of third-party warehousing logistics companies. Logist Eng Manag 33:18–20

The Research on Traffic Flow Simulation and Optimization of Beijing West Railway Station

Zhanping Liu and Xiaochun Lu

Abstract In 1996, Beijing West Railway Station opened and operated officially. In accordance with the original design, the large train transportation hub would work fluently. However, the passengers and vehicles have been suffering the pain of congestion. In this paper, we simulate the traffic flow of Beijing West Railway Station to find out the causes of traffic congestion and provide some optimization proposals through the VISSIM software simulation.

Keywords Beijing West Railway Station • Traffic flow simulation • VISSIM

1 Introduction

Beijing West Railway Station, completed in 1996, is the largest population distributing center and transportation hub in China. So far, the trains have greatly exceeded the maximum capacity. Besides in Beijing West Railway Station, the North plaza has more than 50 bus lines, while the south plaza has only four starting buses. Both of them caused serious traffic jams around north plaza.

The traffic simulation software VISSIM [5] is used in this paper to make traffic simulation around Beijing West Railway station. By simulating we try to find out the causes leading to traffic jams and then provide some optimization ideas.

Z. Liu (✉) • X. Lu
School of Economics and Management, Beijing Jiaotong University, Beijing, China
e-mail: 11120658@bjtu.edu.cn; xclu@bjtu.edu.cn

Z. Zhang et al. (eds.), *LISS 2012: Proceedings of 2nd International Conference on Logistics, Informatics and Service Science*, DOI 10.1007/978-3-642-32054-5_60,

2 Traffic Situation and Underlying Data Analysis

Beijing West Railway Station and the surrounding transportation system include: YangFangDian East Road, YangFangDian Road, YangFangDian West Road from north to south and Lian HuaChi Road from east to west. In addition, a large public transport hub was established in the east of Beijing West Railway Station. Besides, the traffic of YangFangDian East Road is small, so it is not our focus. Currently, the traffic is mainly composed of three types of elements, taxi, bus and private car.

2.1 Present Traffic Situation

The survey data of main traffic routes' traffic flow are as shown in Table 1. In the investigation; it takes 5 min as a time interval.

Furthermore, according to the data in Table 1, we can calculate the relevant distribution of traffic flow.

2.2 Definition of Parking Time Distribution

VISSIM uses the parking time distribution to define the stopping time at the stop sign and bus station [1].

Table 1 The survey data of traffic flow in Beijing West Railway Station

		Quantity (in vehicles)		
Location	Vehicle type	Time 9:30–11:30	Time 16:30–17:30	Proportion (%)
LianHuaChi East Road from east to west	Car	3,010	1,532	83.80
	Bus	564	269	15.37
	Van	35	10	0.83
	Total	3,609	1,811	100
LianHuaChi East Road from west to east	Car	3,617	1,755	83.8
	Bus	665	322	15.37
	Van	36	17	0.83
	Total	4,316	2,094	100
YangFangDian Road	Car	800	401	86.90
	Bus	84	43	9.19
	Van	31	23	3.91
	Total	915	467	100
YangFangDian West Road	Car	574	328	94.98
	Bus	30	8	4
	Van	9	1	1.02
	Total	613	337	100

Parking time distribution of Beijing West Railway Station's traffic system has three major types: (1) the bus stopping time depends on the number of passengers, typically in random distribution of 0–2 min; (2) during the traffic peak period, average stopping time of taxis is about 30 s, however at other time, it only can be measured by average stopping time; (3) in accordance with the schedule, buses run every 10 min, however the reality is the uniform distribution of 8–10 min. In the model, bus stopping time is subject to empirical distribution of 60–360 s.

2.3 Path Decision

When traveling to the crossroads, the vehicles will make path decisions. In VISSIM software, path decisions are made according to different models, so in our investigation, the path decision data is indispensable.

3 Beijing West Railway Station Simulation

3.1 Establish the Model of Road Network and the Traffic Flow Model

Import the background picture into VISSIM software, then build road network of Beijing West Railway Station by ratio [2]. Besides, because the buses must stop at bus station, and buses run in accordance with the dispatching schedule strictly, so comparing with other vehicles, bus arrival rate is special [3]. In VISSIM, the bus lane needs to be set separately.

In VISSIM, the traffic flow model consists of two parameters to form, the traffic flow in different routes and the traffic flow in different periods in the same route. Set the traffic flow by using the table in VISSIM function.

3.2 Establish Vehicle Priority Rules of No Signal Control at the Intersection

At the intersection of no signal control, VISSIM uses priority rules specify the right of going firstly in conflicting traffic flow. In the model of Beijing West Railway Station, the priority rule is set based on the rules that cars go ahead of the rest on the main road and cars going straightly are ahead of ones making a turn [4].

4 Traffic Flow Analysis of Simulation Results

In this paper, we evaluate the traffic system from two aspects, indicators that reflect fluency such as vehicle average queue length, maximum queue length, the total number of vehicles in queue, and indicators that reflect efficiency such as average delay time, average stopping time, average travel time, number of stops.

According to the simulation results, except the two intersections mentioned below, the traffic congestion around Beijing West Railway Station is not so serious.

4.1 The Intersection of LianHuaChi East Road and the Second Floor Waiting Hall in Beijing West Railway Station

By simulation, the average delay time is 6–27 s and average queue length is 22–102 m. The maximum queue length is generally over 110 m and the maximum can be up to 300 m. The highest number of stops can reach 800 times. So it can be seen that congestion here is serious. The main cause is that a large number of cars are driven into the Beijing West Railway Station waiting room on the second floor, so it causes a serious blockage.

4.2 The Intersection of LianHuaChi Road and the Section Where Buses Leaves the Bus Terminal

The average delay time is 5–7 s and average queue length is 4–9 m. The maximum queue length can be up to 140–160 m. In particular departure time, the bus flow and the traffic flow in LianHuaChi East Road generate congestion, which cause a serious blockage by observing the simulation animation.

5 Conclusions and Recommendations for Optimization

In an overall view, to solve the pain of the jam, the general principle is making roads smoother. Then the time of vehicles spending on the road will be reduced, and the number of cars on the road will significantly decrease. With fewer cars on the road, Road is getting clearer, which is a virtuous circle. For example, limit the number of cars driving into the Beijing West railway station. And if the south plaza was enabled, we can guide cars to move to the South square actively. Meanwhile, adjust bus departure time prone to congestion to avoid traffic jam, in the intersection of LianHuaChi East Road from east to west and the second floor waiting hall.

In the light of Beijing rail transit plan, the No.7 and No.9 Metro Line will be set up in this station. After the subway opened, which will help alleviate the traffic pressure.

Acknowledgments The article is supported by the two projects as follows: (1) A joint project of Beijing education commission: A research on simulation optimization system between Beijing railway station and urban transit system. (2) National Natural Science Foundation Project "logistics resource integration and scheduling optimization", item number: 71132008.

References

1. Jianping Yuan, Dongtao Fan (2010) VISSIM application on transport simulation of passenger transport hub in Chang Zhou, vol. 28. J Transp Inf Saf 155:68
2. Junqiang Leng, Yuquan Leng, Yaping Zhang (2009) Study on location of turning left at the intersection based on VISSIM simulation, vol. 34. J KunMing Univ Sci Technol (Nat Sci Ed) 3:58
3. Gaohong An, Tao Xi, Lijing Wang (2007) Prediction of traffic flow simulation, vol. 13. J Tian Jin Inst Urban Constr 3:180–181
4. Jing Gao, Jian Xiong, Qinya Qin (2007) Traffic flow simulation of driving imitation system based on VISSIM, vol. 25. Comput Commun 2:127
5. PTV Planung Transport Verkehr AG (2007) VISSIM user manual-version 4.30 [R]. PTV Corporation

The Risk Assessment of Logistic Finance Based on the Fuzzy Analytic Hierarchy Process

Xiaolong Li and Runtong Zhang

Abstract The information asymmetry between banks and enterprises bring some risks to banks when participate in the logistics and finance business. Based on the fuzzy mathematics theory, this paper sets risk evaluation index system for the pledge risk, financing enterprise credit risk, logistics enterprise risk and regulatory risk, and it can support one or more project risk assessment. Also, this paper indicates the feasibility by an empirical study.

Keywords Logistic finance • Risk prevention • Fuzzy analytic hierarchy process

1 Introduction

Banks should confirm the risk in the logistics finance, which can be divided into four aspects: enterprise reimbursement risk, market risk, regulatory risk and realizable risk. Analyzing from the risks these banks may face, this paper sums up the finance risk classification of logistics, and then sets risk index system and lastly settles on the risk evaluation model, it provide a reference risk evaluation system for banks to carry out the logistics finance business.

2 Logistics Finance Risk Evaluation Index System

Yaodong Bao and Zhang Wuyi [1] used the analytic hierarchy process to assess the risk evaluation system. Chuansong Wang [2] used the fuzzy evaluation method to study the risk. Xiuzhi Zhang [3] used factor analysis method, to calculate the index

X. Li • R. Zhang (✉)
School of Economic and Mangement, Beijing Jiaotong University, Beijing 100044, P.R. China
e-mail: 11120636@bjtu.edu.cn; rtzhang@bjtu.edu.cn

Z. Zhang et al. (eds.), *LISS 2012: Proceedings of 2nd International Conference on Logistics, Informatics and Service Science*, DOI 10.1007/978-3-642-32054-5_61,

Table 1 Logistics finance risk index system

First class index	Second class index			
The pledge risk	Legitimacy	Stability	Marketability	Natural attributes
Financing enterprise risk	Enterprise credit	Executives credit	Management structure	Promising future
Logistics enterprise risk	Enterprise credit	Information platform	Hardware facilities	Business capability
Regulatory ability risk	Warehouse management	Regulations	Warning mechanism	Professional quality

weights and levels of total ranking value and assess the logistics finance risk. Yu Hu and Xia Huijuan [4] proposed a fuzzy complementary judgment matrix scheduling model to control the risk based on risk factors of the fuzzy ordered weighted averaging operator.

The scientificity of factor index can be assessed by reliability analysis. Logistics finance risk index system is set based on risk indicators of combined with the analytic hierarchy process method. Secondary indicators have different weights under the first indicators, and the integration of secondary indicators constitutes first level indicators which are shown in Table 1:

First class index for the upper standards $t_i{}^1(i = 1, 2 \ldots 4)$, second class index for lower standards $t_d{}^2(d = 1, 2 \ldots 16)$. Secondary indexes can increase the index number, summarizing it in one class index, based on the main factors of one class index that level 2 index number.

3 Logistics Finance Risk Evaluation Model

Assuming that N reviewers as $(\mathrm{P1}, \mathrm{P2} \ldots\ldots \mathrm{PN})$, upper criterion for $t_i{}^1(i = 1, 2 \ldots 4)$, Based on the index criteria $t_d{}^2(d = 1, 2 \ldots 16)$ assessing logistics finance projects, the risk evaluation model can assess many projects (λ1, λ2 ... λm).

The upper index weights are determined by the AHP, considering the reviewer's subjective factors which will lead uncertainty and fuzziness to the right decision-making, therefore, this paper uses the triangle fuzzy numbers settle on the upper index weights.

Specific the upper index weight is expressed as:

$$S_i^{\,1} = [A_i, B_i, C_i], \tag{1}$$

$$A_i = \min\{S_{in}\}, \quad B_i = \left\{\prod_{i=1}^{n} S_{in}\right\}^{1/n}, \quad C_i = \max\{S_{in}\} \tag{2}$$

Table 2 The fuzzy numbers of fuzzy variables

Hierarchy variable	Fuzzy numbers
Low	(0, 0.25, 0.5)
Middle	(0.25, 0.5, 0.75)
High	(0.5, 0.75, 1.00)

Among them S_i^1 indicates the weights of upper index, S_{in}^1 indicates the importance of the upper index weights of N-bit reviewer's evaluation.

Reviewers use the weights set H = (low middle high) to evaluate the importance of various criteria, and use the hierarchical collection to evaluate the satisfaction of different criteria. Fuzzy numbers table as shown in Table 2:

The fuzzy weights of the underlying criteria and the satisfaction of many projects are expressed by fuzzy various shown in Table 2, and then integrate evaluators' assessment by means of average algorithm. The integrated approach of fuzzy integration method of underlying criteria weights and satisfaction of the project to be assessed are as follows:

$$S_d{}^2\{\left(\prod_{i=1}^{n} S_{id}\right)^{1/n} \tag{3}$$

$$P_{jd}\left(\prod_{i=1}^{n} P_{jd}\right)^{1/ij} \tag{4}$$

Among them $S_d{}^2$ indicates the geometric mean weight of the underlying criteria, $S_{di}{}^2$ indicates the importance of the underlying index weights of N-bit reviewer's evaluation, P_{jd} indicates the geometric mean fuzzy satisfaction of the project (λ_j) to be assessed based on the underlying criteria $S_d{}^2, P_{jdi}$ indicates the allocation of satisfaction levels of the project (λ_j) of N-bit reviewer's evaluation based on the underlying criteria $S_d{}^2$. Thus the satisfaction of the upper index R_{tj} of the project to be assessed can be integrated by $P_{jd}, S_d{}^2$.

$$R_{tj} = 1/\text{k}\{\left(P_{j1} * S_1{}^2\right) + \left(P_{j2} * S_2{}^2\right) + \cdots \left(P_{jk} * S_k{}^2\right)\} \tag{5}$$

Assuming $P_{jdi} = \left(f_{jdi}, g_{jdi}, h_{jdi}\right), S_{di}{}^2 = (u_{di}, v_{di}, w_{di})$ as triangle fuzzy function,

$$R_j = \left(a_{tj}, y_{tj}, c_{tj}\right), \tag{6}$$

Among them

$$a_{tj} = \sum_{\text{d}=1}^{\text{k}=4} f_{jd} * u_d/k \quad f_{jd} = \left(\prod_{i=1}^{n} f_{jdi}\right)^{1/n} \quad u_d = \left(\prod_{i=1}^{n} u_{di}\right)^{1/n} \tag{7}$$

The other items can be shown for the same reason.

The weight values of the upper index $\left(S_i^{\ 1}\right)$, and the satisfaction of the upper index of the project to be assessed $\left(R_j\right)$, have been inferred, and then the fuzzy evaluation of the project is:

$$w_j = R_j * S_i^{\ 1} = \left(q_j, q_j, t_j\right) \tag{8}$$

4 Empirical Research

This paper uses the Zhongshan branch of the GDB and the financial corporation projects for the case. The benefits of these projects are different.

First, we select four reviewers to evaluate four projects, l the weight evaluation of the upper index is the first step to obtain. Specific satisfaction is shown in Table 3:

Combined with the evaluation of four reviewers, this paper uses the fuzzy algorithm to figure up the fuzzy weights of the upper index, shown in Table 4:

Then, this paper should calculate the weight of the underlying criteria $\left(S_{di}^{\ 2}\right)$ and the satisfaction of the project to be assessed $\left(P_{jdi}\right)$, and then the satisfaction of the underlying criteria $\left(R_j\right)$, can be calculated as follows in Table 5.

Table 3 Evaluator on the project the satisfaction of upper index

	P_1	P_2	P_3	P_4
$S_1^{\ 1}$	0.13	0.10	0.09	0.11
$S_2^{\ 1}$	0.32	0.35	0.29	0.33
$S_3^{\ 1}$	0.21	0.20	0.18	0.20
$S_4^{\ 1}$	0.20	0.19	0.23	0.23

Table 4 The fuzzy weights of the upper index

	Fuzzy weights
$S_1^{\ 1}$	(0.11, 0.10, 0.12)
$S_2^{\ 1}$	(0.33, 0.30, 0.32)
$S_3^{\ 1}$	(0.19, 0.20, 0.21)
$S_4^{\ 1}$	(0.21, 0.20, 0.22)

Table 5 The lower indicators of fuzzy satisfaction index

(0.36, 0.73, 0.92)	(0.36, 0.73, 0.95)	(0.39, 0.76, 0.95)	(0.40, 0.78, 0.90)
(0.46, 0.85, 1.00)	(0.45, 0.83, 0.94)	(0.42, 0.80, 1.00)	(0.39, 0.77, 0.96)
(0.45, 0.83, 0.94)	(0.39, 0.78, 0.92)	(0.40, 0.78, 0.96)	(0.42, 0.80, 1.00)
(0.37, 0.76, 0.93)	(0.38, 0.77, 0.93)	(0.40, 0.79, 0.95)	(0.41, 0.79, 0.99)

Table 6 Project overall fuzzy evaluation form

	Overall fuzzy evaluation
λ_1	(0.36, 0.78, 0.95)
λ_2	(0.40, 0.83, 0.97)
λ_3	(0.30, 0.72, 0.88)
λ_4	(0.28, 0.69, 0.84)

The overall evaluation of these projects can be calculated, as shown in Table 6:

From Table 6, we can see that four projects have different operability ($\lambda 2 > \lambda 1 > \lambda 3 > \lambda 4$), the second project is best for corporation and the last benefits not so well.

5 Conclusion

The risk evaluation index system conforms to the scientific principles. It can assess the risk factors in the logistics finance scientifically. This paper presents a comprehensive evaluation model based on fuzzy math method. It can maximize the benefits and the control of logistics finance risk factors with multiple rating.

Acknowledgments The research was supported by the key project of logistics management and technology lab.

References

1. Bao Yaodong, Zhang Wuyi (2010) Finance risk analysis logistics based on the AHP [J]. China Logist Purch 5:68–69
2. Wang Chuansong (2009) Logistics finance business risk management research [D]. Guangdong Univ Foreign Stud 1:21–28
3. Zhang Xiuzhi (2010) Logistics finance business risk evaluation system and the fuzzy comprehensive evaluation [J]. Logist Technol 1:132–134
4. Hu Yu, Xie Huijuan (2009) The risk of logistics of modern rural finance center don't evaluate and control [J]. China's Circ Econ 5:18–19

A Queueing-Inventory System with Registration and Orbital Searching Processes

Jianan Cui and Jinting Wang

Abstract In this paper, we consider a continuous time queueing-inventory system with customers' registrations and stochastic inventory on the replenishment policy of (s, S). Using the method of matrix analysis, we obtain the steady joint probability distribution of the number of customers in the orbit and the inventory level. Various system performance measures in the steady state are derived and the long-run total expected cost rate is calculated.

Keywords Queueing-inventory system • (s, S) policy • Matrix analytic method • Registration • Retrial customers

1 Introduction

Production-inventory systems have been extensively studied in the area of integrated supply chain management. Traditionally, this topic is usually investigated using queueing networks and multi-echelon inventory models. For details, one can see Berman et al. [1–3] in which some integrated models appeared concerning the problem of how the classical performance measures are influenced by the management of attached inventory and vice versa: How inventory management has to react to queueing of demands and customers, which is due to incorporated service facilities. Recently, Schwarz et al. [4] studied an M/M/1 queueing system with inventory under continuous review and different inventory management policies, and with lost sales.

Artalejo et al. [5] studied inventory policies with positive lead-time and retrial of customers who could not get service during their earlier attempts to access the service station. Krishnamoorthy and Jose [6] took a comparison of inventory systems with service, positive lead-time, loss, and retrial of customers. In their

J. Cui • J. Wang (✉)
Department of Mathematics, Beijing Jiaotong University, Beijing 100044, P.R. China
e-mail: 10121836@bjtu.edu.cn; jtwang@bjtu.edu.cn

Z. Zhang et al. (eds.), *LISS 2012: Proceedings of 2nd International Conference on Logistics, Informatics and Service Science*, DOI 10.1007/978-3-642-32054-5_62,

model, the retrial rate is assumed to be linear with the number of the customers in the orbit. By using the matrix analysis method (Latouche and Ramaswami [7]), they obtain a numeral values of the system. For further details, one can see Yang and Templeton [8] and Falin [9, 10].

In this paper, we will consider a continuous review queueing-inventory system with customers' registrations and the server is required to search for customers in the registration list according to first-come-first-serve (FCFS) discipline. The whole system combines the inventory with the server: the possible stochastic time in the system, such as the customer's inter-arrival time, the service time, the retrial time and the positive lead-time of the replenishment.

2 Problem Formulations and Analysis

We consider a continuous time retrial queueing system with stochastic inventory on the replenishment policy of (s, S). The customers arrive according to a Poisson process with the rate of α. The server is closed when the inventory is out of stock. As the (s, S) replenishment policy, when the on-hand inventory level drops to a prefixed level, say s (>0), an order for Q $(=S - s > s)$ units is placed. The positive lead-time of the replenishment is exponential distribution with the rate β. The server is available to work when there are items in the inventory, the customer at the server will leave the system with one item from the inventory after a random service time, which we assume it is according to an exponential distribution with the rate.

If the arriving customer finds server unavailable, he will leave a message to register the system and then enter the retrial orbit. The server is required to search for these customers once they are available to provide service to them. It means the customers in the orbit will obtain the opportunity of being served after a random time, which is also assumed to be exponentially distributed with the rate θ. Compared to the customer in the orbit, new customer has priority to enter the server when they both attempt to enter the server at the same time; however, the priority is non-preemptive. The customers in the orbit take the retrial mechanism of FCFS, which means he who is the first coming in to the orbit, who is the first coming out to take the retrial chance. If he encounter the server is available to work, he will be served immediately, and otherwise he will return to the orbit.

To describe this system, at any moment t, let $N(t)$ denotes the number of customers in the orbit, $S(t)$ denotes the state of server 0, 1, 2, respectively, corresponding to the server is closed, idle, or at working and $I(t)$ denotes the on-hand inventory level. The possible values of the above variables are: $N(t) \in \{0, 1, 2, \ldots\}, I(t) \in \{0, 1, 2, \ldots, S\}$. We assume that the above three random variables are independent of each other; we can conclude that the stochastic process $\{N(t), S(t), I(t)\}_{t \geq 0}$ is a three-dimensional Markov process. Its state space is listed as follows:

$\Omega = \{(i, 0, 0) : i \in N\} \cup \{(i, 1, m) : i \in N, m = 1, 2, \ldots, S\} \cup \{(i, 2, m) : i \in N, m = 1, 2, \ldots, S\}$. The infinitesimal generator is $A = (a((i,j,k), (i',j',k')))$, (i,j,k)

$\in \Omega(i',j',k') \in \Omega$. By ordering the sets of state space as lexicographically, the infinitesimal generator A can be expressed in a block partitioned matrix with entries:

$$A = \begin{matrix} 0 \\ 1 \\ 2 \\ \vdots \end{matrix} \begin{bmatrix} \widehat{A_0} & C & & \\ D & B & C & \\ & D & B & C \\ & \ddots & \ddots & \ddots \end{bmatrix} \quad (1)$$

It readily seen that $\{N(t), S(t), I(t)\}_{t \geq 0}$ is a level-independent QBD process.

3 Steady State Analysis

Under the stability condition, we can consider the steady probability of the system. Let $\pi_{i,j,k} = \lim_{t \to \infty} \Pr\{N(t) = i, S(t) = j, I(t) = k\}$ and let

$$\Pi = (\Pi_0, \Pi_1, \Pi_2, \ldots), \Pi_i = (\pi_{i,1}, \pi_{i,2})$$
$$\pi_{i,1} = (\pi_{i,0,0}, \pi_{i,1,1}, \pi_{i,1,2}, \ldots, \pi_{i,1,S}), \pi_{i,2} = (\pi_{i,2,1}, \pi_{i,2,2}, \ldots, \pi_{i,2,S}), (i = 0, 1, 2, \ldots)$$

The vector Π satisfies:

$$\begin{cases} \Pi A = 0 \\ \Pi e = 1 \end{cases} \quad (2)$$

From the well-known result on matrix-geometric methods (see [7]); the steady-state probability vector Π can be given by:

$$\Pi_i = \Pi_0 R^i, (i = 1, 2, \ldots), \quad (3)$$

where the matrix R satisfies the matrix quadratic equation:

$$R^2 D + RB + C = 0, \quad (4)$$

and the vector Π_0 can be obtained by solving the following equations

$$\begin{cases} \Pi_0 \left(\widehat{A_0} + RD \right) = 0 \\ \Pi_0 (I - R)^{-1} e = 1 \end{cases} \quad (5)$$

From the matrix quadratic Eq. (4), due to the special structure of matrix C, the matrix R has the follow structure:

$$R = \begin{bmatrix} 0_{(S+1)\times(S+1)} & 0_{(S+1)\times S} \\ R_1 & R_2 \end{bmatrix}, R_1 = (r_{i,j})_{S\times(S+1)}, R_2 = (t_{i,j})_{S\times S}$$

Then, we can get

$$R^i = \begin{bmatrix} 0_{(S+1)\times(S+1)} & 0_{(S+1)\times S} \\ {R_2}^{i-1}R_1 & {R_2}^i \end{bmatrix}, (i = 0, 1, 2, \ldots)$$

Substituted it into the (4), the matrix quadratic equation can be re-written as:

$$\begin{bmatrix} 0_{(S+1)\times(S+1)} & 0_{(S+1)\times S} \\ R_1B_{11} + R_2B_{21} & R_2R_1D_1 + R_1B_{12} + R_2B_{22} + \alpha I_{S\times S} \end{bmatrix} = 0_{(2S+1)\times(2S+1)} \quad (6)$$

We can get a series of non-linear equations. Then, for any actual model, once the parameters are given, we can use some numeral iterative method, such as Gauss-Seidel method, to solve the numeral equations relevant to (5). All the probability vectors Π_i can be described as follow:

$$\begin{aligned} \Pi_i = \Pi_0 R^i &= \left(\pi_{0,1}, \pi_{0,2}\right) \begin{bmatrix} 0_{(S+1)\times(S+1)} & 0_{(S+1)\times S} \\ {R_2}^{i-1}R_1 & {R_2}^i \end{bmatrix} \\ &= \left(\pi_{0,2}{R_2}^{i-1}R_1, \pi_{0,2}{R_2}^i\right), (i = 1, 2, \ldots) \end{aligned}$$

We can partition the vector Π_i as

$$\Pi_i = \left(\pi_{i,1}, \pi_{i,2}\right), \text{where} \begin{cases} \pi_{i,1} = \pi_{0,2}{R_2}^{i-1}R_1 \\ \pi_{i,2} = \pi_{0,2}{R_2}^i \end{cases}, (i \geq 1)$$

We can also use an alternative iterative method to solve the rate matrix R. Let

$$\begin{cases} U = B + C \cdot G \\ G = (-U)^{-1} \cdot D \\ R = C \cdot (-U)^{-1} \\ U = B + R \cdot D \end{cases}, \text{(see Latouche and Ramaswami [7])}$$

We can use them iteratively to obtain the rate matrix R.

Algorithm 1.

$$\begin{aligned} & G := (-B - C) \cdot D; \\ & \mathit{repeat} \\ & \begin{cases} G_{old} := G; \\ U := B + C \cdot G; \\ G := G = (-U)^{-1} \cdot D \end{cases} \\ & \mathit{until} \|G - G_{old}\| \leq \varepsilon \\ & R := C \cdot (-U)^{-1} \end{aligned}$$

Then, combined with (5), we can get a special non-zero solution vector Π_{0*} by solving the first matrix equation: $\Pi_0\left(\widehat{A_0} + RD\right) = 0$; furthermore, together with the second equation: $\Pi_0(I-R)^{-1}e=1$, we can get the boundary probability vector Π_0:

$$\Pi_0=\left[\Pi_{0*}\cdot(I-R)^{-1}\cdot e\right]^{-1}\cdot\Pi_{0*}$$

Having the R and Π_0, we can obtain the rest probability vectors by the formula:

$$\Pi_i=\Pi_0 R^i,\ (i=1,2,\ldots)$$

4 System Performance Measures

1. Expected Inventory Level: $\zeta_I=\sum_{i=0}^{\infty}\sum_{k=1}^{S}k\left(\pi_{i,1,k}+\pi_{i,2,k}\right)$
2. Expected Reorder Rate: $\zeta_R=\mu\sum_{i=0}^{\infty}\pi_{i,2,s+1}$
3. Expected Number of Customers in the Orbit: $\zeta_0=\sum_{i=1}^{\infty}i\Pi_i e=\Pi_0\sum_{i=1}^{\infty}iR^i e$
4. Overall Rate of Losing Customers: $\zeta_L=\alpha P_{OFF}=\alpha\sum_{i=0}^{\infty}\pi_{i,0,0}$
5. Overall Rate of Retrial: $\zeta_{OR}=\theta\cdot\sum_{i=1}^{\infty}\sum_{k=1}^{S}\left(\pi_{i,1,k}+\pi_{i,2,k}\right)=\theta\cdot\sum_{i=1}^{\infty}\left(\Pi_i\cdot e-\pi_{i,0,0}\right)$
6. Successful Rate of Retrial: $\zeta_{ORS}=\theta\sum_{i=1}^{\infty}\sum_{k=1}^{S}\pi_{i,1,k}=\theta\left[\pi_{0,2}(I-R_2)^{-1}R_1 e-P_{OFF}\right]$
7. Fraction of Successful Rate of Retrial: $\zeta_{FORS}=\frac{\zeta_{ORS}}{\zeta_{OR}}$

5 Cost Analysis

Let $TC(s,S)$ denote the long-run expected cost rate under the following cost structure.

c_O: Waiting cost of a customer in the orbit per unit time;

c_I: The inventory carrying cost per unit item per unit time;

c_R: Setup cost per order.

Then we have $TC(s,S)=c_O\zeta_O+c_I\zeta_I+c_R\zeta_R$

Based on the equations obtained in Sect. 4, we get that:

$$TC(s,S)=c_O\Pi_0 R(I-R)^{-2}e+c_I\sum_{i=0}^{\infty}\sum_{k=1}^{S}k\left(\pi_{i,1,k}+\pi_{i,2,k}\right)+c_R\mu\sum_{i=0}^{\infty}\pi_{i,2,s+1}$$

Acknowledgment This work is supported by National Natural Science Foundation of China (No.11171019), Program for New Century Excellent Talents in University (NCET-11-0568) and the Fundamental Research Funds for the Central Universities (No.2011JBZ012).

References

1. Berman O, Kim EH, Shimshak DG (1993) Deterministic approximation for inventory management for service facilities. IIE Trans 25:98–104
2. Berman O, Kim EH (1999) Stochastic inventory policies for management for service facilities. Stoch Model 15:695–718
3. Berman O, Sapna KP (2000) Inventory management at service facilities for systems with arbitrarily distributed service times. Stoch Model 16:343–360
4. Schwarz M, Sauer C, Daduna H, Kulik R, Szekli R (2006) M/M/1 Queueing systems with inventory. Queueing Syst 54:55–78
5. Artalejo JR, Krishnamoorthy A, Lopez-Herrero MJ (2006) Numerical analysis of (s, S) inventory systems with repeated attempts. Ann Oper Res 141:67–83
6. Krishnamoorthy A, Jose KP (2007) Comparison of inventory systems with service, positive lead-time, loss, and retrial of customers. J Appl Math Stoch Anal, Article ID 37848, 2007:23
7. Latouche G, Ramaswami V (1999) Introduction to matrix analytic methods in stochastic modeling. American Statistical Association and Society for Industrial and Applied Mathematics, Philadelphia
8. Yang T, Templeton JGC (1987) A survey of retrial queues. Queueing Syst 2:201–233
9. Falin GI (1990) A survey of retrial queues. Queueing Syst 7:127–167
10. Falin GI, Templeton JGC (1997) Retrial queues. Chapman & Hall, London

Study on Warehousing Management System Information Performance Based on Analytic Hierarchy Process-Fuzzy Comprehensive Evaluation Method

Bing Han, Bing Zhu, and Li Huang

Abstract The level of information in warehousing management takes an important role in a logistics company. This paper builds an evaluation model based on Analytic Hierarchy Process-Fuzzy Comprehensive Evaluation so as to make assessment of information performance in a warehousing management information system. A case study is delivered showing the feasibility and credibility of this model.

Keywords AHP • FCE • WMS • Information performance

1 Introduction and Description of the Problem

Warehousing management system is an important part of modern logistics. Effective warehousing management can achieve the integration of the products, adjust the supply and demand, it can make the whole logistics system work fast and correctly. Storage is always the key section in logistics system. Information performance in warehousing management system can make full use of various resources in Logistics Company and thus it can dig out a greater profit space for logistics enterprise.

In the past decades the warehousing management information system in our country have already rapid developed, but compared with the developed countries, the popularization, the depth and level of application still have a long way to go [1]. In recent years nearly all the logistics enterprise have been start using warehousing information management system [2], but the application of information system is still in a low level [3]. This paper analyzes the actual situation of the corporate performance and gives out the result.

B. Han (✉) • B. Zhu • L. Huang
School of Economics and Management, Beijing Jiaotong University,
No.3 Shang Yuan Cun, Hai Dian District, Beijing, P.R. China 100044,
e-mail: hb200507@163.com; bzhu@bjtu.edu.cn; 11120688@bjtu.edu.cn

Z. Zhang et al. (eds.), *LISS 2012: Proceedings of 2nd International Conference on Logistics, Informatics and Service Science*, DOI 10.1007/978-3-642-32054-5_63,

2 Level Analysis of a Fuzzy Comprehensive Evaluation Model

There are two parts in this model: First, using the analytic hierarchy process to determine the weight of each indicator in the evaluation system, and then on this basis the use of fuzzy evaluation method to evaluate.

2.1 *Analytic Hierarchy Process and Fuzzy Comprehensive Evaluation Method*

The Analytic Hierarchy Process is a method of a decision; it put forward by the U.S. operations research professor T.L. Saaty in the early 1970s. It helps decision-makers to do the event level analysis [4]. The basic idea of fuzzy and comprehensive evaluation is to using the principle of linear transformation and maximum membership function, make a comprehensive consideration on evaluation objects, and then make a scientific evaluation on these elements [5].

2.2 *Principles and Procedures*

The step of using analytic hierarchy process modeling is generally as follows: I. establishes an index system about the warehousing management system. II. constructs all judgment matrixes in all levels. Make elements in an above as a standard, the next level of elements A_1, $A_2 \ldots A_{ii}$ dominant relationship, according to the standard we can give A_1, $A_2 \ldots A_{ii}$ corresponding weights through their relative importance. $a_{ij} > 0$, $a_{ij} = 1/a_{ij}$, $a_{ij} = 1$. III. Calculate the single criterion of the relative weight, and make the consistency examination. IV. Consistency test according to hierarchy combination sort, each element in next level with the calculation results of CI, is relative to present level, the corresponding standardize is RI, thus the consistency rate in next level is $CR = CI/RI$, when $CR < 0.1$, the consistency of the layer of k is acceptable, otherwise, you must adjust the comparison matrix.

Fuzzy and comprehensive evaluation method make the character of fuzziness be quantified by means of membership function, thus can use traditional mathematical calculation method for processing.

3 The Empirical Analysis

In this paper, the author chooses two warehousing companies to give them a comprehensive evaluation (Fig. 1 and Tables 1, 2, 3, 4, 5, 6, 7, 8, 9, and 10).

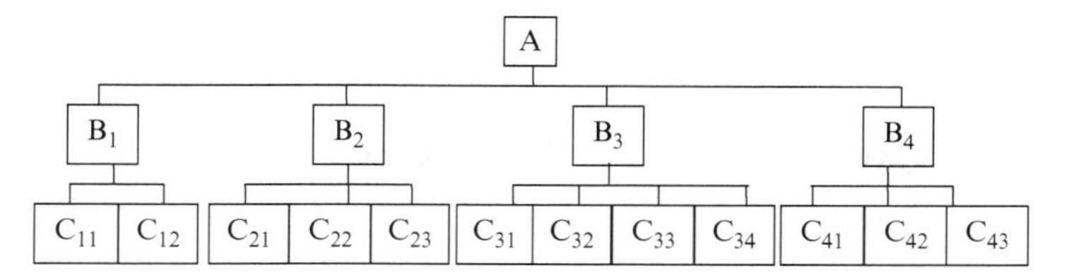

Fig. 1 The structure of logistics enterprise
A: the performance of information in Logistics Company, B_1: customer index, B_2: financial risk index, B_3: internal operation index, B_4: learning and growth index, C_{11}: the satisfaction of customer, C_{12}: contain of customer, C_{21}: occupy rate of inventory capital, C_{22}: return on net assets, C_{23}: turnover rate of funds, C_{31}: inventory turnover rate, C_{32}: order processing accuracy rate, C_{33}: constituent structure is optimized, C_{34}: order on time delivery rate, C_{41}: the population of information, C_{42}: satisfaction of stuff, C_{43}: IT team skills. Using the method of the experts' interviews and list all levels in firm A and B of judgment matrix according to expert, as the following

Table 1 The judgment matrix of the layer A-B in firm A

A	B_1	B_2	B_3	B_4
B_1	1	2	5	4
B_2	1/2	1	3	2
B_3	1/5	1/3	1	1/2
B_4	1/4	1/2	2	1

$w_A = (2.035, 1.061, 1.346, 0.574)$
$\lambda_{max} = 4.021$, CR $= 0.008 < 0.1$, meets the requirements of consistency

Table 2 The judgment matrix of the layer A-B in firm B

A	B_1	B_2	B_3	B_4
B_1	1	2	4	2
B_2	1/2	1	4	2
B_3	1/4	1/4	1	1/3
B_4	1/2	1/2	3	1

$w_A = (0.423, 0.299, 0.081, 0.197)$
$\lambda_{max} = 4.081$, CR $= 0.03 < 0.1$, meets the requirements of consistency

Table 3 The judgment matrix of the layer B_1-C in firm A

B_1	C_{11}	C_{12}
C_{11}	1	2
C_{12}	1/2	1

$w_{B1} = (0.667, 0.333)$
$\lambda_{max} = 2$, CR $= 0 < 0.1$, meets the requirements of consistency

Table 4 The judgment matrix of the layer B_1-C in firm B

B_1	C_{11}	C_{12}
C_{11}	1	1/3
C_{12}	3	1

$w_{B1} = (0.250, 0.750)$
$\lambda_{max} = 2$, CR $= 0 < 0.1$, meets the requirements of consistency

Table 5 The judgment matrix of the layer B_2-C in firm A

B_2	C_{21}	C_{22}	C_{23}
C_{21}	1	1/4	1/2
C_{22}	4	1	3
C_{23}	2	1/3	1

$w_{B2} = (0.137, 0.625, 0.238)$
$\lambda_{max} = 3.018$, CR = 0.016 <0.1, meets the requirements of consistency

Table 6 The judgment matrix of the layer B_2-C in firm B

B_2	C_{21}	C_{22}	C_{23}
C_{21}	1	1/3	1/2
C_{22}	3	1	2
C_{23}	2	1/2	1

$w_{B2} = (0.163, 0.540, 0.297)$
$\lambda_{max} = 3.009$, CR = 0.009 <0.1, meets the requirements of consistency

Table 7 The judgment matrix of the layer B_3-C in firm A

B_3	C_{31}	C_{32}	C_{33}	C_{34}
C_{31}	1	2	2	2
C_{32}	1/2	1	3	1/2
C_{33}	1/2	1/3	1	2
C_{34}	1/2	1/3	1/2	1

$w_{B3} = (0.398, 0.220, 0.255, 0.127)$
$\lambda_{max} = 4.224$, CR = 0.083 <0.1, meets the requirements of consistency

Table 8 The judgment matrix of the layer B_3-C in firm B

B_3	C_{31}	C_{32}	C_{33}	C_{34}
C_{31}	1	2	2	2
C_{32}	1/2	1	2	2
C_{33}	1	1/2	1	3
C_{34}	1/2	1/2	1/3	1

$w_{B3} = (0.333, 0.280, 0.261, 0.126)$
$\lambda_{max} = 4.241$, CR = 0.089 <0.1, meets the requirements of consistency

Table 9 The judgment matrix of the layer B_4-C in firm A

B_4	C_{41}	C_{42}	C_{43}
C_{41}	1	2	6
C_{42}	1/2	1	4
C_{43}	1/6	1/4	1

$w_{B4} = (0.588, 0.322, 0.090)$
$\lambda_{max} = 3.009$, CR = 0.009 <0.1, meets the requirements of consistency

Table 10 The judgment matrix of the layer B_4-C in firm B

B_4	C_{41}	C_{42}	C_{43}
C_{41}	1	2	5
C_{42}	1/2	1	3
C_{43}	1/5	1/3	1

$w_{B4} = (0.627, 0.254, 0.118)$
$\lambda_{max} = 3.052$, CR = 0.009 <0.1, meets the requirements of consistency

There are four levels in the performance of information of the evaluation in Logistics Company in this paper: $V = \{v_1(\text{excellent}), v_2(\text{good}), v_3(\text{general}), v_4(\text{poor})\}$, according to opinion of experts on A company,judgment matrix shown as follows:

$$R_{B1} = \begin{bmatrix} 0.1 & 0.6 & 0.3 & 0 \\ 0 & 0.7 & 0.3 & 0 \end{bmatrix} \qquad R_{B2} = \begin{bmatrix} 0.1 & 0.5 & 0.4 & 0 \\ 0.2 & 0.4 & 0.4 & 0 \\ 0.3 & 0.5 & 0.2 & 0 \end{bmatrix}$$

$$R_{B3} = \begin{bmatrix} 0.1 & 0.4 & 0.5 & 0 \\ 0.2 & 0.4 & 0.4 & 0 \\ 0.1 & 0.3 & 0.6 & 0 \\ 0.3 & 0.3 & 0.4 & 0 \end{bmatrix} \qquad R_{B4} = \begin{bmatrix} 0.1 & 0.3 & 0.5 & 0.1 \\ 0 & 0.5 & 0.5 & 0 \\ 0.2 & 0.3 & 0.4 & 0.1 \end{bmatrix}$$

The comprehensive evaluation of the elements in middle layer of firm A as follows:

$Q_1 = W_{B1} \cdot R_{B1} = (0.067, 0.633, 0.210, 0)$
$Q_2 = (0.219, 0.953, 0.311, 0)$
$Q_3 = (0.480, 0.362, 0.491, 0)$,
$Q_4 = (0.239, 0.364, 0.491, 0.068)$
$R_A = (Q_1 Q_2 Q_3 Q_4)$, $Q_A = W_A R_A = (0.167, 0.656, 0.301, 0.010)$.

Use the same method we can get the comprehensive evaluation of the elements in middle layer in firm B as follows:

$Q_1 = (0.025, 0.675, 0.300, 0)$
$Q_2 = (0.200, 0.408, 0.417, 0.030)$
$Q_3 = (0.187, 0.435, 0.379, 0)$
$Q_4 = (0.386, 0.350, 0.463, 0.019)$
$Q_B = (0.162, 0.512, 0.373, 0.013)$.

And then calculate the value of comprehensive evaluation and determine the Level of the evaluation, first, give the score of the set of evaluation according to the hundred-mark system, thus we can get the data of the set of evaluation by assign values: $K = \{95, 85, 70, 50\}$, finally, got the scores of comprehensive evaluation of firm A and B as follows: $E_A = Q_A \cdot K = 93.195$, $E_B = Q_B \cdot K = 85.670$

E_A is greater than E_B, so we think that the performance of the information in firm A is greater than firm B.

4 Conclusions

This paper analyzes the problem of warehousing management system information performance in logistics enterprise and establishes an evaluation modeling—level analysis of a fuzzy comprehensive evaluation model specialized for actual operation in China. We evaluate the information performance of two logistics enterprises which are similar in nature and size. The final conclusion is that the evaluation results of information performance which obtained by the model in this paper tally with the actual situation of the corporate performance. It is proved that the evaluation of this model is feasible and credible; this method has a great significance for logistics industry in China.

References

1. Jiang Chaofeng (2011) WMS information performance present situation and development prospect in our country. China Logist Time 5:58–59
2. Zhang Jiaoyan (2010) Research on the building and application of index system in logistics information performance evaluation. Dongbei University of Finance and Economics, Dalian, p 12
3. Hubei 337 Department of National Reverse Goods Administration (2010) Construction of information performance and the practice of warehousing management. Logist Technol 29(18):60
4. Wang Yizhong (2009) Warehousing information technologies urgently need to upgrade. China Logist Parching 8:60
5. Yan Huashi (2010) A study on the application of APH in the investment decision-making of South-International Petroleum-Chemical Storage Project. South China University of Technology, Guangzhou, p 10

Part III
Service Management

Study on Simulation Method for Intersection Hybrid Traffic Flow

Yalong Zhao, Xifu Wang, Hongfeng Li, and Tingting Zhu

Abstract In hybrid traffic conditions at intersections, the speed of various transport modes is different and the traffic conflicts occur frequently, which lead to traffic chaos. Reasonable simulation method is very important for improving intersection state and capacity. The process and method of microscopic traffic simulation are studied, this paper considers the impacts of pedestrians and non-motorized vehicles, then designs a simulation process for intersection hybrid traffic flow. Taking an orthogonal intersection as an example, through data survey, problems analysis, optimization scheme design and simulation evaluation, the validity of the proposed method is verified.

Keywords Traffic engineering • Hybrid traffic flow • Microscopic traffic simulation • Intersection • Traffic organization and optimization

1 Introduction

As an effective mean and auxiliary tool of traffic planning and designing, microscopic traffic simulation has been widely applied in developed countries [1]. However, the technology of simulating urban hybrid traffic flow is not mature in china. Accurate data collection and analysis and model validation are often neglected in traffic simulation, the model can't reflect the actual situations completely when mock the real system, which may influence the authenticity and reliability of simulation results [2, 3]. Urban road intersection is a complex system, if the simulation method is not reasonable, the efficiency and results will be difficult be guaranteed.

Y. Zhao (✉) • X. Wang • H. Li • T. Zhu
School of Traffic and Transportation, Beijing Jiaotong University, Beijing 100044, China
e-mail: ylzhaobjtu@163.com; xfwang@bjtu.edu.cn; hfli@bjtu.edu.cn; 11121071@bjtu.edu.cn

Z. Zhang et al. (eds.), *LISS 2012: Proceedings of 2nd International Conference on Logistics, Informatics and Service Science*, DOI 10.1007/978-3-642-32054-5_64,

2 Simulation Process Design for Hybrid Traffic Flow

2.1 *Process Design*

This paper divides microscopic traffic simulation process into five parts: determination of simulation target, traffic data survey, traffic data analysis, simulation modeling, design and evaluation of optimization schemes. The simulation process is shown in Fig. 1.

2.2 *Traffic Data Survey and Analysis*

As the foundation of model parameters calibration and traffic simulation, basic traffic data must be surveyed accurately and comprehensively, which directly affects the fidelity of models and the reliability of schemes evaluation [4]. The collection traffic date mainly include static flow of the various traffic means, vehicle types, desired speed of different type vehicles, pedestrian and non-motorized vehicle flow, traffic management data, bus information and detailed geometric data, etc.

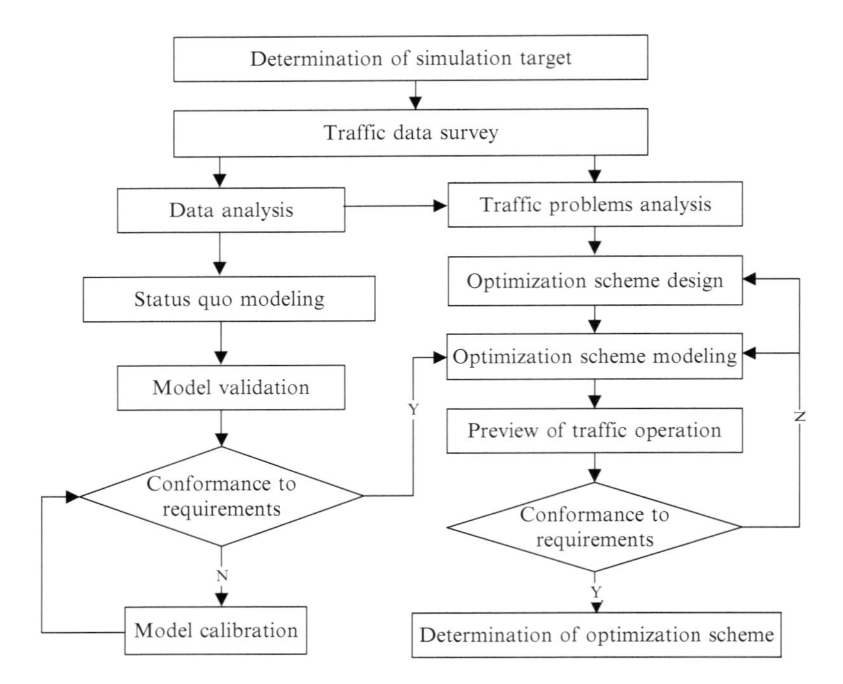

Fig. 1 Simulation process of hybrid traffic flow

The traffic flow not only changes over time, but also with the spatial variation. By analyzing the spatial and temporal distribution and variation tendency of the traffic flow, we can determine the simulation period, meanwhile, the traffic demand change during the entire simulation period is reflected. Through the comprehensive analysis of traffic data, the major traffic problems of the intersection are determined, which provides the basis for designing optimization schemes.

2.3 Design and Evaluation of Optimization Schemes

On the basis of analyzing the principal contradiction, this paper optimizes the intersection from several aspects, including channelization, traffic organization optimization and signal timing. After the optimization schemes are designed, it is necessary to employ simulation software to test them, the final optimization schemes should be implemented conveniently and solve the traffic problems effectively.

There exist a certain degree of similarities among different intersections, for improving the efficiency of the program design, we should classify different kinds of traffic problems in accordance with certain characteristics, such as signal timing, violation and channelization, etc. According to the characteristics of different problems, we propose corresponding solutions and establish scheme library [5]. This method can greatly improve the efficiency of optimization scheme design when optimize a large number of intersections.

3 Instance Application

3.1 Field Data Survey and Analysis

This paper takes Zhixinqiaoxia intersection as an instance, uses micro-simulation software VISSIM to analyze it, and then studies the validity of the method. Zhixinqiaoxia intersection is a signalized crossing, Fig. 2 shows its geometric characteristics. The intersection adopts four-phase signal control program, the cycle length during peak hour is 180 s, Fig. 3 shows the signal timing scheme.

Selecting 3 working days and 1 weekend day of a week, this paper determines the peak hour by comparing the different period traffic flow, the results show that Tuesday morning 7:30–8:30 traffic flow is maximum, which is 6306pcu. The peak hour is the most critical period of traffic running in a day, it requires greater traffic capacity. Due to traffic flow of peak hour is very large, traffic problems are prominent particularly.

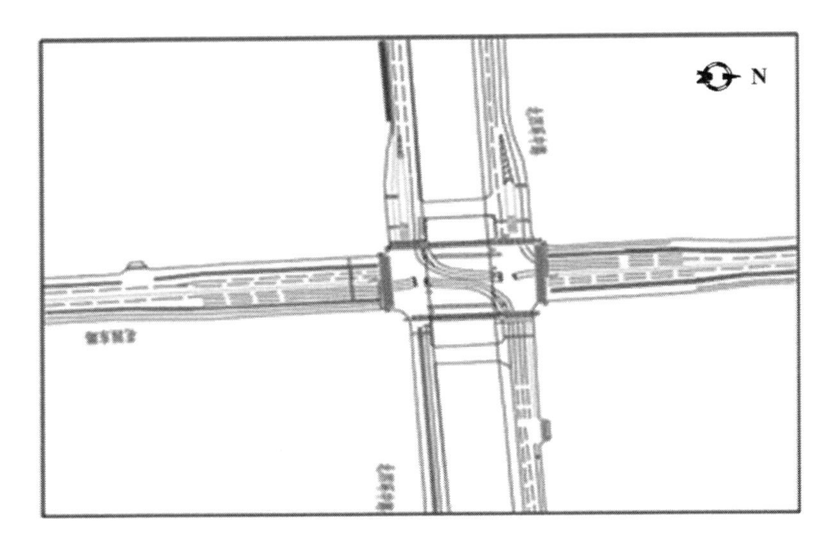

Fig. 2 Geometric diagram of Zhixinqiaoxia intersection

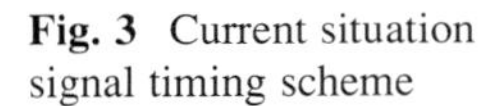

Fig. 3 Current situation signal timing scheme

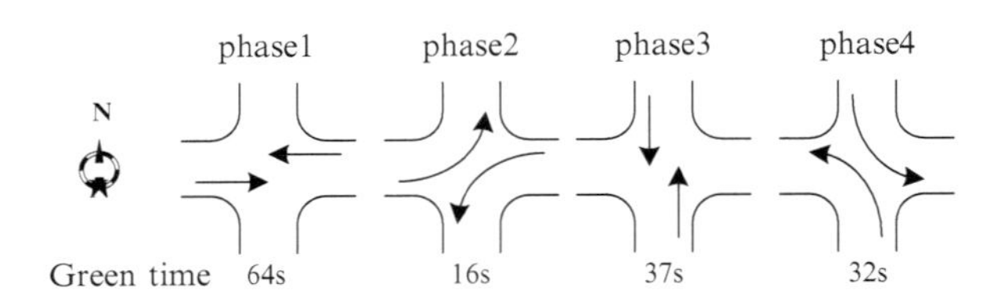

Pedestrian and bicycle analysis. The number of pedestrians and bicycles during peak hour is 1,673 and 2,528 respectively. The violation rate reach 6.7 and 7.1% particularly, which serious interfere the normal operation of motor vehicles.

Motor vehicle traffic analysis. As is revealed in Fig. 4, straight flow of east entrance is maximum, which is the main flow. From the perspective of vehicle type, small vehicles account for the vast majority, mid-size vehicles and heavy vehicles account for a small percentage. Peak hour factor of the intersection is 0.91, which manifests that traffic flow changes small at different periods of peak hour.

Comprehensive traffic problem analysis. Input lanes of east entrance and underground passage of west entrance are unreasonable. The main contradictions of the east entrance are saturation of straight lanes is too high and queue length is too long, however, the utilization rate of right-turn lane is low. West side of the intersection set an underground passage, the utilization rate of it is low, which peak hour pedestrian flow is only 60. Entrance and exit of the underground passage are set between right-turn lane and straight lane, they take up the intersection area and badly influence the capacity of west entrance and lead to traffic congestion during rush hour.

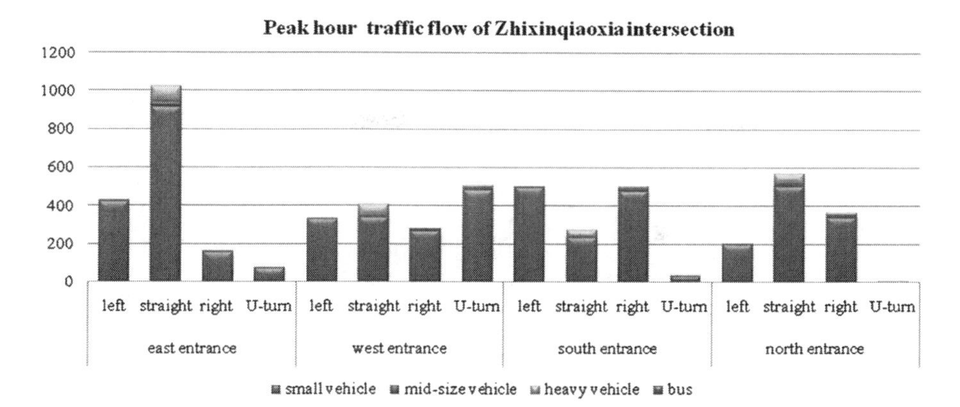

Fig. 4 Peak hour traffic flow diagram

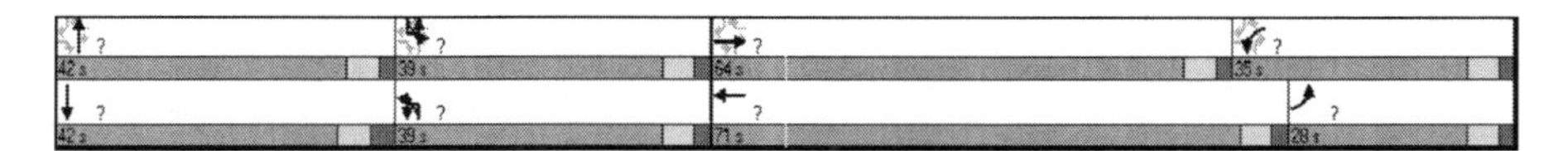

Fig. 5 The signal timing plan of split optimization

3.2 Optimization Scheme Design

This paper takes traffic conflict, signal timing and channelization into consideration, designs a comprehensive optimization scheme. The program compresses the width of east entrance bicycle lane and moves it outside, adds a straight motor vehicle lane; moves the entrance and exit of the underground passage outside of bicycle lane, adds a left-turn lane. On the basis of above channelizing scheme, keep the signal cycle unchanged, adopt traffic signal timing optimization software SYNCHRO to optimize the split. The new signal control scheme is shown as Fig. 5.

3.3 Simulation Analysis

Current situation simulation. The mission of current situation simulation is to mock real traffic phenomena as much as possible, by comparing simulation data and field date, analyzes the authenticity of the simulation. The range of this simulation is intersection stop line back within 300 m of road. Establishing the current model of intersection based on fundamental data, input the traffic volume, signal timing scheme and traffic operation parameters, etc.

Optimization scheme simulation and evaluation. Establishing the optimization scheme model based on the status quo model, testing the traffic operation condition

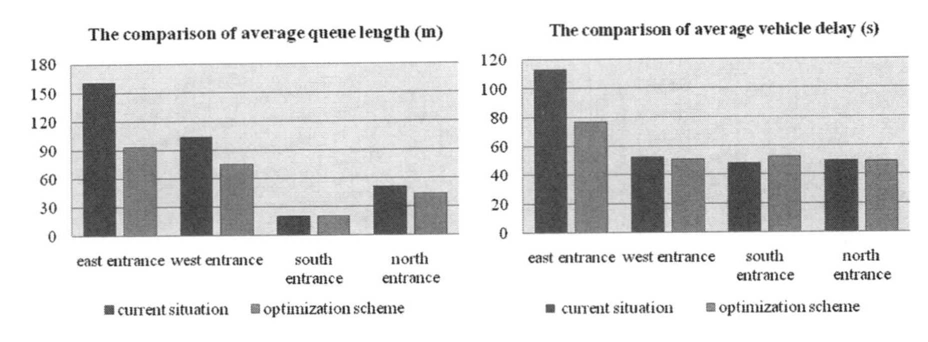

Fig. 6 The comparison of simulation results

and outputting the indicators. Using average queue size and average vehicle delay to evaluate the optimization scheme. The simulation results as illustrated in Fig. 6.

The overall average queue size and average vehicle delay of Zhixinqiaoxia intersection decreased by 21 and 12.8% respectively. Among them, the ameliorative effect of east entrance and west entrance are most obvious. The optimization scheme is feasible due to the intersection has channelization space.

4 Summary

This paper designs a simulation process for intersection hybrid traffic flow, and describes the specific method in detailed. Taking Zhixinqiaoxia intersection as an instance, a combinatorial optimization scheme is designed based on main traffic problems, employing VISSIM microscopic simulation software to simulate and analyze the optimization scheme, the results show that the method has practical value.

References

1. Yang Jia, Sun Jian (2010) Application of microscopic traffic simulation. Urb Transport China 8(5):79–83
2. Sun Jian, Yang Xiao-guang, Liu Hao-de (2007) Study on microscopic traffic simulation model systematic parameter calibration. J Syst Simul 19(1):48–50
3. Sun Jian, Li Ke-ping (2010) Credibility evaluation for micro traffic simulation model. Comput Simul 27(1):276–280
4. Wei Ming, Yang Fang-ting, Cao Zheng-qing (2003) A review of development and study on the traffic simulation. J Syst Simul 15(8):1179–1183
5. Li Mei-ling, Rong Jian, Sun Zhi-yong (2009) Mixed traffic simulation decision system of signalized intersection. Commun Stand 23:107–110

Study on Train Dispatching Model During Holidays

Fei Dou, Limin Jia, Jie Xu, Yangfan Zhou, and Li Wang

Abstract As the passenger flow volume fluctuates significantly during holidays and in case of an unexpected large passenger flow, it's important to arrange train assignment, make temporary passenger train operation scheme and adjust the train operation plan to assure the normal operation of high-speed railway. This paper aims to ease a large passenger flow condition during holidays caused by imbalanced passenger flow. The process of vehicle deployment always leads to the phenomenon that there are no passengers on the round-trip train, so that the railway transportation cost increases. To solve this problem, this paper puts forward the holiday train assignment model, and apply corresponding algorithm to do numerical simulation for solving the train dispatching model by case analysis. It is proved that this train dispatching optimization model and its algorithm can meet the actual demand reasonably.

Keywords Railway transportation • Train dispatching • Optimization model • Nonlinear complementarity

1 Introduction

With the rapid development of national economy and the accelerated process of urbanization, passenger dedicated line passenger demand between the regions is also increasing, especially the sudden increase of passenger volume during the

---

F. Dou • Y. Zhou • L. Wang
School of Traffic and Transportation, Beijing Jiaotong University, Beijing, China
e-mail: doufei911@163.com; 11114216@bjtu.edu.cn; wangli298@gmail.com

L. Jia (✉) • J. Xu
State Key Laboratory of Rail Traffic Control and Safety,
Beijing Jiaotong University, Beijing, China
e-mail: jialm@vip.sina.com; jxu1@bjtu.edu.cn

Z. Zhang et al. (eds.), *LISS 2012: Proceedings of 2nd International Conference on Logistics, Informatics and Service Science*, DOI 10.1007/978-3-642-32054-5_65,

holiday season. As peak passenger flow period is often in the holidays before and after the period of time, a substantial increase in the traffic volume bring overload effect to passenger dedicated line. Under sudden large passenger flow conditions, the normal organization of the passenger has been unable to meet the passenger's demand, so it is necessary to assign the train and adjust train operation plan.

The train dispatching scheme is a complex combinatorial optimization problem. But recently more and more researches have the train dispatching model and optimization algorithm [1–4], such as linear programming, nonlinear programming and genetic algorithm, have been applied at present.

2 Model of the Train Dispatching

The whole procedure of the model needs to judge the passenger flow whether to meet the traffic demand. If it meets the traffic demand, the operation remains unchanged, or the operation would be dispatched for passenger flow assignment, and attains the traffic demand.

2.1 Notation

Input data

- l: The train type.
- k: The railway line type. There are two kinds of railway lines in China, high-speed railway and existing railway.
- F_l: The variable operating cost for train l running 1 km.
- F_l': The variable operating cost for train l empty running 1 km.
- L_{ij}: The distance between stations i and j.
- N_{ij}^k: The required number of trains between stations i and j on the railway line k.
- C_{ij}^k: The carrying capacity between stations i and j on the railway line k.
- t_l: The temporal summation of the time span of vertical skylights and stopping time of train l.
- $T_{i,j,l}$: The running time of train l between stations i and j.

Decision variables

- $x_{i,j,l}^k$: The number of trains l between stations i and j on the railway line k.
- $y_{i,j,l}^k$: The number of the empty running trains l between stations i and j on the railway line k.

2.2 Objective Function

Transportation cost is minimized as the objective function to meet the traffic demand. The variable operating cost includes variable operating cost and variable operating cost for the train empty scheduling, therefore, the objective function of the model is

$$\min \quad z = \sum_k \sum_l \sum_{i \neq j} \sum_j F_l \cdot L_{ij} \cdot x^k_{i,j,l} + \sum_k \sum_l \sum_i \sum_j F_l' \cdot L_{ij} \cdot y^k_{i,j,l}$$

2.3 Constraint Conditions

1. The required number of trains' restrictions:
   The number of train l round trip running between stations i and j during day and night is $t_{i,j,l} = \frac{24 - t_l}{2T_{i,j,l}}$, therefore,

$$\sum_k \left(x^k_{i,j,l} + y^k_{i,j,l} \geq \frac{N^k_{ij}}{t_{i,j,l}} \right).$$

2. Carrying capacity restrictions:

$$\sum_l \left(x^k_{i,j,l} + y^k_{i,j,l} \right) \leq C^k_{ij}.$$

3. Train dispatching restrictions:
   If the round-trip trains are no-load between stations i and j, it can not satisfy the condition of the burst of passenger flow between stations i and j. Therefore, $y^k_{i,j,l}$ or $y^k_{j,i,l}$ must be zero,

$$y^k_{i,j,l} \cdot y^k_{j,i,l} = 0, i \neq j$$

2.4 Model of the Train Maintenance Optimization

We could obtain the following optimization model of the train dispatching.

$$\min \quad z = \sum_k \sum_l \sum_{i \neq j} \sum_j F_l \cdot L_{ij} \cdot x^k_{i,j,l} + \sum_k \sum_l \sum_i \sum_j F_l' \cdot L_{ij} \cdot y^k_{i,j,l}$$

$$s.t.\quad \sum_k \left(x_{i,j,l}^k + y_{i,j,l}^k \geq \frac{N_{ij}^k}{t_{i,j,l}} \right)$$

$$\sum_l \left(x_{i,j,l}^k + y_{i,j,l}^k \right) \leq C_{ij}^k$$

$$y_{i,j,l}^k \cdot y_{j,i,l}^k = 0,\ i \neq j$$

$$x_{i,j,l}^k \geq 0$$

$$y_{i,j,l}^k \geq 0$$

Among which train dispatching restrictions are nonlinear complementarity constraints. We can use the corresponding algorithm to solve this problem; an example analysis is given in the following section.

3 Example Analysis

Many different algorithms for large scale nonlinear programming have been put forward. An interior point algorithm by R.H. Byrd, M.E. Hribar, and J. Nocedal was proposed to solve nonlinear programming [5]. And the optimization methods and software by R. Fletcher and S. Leyffer was put forward in order to solve the problem that is the nonlinear program with complementarity constraints [6]. Here, we can use the interior-penalty algorithm [7, 8] for train dispatching model in condition that the constraint of this train dispatching optimization model is nonlinear complementarity constraint. We complete a case studies with simulation data, and analyze it by the results. Assume that there are two railway lines, and four stations: station A,B,C,D, thus, $i = \{1,2,3,4\}$ and $j = \{1,2,3,4\}, i \neq j$, in addition, there is the only one type of the train, $\frac{N_{12}}{t_{12}} = \frac{N_{21}}{t_{21}} = 4$, $\frac{N_{13}}{t_{13}} = \frac{N_{31}}{t_{31}} = 6$, $\frac{N_{14}}{t_{14}} = \frac{N_{41}}{t_{41}} = 8$, $\frac{N_{23}}{t_{23}} = \frac{N_{32}}{t_{32}} = 2$, $\frac{N_{24}}{t_{24}} = \frac{N_{42}}{t_{42}} = 4$, $\frac{N_{34}}{t_{34}} = \frac{N_{43}}{t_{43}} = 3$, the carrying capacity of the existing railway is $C_{ij}^1 = C_{ji}^1$, that are $C_{12}^1 = C_{21}^1 = 10$, $C_{13}^1 = C_{31}^1 = 9$, $C_{14}^1 = C_{41}^1 = 10$, $C_{23}^1 = C_{32}^1 = 12$, $C_{24}^1 = C_{42}^1 = 11$, $C_{34}^1 = C_{43}^1 = 8$, also assume that carrying capacities of the high-speed railway are $C_{12}^2 = C_{21}^2 = 8$, $C_{13}^2 = C_{31}^2 = 8$, $C_{14}^2 = C_{41}^2 = 7$, $C_{23}^2 = C_{32}^2 = 8$, $C_{24}^2 = C_{42}^2 = 7$, $C_{34}^2 = C_{43}^2 = 5$. In the process of train dispatching between stations i and j, assume that the operating cost for running is $F_l \cdot L_{ij}$, that are $F_l \cdot L_{12} = 20000$, $F_l \cdot L_{13} = 30000$, $F_l \cdot L_{14} = 40000$, $F_l \cdot L_{23} = 10000$, $F_l \cdot L_{24} = 10000$, $F_l \cdot L_{34} = 10000$ (unit: yuan), and the variable operating cost $F_l' \cdot L_{ij}$ for empty running are $F'_l \cdot L_{12} = 30000$, $F'_l \cdot L_{13} = 40000$, $F'_l \cdot L_{14} = 50000$, $F'_l \cdot L_{23} = 20000$, $F'_l \cdot L_{24} = 30000$, $F'_l \cdot L_{34} = 30000$ (unit: yuan).

Table 1 The number of the train on the existing railway

Station \ Station Number	A	B	C	D
A	—	5	4	4
B	5	—	3	4
C	4	3	—	2
D	4	4	2	—

Table 2 The number of the train on the high-speed railway

Station \ Station Number	A	B	C	D
A	—	2	3	4
B	2	—	1	2
C	3	1	—	2
D	4	2	2	—

We can calculate that the minimum cost of the train dispatching is 1.98 million yuan, the number of the train on the existing railway x_{ij}^1 is shown in Table 1, and the number of the train on the high-speed railway x_{ij}^2 is shown in Table 2.

The number of the train on the existing railway is $y_{ij}^1 = 0$, and the number of the train on the high-speed railway is $y_{ij}^2 = 0$. Therefore, we can determine that there are not the round-trip no-loading trains from the result, and the train dispatching scheme is achieved maximum utilization.

4 Conclusions and Future Work

This paper deals with the problem of train dispatching which is a discrete nonlinear program. There are some aspects to explore concerning the proposed model. Firstly, the train assignment model only considers the assignment of transportation costs, at the same time, the necessary evacuation time should also be considered. Secondly, we need to consider more constraint conditions when passenger flow increases during holidays. Finally, origins and destinations passenger flow forecast should be done before the formation of the model. The train operation plan adjustment is based on the difference in passenger flow between origins and destinations. For complex examples, the effectiveness of the train dispatching model and corresponding algorithm needs further study.

Acknowledgment This work has been supported by the National Natural Science Foundation of China (Grant: 61074151), the National Key Technology R&D Program (Grant: 2009BAG12A10), the Research Fund of the State Key Laboratory of Rail Traffic Control and Safety (Grant: RCS2008ZZ003, RCS2009ZT002), and the Research Fund of Beijing Jiaotong University (Grant: 2011YJS035).

References

1. Beaujon GJ, Turnquist MA (1991) A model for fleet sizing and vehicle allocation. Transport Sci 25:19–45
2. Chen Yan-Ru, Peng Qi-Yuan, Jiang Yang-sheng (2003) Research on a model for adjusting train diagram on double track railway with satisfactory optimization. J China Rail Soc 25(3):8–12
3. Dejax PJ, Crainic TG (1987) A review of empty flows and fleet management models in freight transportation. Transport Sci 21(4):227–247
4. Misra SC (1972) Linear programmings of empty wagon disposition. Rail Int 3(3):151–180
5. Byrd RH, Hribar ME, Nocedal J (1999) An interior point algorithm for large scale nonlinear programming. SIAM J Optim 9(4):877–900
6. Fletcher R, Leyffer S (2004) Solving mathematical program with complementarity constraints as nonlinear programs. Optim Method Softw 19(1):15–40
7. Doufei (2011) Study on train deployment model and algorithms during holidays. Beijing Jiaotong Univ 6:15–35
8. Hu XM, Ralph D (2004) Convergence of a penalty method for mathematical programming with complementarity constraints. J Optim Theory Appl 123(2):365–390

How Can High Speed Railway Survive the Competition from Civil Aviation?

Hongchang Li and Xujuan Kuang

Abstract With the implementation of Medium and Long Term Railway Network Plan 2008, Chinese high speed railway gradually came into being. In order to cope with competition from high speed railway, civil aviation put forward comprehensive strategies including cost, big-client, on-web check-in, price and etc. Under such a circumstance, it's of great importance for high speed railway to adopt corresponding strategies to improve its core competence. High speed railway and civil aviation have their comparative advantages as to distance, speed, price, safety and energy consumption and etc. Time model shows that, within 1,000 km, high speed railway dominates market share because of its time saving characteristics, while above 1,000 km, civil aviation has comparative advantage. In order to improve consumer's surplus and gain market competition edge, high speed railway can take strategies of product, price, place, promotion, resource, competition and cooperation, knot and informationization.

Keywords High speed railway • Civil aviation • Market competitiveness

1 Introduction

With the rapid development of highway and civil aviation in China, market share of high speed railway dropped significantly during the last two decades, see Table 1.

H. Li (✉)
School of Economics and Management, Beijing Jiaotong University,
Shangyuancun 3, District Haidian, Beijing 100044, P.R. China
e-mail: hc_li001@yahoo.com.cn

X. Kuang
Department of Economics and Management, Civil Aviation Management Institute of China,
Beijing 100044, P.R. China
e-mail: kuang.juan@126.com

Z. Zhang et al. (eds.), *LISS 2012: Proceedings of 2nd International Conference on Logistics, Informatics and Service Science*, DOI 10.1007/978-3-642-32054-5_66,

Table 1 Chinese railway market share

Year	Total passengers transported (billion persons)	Railway (%)	Highway (%)	Water (%)	Civil aviation (%)
1991	8.06	11.80	84.70	3.24	0.27
1992	8.609	11.58	85.00	3.08	0.34
1993	9.966	10.58	86.37	2.72	0.34
1994	10.929	9.95	87.29	2.39	0.37
1995	11.726	8.76	88.76	2.04	0.44
1996	12.454	7.61	90.10	1.84	0.45
1997	13.261	7.04	90.84	1.70	0.42
1998	13.787	6.90	91.20	1.49	0.42
1999	13.944	7.18	91.01	1.37	0.44
2000	14.786	7.11	91.13	1.31	0.45
2001	15.341	6.85	91.44	1.22	0.49
2002	16.082	6.57	91.73	1.16	0.53
2003	15.875	6.13	92.24	1.08	0.55
2004	17.675	6.32	91.91	1.08	0.69
2005	18.47	6.26	91.90	1.10	0.75
2006	20.242	6.21	91.91	1.09	0.79
2007	22.278	6.09	92.05	1.03	0.83
2008	28.679	5.10	93.52	0.71	0.67
2009	29.769	5.10	93.52	0.71	0.67
2010	32.695	5.13	93.37	0.68	0.82

According to the Medium and Long Term Railway Network Plan 2008, by the end of 2012, Chinese railway network will reach 110,000 km, double track and electrification ratio will be over 50%, high speed railway with the operational speed over 200 km/h will reach 13,000 km. By 2020, the total length of high speed railway will be over 17,000 km, and the total length of railway will be more than 120,000 km, see Fig. 1.

Chinese civil aviation put forward comprehensive measures to cope with the competition from high speed railway. It's imperative for high speed railway to adopt counter measures to survive the competition from high speed railway.

2 Literature Review

Transportation resource serves as the foundation [1], transportation product as the carrier, marketing as the method to achieve market competence, see Fig. 2.

2.1 Transportation Resources

Transportation resource includes fixed and movable transportation resources, and soft ones such as labor, information, organization and management and etc. Railway transportation resource see Table 2.

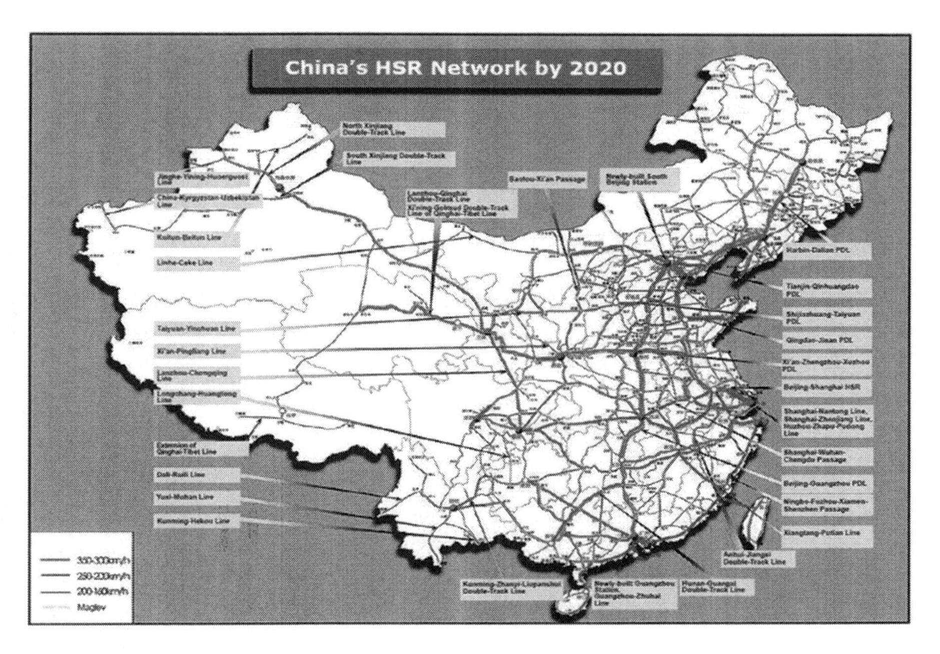

Fig. 1 Long and medium term DPL plan by 2020

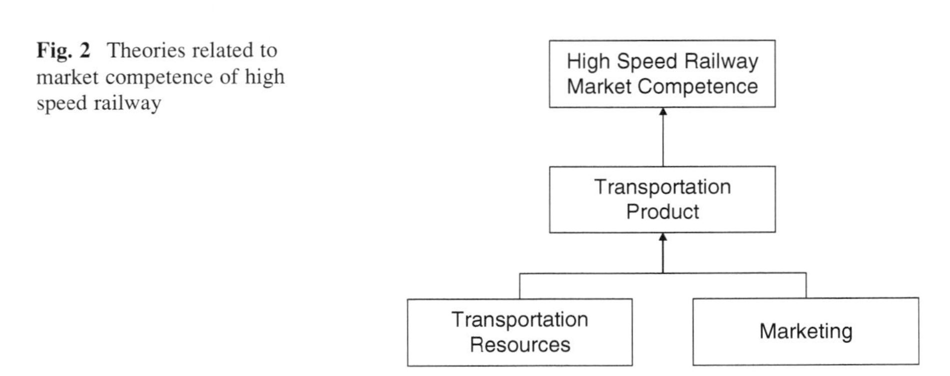

Fig. 2 Theories related to market competence of high speed railway

Railway transportation resources consist of locomotive, wagon, station, communication and signal equipment and etc., and form the foundation for transportation product.

2.2 *Transportation Product*

Product refers to the commodity that can satisfy consumer's need and bring about utility [2]. Product has at least three layers, that is, core product, formal product and

Table 2 Railway transportation resources

Resource type	Resource forms	
Movable resource	Locomotive	1. Steam
		2. Diesel
		3. Electrified
	EMU	
	Ferry	
	Wagon	1. Passenger
		2. Cargo
	Container	1. 20 in.
		2. 10 tons
		3. 5 tons and others
Fixed resource	Railway lines	1. Track bed
		2. Track
		3. Switch
		4. Crossing
		5. Bridge
		6. Tunnel
		7. Culvert
		8. Others
	Station	1. Station routes
		2. Freight facility
		3. Passenger facility
	Telecommunication and signal	
	Water and electricity	
	Other fixed transportation resources	
Soft resource	Human being	
	Capital	
	Organization	
	Operational management	

additional product [3]. Of which, core product is the direct benefit and utility given to the consumers; formal product is the external appearance and characteristics of product including shape, mark and package; additional product is the value added services including installation, maintenance, financing, logistics and etc. It's very important for transportation industries to provide transportation product with high service quality, see Fig. 3.

Other factors keep constant, travel time will be the determinant affecting people's transportation choice.

2.3 Marketing

In 1953, Neil Borden brought up the idea of marketing mix, and maintained that market demand is influenced, to some extent, by so-called marketing factors

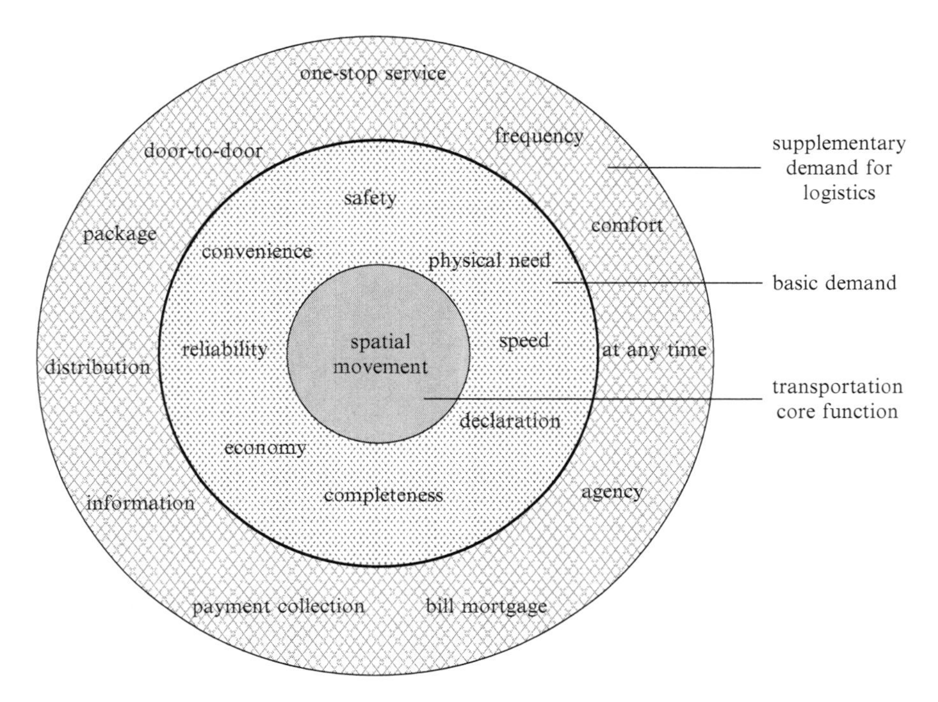

Fig. 3 The importance of speed for transportation product

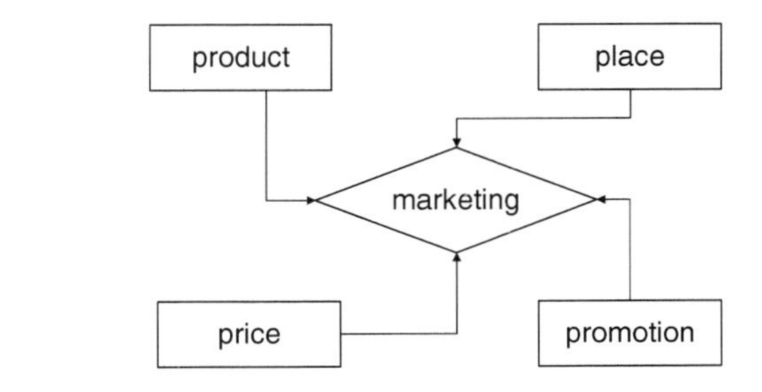

Fig. 4 4Ps

including market demand, cost, price place and promotion. In 1960s, based on marketing mix theory, E. Jerome McCarthy proposed the marketing theory of 4Ps, that is, product, price, place and promotion, see Fig. 4.

Through marketing strategies, high speed railway can reinforce consumers' loyalty and make high speed railway product a popular and welcome transportation service image and gain market competiveness.

3 SWOT Analysis of High Speed Railway and Its Basic Competitive Distance

We first make SWOT analysis of high speed railway and maintain that high speed railway's advantage outfits its disadvantage and opportunity exceeds its challenge.

3.1 SWOT Analysis

SWOT analysis of high speed railway is shown by Table 3.

3.2 Basic Competitive Distance of High Speed Railway

Suppose the operational speed of high speed railway is V_r while the speed of civil aviation V_a. ΔT is the time saving of high speed railway because of short travel distance of high speed railway from home to railway station and from railway station to home, and simple and convenient checking procedure at railway stations. Then, the equilibrium distance is as follows [4]:

$$S^* = \frac{\Delta T \times V_a \times V_r}{V_a - V_r} \tag{1}$$

Of which, S^* is the equilibrium distance at which high speed railway and civil aviation spend the same time and cover the same travel range. Consequently, when travel distance is below S^*, from time saving perspective, high speed railway is superior to civil aviation. When travel distance is above S^*, civil aviation has outstanding competitive advantage, see Fig. 5.

4 Strategies for High Speed Railway to Compete with Civil Aviation

In order to compete with civil aviation, high speed railway of China can take comprehensive strategies.

1. **Price strategy**. High speed railway can set price according to the market segment, peak and trough time period and quantity demanded.
2. **Product strategy**. High speed railway can provide the market with diversified product service, see Table 4.

Table 3 SWOT analysis of high speed railway

Internal environment		Score	Weight	Weighted score	External environment		Score	Weight	Weighted score
Advantage	1. High speed	5	0.10	0.50	Opportunity	1. Strong policy support	4	0.15	0.60
	2. Capable management team	3	0.05	0.15		2. Implementation of medium and long term railway network plan 2008	4	0.25	1.00
	3. Market oriented operational mechanism	2	0.05	0.10		3. Local government supports high speed railway financially	5	0.20	1.00
	4. Time saving within 1,000 kms	4	0.20	0.80		4. Promising market demand	5	0.15	0.75
	5. Stations adjacent to downtown	5	0.20	1.00		5. Reform of ministry of railways	4	0.15	0.60
	6. Harmony brand's popularity	5	0.15	0.75		6. Increasing traffic volumes	4	0.10	0.40
	7. Professional personnel	4	0.05	0.20					
	8. Nationwide agencies	3	0.05	0.15					
	9. Perfect information system	3	0.10	0.30					
	10. Dominant market share in some regions	3	0.05	0.15					
Sub-total			**1.00**	**4.10**	**Sub-total**			**1.00**	**4.35**
Disadvantage	1. Product structure needs to be optimized	−5	0.25	−1.25	Challenge	1. Civil aviation takes comprehensive competing measures	−5	0.3	−1.50
	2. Lack of operational experiences	−3	0.20	−0.60		2. Increasing cost pressure	−5	0.25	−1.25
	3. EMU accident of 2011	−4	0.20	−0.80		3. Inter-mode competition from highway	−3	0.2	−0.60
	4. Imperfect station infrastructure	−3	0.10	−0.30		4. Construction downsize and speed reduction	−3	0.15	−0.45
	5. Over speed of high speed railway and high cost	−5	0.15	−0.75		5. Debt payment pressure	−3	0.1	−0.30
	6. Diversified operation should be developed	−3	0.10	−0.30					
Sub-total			**1.00**	**−4.00**	**Sub-total**			**1.00**	**−4.10**
Sum of advantage and disadvantage				**0.10**	**Sum of opportunity and challenge**				**0.25**

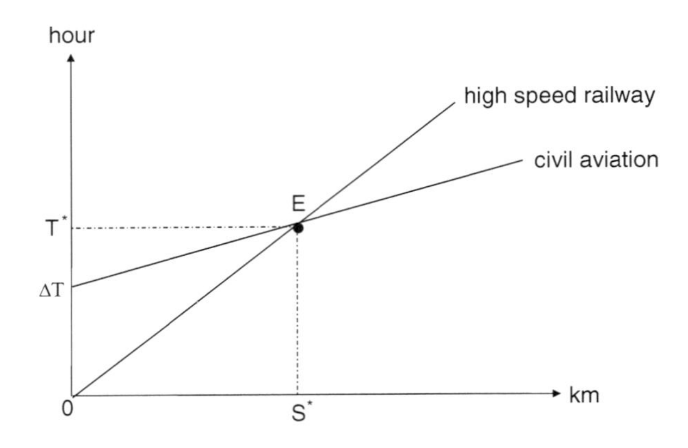

Fig. 5 Equilibrium time and distance of high speed railway and civil aviation

Table 4 Product strategy of high speed railway

Order	Product: Core	Additional	
1	EMU operational plan	Entertainment	Door to door ticket delivery
2	Time	Food and beverage	Guest services
3	Connection with city transportation	Office work	Hotel reservation
4	Ticket booking	Business work	
5	Standardized product	Internet	
6		Tourism	

3. **Place strategy**. High speed railway has three ways to have tickets sold, that is, railway station, ticket selling agencies and Internet.
4. **Promotion strategy**. High speed railway product can be known by consumers through publicity and promotion activities including advertisement and sponsorship.

5 Conclusion

Other parameters kept constant, high speed railway has competitive advantage within 1,000 km. In order to survive the competition of civil aviation, high speed railway has to improve the quality of transportation product and increase consumer's utility. Chinese high speed railway can take strategies to reinforce its market competitiveness by adopting price, product, place and promotion strategies.

Acknowledgments Thank railway expert of the World Bank, Mr. David Burns for his helpful advice. The paper is sponsored by National Science Foundation Project (No.: 41171113) and the key project of logistics management and technology lab.

References

1. Antonio Estache, Gines De Rus (2000) Privatization and regulation of transport infrastructure. The World Bank, Washington, D.C
2. Daughety AF (1985) Analytical studies in transport economics. Cambridge University Press, Cambridge
3. Rong Chaohe (2002) Western transportation economics. Economic Science Press, Beijing
4. Zhao Jian, Su Hongjian (2010) Value of travel time saving and transportation mode choice. Compr Transp 32:60–65

Model of Passenger Route Choice in the Urban Rail Transit Network

Qiao Ke, Zhao Peng, and Qin Zhi-peng

Abstract Passenger travel route choice behavior is becoming diversified with the expansion of urban rail transit network. The paper proposes the passenger route choice model and algorithm on the basis of passenger classification. Calculation and analysis of the local network of Beijing rail transit proved that the results are more realistic after the passenger classification.

Keywords Urban rail transit • Route choice • Travel behavior • Travel impedance

1 Introduction

With the expansion of urban rail transit network, passenger route choice has become diversified. Automatic fare collection system (AFC) can accurately acquire the passenger entry and exit information, and OD through the subway. But AFC can not accurately acquire the passenger travel route, and this flaw brought difficulties to the task based on passenger flow such as network operation, emergency management and fare sorting. Thus, to analyze the passenger route choice process in detail and to calculate route choice accurately are the basis to solve the range of issues related to passenger flow.

The existing researches [1–5] have considered the route choice influencing factors including quantitative factors such as travel time and transfer time and qualitative factors such as crowd and transfer convenience. Different route choice

Q. Ke (✉) • Z. Peng
School of Traffic and Transportation, Beijing Jiaotong University,
Beijing 100044, P.R. China
e-mail: 09114218@bjtu.edu.cn; pzhao@bjtu.edu.cn

Q. Zhi-peng
Transportation Planning Division of The Survey & Design
Institute of Railway Ministry, Tianjin 300142, P.R. China
e-mail: 09125157@bjtu.edu.cn

Z. Zhang et al. (eds.), *LISS 2012: Proceedings of 2nd International Conference on Logistics, Informatics and Service Science*, DOI 10.1007/978-3-642-32054-5_67,

model and algorithm were also proposed. But these researches didn't consider the passenger familiarity on the network. In order to make the results more match the reality and improve the accuracy of models, this paper proposed the passenger route choice model and algorithm based on the passenger classification. The results were verified having the local network of Beijing rail transit for example.

2 Passenger Integrated Travel Impedance

2.1 Passenger Classification

Based on the passengers' familiarity on the network and selection basis, passengers were divided into familiar type and strange type. Familiar passengers have multiple travel experience and select routes by experience, such as commuters. Strange passengers rarely take the route and select route by network sketch map, such as tours or shoppers.

2.2 Integrated Travel Impedance of Familiar Passengers

Familiar passengers integrated travel impedance can be expressed by generalized travel time function including train running time, transfer time, crowd, etc. According to the existing research, this paper refined transfer coefficient and considered additional entry and exit time. The travel impedance is expressed as:

$$C_r^{a,b} = (1 + Q(\delta))\left(\sum T_{i,j} + \sum T_s\right) + \alpha \sum T_{k,walk}^{m,n} + \beta \sum T_{k,wait}^{m,n} + \alpha(\Delta T_a + \Delta T_b) \quad (1)$$

where, $C_r^{a,b}$ is the r-path travel impedance from a to b of familiar passengers; $T_{i,j}$ is the train running time; T_s is the dwelling time; $Q(\delta)$ is the crowd coefficient; $T_{k,walk}^{m,n}$ is the transfer walking time, α is the amplification coefficient; $T_{k,wait}^{m,n}$ is the transfer waiting time, β is the amplification coefficient; ΔT_a is the additional entry time at original station; ΔT_b is the additional exit time at terminal station.

2.3 Integrated Travel Impedance of Strange Passengers

The strange passengers judge and select routes by the rail transit network sketch map, so factors affecting their judgment are the straight line distance, line curve distance, number of stations and number of transfer stations in the map. The travel impedance is expressed as:

$$T_r^{a,b} = \gamma \sum L_{i,j} + \lambda \sum H_{i,j} + \mu N_r + \theta M_r \tag{2}$$

where, $T_r^{a,b}$ is the r-path travel impedance from a to b of strange passengers; $L_{i,j}$ is the straight line distance between two adjacent stations of r-path in the map; $H_{i,j}$ is the line curve distance between two adjacent stations of r-path in the map; N_r is the number of stations except transfer stations of r-path in the map; M_r is the number of transfer stations of r-path in the map; γ, λ, μ, θ are the convert coefficient.

3 Route Choice Model and Algorithm

Improved Logit model and probability distribution model based on normal distribution are mainly used to take passenger route choice. Qin Luo proposed a ratio correction method of multi-route distribution model [5], in this paper the model was improved, firstly the initial choice probability was determined by time-based travel impedance, then the probability was corrected by transfer times, crowd and demand for seats.

3.1 *Route Choice Probability Distribution Model*

According to integrated travel impedance, probability distribution model was used to take the initial passenger route choice, the model is expressed as:

$$\begin{cases} x = \dfrac{T_i - T_{\min}}{\min\{T_{\min} \times f, f_{\max}\}} \\ s_i = e^{-(x-a)^2/2\sigma^2} \\ p_i = \dfrac{s_i}{\sum s_i} \end{cases} \tag{3}$$

Where, T_i is the integrated travel impedance of i-route; $T_{\min}$ is the integrated travel impedance of the shortest route; f is the maximum allowable increasing coefficient of travel impedance; $f_{\max}$ is the maximum allowable increasing value of travel impedance; p_i is the choice probability of i-route.

3.2 The Concrete Algorithm

The algorithm is as follows:

Step 1: build urban rail transit network and determine the proportion of familiar passengers and strange passengers based on passenger surveys
Step 2: calculate the integrated travel impedance of familiar and strange passengers according to Eqs. 1 and 2
Step 3: obtain OD feasible routes according to the K-short path search algorithm and determine effective route by judgment conditions. If travel impedance of a route $T_k > \min\{T_{\min}(1+f), T_{\min}+f_{\max}\}$, the route is ineffective
Step 4: determine the initial choice probability of each effective route according to Eq. 3
Step 5: correct the probability through transfer times correction coefficient X_{change}. Compare transfer times between shortest route and other routes, then the initial choice probability of other routes multiply X_{change} to obtain the corrected probability and ensure the sum of probabilities is 1 at the same time
Step 6: correct the probability through crowd correction coefficient $X_{congestion}$. The correction method is similar to X_{change}, but this coefficient is considered only for the familiar passengers, because only it may be sufficiently familiar with the OD, it can distinguish the crowd difference between the different routes
Step 7: correct the probability through demand for seats correction coefficient X_{seats}. The correction method is as above. This coefficient is for long-distance passengers, seat is an important indicator of passenger travel comfort in the long-range condition, so the passengers will first choose a route that has seats under certain threshold.

4 Computing Examples

The model was proved having the Beijing local rail transit network for example. The OD from Yonganli to Jiandemen was calculated. Basis date of network derived from actual data.

4.1 Parameter Value

Parameters of familiar passenger are shown in Table 1, they are from literature [6].

According to the results of the actual survey, the paper set that strange passenger ratio is 33% and familiar passenger ratio is 67%, and parameter values of strange passenger are shown in Table 2.

Table 1 Parameter value of familiar passenger

Parameter	α	β	f_{max}
Value	1.9	1.5	12

Table 2 Parameter value of strange passenger

Parameter	μ	γ	λ	θ
=	0.4	2.5	2	2.9

Table 3 Route choice probability results

Route	Route detail	Passenger type	Passenger ratio (%)	Computing result (%)	Final probability (%)
1	Yonganli-Dongdan-Huixinxijienan kou-Jiandemen	Familiar	67	10.42	17.22
		Strange	33	31.03	
2	Yonganli-Guomao-Jiandemen	Familiar	67	89.58	71.63
		Strange	33	42.5	
3	Yonganli-Jianguomen-Yonghegong-Huixinxijienankou-Jiandemen	Familiar	67	0	7.49
		Strange	33	22.71	
4	Yonganli-Jianguomen-Dongzhimen-Shaoyaoju-Jiandemen	Familiar	67	0	1.24
		Strange	33	3.76	

4.2 Computing Result

According to the impedance of effective route, the initial choice probability was calculated by probability distribution models. This case mainly considered the correction of transfer times. The final passenger route choice probability is shown in Table 3.

It can be seen Route 3 and Route 4 are ineffective for familiar passengers, but are effective for strange passengers who judge the route by map. Indeed there will be a certain percentage of passengers choosing Route 3 and Route 4, this showed taking passengers classification can get more comprehensive results, it is correct to take passenger classification.

5 Conclusions

According to passengers' familiarity on the network, passengers were divided into two types: familiar type and strange type. The different integrated travel impedance function for two types of passengers was set up. For the familiar passengers, the travel impedance increased additional entry and exit time; the transfer amplification coefficient was refined to make the function more realistic. For the strange passengers, the travel impedance was mainly based on the rail transit network map. The ratio correction method of multi-route distribution model was used to take route choice. The results can be corrected by coefficient of crowd, coefficient of transfer times and coefficient of demand for seats. The case of Beijing local rail transit network showed that the passenger route choice behavior can be depicted accurately through classifying the passengers.

Acknowledgments This research was supported by the Fundamental Research Funds for the Central Universities (2011YJS037)

References

1. Wu XY, Liu CQ (2004) Traffic equilibrium assignment model specially for urban railway network. J Tongji Univ(Nat Sci) 32(9):1158–1162
2. Si BF, Mao BH, Liu ZL (2007) Passenger flow assignment model and algorithm for urban railway traffic network under the condition of seamless transfer. J China Rail Soc 29(6):12–18
3. Xu RH, Luo Q, Gao P (2009) Passenger flow distribution model and algorithm for urban rail transit network based on multi-route choice. J China Rail Soc 31(2):110–114
4. Liu JF, Sun FL, Bo Y (2009) Passenger flow route assignment model and algorithm for urban rail transit network. J Transp Syst Eng Inf Technol 5:131–133
5. Luo Q (2009) Theory and simulation analysis of passenger flow distribution based on network operation for urban mass transit. Tongji University, Shanghai
6. Lai SK (2008) Research on income distribution model of urban rail transit. Beijing Jiaotong University, Beijing

Modeling on Dynamic Passenger Flow Distribution in Urban Mass Transit Network

Xiang-ming Yao, Peng Zhao, Ke Qiao, and Wei-jia Li

Abstract The spatial and temporal characteristics of passenger flow distribution in urban mass transit network are foundation and core for cooperative transportation organization. However urban mass transit is a complex and dynamic system, which is difficult to be described in a global mathematical model. In this thesis we put forward a simulation model based on multi-agent approach to analyze passenger flow distribution. Using real passenger flow data simulated in this system we verify accuracy and computation efficiency of the model. The simulation results show that it well depicts passengers' whole travelling process and can be used to evaluate the strategies and status of network.

Keywords Urban mass transit network • Dynamic passenger flow distribution • Simulation modeling

1 Introduction

Urban mass transit has played a key role to people mobility. With expansion of network, it is more important for providing a convenient, secure and economical transport by collaborative transportation organization, especially in peak hour and emergency. However, this system is so large-scale and complex that the passenger flow characteristics is temporal and spatial dynamic, so it is very difficult or even impossible for making new transportation solutions. In order to figure out the problem effectively and fast, a comprehensive method must be found.

With the development of computer simulation, a growing number of simulation models have been applied to traffic analysis and evaluation. Shang Lei, Lu Hua-pu

X. Yao (✉) • P. Zhao • K. Qiao • W. Li
School of Traffic and Transportation, Beijing Jiaotong University, Beijing 100044, China
e-mail: 10114232@bjtu.edu.cn; pzhao@bjtu.edu.cn; 09114218@bjtu.edu.cn; 11120863@bjtu.edu.cn

Z. Zhang et al. (eds.), *LISS 2012: Proceedings of 2nd International Conference on Logistics, Informatics and Service Science*, DOI 10.1007/978-3-642-32054-5_68,

put forward a micro simulation model of road traffic to analysis the network through single vehicle moving and intersecting [1]. David Meignan adopt a multi-agent approach to describe the global system of bus network as behaviors of numerous autonomous entities such as buses and travelers, through interaction between the main components to reproduce the passengers' travelling process, from passenger load of network and waiting time to evaluate the operational efficiency [2]. However, using simulation to study the dynamic passenger flow distribution for urban mass transit network is still lacking. In this paper, we build a multi-agent simulation model to describe the entire subway system of passenger travelling and train running. The core thinking is obtaining the flow distribution characteristics through a large number of individual passenger agents travelling on network to evaluate and optimize the strategies of transportation organization.

2 Simulation Modeling

Urban mass transit is a complex and dynamic system, including multiple components: passengers, stations, trains, sections, etc. This complexity is due to the interactions between different components. The multi-agent approach is a well suited method to design such complex system. We see every component as autonomous entity, which called agent and can interact to deal with local or global tasks.

2.1 Framework of Simulation Model

A multi-agent system (MAS) is defined as a set of interaction agents, two types of MAS can be defined following the agent architecture which is used: deliberative (or cognitive) and reactive architecture. Deliberative agents have generally a symbolic representation of their environment and cooperate thanks to high level communication protocols [3]. At the opposite, reactive agents do not have representation of their environment. They act following their perceptions, which are very limited. Reactive agents can cooperate and communicate through their interactions with the environment (called indirect communication). As a consequence, such reactive systems present some global intelligent behavior that result from the numerous interactions between agents and their environment [4]. The model we presented for urban mass transit network relies on the reaction approach. Passengers and trains are considered as reactive entities and other components such us stations and sections are describe as environment.

The *environment*, where passengers and trains move, is the composition of train running network, passenger travelling network. The main role of the environment is to constraint perceptions and interactions of agents. For example, passenger agent and train agent can interact only when they are located at the same station-stop.

The *passenger agent* is the core entity of this system, which has a strong adaptability and intelligence. It can achieve all process of a passenger travelling on network, such as entering station, waiting for train, and transferring from one line to another; even they can adjust their behavior under traffic policy changing or network changing.

The *train agent* also is a move agent in system, which runs according with operation routes and timetables. The main purpose is to load passengers from one station to another, control process of passengers' boarding and alighting.

2.2 Passenger Route Choice Model

Passenger route choice is divided into two processes: a feasible path searching and path selection. It is difficult to achieve real-time path searching for single travelers by restrictions of rail network size and number of travelers, so this article combines off-line path searching and passenger path selecting dynamically. When initialize the simulation feasible paths of each OD are searched in the network. When the passengers generated, they can find their own path in the path collection to get their own path according to their travel information and attributions. Under normal circumstances, once a passenger selects his own path, the path will not change through the entire travel process. If there is a failure of network or transport organization change caused by emergencies, the affected passenger will re-select his route, and then path changing behavior occurred.

Generally passengers choose the route with least cost. However, due to the complexity of network, the judgment of "shortest path" may appear differently. Route choice behavior has a certain degree of randomness, in particular increase of loop line. The route choice model based on probability fits more with the behavioral characteristics of travelers. Therefore in this thesis we assume that the random error obeys the distribution of *Gumbel*. We build a route choice model based on *Logit* model; the probability of a passenger to select the n_{th} path in OD pair rs is as follows:

$$p_n^{rs}(a) = \frac{\exp[V_n^{rs}(a)]}{\sum_{i=1}^{K} \exp[V^{rs}{}_i(a)]} \tag{1}$$

$$V_n^{rs}(a) = C_n^{rs} + \varepsilon_n^{rs} + Y_n^{rs}(a) \tag{2}$$

p_n^{rs}: The probability of a passenger to select the n_{th} path in OD pair rs;
V_n^{rs}: The travel utility value of the n_{th} path in OD pair rs;
C_n^{rs}: The comprehensive cost of the n_{th} path of OD pair rs;
ε_n^{rs}: The random error of the n_{th} path in OD pair rs;
$Y_n^{rs}(a)$: The additional cost of the n_{th} path because of congestion and a is a parameter of crowd degree.

Once stations and intervals suffer a temporary breakoff due to emergencies, passenger's route choice behavior will change greatly. According to different situations, the affected passengers are often divided into three categories: delay passengers, detour passengers and loss passengers. Delay passengers and detour passengers will continue to travel in network, the loss passengers will give up to ride because delay time or bypassing distances is too long [5].

Considering the complexity of passenger behavior in emergencies, we simplify that delay passengers and detour passengers change to re-select the "shortest" path from current station to their destination. Loss passengers get out from current station directly.

2.3 Passenger Transfer Queue Model

The behavior of passenger walking in station is very complex, there have lots of researches using micro-simulation approach [6]. Our work mainly focus on passenger flow distribution of network and is difficult to take into account passenger micro-behaviors in station, so we propose a simple queue model to describe passenger walking in station. The queues include: entering queue, waiting queue, stranded queue (failure to board passengers), getting off queue, transferring queue (if station is a transfer station), and exiting queue. The process of passenger walking in station is simplified as tourists transfer from one queue to another queue. Transfer queue model is as follow:

$$Q_j^t = Q_j^{t-1} + \Delta t \times v_{ij} \tag{3}$$

$$Q_{ij} = \Delta t \times v_{ij} \geq 0 \tag{4}$$

Q_j^{t-1}: The passengers of queue j at time t;

Q_{ij}: The passengers transferring from queue i to queue j;

Δt: The time step;

v_{ij}: The speed of transferring, which reflects the passing capacity of a routeway (from node i to j), such as entrance channel or exit channel.

2.4 Passengers Get On and Off Model

When a train stops at platform then train agent will "inform" on-aboard passengers of the stop information, such as station name and station type. Then on-board passengers decide whether to get off or not according to their OD information and selected path. At the same time, waiting passengers get train information, such as direction and next stop and then decide whether to get on or not. These two processes are complied with the principle "first alight then aboard".

1. Passengers get off model

   Step 1: Train agent "informs" on-board passengers of the stop information
   Step 2: The passengers determine whether to get off or not according to their OD information and selected path
   Step 3: Transfer passengers who will transfer from getting off queue to exiting queue or transferring queue according to their purpose
   Step 4: The process of passenger getting off is ended.

2. Passengers get on model

   Step 1: After train stops, the train agent will "informs" waiting passengers on platform of the train running information (direction and next-stop) in order to help travellers decide whether to get on or not
   Step 2: The passengers who were failure to get on last train will have a priority to get on train, so they are added to queue in train under the capacity constraint
   Step 3: Add other waiting passengers who will get on to passenger queue in train also under the capacity constraint
   Step 4: Transfer passengers who can't get on from waiting queue to stranded queue, and need to wait for next train.
   Step 5: The process of passenger getting on is ended.

3 Model Application

The proposed model has been entirely implemented in a simulation tool called URT-DPSS (Fig. 1), which is dedicated to analyze and evaluate the operation effectiveness of rail transit network.

Using Beijing subway network (Fig. 2) we have verified the accuracy and computation efficiency of the model. By the year of 2011 the transit network includes 12 lines with about 300 km. We select a day of December, 2011 to simulate; passenger flow data comes from the AFC system, which includes individual traveller's information. Simulation real time is from 7:00 to12:00 pm, and the loaded passengers of network are about 1.8 million.

A significant number of outputs have been produced by this simulation system to analyze the spatial and temporal characteristics of passenger flow distribution, which include entry, exit, transfer and detention passenger flow volume of station by periods, real-time and historic passenger flow of sections, and fully loaded rate of trains, etc. We will show some passenger flow data of one train for example.

Figure 3 shows the number of on-board passengers of a train (train number is 1069) in each section of Line1, which reflects the utilization of train's transport capacity, also can be used to locate overload trains and unused trains. Then, some

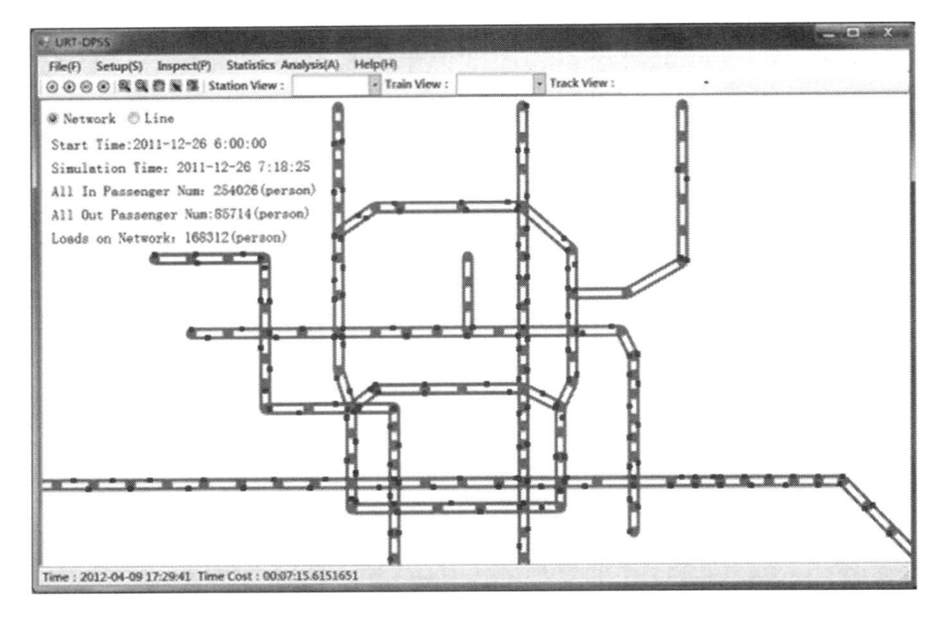

Fig. 1 Screen shot of URT-DPSS

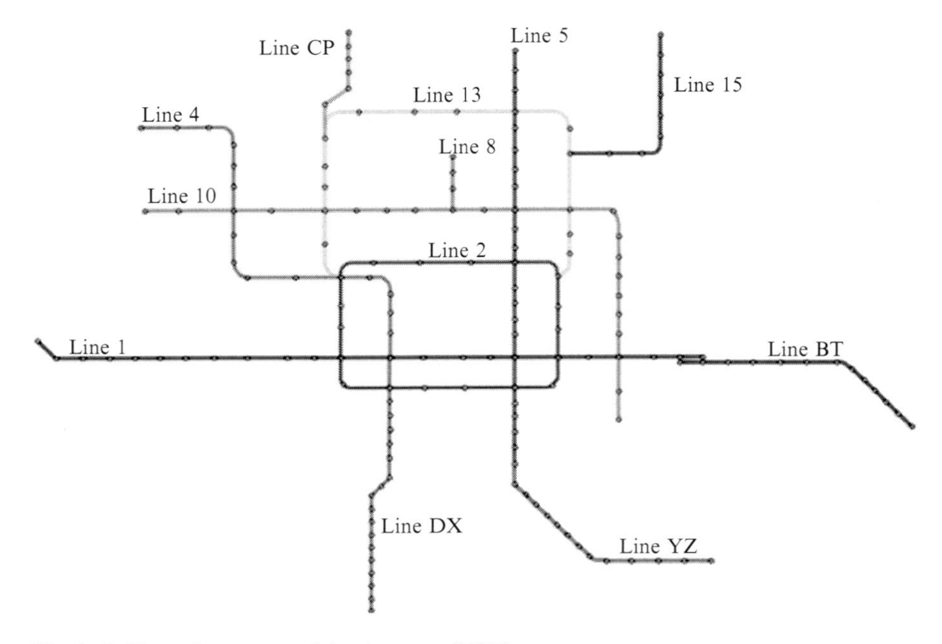

Fig. 2 Beijing subway network by the year of 2011

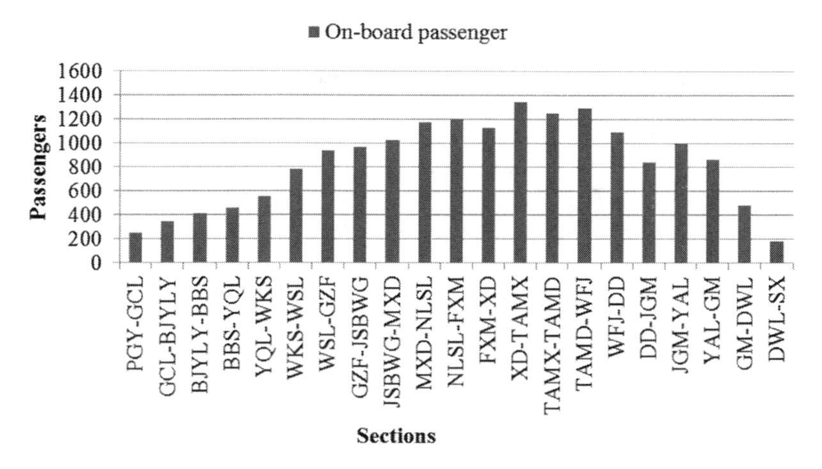

Fig. 3 On-board passengers

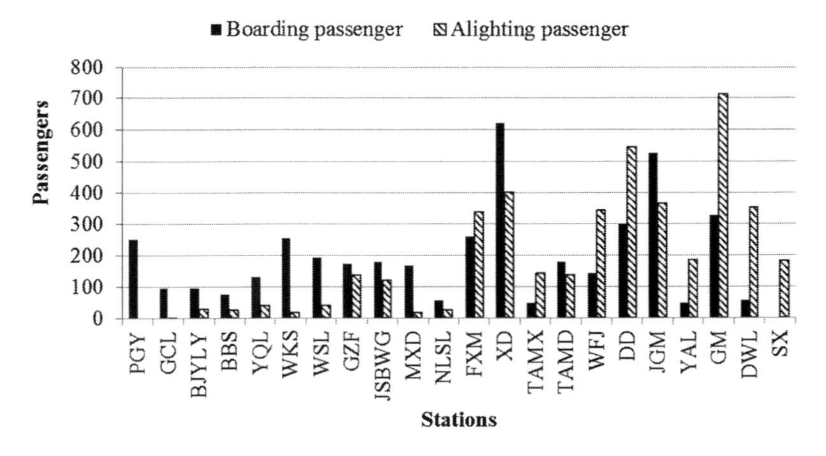

Fig. 4 Boarding and alighting passengers of a train

adjustment of train operation plan can be adapted to reduce or avoid overload problems, such as adjustment of the train running routes and train numbers.

Figure 4 shows the number of boarding and alighting passengers of a train (train number is 1069) at each station in Line1. We find that passengers at FXM station, XD station, DD station and JGM station are more than others, because they are transfer stations. This kind of statistical data can provide the very basis for dispatchers to make the stopping station plan of trains operation in emergency.

4 Conclusions

In this paper, a multi-agent simulation model for urban rail transit network has been presented. We have shown that an agent-based approach allows designing such autonomous, dynamic and interacting system. The model combines the process of passenger walking in station and travelling in network, which can well describe passengers' whole travelling process in network to obtain the spatial and temporal characteristics of passenger flow.

Using real passenger flow data of Beijing subway network to take a case study, results show that our simulation tool can be used to analyze and evaluate the operation effectiveness of network well. The main perspective of this work is to provide decision support for cooperative organization operators and evaluate operation effectiveness of the network. So they are useful to make train operation plan and dispatching measures under emergence. Modeling and measuring the efficiency of these strategies is a challenge. Forthcoming works will consider how to give decision support for operators.

Acknowledgment This research was supported by the Fundamental Research Funds for the Central Universities (2012YJS071).

References

1. Lei Shang, Hua-pu Lu (2006) Urban microscopic traffic simulation system and its application. J Syst Simul 18:221–224
2. David M, Olivier S (2007) Simulation and evaluation of urban bus-networks. Simul Model Pract Theory 15:659–671
3. Rao AS, Georgeff MP (1995) Bdi agents: from theory to practice. Technical report. Australian Artificial Intelligence Institute, Melbourne
4. Ferber J (1999) Multi-agent systems: an introduction to distributed artificial intelligence. Addison Wesley, Harlow
5. Ling Hong, Rui-hua Xu (2011) Calculation method of emergency passenger flow in urban rail network. J Tongji Univ 10:1485–1489
6. De-wei Li (2007) Modelling and simulation of microscopic pedestrian flow in MTR hub. Beijing Jiaotong University, Beijing

Semiotics-Oriented Method for Generation of Clinical Pathways

Jasmine Tehrani, Kecheng Liu, and Vaughan Michell

Abstract Large numbers of people continue to be successfully cared for and treated in the National Health Service, but a significant number of errors and other forms of harm occur. These errors are mainly a result of human error or occur in consequence of poor process, procedure and control design. In this paper, we present a semiotic approach to generating clinical pathways by capturing knowledge from syntactic level to social level and guiding the modelling of clinical pathways using Business Process Modelling Notation (BPMN) best practice to visualise the process. This will result in more rigorous control and visibility of the care process ensuring completeness, consistency and patient safety by enabling the mapping of formal and informal/ safety controls into clinical pathways.

Keywords Healthcare informatics • Clinical pathways • Process modeling • Organizational semiotics • Knowledge management • Information system

1 Introduction

Large numbers of people continue to be successfully cared for and treated in the National Health Service, but a significant number of errors and other forms of harm occur [1]. Increasing costs of health care, fuelled by demand for high quality, cost-efficient health care has propelled hospitals to restructure their patient care delivery systems. One such systematic approach is the adaptation of an engineering project management methodology, the critical path method (CPM), as a tool to organise, standardise and improve the quality of healthcare delivery and hence patient outcomes [2]. However, the application and adaptation of CGs in a local hospital setting, inevitably has some limitations of process management in practice as they

J. Tehrani (✉) • K. Liu • V. Michell
Informatics Research Centre, The University of Reading, Reading RG6 6AH, UK
e-mail: g.tehrani@pgr.reading.ac.uk; k.liu@henley.reading.ac.uk; v.a.michell@reading.ac.uk

Z. Zhang et al. (eds.), *LISS 2012: Proceedings of 2nd International Conference on Logistics, Informatics and Service Science*, DOI 10.1007/978-3-642-32054-5_69,

often fail to offer a clear description of activities, conditions, sequence and authorities of action of a care process [3]. The proposed modeling approach to generate clinical pathways (CPs) adopts organisational semiotics to capture and represent the CG knowledge by determining the underlying semantics and the relationship between agents and their patterns of behaviour. We use Norm Analysis Method (NAM) to extract and analyse patterns of care activities and informal safety norms that affect patient safety outcomes. The proposed method enables the generation of CP from a semiotic perspective by capturing all necessary knowledge from syntactic level to social level and guiding the modeling of clinical pathways using Business Process Modeling Notation (BPMN) best practice and adopts a socio-technical approach to map informal safety norms in to clinical pathways.

2 Related Work in Clinical Pathways

Clinical pathways (CPs) are evidence-based patient care algorithms that describe the process of care for specific medical condition within a localized setting. A large and growing body of literature has investigated the modeling of clinical pathways. Abidi and Chen [4] present a semantic web framework and rendered the technical basis for a services-oriented architecture to generate and orchestrate patient-specific healthcare plans. Recently, various methods are proposed to represent the clinical guidelines which were originally paper based as Computer interpretable guidelines, (CIG) most of which can be visualised in flowcharts [5]. Other authors represented clinical pathway knowledge as a clinical pathway ontology which offers a detailed ontological model describing the structure and function of clinical pathways [6].

The above methods describe clinical pathways from a structural aspect comprising concepts, relationships between concepts, and properties that describe the concept, which provide foundations for the work conducted in our project. However, these methods lack a mechanism for describing possible patterns of human behaviour and the conditions under which the behaviour will occur. This mechanism is crucial for conceptualizing and generating clinical pathway. To overcome this weakness, a semiotic approach, which adopts Organizational Semiotics (OS) methods is adopted [7]. The clinical pathway generated from SAM together with norms will semantically enrich the knowledge conceptualisation of the domain as to analyse the patterns of behaviours of various agents in healthcare setting.

3 Semantics-Oriented Method for Generation of Clinical Pathways (SOG-CP)

Semantics-Oriented Method for Generation of Clinical Pathways (SOG-CP) is a method for generating clinical pathways. SOG-CP adopts organisational semiotics methods including Semantic Analysis Method (SAM) and Norm Analysis Method

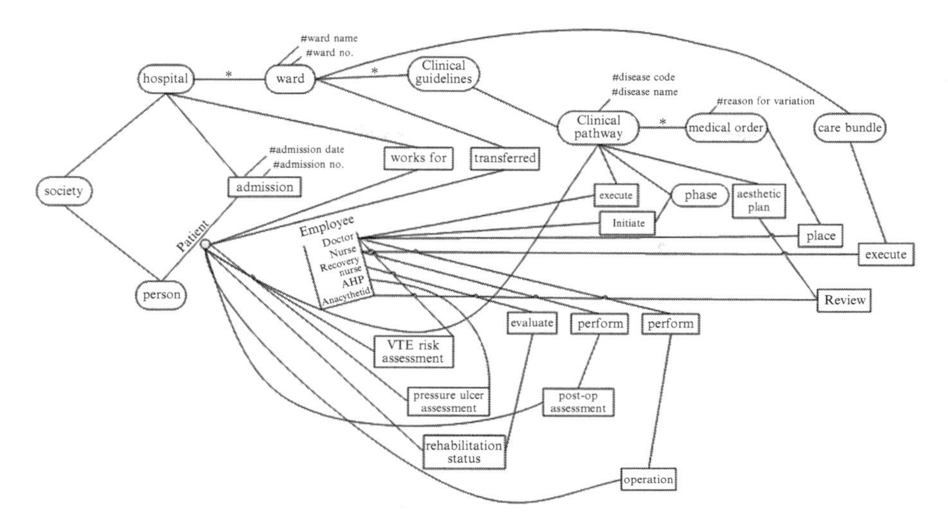

Fig. 1 Semantic analysis for modelling of agents, patterns of behaviour and their relationships

(NAM) to explicitly represent the semantics of the concepts and their relationships, patterns of behaviour and norms governing the action taken. SOG-CP is carried out in three main stages.

Stage 1: A knowledge management approach is taken to represent CG knowledge by ontologically modeling clinical guidelines in terms of possible patterns of behaviours, authorities of action, semantic units and their relationships. **Stage 2**: Norm analysis is conducted to identify rules that govern the actions identified on the ontology chart. **Stage 3**: Information collected during SA and NA are integrated to guide the generation of clinical pathways. The initial phase of this stage is to generate process models using BPMN.

3.1 Semantic Analysis for Business Domain Modeling

The semantic analysis method (SAM) is a method for conceptualizing knowledge of a problem domain and to analyse the patterns of behaviours of various agents (i.e., physician or doctors) in an organization [8]. The result of semantic analysis are provided in a graphical format, using what is called 'ontology chart'. The ontology chart models concepts and the responsible authorities of actions in the problem domain and captures domain knowledge supported by semantic units and ontological dependencies [9]. Figure 1 is the result of SAM conducted for a major gynecology surgery guideline. An affordance is a possible pattern of behaviour available to a member of society, represented as rectangles. In addition, a concept may have properties prefixed by the hash sign (#) that capture descriptive information about

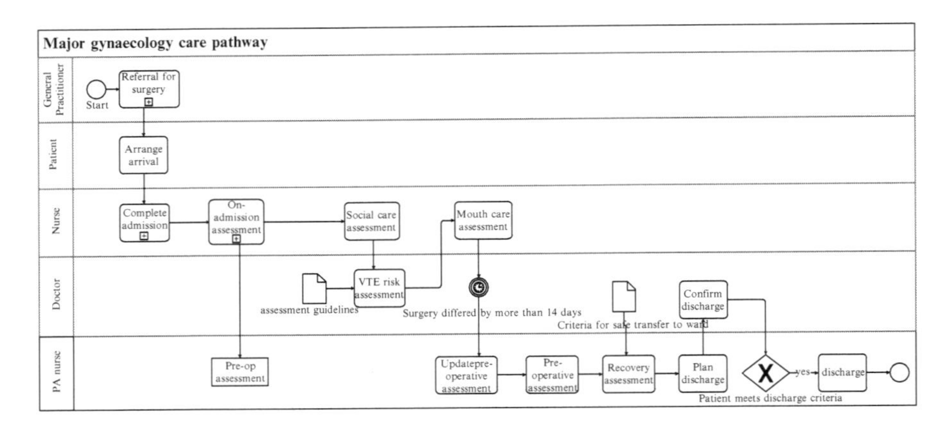

Fig. 2 Part of CP for major gynaecology surgery

the concept. For example, the concept of Ward has the properties of '#ward name' and '#ward number' in the following charts which are called determiners. Agents are represented as ellipses and represent those who can take responsibility for their actions. It could be an individual, a group of people e.g., 'society', 'hospital' and 'person'. Through SAM, possible patterns of behaviours in CP and their relationships are represented in an ontology chart which delineates the boundary of concern in the analysis and defines the meaning of terminology used in the clinical pathway model.

3.2 *Generation of Clinical Pathways*

The information collected during semantic analysis and norm analysis of clinical guidelines is integrated to guide the generation of clinical pathways. We adopt process-modeling techniques because it gives a rational means of organising information that is processed to perform care activities. Figure 2 shows part of a Major Gynecology pathway modeled using BPMN techniques.

3.3 *Extending Clinical Pathways by Capturing Business Dynamics Using Norms*

Semantic analysis method captures all possible behaviours in a CP which need further description by norms that govern these behaviours. Norms, in addition to the knowledge represented in the semantic model, specify the details of the possible patterns of behaviors; e.g., the conditions where certain clinical actions must

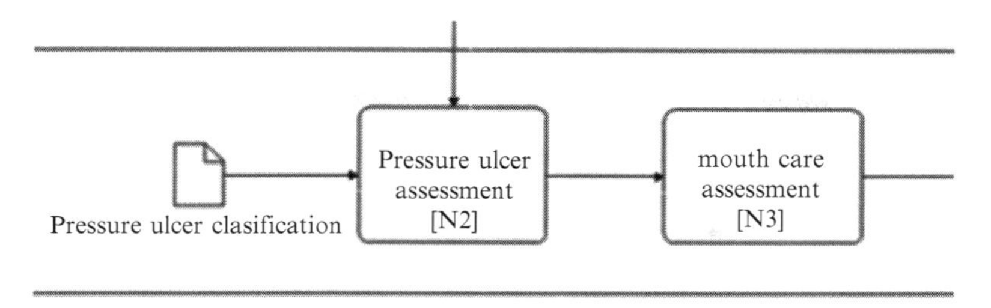

Fig. 3 Extension of clinical pathway for major gynaecology surgery with norm

happen or where they are actually impossible. The norms are described as the following format [8]: **Whenever** < condition > **If** < state > **Then** < agent > **Is** < deontic operator > **To** < action>. "Condition" and "state" are defined from the determined pre-conditions. "Agent", as an authority, is responsible for executing defined action(s). The entitlement is defined by the Deontic operators. "Deontic operator" is derived from Deontic Logic and can be one of the following: "obliged", "permitted" and "prohibited" which prescribe what people must, may, and must not do. The extension is carried out by incorporating norms into the business process diagram. In the diagram, each control condition is labelled as [N#] where # is the number for identification. The labels are then elaborated in the norm specifications to indicate the condition, the actor and action to be undertaken. Figure 3 depicts the extended care process for applying water low pressure ulcer assessment.

4 Discussion and Conclusions

The proposed approach for generation of clinical pathways, can addresses social and informal/safety factors which conspire together to influence the outcome of patient safety. A unique contribution of this paper is a semiotic method for generation of clinical pathways which can addresses social and informal/safety factors which conspire together to influence the outcome of patient interaction and safety. This is achieved through modeling clinical pathways using SAM and NAM. Furthermore, a semiotic approach for mapping of safety/informal factors in clinical pathways is presented. The strength of this modeling approach results from two base methods, both of which are sound and well tested. The BPMN is rigorous method that provides a rich set of techniques and notations for process modeling. Norm analysis enables one to specify norms in the problem domain. Hence, allows the recognition of human responsibilities and obligations and more importantly, the ultimate power of decision making in exceptional circumstances.

Future work will aim at applying the prosed method of generating care pathways to other instances of clinical guidelines and validate the approach by implementing the pathways in a hospital setting. We also emphasized that the initial findings of this research has been acknowledged by clinical experts from a UK hospital.

Acknowledgement We thank Dr. Hester Wain and Dr. Debbie Rosenorn-lanng for their support and contribution to this research.

References

1. Chang A et al (2005) The JCAHO patient safety event taxonomy: a standardized terminology and classification schema for near misses and adverse events. Int J Qual Health Care 17(2):95–105
2. Yang H, Liu K, Li W (2010) Adaptive requirement-driven architecture for integrated healthcare systems. J Comput 5(2):186–193
3. Carthey J, Clarke Julia Field, Campaign; Associate (Safer Care Priority Programme), NHS Institute for Innovation and, Improvement (2010) Leadership for safety: implementing human factors in healthcare
4. Abidi SSR, Chen H (2006) Adaptable personalized care planning via a semantic web framework. In: 20th international congress of the European Federation for medical informatics, Maastricht, 2006
5. Sonnenberg F, Hagerty C (2006) Computer-interpretable clinical practice guidelines. Where are we and where are we going. Yearb Med Inform 45:145–158
6. Hurley KF, Abidi SSR (2007) Ontology engineering to model clinical pathways: towards the computerization and execution of clinical pathways. IEEE
7. Salter A, Liu K (2002) Using semantic analysis and norm analysis to model organizations. Citeseer: http://citeseerx.ist.psu.edu/viewdoc/summary?doi:10.1.1.5.4577
8. Liu K (2000) Semiotics in information systems engineering. Cambridge University Press, Cambridge/New York, xii, 218 p
9. Liu K (2002) Organizational semiotics: evolving a science of information systems: IFIP TC8/WG8.1 working conference on organizational semiotics, evolving a science of information systems, 23–25 July 2001, Montreal, Quebec, Canada. Ifip. 2002. Kluwer Academic Publishers, Boston, xxii, 308 p

Improving Mental Models Through Learning and Training – Solutions to the Employment Problem

Xiang Kaibiao and Xian Heng

Abstract The increasing flat, diverse and complex and has now entered the era of knowledge economy in the world, whether business or personal, in order to maintain long-term competitiveness and vitality, it must be continuing to improve mental models, keep the change corresponds to a flexible and open, the "spiritual conversion". With the introduction of the concept of mental models, people pay more attention to the psychological factors which the management. By analyzing the meaning and connotation of mental models, learning and training employees to improve their mental models to meet the employment requirements of the development of the times, through two levels: the enterprise level with the staff at the individual level to improve the mental models, to enhance corporate and individual competitiveness.

Keywords Mental models • Learning • Training • Employment problem

1 Introduction

With the rapid development of the global economy, today's businesses or countries in order to have a relatively long-term competitiveness must be constantly learning, active learning, effective learning, the only way companies or countries in the midst of an environment to be able to maintain a sustainable living and healthy and harmonious development. This study, it is not only to obtain new knowledge and information and, more importantly, changes in the way of thinking and values of

---

Department of Education Academic Humanities and Social science research projects, Guizhou. No.: 11FDY009.

X. Kaibiao (✉)
School of Economics and Management, Beijing Jiaotong University, Beijing, China
e-mail: xiangkaibiao@vip.163.com

X. Heng
The Materials and Metallurgical College, Guizhou University, Guiyang, China

Z. Zhang et al. (eds.), *LISS 2012: Proceedings of 2nd International Conference on Logistics, Informatics and Service Science*, DOI 10.1007/978-3-642-32054-5_70,

people's soul. The transformation of this way of thinking and values that people improve their own mental models. Mental models to be effectively improved, people will really take the initiative and lasting learning to understand and transform the world around, constantly self-transcendence.

Mental models concept was first put forward by the Scottish psychologist Kenneth Craik in 1943, refers to those deep-rooted understanding of the world affect how people interpret the world, to face the world and how to take action in the minds of many of the assumptions, stereotypes, or impression. Been widely recognized in the literature, the most cited Peter Senge on the definition of mental models. Peter Senge [1] in his book "The Fifth Discipline" mental models are deeply ingrained in the mind, affect how people understand the world how to take action many of the assumptions, stereotypes, or images, impression. Ning Xiaoyong and Cui Yueming [2], the mental model is due to the living environment of the past, life experience, professional background, knowledge, literacy and other aspects of certain values in the long life, ways of thinking, behavior, and it determines people how to look, think, solve problems, is a reflection of people of ability and accomplishment.

The notion of shared mental models (or collective mind) has received considerable attention in management and organizational psychology literature (e.g., [3, 4]). Holyoak [5] defined a mental models a "...psychological representation of the environment and its expected behavior." Rousse and Morris [6] further noted that if a group shares a mental model, it serves as the basis for future event prediction and choice regarding courses of action. Such diagnoses and decisions are all fundamental to the safety process in any organization. For example, Weick and Roberts [4] examined aircraft carrier processes and demonstrated that groups with shared models perform more effectively than those without a "collective mind." Shared mental models are the result of selection, training, and experience, and they are more likely to exist under conditions of cohesiveness and membership stability. Furthermore, they are particularly important in environments requiring nearly continuous operating reliability: "Organizations concerned with reliability enact mental processes that are more fully-developed than those found in organizations concerned with efficiency".

2 The Company

2.1 Communication in Team

Improve the mental models is to expand the organization internal communication within the organization each member of each other to open their hearts, "face to face" in-depth exchanges. Only in this way, people can make each other understand their own mental models, exposing the defects of which the other party will also calm voice their opinions, allowing people to enrich their knowledge by learning from different people's thoughts, perspectives and views of improving mental models. By the real depth of dialogue, the group can enter an individual alone can not enter the larger "common sense brings together", enables organizations of every member to be a winner.

2.2 Create a Continuous Learning Environment

To improve the organization's mental models, the first and very important point is to create the atmosphere of continuous learning, and each member of the organization to open their inner world through the "learning" to constantly improve their own mental models. In an organization which the leader is not only the interests of the creators, but also organized the founder of the cultural. Therefore, to improve the leader's mental model to create a culture of continuous learning which plays a key role. Mental models of the so-called leader is the leader in the formation of long-term live, work and practice, a way of thinking, behavior and values. The leader's mental model is not only a direct impact on their own leadership style and organizational decision-making, but also affect the cultural atmosphere of the entire organization and organizational innovation. If the leader of an organization on mental models has serious shortcomings, the heart which does not attach importance to organizational learning, mental models of the organization there must be a big problem, let alone to create a strong learning culture. Conversely, if the leader of an organization can continue to improve their own mental models, a good grasp of their behavior, focus on learning from the heart of them, to promote learning, lifelong learning in the organization to set a good example, then this can not only improve the leader management capabilities, and improve an organization's mental models, in particular, to create a culture of continuous learning will play a very active role.

2.3 Basing on the Employees

Xiang Kaibiao [7] Companies should treat the employees as important assets, as they can create value for the company and outstanding employees can create more value. Some companies reduce costs through cutting the number of employees, which will have effects in a short term, but can also bring unpredicted results. For example, existing or potential customers may leave, suspicions from inside and outside the companies will enlarge the influence of the crisis, simple and crude layoff won't help reduce the burden on companies, but may cause loss of technicians and skilled managers. What is the most important, the corporate culture and spirit brought up in years' work may collapse, and the excellent employees may leave. In this way, to companies, it is better to take flexible measures against the

Table 1 Second-order confirmatory factor analysis of organizational knowledge

	First-order			Second-order	
First-order construct	Indicator	Loading	t-value	Loading	t-value
Knowledge acquisition	KA1	0.58	–[a]	0.49	9.69
	KA2	0.7	10.56		
	KA3	0.83	11.12		
Knowledge distribution	ID1	0.72	–[a]	0.69	14.02
	ID2	0.76	13.92		
	ID3	0.73	13.53		
Knowledge interpretation	II1	0.81	–[a]	0.61	14.28
	II2	0.76	15.01		
	II3	0.62	12.47		
Organizational memory	OM1	0.64	–[a]	0.36	8.11
	OM2	0.83	14.49		
	OM3	0.8	14.08		
	OM4	0.92	15.33		

From Daniel Jiménez-Jiménez and Raquel Sanz-Valle [15]
Fit statistics for measurement model of 13 indicators for four constructs: χ^2 (61) = 117.19; GFI = 0.96; RMSEA = 0.044; CFI = 0.98; TLI (NNFI) = 0.97
[a]Fixed parameter

crisis, instead of layoffs, to stabilize the confidence of the employees. Without confidence, a company hardly can walk out the difficulty.

For example, Ningxia Tongda Coal Group Company doesn't take measures like production cuts or layoffs, or even reducing salaries. The strategy of the company is to keep production, if there is enough cash flow, as the company will have a faster development than other companies when the fundamental of the market changes. The decision by the management team was welcomed by the employees, and their loyalty and the team-spirit are improved.

2.4 Organizational Learning

The paper focuses on the Huber's model of organizational learning. After reviewing the theoretical literature [8, 9] and empirical research [10–13], the present study adopts the organizational learning scale of Perez Lopez et al. [14]. This thesis quoted Daniel and Raquel [15] research measures organizational learning as a single construct, made up of the four behavioral dimensions of this process. A second order factor analysis demonstrates that the four dimensions reflect a higher-order construct (Table 1). They uses LISREL 8.50 to estimate the measurement model. The results suggest a good fit of the second-order specification for the measure of organizational learning (χ^2 = 117.19, df = 61; goodnessof-fit index [GFI] = 0.96; root mean square error of approximation [RMSEA] = 0.044; comparative fit index [CFI] = 0.98; Tucker–Lewis index [TLI] = 0.97; incremental fit index [IFI] = 0.98). The GFI, CFI, TLI and IFI statistics exceed the recommended threshold level of 0.90 [16]. The RMSEA is nearly 0.050 and the root mean square residual [RMR] and standardized RMR are 0.029 and 0.035, respectively, which indicates an acceptable fit.

2.5 *Innovation and Research*

Though the oversea is weak and orders decrease, some strong export-oriented companies managed to reduce the bad effects of the financial crisis, through boosting self-development and research, innovating the technology, improving product mix for export and so on.

Fujian Hongyuan Group is a company focusing on exporting textile fabrics. Due to the international financial crisis, the company had an export value in 2008 16.0 plus million US dollars lower from a year ago, two production lines were idled, and more than 400 workers laid off. All these meant the company was in danger. But, in the later half of 2008, Hongyuan Group and Institute of Chemistry, the Chinese Academy of Sciences together successfully developed "new renewable bamboo fiber textile material", which has a leading technology in the world, and got independent intellectual property rights. During the 2nd Asian Sporting Goods Exhibition held in Hong Kong in Nov 2008, this new material interested more than 220 oversea customers, and brings back the orders. In 2009, the company opens a new production line for the new renewable bamboo fiber textile material, which provides 500 plus job positions, meaning Hongyuan employs 100 plus new workers.

3 For the Individual

People should do something when facing unemployment, as people should improve their skills for the salaries all the time. Senge [17], in his book Schools that Learn, simply stated that when individuals become aware of the sources of their thinking they begin to understand their mental models. Jonassen [18] Mental models are complex and inherently epistemic (and that) they form the basis for expressing how we know what we know. Because mental models are epistemic, they are not readily known to others and, in fact, not necessarily comprehended by the knower. Mental models, like all knowledge, must be inferred from performance of some sort.

How does one transform mind-sets? Often one's mind-set is the product of concepts and contexts one has grown up with and developed over a significant portion of one's life [19, 20]. International business research has much to say about the global mind-sets [21–27]. To change a mind-set means to motivate an individual to challenge the status quo, search for alternatives, provide concepts and contexts for new cognitive structures, and provide sufficient logic and reasoning so the participants replace existing structures with new ones [28].

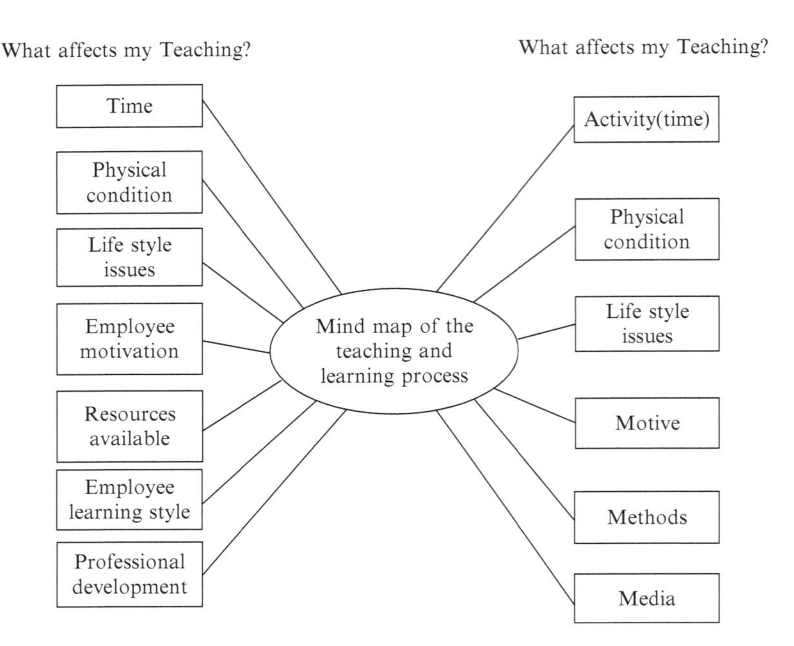

Fig. 1 Participant #1's (PI) mind map on the teaching and learning process (From Bogner Leonard [40])

3.1 *Learning Model*

Greenaway [29], explained that learning models are theories about how people learn. Greenaway researched how to create learning cycles (models) based on different theories of experiential learning and how to apply them to teaching and training. He went on to state that experiential learning can also be applied to almost any type of learning that occurs through experience, and that it is often used by instructors in a structured learning procedure that follows a cyclical model. Neill [30] believes, there has been extensive research conducted on learning models called experiential learning cycles. Experiential learning cycles are models for understanding how the process of learning works (Fig. 1).

3.2 *Training Strategies*

Researchers have sought to determine effective training strategies to convey appropriate content [31]. Conceptual training uses metaphorical techniques to convey the workings of a system. This may take the form of describing the overall structure and integrated workflow of the system and instructing end-users on a new system by drawing analogies in terms of a system they are already familiar with (i.e., database

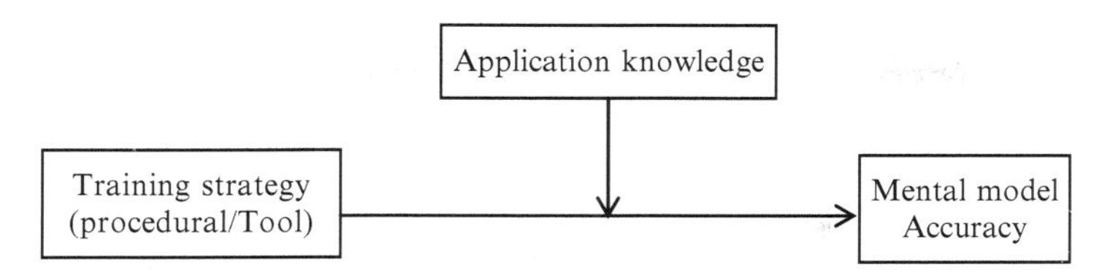

Fig. 2 Application knowledge influence on training methods and mental models (From Coulson Antony [41])

software), in terms of the training strategy framework, this approach would increase the knowledge-levels of training to add a tool conceptual component.

Many employees have examined the impact of application knowledge on mental model development, as measured by mental model accuracy. Specifically, these employees have reinforced the concept that end-user trainees with specific application knowledge have improved strategies and performance outcomes, suggesting an accurate mental model [32–39] (Fig. 2).

3.3 Improve Personal Competing Ability

Personal competing ability refers to the ability that keeps the oneself advancing others in the same competing conditions through continuously strengthening and improving oneself's skills. In this way, you can have more chances and can create more value than others.

For example, we notice that more companies in the talent market seek for technicians or skilled workers, which means experienced workers often have more chances and better salaries and benefits than the common workers, the competing advantages of experienced workers on common ones.

Microsoft found that in the statistics of seven main layoffs since 1996, employees in technology and sale departments have a very low unemployment ratio, never exceeding 10%, as these departments are where the core competeting ability comes from. Now many high officials from companies are trying to improve their knowledges in Chinese and Chinese Culture, learning a foreign language, or having a college class, in order to keep their competing advantages. Now the demand for small languages experts is strong, but the reserve is not that large. Learning or improving the skill of second foreign language become popular among the college students and the white-collar.

3.4 Adjust the Idea of Employment

Facing the talent competition, employees should get a clear understanding of the job market, and adjust the ideas of employment accordingly, form a positive attitude towards the life and the future, and improve the skills. Meanwhile, the employees should initiatively participate in training projects, get experiences and improve skills, planning for self-own carrier and development plan.

Change the old idea of employment, and form a pragmatic and diversified attitude. No hurry in finding a perfect job in one-step, and seek for a development with a position. Now, stay calm and learn some basic skills, making preparation for future opportunities.

3.5 Create Self-Own Businesses

Facing the talent competition, many people have lost jobs, but it is an important way to solve the problem of unemployment to set up self-own businesses, as it not only can solve the problem of selfown, but also create more positions. It is estimated that one new business can add 3–4 new positions.

The owners of Happy Pot Boiled Food, a food brand in Shenyang City, Miao Guoqing and Zhao Qingxiang, are two former army officials, who served in the army for 26 years. Since 18th June the first shop opened, the brand has had 14 chain shops, and paid more than one million RMB of tax. Now the company is planning to establish a national brand. This is a good example.

4 Conclusion

In this paper, through the literature review to explain the learning and training to upgrade and amends employee's mental model to deal with the future talent competition, it is the main way through the company and individual two aspects to solve the problem.

References

1. Peter Senge (1990) The fifth discipline: the art and practice of the learning organization. Doubleday, New York
2. Ning Xiaoyong, Cui Yueming (2009) On improving mental models and build a learning organization. Econ Res Guide 17(55):255–256
3. Klimoski R, Mohammed S (1994) Team mental models: construct or metaphor? J Manag 20:403–437
4. Weick KE, Roberts KH (1993) Collective mind in organizations: heedful interrelating in flight decks. Adm Sci Q 38:357–381
5. Holyoak KJ (1984) Mental models in problem solving. In: Anderson JR, Kosslyn SM (eds) Tutorials in learning and memory. W.H. Freeman, New York, pp 193–218
6. Rousse WB, Morris NM (1986) On looking into the black box: prospects and limits in the search for mental models. Psychol Bull 100:349–363

7. Xiang Kaibiao (2009) Solutions to the employment problem. China's Foreign Trade 5:31–32
8. Lei D, Slocum JW, Pitts RA (1999) Designing organizations for competitive advantage: the power of unlearning and learning. Organ Dyn 27(3):24–38
9. Slater SF, Narver JC (1993) Product–market strategy and performance: an analysis of the Milesy Snow strategy types. Eur J Mark 27(10):33–51
10. Baker WE, Sinkula JM (1999) The synergistic effect of market orientation and learning orientation on organizational performance. J Acad Mark Sci 27(4):411–427
11. Hurley RE, Hult GTM (1998) Innovation, market orientation and organizational learning: an integration and empirical examination. J Mark 62:42–54
12. Jerez-Gomez Pilar, Cespedes-Lorente José, Valle-Cabrera Ramón (2005) Organizational learning capability: a proposal of measurement. J Bus Res 58:715–725
13. Tippins MJ, Sohi RS (2003) IT competency and firm performance: is organizational learning a missing link. Strateg Manag 24(8):745–761
14. Perez Lopez S, Montes Peon JM, Vazquez Ordas CJ (2004) Managing knowledge: the link between culture and organizational learning. J Knowl Manag 8(6):93–104
15. Daniel Jiménez-Jiménez, Raquel Sanz-Valle (2011) Innovation, organizational learning, and performance Original Research Article. J Bus Res 4(4):408–417
16. Hoyle RH, Panter AT (1995) Writing about structural equation modeling. In: Hoyle RH (ed) Structural equation modelling. Sage, Thousand Oaks, pp 158–176
17. Senge P (2000) Schools that learn. A fifth discipline fieldbook for educators, parents, and everyone who cares about education. Doubleday/Currency, New York
18. Jonassen DH (2000) Operationalizing mental models: strategies for assessing mental models to support meaningful learning and design-supportive learning environments. Retrieved 8 Dec 2004. From http://www, ittheory.com/j onassen2.htm
19. Craik K (1943) The nature of explanation. Cambridge University Press, Cambridge
20. Johnson-Laird PN (1983) Mental models: towards a cognitive science of language, inference and consciousness. Cambridge University Press, Cambridge
21. Gupta A, Govindarajan V (2002) Cultivating a global mindset. Acad Manag Exec 16 (1):116–126
22. Harveston PD, Kedia BL, Davis PS (2000) Internationalization of born global and gradual globalizing firm: the impact of the manager. Adv Compet Res 8(1):92–99
23. Jeannet J-P (2000) Managing with a global mindset. Financial Times/Prentice Hall, London
24. Levy O (2005) The influence of top management team attentional patterns on global strategic posture of firms. J Organ Behav 26(7):797–819
25. Lobel SA (1990) Global leadership competencies: managing to a different drumbeat. Hum Resour Manage 29(1):39–48
26. Murtha TP, Lenway SA, Bagozzi RP (1998) Global mindsets and cognitive shift in a complex multinational corporation. Strateg Manag J 19:97–114
27. Perlmutter HV (1969) The tortuous evolution of the multinational corporation. Columbia J World Bus 4(1):9–18
28. Muñoz CAC, Mosey S, Binks M (2011) Developing opportunity-identification capabilities in the classroom: visual evidence for changing mental frames. Acad Manag Learn Educ 10 (2):277–295
29. Greenaway R (2002) Experiential learning cycles. Retrieved 13 Apr 2006. From http://reviewing.co.uk/research/learning.cycles.htm
30. Neill James (2004) Experiential learning cycles: overview of 9 experiential learning cycle models. Retrieved 13 Apr 2006. From http://www.wilderdom.com/experiential/elc/Experiential Learning Cycle.html
31. Olfman L, Mandviwalla M (1994) Conceptual versus procedural software training for graphical user interfaces: a longitudinal field experiment. MIS Quart 18(4):405–426
32. Mack RL, Lewis CH et al (1983) Learning to use word processors: problems and prospects. ACM Trans Inform Syst 1(3):254–271

33. Polson PG, Kieras DE (1985) A quantitative model of learning and performance, of text editing knowledge. Conference on human factors in computing systems. ACM Publications, San Francisco
34. Karat J, Boyes L et al (1986) Transfer between word processing systems. Conference on human factors in computer systems. ACM Publications, Boston
35. Polson PG, Muncher E et al (1986) A test of a common elements theory transfer. Conference on human factors and computing systems. ACM Publications, Boston
36. Ziegler J, Hoppe H et al (1986) Learning and transfer for text and graphics editing with a direct manipulation interface. CHI 1986 conference on human factors in computer systems. ACM Publications, Boston
37. Black JB, Bechtold JS (1989) On-line tutorials: what kind of inference leads to the most effective learning. Conference on human factors in computer interaction. ACM Publications, Austin
38. Igbaria M (1993) User acceptance of microcomputer technology: an empirical test. Omega 21(1):73–90
39. Satzinger JW, Olfman L (1998) User interface consistency across end-user application: the effects of mental models. J Manag Inf Syst 14(4):167–194
40. Bogner Leonard A (2007) Emerging mental models of teaching and learning. Using lesson study in acareer and technical education course. Ed.D., University of Minnesota
41. Coulson Antony (2002) ERP training strategies The role of knowledge-levels in the formation of accuratemental models. Ph.D., The Claremont Graduate University

Modeling and Analyzing of Railway Container Hub Scheduling System Based on Multi-agent System

Wang Li, Zhu Xiaoning, and Xie Zhengyu

Abstract Railway container hub scheduling system is a typical discrete element dynamic system (DEDS) which has the feature of high randomness, low flexibility, and uncertainty of operation time and high coordination demand of facilities. Multi-agent systems (MAS) have demonstrated their potential for solving complex problems which are asynchronism, concurrency, distributivity, parallel and overcome the disadvantages when faces complex discrete element dynamic system. According to the analysis of layout, establishment, facility and operation procedure of railway container hub, this paper chooses MAS as the basis to build the railway container hub scheduling system model, analyzes the function of each agent, and describes the schedule procedure analysis of agents.

Keywords Container • Railway container hub • Multi-agent system • Scheduling

1 Introduction

Railway container hub scheduling system (RCHSS) is a complex discrete element dynamic system (DEDS) composed by transportation systems, dispatching systems and information systems. It is related to the modeling and optimizing of logistic, dispatch and information. The operations of system have features of parallel, coordination and competition. Because of the high randomness, low flexibility, uncertainty

W. Li (✉) • Z. Xiaoning
School of Traffic and Transportation, Beijing Jiaotong University, Beijing 100044, P.R. China
e-mail: liking_bjtu@126.com; xnzhu@bjtu.edu.cn

X. Zhengyu
School of Traffic and Transportation, Beijing Jiaotong University, Beijing 100044, P.R. China

State Key Laboratory of Rail Traffic Control and Safety, Beijing Jiaotong University, Beijing 100044, P.R. China
e-mail: silinsherwin@126.com

Z. Zhang et al. (eds.), *LISS 2012: Proceedings of 2nd International Conference on Logistics, Informatics and Service Science*, DOI 10.1007/978-3-642-32054-5_71,

of operation time and high coordination demand of facilities in RCHSS, the system model description has features of concurrency, flexibility and dynamic.

Multi-agent systems (MAS) have demonstrated their potential for solving complex problems in various domains. A multi-agent system (MAS) is defined as a loosely coupled network of problem solvers that work together to solve problems that are beyond the individual capabilities or knowledge of each problem solver [1]. The increasing interest in MAS research is due to significant advantages inherent in such systems, including their ability to solve problems that may be too large for a centralized single agent, provide enhanced speed and reliability and tolerate uncertain data and knowledge.

For the advantages of MAS, it is largely applied in operation and schedule of container. Many studies which focus on the management of port container terminal, schedule of the container yard, distribution of berth, etc. have got several achievements. Degano and Pellegrino [2] apply the MAS to the chain of terminal operations and take Italy Voltri terminal for simulation. Gambardella and Rizzoli [3] combine operational research and MAS to study the scheduling and loading-unloading process. M. Yu et al. [4] use MAS as the basis for an intelligent terminal schedule system and proposes its framework, communication mechanism and negotiation mechanism between agents. Li and Li [5] present a modeling framework of container terminal logistics system which based on multi-agent and the successful algorithms in the computer domain and use AnyLogic platform to simulate for Shanghai harbor. Zhang and Yan [6] use MAS as the basis to build a container rail-sea intermodal transportation model. X.J. Wei [7] builds a container backup yard simulation model and makes this agent system can self study and autonomy. B. Sun proposes the MAS based architecture model of the operational scheduling system of logistics in container terminal and builds a robust berth allocation model based on ant colony optimization.

According to the studies mentioned above, we can find the applications of MAS in container logistic system most focus on port container terminals. The study about the application of MAS in RCHSS is few. As an important hinge of inland container transportation, railway container hub is different from port container terminal in distribution, facility, operation procedure and dispatch. So the study about the application of MAS in RCHSS is necessary. In this paper, we will use MAS for the modeling and analysis of schedule in RCHSS.

The rest of paper is organized as follows. Section 2 introduces the layout and facility of railway container hub. Section 3 describes operation procedure of railway container hub. Section 4 is the modeling and analysis of schedule in RCHSS. Section 5 concludes the paper and provides some directions for future work.

2 Layout of Railway Container Hub

Railway container hub can be divided into eight function areas which include operation area, main container yard, auxiliary container yard, security inspection area, intelligent door, control tower, parking area and container service area. The functions of area are described as follow:

1. Operation area
   Operation area is the core function area of railway container hub. The main loading and unloading processes of railway container hub are in this area
2. Main container yard
   Main container yard is the warehouse of arrival containers (temporary storage), transit containers and departure containers. If the container turnover volume of railway container hub is low, all common containers will be stored in this area.
3. Auxiliary container yard
   Auxiliary container yard is the auxiliary of main container yard. The special containers, refrigerated containers and empty containers are storaged in this area. When storage content of main container yard cannot meet the demand of container storage, railway container hub need transfer the demurrage arrival containers, transit containers and departure containers from the main container yard to the auxiliary container yard by container trucks.
4. Security inspection area
   Security inspection area is responsible for security checking of containers which are carried in railway container hub by container trucks.
5. Intelligent door
   Intelligent door can automatically recognize the number of container and container truck, and match truck and container information with electronic data to decide which trucks are allowed entrance or exit.
6. Control tower
   Control tower is responsible for supervising and managing the real-time status of containers and facility, giving task instructions for loading-unloading facilities.
7. Parking area
   Parking area is a parking lot of internal container trucks and reaches stackers.
8. Container service area
   Container service area is responsible for cleaning and maintaining of containers. The containers which need service are transported by internal container trucks from main container yard and carried by reach stackers from auxiliary container yard.

   Layout of railway container hub is shown in Fig. 1:

3 Operation Procedure of Railway Container Hub

Operation procedure of railway container hub can be divided into two parts. First is the operation procedure of container export which is shown in the right of Fig. 2. This part operation mainly focuses on the empty containers, heavy containers, special containers and transit containers loading. Another part is the operation procedure of container import which is shown in the left of Fig. 2. This part operation mainly focuses on the arrival container train unloading. The whole operation procedure of railway container hub is shown in Fig. 2:

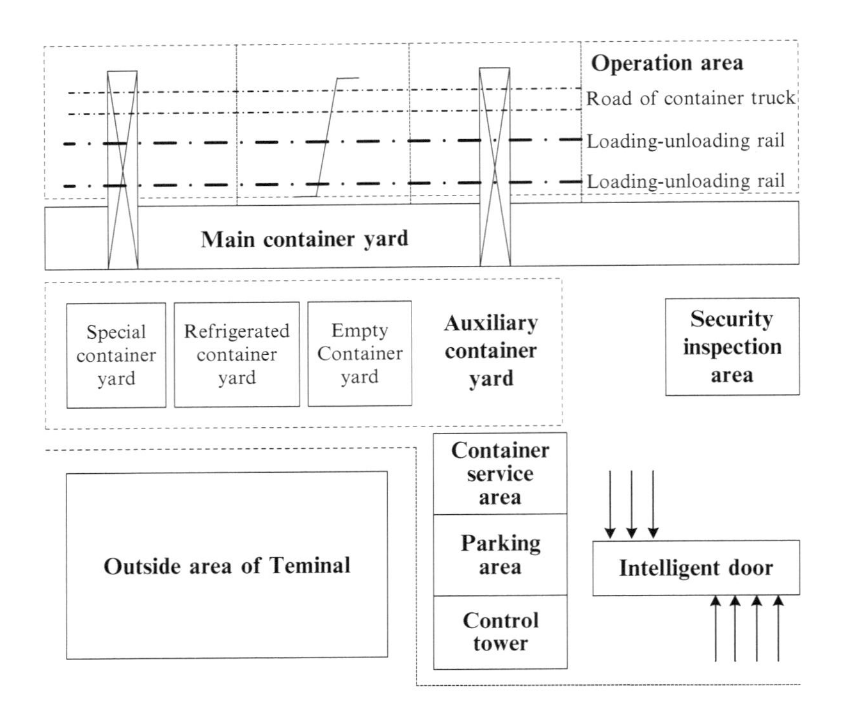

Fig. 1 Layout of railway container hub

4 Modeling and Analyzing of RCHSS Based on MAS

4.1 Modeling of RCHSS Based on MAS

According to the layout and operation procedure of railway container hub mentioned above and the characteristics of RCHSS, we use the methodology of MAS-commonKADS, comprehensively consider the operation type, operation organization, communication mode and consultation mechanism, adopt the mixed distribution pattern, build the model of RCHSS.

The model is composed by ten agents include rail designation agent, container yard distribution agent, gantry crane dispatch agent, reach stacker dispatch agent, container truck dispatch agent, rail agent, container yard management agent, gantry crane agent, reach stacker agent and container truck agent.

In the ten agents, rail designation agent and container yard distribution agent belong to the establishment control agent class. Gantry crane dispatch agent, reach stacker dispatch agent and container truck dispatch agent belong to the facility control agent class. Gantry crane agent, reach stacker agent and container truck agent belong to facility implement agent class. Container yard management agent is responsible for the management of main and auxiliary container yards. The model of RCHSS based on MAS is shown in Fig. 3:

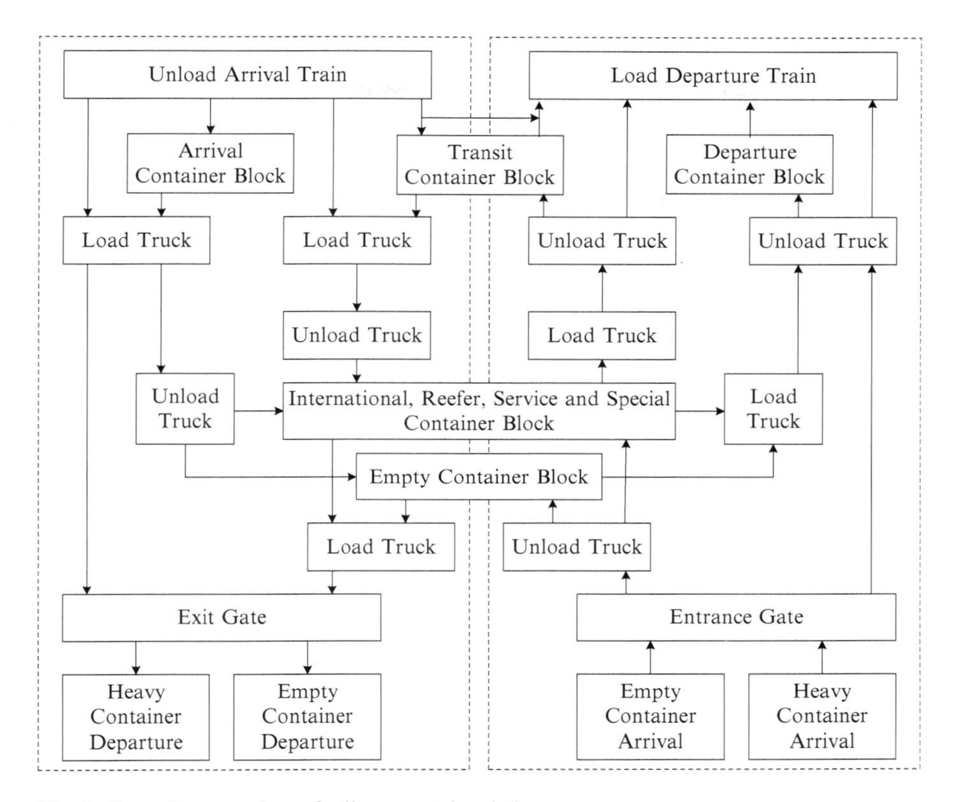

Fig. 2 Operation procedure of railway container hub

In the model, we divide the schedule problem of railway container hub into several sub-problems, and each sub-problem is solved by specific agent. These agents are relatively independent between each other. There are consultation mechanism and communication mode among agents.

4.2 *Function Analyzing of Agents*

Functions of ten agents are analyzed as follow:

1. Rail designation agent
   This agent is responsible for the distribution of loading-unloading rail, planning the service time and coordinating the utilization of rail.
2. Rail agent
   This agent is responsible for the arrival and departure of container trains and sending service application for gantry crane dispatch agent.

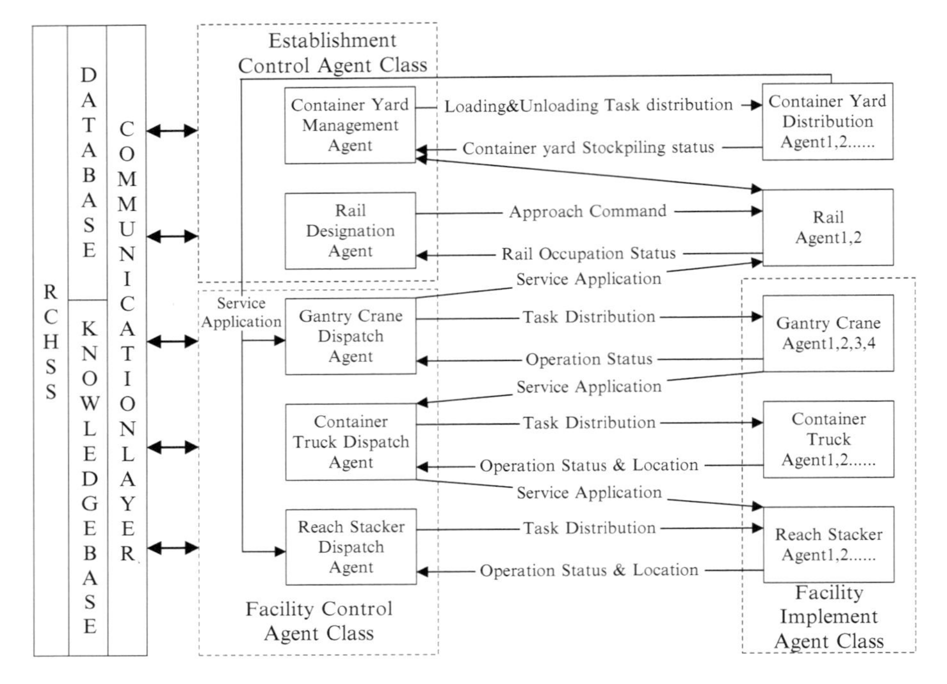

Fig. 3 Model of RCHSS based on MAS

3. Gantry crane dispatch agent
   According to the service application from rail agent, this agent is responsible for ensuring the loading-unloading order of containers in main container yard and dispatching gantry crane to accomplish loading-unloading.
4. Gantry crane agent
   This agent is responsible for generating loading-unloading sequence on the basis of gantry crane dispatch agent order and container attribute, sending service application to container truck agent.
5. Reach stacker dispatch agent
   This agent is responsible for the loading-unloading operation of container in auxiliary container yard and dispatching reach stacker to accomplish loading-unloading.
6. Reach stacker agent
   This agent is responsible for generating loading-unloading sequence on the basis of reach stacker dispatch agent order and container attribute, sending service application for container truck agent.
7. Container truck dispatch agent
   According to the service application from gantry crane agent and reach stacker agent, this agent is responsible for generate loading-unloading sequence and routes, and cooperating with gantry crane and stacker to accomplish loading-unloading.

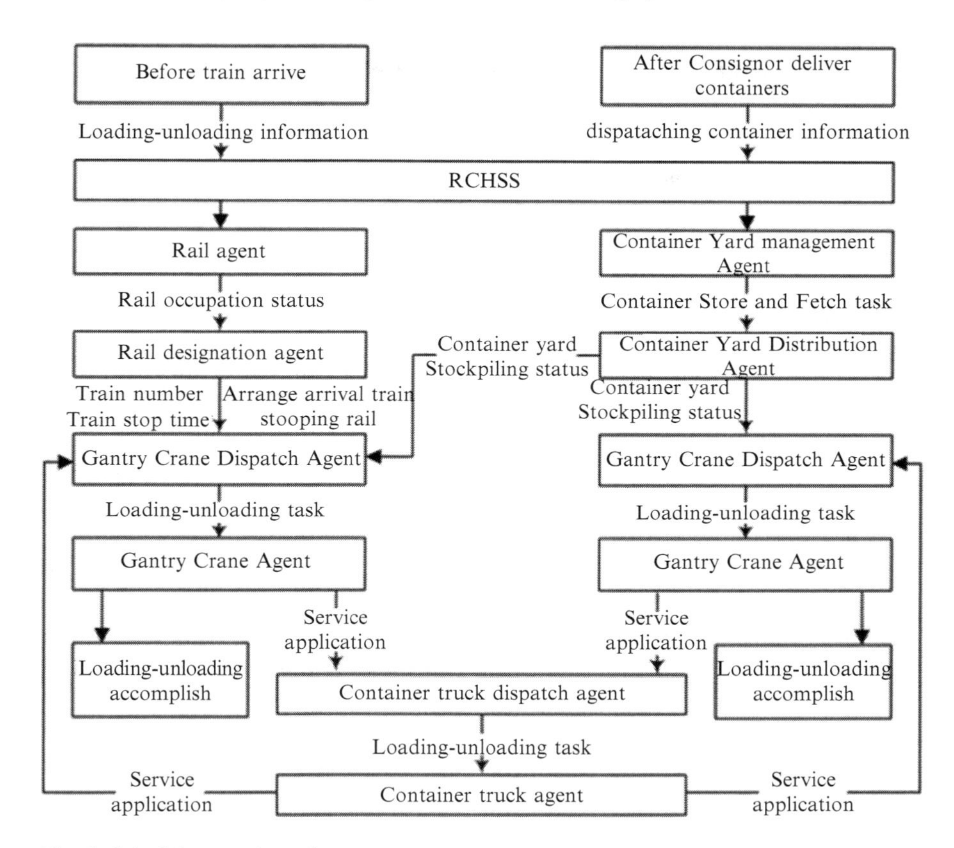

Fig. 4 Schedule procedure of agents

8. Container truck agent
   This agent is responsible for sending status and feedback information to container truck dispatch agent.
9. Container yard distribution agent
   This agent is responsible for ensuring distribution quantity and stack position of import containers in main and auxiliary container yard.
10. Container yard management agent
    This agent is responsible for ensuring the container stack plan, and the position of import and export container.

4.3 Schedule Procedure Analyzing of Agents

The RCHSS model is composed by several agents who have different knowledge are joined by communication network. The agents make decision by the message which is sent from other agents. Schedule procedure of agents is shown in Fig. 4:

5 Conclusion

In this paper, we introduce the layout, establishment, facility, and operation procedure of railway container hub. According to the characteristics of railway container hub, we use MAS as the basis to build the railway container hub schedule system model, analyze the functions of each agent, and describe the schedule procedure analysis of agents. The study about communication mechanism and negotiation mechanism among RCHSS agents will be the future direction.

Acknowledgments The authors wish to thank the National Natural Science Foundation of China for contract 60870014 under which the present work was possible.

References

1. Durfee EH, Lesser V (1989) Negotiating task decomposition and allocation using partial global planning. Distrib Artif Intell 2:224–229
2. Degano C, Pellegrino A (2002) Multi-agent coordination and collaboration for control and optimization strategies in an intermodal container terminal. In: IEEE international engineering management conference, Cambridge, 2002, pp 590–595
3. Gambardella LM, Rizzoli AE (ed) (1998) Simulation and planning of an intermodal container terminal. Simulation 71:107–116
4. Yu M (2007) Research on container terminal scheduling system based on multi-agent. WUHAN University of technology, Wuhan
5. Li Bin, Li Wenfeng (ed) (2007) Modeling of container terminal logistics system based on multi-agent. J Southeast Univ 23:146–150
6. Zhang R, Yan P (2007) Information system model of rail-sea intermodal transportation based on multi-agent. J Tonggi Univ 35:72–76
7. Wei X (2007) Research on container backup yard machine scheduling simulation system based on agent. Dalian Maritime University, Dalian

An Integration Framework for HTML5-Based Mobile Applications

Donghua Chen and Xiaomin Zhu

Abstract More HTML5-based mobile apps provide users with better user experience for app services. However, faced with difficulties of integrating massive existing Internet-based resources rapidly and creatively into Mobile Internet on HTML5 to better service in some areas like enterprise management, this paper introduces a new integration framework including its models helping service enterprises better integrate and reuse their Internet-based resources. Finally it introduces our prototype integration platform and proves its high feasibility of integration to promote the Mobile Internet.

Keywords HTML5 • Mobile application • Resource integration • Mobile Internet

1 Introduction

With HTML5 developing and Mobile Internet becoming mature in recent years, increasing mobile users acquire app services from service fields. However, they always fail to get quite satisfactory services because of incompatibility for Internet-based resources integrated into Mobile Internet through mobile ends [1–3]. This paper will introduce an Internet-based resource integration framework helping enterprises quickly reconstruct and redeploy HTML5-based app services based on their existing information resources, which promotes information resource reuse and enterprise value rediscovery in the rapid developing Mobile Internet.

This paper firstly reviews related work in Sect. 2. And then we discuss about two integration models of our framework in Sect. 3. Then a prototype platform will be tested to show its feasibility. At last, we have an outlook of our further work.

D. Chen (✉) • X. Zhu
School of Mechanical, Electronic and Control Engineering,
Beijing Jiaotong University, Beijing 100044, P.R. China
e-mail: tungwahchan@gmail.com; xmzhu@bjtu.edu.cn

Z. Zhang et al. (eds.), *LISS 2012: Proceedings of 2nd International Conference on Logistics, Informatics and Service Science*, DOI 10.1007/978-3-642-32054-5_72,

2 Related Work

Since the HTML5 standard [4] was introduced, it has made significant influence on mobile ecosystem. Though it is a general-purpose web standard, many of the new features are aimed squarely at making the Web a better place for desktop-style web applications [5]. So many researchers have gained fruitful consequences in HTML5 studies. Yang and Zhang [6] introduced interactive 3D graphics toward HTML5, getting rid of browser plug-ins for which they thought its future in interactive 3D graphics is bright. As more HTML5-based researches are successful, it is indicating a bright future of HTML5-based Mobile Internet.

With respect of integration framework, Paul et al. [7] once introduced a system called Cabana using diagrams to develop HTML5-based mobile apps, while it may be difficult to handle with problems of complicate mobile app logic integration as here tries to solve. The similar jobs are like TouchDevelop [8]. Michael, et al. [9] proposed a large-screen web content adaptative solution to enhance user experience by powerful HTML5. However, we think that their research in limit devices is more practical. Also, some other papers [10–13] discussed about mobile application integration, but most of them as we think just focus on how to make mobile development more quick and easy. Few turns back on the mature Internet containing massive information resources to review the value of the existing Internet-based resources and creatively apply them to Mobile Internet on HTML5 to provide better mobile app services for mobile users, which need to be discussed and solved in this paper.

3 The Integration Model for Internet-Based Resources

Faced with problems above, we introduce an app construction model as Fig. 1 shown.

This model can enable us to use a more structured mechanism to solve the integration issues. It contains five layers and five strategies to enrich the model in its entire implementation. They are namely HTML5-based View, App View, App Page, App Controller and App Metadata. The Internet-based resources include tagged resources, binary resources, database resources and script resources. Thereafter, a HTML5-based mobile app will be created by series of transformation based on these existing Internet-based resources. In the model, we also define five strategies to ensure its high flexibility and robustness in Fig. 1. All these specific strategies will enrich this model, making each mobile app as an instance of this model.

Based on the models above, this section will introduce our prototype platform that provides integral functions of resources integration, shown in Fig. 2.

Let $S = \{S_i\}$ be a set of data sources, $F = \{F_i\}$ be a set of integral functions, $B = \{B_i\}$ be a set of business modules, $A = \{A_i\}$ be a set of apps and $E = \{E_i\}$ be a

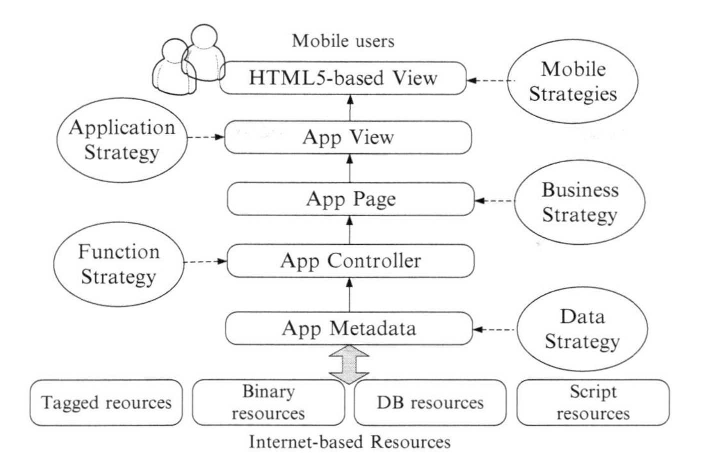

Fig. 1 A HTML5-based mobile app construction models

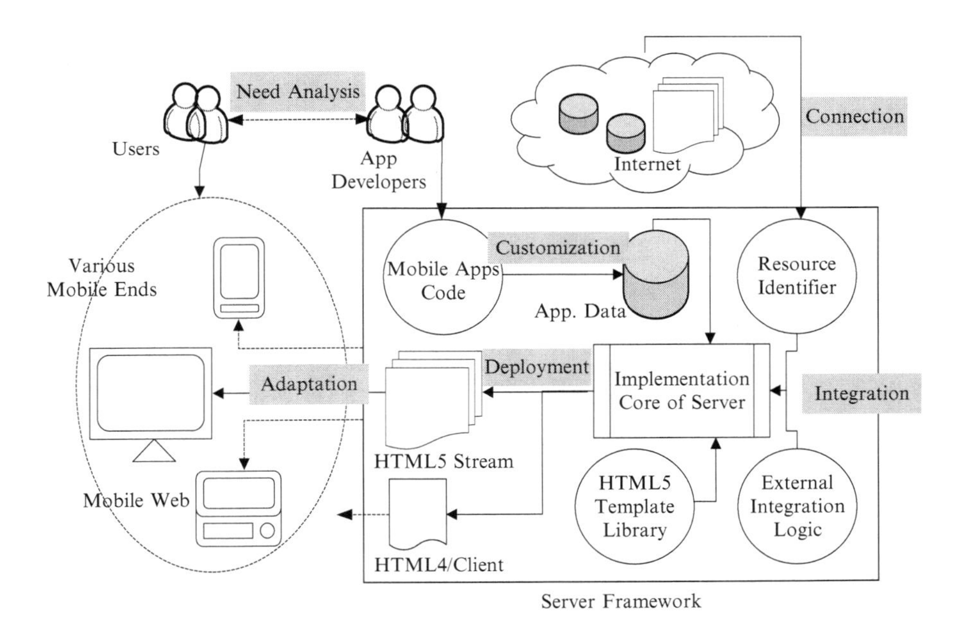

Fig. 2 The framework of our mobile HTML5-based app platform

set of different environments. The final view of the app that mobile can operate is denoted as M. And then we can get equations as followed in

$$M = E\big(a \in \big\{\big(\big\{\big(b \in \big\{\big(f \in \{(S_i)\}, L_f\big)\big\}, L_b\big)\big\}, L_a\big)\big\}, L_e\big) \tag{1}$$

In (1), L_f stands for combination method of data sources, L_b stands for combination method of functions, L_a stands for combination method of business logic and L_e will be the environment factors. $f \in \{(S_i)\}$ indicates that f is a subset of S. Finally, we establish a user model to help us to identify the behavior of mobile users, which will be very useful to the app construction model.

4 Efficiency Test of Our Prototype Integration Platform

The platform is in charge of integrating resources and deploying mobile app services, which means its comprehensive efficiency is extremely important. We implement some tests.

Firstly we select 25 HTML single page resources of small size between 1–30 KB to integrate them through our platform, where we collected key parameters like time cost of page loading in the process. Finally, we can visit the integral mobile app service through an android app service client we developed. We collected time cost of app page loading through both our client and native mobile browsers. This experimental result is as Fig. 3 shown below.

In Fig. 3, the time cost of some page loading is decreasing when pages are reconstructed and integrated into mobile app services that can be visited through our client compared with that of directly visiting related pages in native mobile browsers. Thus, we can draw two conclusions. Considering the time cost in android client against our native mobile browsers, we find over 50% of pages will shorten their loading time in our client, which plays a significant role in trying to integrate

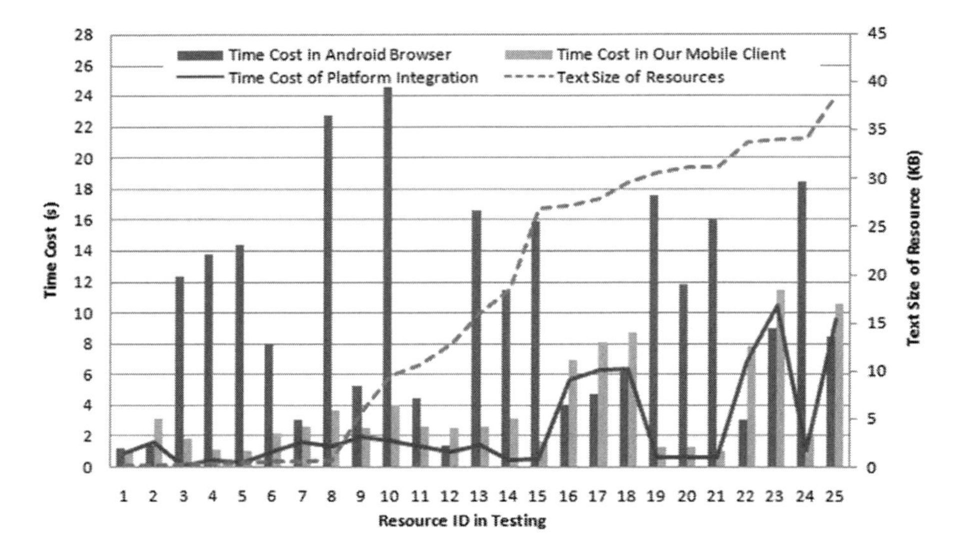

Fig. 3 The efficiency test of integration mechanism in our platform

and deploy hundreds of page resources online rapidly and also provide us with the more specific information than directly visiting web pages. It shows its high efficiency of information integration compared with browses. In addition, with rapidly increasing page size, it seems that time cost didn't increase a lot following its size, proving that it's able to integrate rather a huge amount of resources without increasing too much extra time to some degree.

However, we also make another test on whether the platform can deal with a single text resource of a huge size. Based on our data collected, the time cost of visiting app services in ours has exceeded that in native mobile browsers rapidly when the size was larger than 250 KB. It seems that when visiting a huge HTML page resource, the platform will have difficulty in turning it rapidly into a visual view of app service. In conclusion, we think that our platform still needs more improvement in integrating text resources of a huge size.

5 Conclusion

This paper introduces our complete Internet-based resource integration architecture aiming to apply to Mobile Internet based on HTML5. In this architecture, an app construction model is discussed to achieve the efficient integration mechanism, making it more powerful faced with rapid reuse and creative integration toward Mobile Internet on HTML5. And based on these research works, we had developed a prototype platform that achieved high flexible mechanism of integration. Based on our experimental results, we proved the feasibility of the architecture and models discussed above. We should see its bright outlook in the future.

Acknowledgments This paper is partially supported by a Key Project of National Natural Science Foundation of China (Contract No.: 71132008) and a Project of Basic Scientific Research Program for Central Universities (Contract No.: 2011JBM365).

References

1. Fling B (2010) Mobile design and development. O' Reilly Media, Inc/Publishing House of Electronics Industry, Beijing
2. Golding P (2008) Next generation wireless applications in web 2.0 and mobile 2.0 world. Wiley, Chichester
3. Hemel Z, Visser E (2011) Declaratively programming the mobile web with Mobl, In: 2011 ACM international conference on object oriented programming systems languages and applications, Portland, pp 695–713
4. Mark Pilgrim. "Dive into HTML5." http://diveintohtml5.info. Accessed on 14 Jan 2013
5. Anttonen M, Salminmen A, Mikkonen T et al (2011) Transforming the web into a real application platform new technologies, emerging trends and missing pieces, In: SAC'11, TaiChung, Taiwan, 2011, pp 800–808

6. Yang Jianping, Zhang Jie (2010) Towards HTML5 and interactive 3D graphics. In: International Conference on Educational and information Technology(ICEIT), Chongqing, China, vol 1, pp 522–527
7. Dickson PE, Hampshire College (2012) Cabana: a cross-platform mobile development system. In: The 43rd ACM technical symposium on computer science education, Raleigh, 2012, pp 529–534
8. Tillmann N, Moskal M, de Halleux J et al (2011) TouchDevelop: programming cloud-connected mobile devices via touchscreen. In: ONWARD'11, Portland, 2011, pp 49–61
9. Nebeling M, Matulic F, Streit L et al (2011) Adaptive layout template for effective web content presentation in large-screen contexts. In: The 11th ACM symposium on document engineering, California, 2011, pp 219–229
10. Stuedi P, Mohomed I, Terry D (2010) WhereStore: location-based data storage for mobile devices interacting with the cloud. In: The 1st ACM workshop on mobile cloud computing & services: social networks and beyond, San Francisco
11. Aghaee S, Pautasso C (2010) Mashup development with HTML5. In: The 3rd and 4th international workshop on web APIs and services Mashups, Ayia Napa, Cyprus, 2010
12. Mansfield-Devine S (2010) Divide and conquer: the threats posed by hybrid apps and html5. Netw Secur 3:4–6
13. Castano S, Ferrara A, Montanelli S (2012) Structured data clouding across multiple webs. Inform Syst 37:352–371

Investment Risk Prediction Based on Multi-dimensional Tail Dependence Empirical Study

Wang-Xiaoping and Gao-Huimin

Abstract Risk prediction plays a very important role in avoiding capital risk of investors, while tail dependence analysis is vital to risk prediction. Adopting D-vine model, with the data of the weekly-closing-price of China stock market and the stock market of neighboring countries and regions, this paper puts forward empirical distribution fit marginal distribution by using the *t*-copula, Clayton copula and Joe-Clayton copula to decompose the multivariate density function and analyze the tail dependence in multi-dimensional case. The experiments show that the pair copula model surely can be used to solve the tail dependence in multi-dimensional case efficiently.

Keywords Risk Prediction • Tail Dependence • D-vine Model • Stock Market

1 Introduction

With the fast development of capital market globalization, Global financial markets become increasingly closer and even more complicated in their relations. It is very important to analyze the tail interrelation between different financial markets. Definitions of tail dependence for multivariate random vectors are mostly related to their bivariate marginal distribution functions. Loosely speaking, tail dependence describes the limiting proportion that one margin exceeds a certain threshold given that the other margin has already exceeded that threshold. For instance, the tail

---

Wang-Xiaoping
School of Business, Jiaxing University, Jiaxing 314001, P.R. China

Gao-Huimin (✉)
Mechanical and Electrical Engineering College, Jiaxing University, Jiaxing 314001, P.R. China
e-mail: humorgao@sohu.com

Z. Zhang et al. (eds.), *LISS 2012: Proceedings of 2nd International Conference on Logistics, Informatics and Service Science*, DOI 10.1007/978-3-642-32054-5_73,

correlation coefficient between Shanghai and Shenzhen Stock Market can reflect the possibility of the two markets' major fluctuations simultaneously.

Copula function is introduced herein to the financial risk management by Embrechts et al. [1], which provided an easy and simple intelligent method for handling the correlation between variables, not only measuring nonlinear, asymmetric correlation, but also capturing tail dependence. Juri and Wutrich [2] proposed copula convergence theory of tail dependence, proved tail correlation between variables. Joe [3] presented a method to build two-dimensional copulas and put it into application, consequentially, great progresses have been made. Recently the n-variate tail distribution problem was studied by Klüppelberg et al. [4], Schmidt [5]. Vine copula [6] is applied in this paper, and its property is used in improving the tail dependence, which can be decomposed two-dimensional copula and marginal distribution function, make multivariate statistical analysis come into being.

The goal of this paper is to assess the investment risk based on dependence structure of international financial markets. This paper is organized as follows; Sect. 2 introduces some basic notions of vine copula. Parameter estimate method is given in Sect. 3, while the empirical study on multivariate vine copula model in SSEC, HIS, N225 and STI Stock is in Sect. 3. Finally, discussions and final remarks will be given in Sect. 4.

2 Basic Notions of Vine Copula

Recent developments in the domain of multivariate modeling are hierarchical or copula based structures. One of the most promising of these structures is the vine construction. The vine was originally proposed by Joe [7] and has further been explored by Bedford and Cooke [8, 9] and Kurowicka and Cooke [10]. N-Dimensional density function of canonical vine and D-vine are presented as formula (1) and (2).

1. Canonical vine

$$\prod_{k=1}^{n} f(x_k) \prod_{j=1}^{n-1} \prod_{i=1}^{n-j} c_{i,i+j|i+1,\ldots,i+j-1}\left(F\left(x_i \middle| x_{i+1}, \ldots x_{i+j-1}\right), F\left(x_{i+j} \middle| x_{i+1}, \ldots x_{i+j-1}\right)\right) \quad (1)$$

2. D-vine

$$\prod_{k=1}^{n} f(x_k) \prod_{j=1}^{n-1} \prod_{i=1}^{n-j} c_{j,j+i|1,\ldots,j-1}\left(F\left(x_j \middle| x_1, \ldots x_{j-1}\right), F\left(x_{j+i} \middle| x_1, \ldots x_{j-1}\right)\right) \quad (2)$$

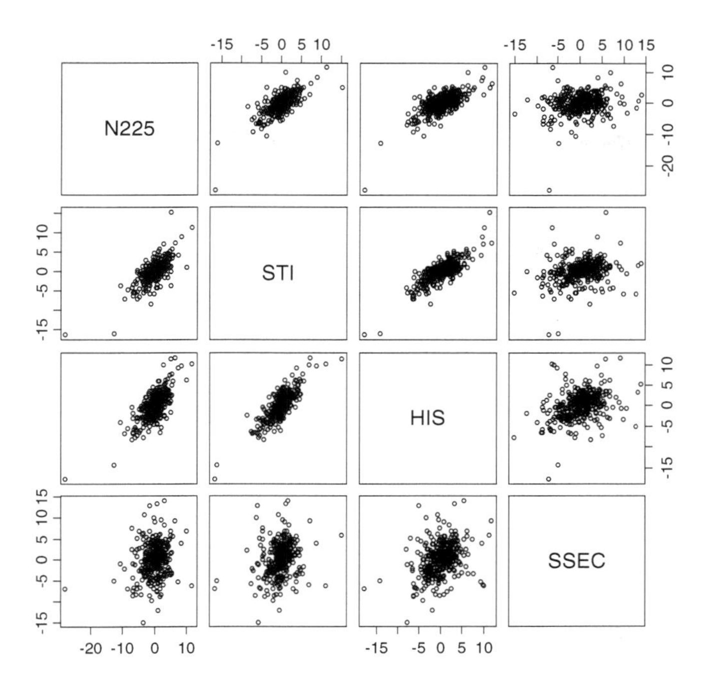

Fig. 1 Scatters diagrams of four stock markets

3 Empirical Analysis

In this section, we study weekly closing price of stock market from the period Jan. 01, 2005 to Jan. 01, 2011. Taking SSEC (Shanghai Stock Exchange), HIS (Hong Kong Hang Seng index), N225 (NIKKEI 225) and STI (Singapore Straits Times index) as samples to conduct tail dependence analysis. The data set were downloaded from http://finance.yahoo.com. Before further processing, we set the rate of return $R_{i,t}$ as formula (3).

$$R_{i,t} = 100\left(\ln P_{i,t} - \ln P_{i,t-1}\right) \tag{3}$$

Where $P_{i,t}$ is weekly closing price of ith stock.

Scatter diagrams (as shown in Fig. 1) show that N225 and STI are positively correlated with HIS. In addition, as the results of correlate analysis, SSEC is not closely related to other three markets.

We fitted a *4*-dimensional D-vine, with *t*-copula for all pairs. We use the empirical distributions as the conditional distribution. The return vectors are converted to uniform pseudo-observations before further modeling.

The tail dependence coefficients from our case are shown in Table 1. The largest one correspond to the largest degrees of freedom values. Based on this table, we

Table 1 Degrees of freedom of t-copula

	N225	HSI	SSEC
STI	6.665653	7.412836	7.989282
N225		249.5002	13.38285
HSI			6.88067

Table 2 Estimated parameter of vine copula

	t-copula		Clayton copula	Joe-Clayton copula	
	υ	ρ	θ	σ	δ
C_{12}	6.665553	0.670086	1.412931	1.219078	1.287829
C_{23}	7.359852	0.767640	1.994341	1.312952	1.843417
C_{34}	6.880670	0.375336	0.612739	1.012465	0.603943
$C_{13\|2}$	149.5001	0.266722	0.307240	1.074209	0.289747
$C_{24\|3}$	300	0.032291	0.002733	1.001000	0.001000
$C_{14\|23}$	300	−0.070987	−0.042561	1.011088	0.002813
AIC	−566.9045		−551.393	−571.6973	
BIC	−521.0279		−528.4547	−525.8207	

The tail dependence coefficient is obtained, as shown in Table 3

Table 3 The tail dependence coefficient

	t-copula	Clayton copula	Joe-Clayton copula	
ρ_{12}	0.255016	0.612275	0.234237	0.583781
ρ_{23}	0.323805	0.706412	0.304580	0.686594
ρ_{34}	0.095832	0.322637	0.016995	0.317365
$\rho_{13\|2}$	0.000000	0.104764	0.093512	0.091424
$\rho_{24\|3}$	0.000000	0.000000	0.001384	0.000000
$\rho_{14\|23}$	0.000000	0.000000	0.015145	0.000000

choose C12 as the copula of N225 and STI, C23 as the copula of HSI and SSEC, and C34 as the copula of STI and HSI.

Based on previous studies, three kinds of copula-Clayton, Joe-Clayton and t-copula are selected. The parameters of the vine copula are estimated by the non-parameter method, see Algorithm 4 in Aas et al. [11]. Results are shown in Table 2. Where υ is degree of freedom of t-copula, ρ is correlation coefficient of t-copula, θ is parameter of Clayton copula, σ and δ are upper, lower tail dependence parameter of Joe-Clayton copula respectively. To evaluate whether a construction appropriately fits the data, a goodness-of-fit test is necessary. We use the AIC, BIC testing method in this paper.

From the Table 3, we can see the goodness-of-fit test does not reject the Student vine copula for the returning data. The estimated correlation coefficient of N225 and STI is 0.255016, the one between STI and HSI is 0.323805, and indicating strong correlations between stocks in the same group are existed. In contrast to above two group, ρ_{34} is 0.095832, hence significantly weaker dependence between HSI and SSEC. Moreover, these four stock markets also have no obvious return condition relationship.

And some useful conclusions have been drawn from the analysis of *t*-copula, such as investing all three stock markets at the same time provides more diversification risk anyway. However, *t*-copula cannot measure tail dependence.

In addition, conclusions we can make from the correlation coefficient of Clayton copula, is that there is very significant lower tail dependence between N225 and STI, STI and HIS, HIS and SSEC respectively. Even under the hypothesis of conditional correlation, lower tail dependence, among N225, HSI and STI, still exist. Therefore, investors concerned with avoiding risk should not choose these three stock markets simultaneously.

At the same time, an analysis of the obtained results by Joe-Clayton copula show that the lower tail correlation coefficients between N225 and STI, STI and HIS, HIS and SSEC are higher compared with upper one in the context of non-conditional correlation. This is in accordance with the truth that the fluctuation extent of yield in bear market is more intense than the fluctuation extent of yield in the bull market. The result is different on the premise of conditional correlation, which displays the upper tail dependence and the coherence degree is decreased.

4 Conclusion

We have studied how *n*-dimensional variables exhibiting complex patterns of dependence in the tails can be modeled using vine copula. This construction is hierarchical in nature, the various levels standing for growing conditioning sets, incorporating more variables. This differs from traditional *2*-dimensional copula models, where levels depict conditional independence. The fit of the constructions has been tested on weekly Return Ratios of *4*-dimensional Stock Market, which enable better tail dependence being understood, problems in *multi*-dimensional variable tail dependence have been solved. Risk manager and venture capitalists can measure invest risk and make correct decision based on analysis of the tail relativities among some stock markets. Consequently, a new thought to improve the prediction accuracy of the invest risk is presented in this dissertation, when you are ready for doing pluralism investment.

Acknowledgments This work was supported in part by the Natural Science Foundation of Shanxi Province (No. 2009011011-3), Humanity and Social Sciences Planning Fund of China Education Ministry (No. 11YJA630082), the Soft Science Foundation of Shanxi Province (No. 2011041001-02), and the Research Project Supported by Shanxi Scholarship Council of China (No. 2011-078).

References

1. Embrechts P, Mcneil A, Straumann D (1999) Correlation and dependence in risk management: properties and pitfalls. In: Dempster M (ed) Risk management: value at risk and beyond. Cambridge University Press, Cambridge, pp 176–223
2. Juri A, Wutrich MV (2002) Copula convergence theorems for tail events. Insur Math Econ 30:405–420

3. Joe H (1997) Multivariate models and dependence concepts. Chapman and Hall, London
4. Klüppelberg C, Kuhn G, Peng L (2007) Estimating the tail dependence function of an elliptical distribution. Bernoulli 13:229–251
5. Schmidt R (2003) Credit risk modeling and estimation via elliptical copulae. In: Bohl G, Nakhaeizadeh G, Rachev ST, Ridder T, Vollmer KH (eds) Credit risk measurement, evaluation and management. Physica-Verlag, Heidelberg, pp 267–289
6. Kurowicka D, Cooke R (2006) Uncertainty analysis with high dimensional dependence modelling. Wiley, New York
7. Joe H (1996) Families of m-variate distributions with given margins and m(m-1)/2 bivariate dependence parameters. In: Ruschendorf L, Schweizer B, Taylor MD (eds) Distributions with fixed marginals and related topics. Institute of Mathematical Statistics, Hayward
8. Bedford T, Cooke RM (2001) Monte Carlo: simulation of vine dependent random variables for applications in uncertainty analysis. In: Proceedings of ESREL2001, Turin, 2001
9. Bedford T, Cooke RM (2002) Vines: a new graphical model for dependent random variables. Ann Stat 30(4):1031–1068
10. Kurowicka D, Cooke RM (2005) Sampling algorithms for generating joint uniform distributions using the vine-copula method. In: 3rd IASC world conference on computational statistics & data analysis, Limassol, Cyprus, 2005
11. Aas K, Czado C, Frigessi A et al (2007) Pair-copula constructions of multiple dependence. Insur Math Econ 44:182–198

Ticket Pricing Model for Group Passenger Based on Dynamic Programming

Gao Ronghuan

Abstract Ticket pricing is a key issue in airline's revenue management. The whole paper can be divided into two main parts. First, according to the characteristic of air seat, a discount rate for the group is calculated, and a discrete-time dynamic programming model for group passenger optimal pricing is established. Second, establishing pricing model and the optimal price policy of group passenger is given by proposed model. Finally, the practicability of the model is illustrated by using a numerical example.

Keywords Revenue management • Airlines transportation • Group passenger • Ticket pricing

1 Introduction

Ticket pricing problem is the short-term seats for sales that correspondence with reservation request, how to control individual fares and reservations [1, 2]. In order to increase revenue, the airline should try to sell a high level of air tickers, but the various levels of inventory request arise time is inconsistent, thus how to determine the number of sales at various levels of inventory is the key factors affecting the airline income [3, 4]. Expected Marginal Seat Revenue (EMSR),a seat inventory control method based on the principle of marginal revenue, has become a classic airline seat control method [5].

Since the low proportion of group of passengers in foreign aviation market, typically less than 10%, there is rarely literature about group seat pricing. But due to national identity, at the domestic market or the Southeast Asian markets, group passengers account for a large proportion of the sales volume, on some routes

G. Ronghuan (✉)
College of Airport, Civil Aviation University of China, Tanjing 300300, P.R. China
e-mail: grh0501@yahoo.com.cn

Z. Zhang et al. (eds.), *LISS 2012: Proceedings of 2nd International Conference on Logistics, Informatics and Service Science*, DOI 10.1007/978-3-642-32054-5_74,

accounted for 30–40% of the annual sales [6]. Group price is a problem worthy of study.

Air seat is perishable that has the following feature: (1) Products cannot be stored. (2) Products can be classified with differential price. (3) Demand is fluctuations.

Groups are different from individual, and often enjoy more discounts treatment by the reason of great demand for seats. Accept the group means squeezing out some of the high-fare passengers. But at the same time due to the perishable of air transport products, that if a flight seat is not completely sold before takeoff, the remaining seats will be simply wasted. So the premise of pricing is to identify a reasonable discount for groups. Then, using discount rate to establish pricing model.

2 Discount Formula

Since the air tickets are void value when aircraft departure, the tickets selling process can be assumed as value – index of time varying function. The ticket price is fluctuations around the value of the ticket. The value of changes is continuous, while the price is impossible to continuous change.

Therefore, in order to get a discrete price list, ticket sales cycle is evenly divided into equal time intervals, and then follow the principle that the total value of the ticket are the same at each selling period, calculate the average value of the ticket at each time, and dividing the result by tickets fare. Then the discount rate is obtained, and it is various in different phases.

2.1 Notations and Assumptions

$D(t)$ indicates customer demand at t; t indicates the number of days, a period from the flight ticket sales to the flight departure. $t = 1$ indicates the first day of sale for the flights, $t = T$ indicates the day that flight took off. And so on, generally the value of T is determined according to the actual situation; $P_V(t)$ indicates the value of the ticket at t; P_{Vi} indicates average value of the ticket sold during period $i; P_m$ represents the highest passenger ticket fares; α indicates the deteriorated rate of the ticket value; a_i indicates the ticket discount rate during period i.

Assume that:

1. only consider a single flight;
2. the passengers demand was as the law of normal distribution function, D: the number of forecast passengers $D = (d_1, d_2 \cdots d_n)$, $d_i(t)$ is passengers demand when ticket charge with j; following a normal distribution, $d_j \sim \mathrm{N}(\mu_j, {\sigma_j}^2)$;

3. the value of ticket changes in a decreasing function of time index; $P_V(t) = P_m e^{-\alpha t}$;
4. the time of selling tickets start with T;
5. the demand of passengers does not vary with price changes.

2.2 Discount Formulas

Since the total value of tickets at each period is constant, the average value of the tickets during i period can be solved with the following expression:

$$\int_{t_{i-1}}^{t_i} P_{vi} D(t) dt = \int_{t_{i-1}}^{i} P_v(t) D(t) dt \tag{1}$$

Since $t_i = i * \frac{T}{n}$, the discount rate formulas are variant as follows:

$$a_i = \frac{\int_{t_{i-1}}^{t_i} P_v(t) D(t) dt}{P_m} \tag{2}$$

3 Dynamic Optimization Pricing Model

The discount rate formula is only for group travelers. In fact, however, the same flight not only supply service for group tourists, but also considering the individuals. So, the next research is about seat allocation to individual as well as group passengers in the same flight, establishment the reverse of dynamic pricing model which make it possible for the group travelers to "fill" the seat that not fully occupied by individuals in a low price. Meanwhile, make sure the group travelers do not crowd out the high fare passengers, and effectively increase the flight proceeds.

3.1 Symbols and Assumptions

$j = 1, 2, 3 \ldots k$ indicates several fare class, $j = 1$ represents the top fare class, while $j = k$ represents the lowest fare class. $n = 0, 1, \ldots N$ indicates decision-making stage; $n = 0$, means the takeoff time; $m = 1, 2, \ldots M$, the number of groups, $m \geq 10$; s: number of available seats for sale; apparently, $M \leq s$; F_j, the fare at level j, α_{ij}: j class fare discounts during i period; P_j^n indicates the probability of seat reserved for the j level at the n-th decision-making stage; R_s^n : the expected total revenue when the number of seats is s at the n-th stage; $\sigma_m(n, s)$ expected marginal revenue for the tickets sold during the n-th stage.

Assumptions:

1. the aircraft capacity is unchanged in a flight
2. do not sale tickets more than aircraft capacity, and the reservation cannot canceled.
3. the reservation requests following the Poisson distribution.
4. at the same decision-making stage, the time of seat requested for groups is different from individual at the same class level.

3.2 Group Passengers Optimization Pricing Model

When aircraft has no seat for sale, there is no need to make any decisions, so the total expected revenue should be equal to the income of the last decision-making stage so:

$$R_s^n = R_0^{n-1}(n \succ 0, s = 0) \tag{3}$$

When $n \succ 0, 0 \prec s \prec c$, seat request are made by individuals:

$$R_s^n = P_0^n R_S^{n-1} + \sum_{j=1}^{k} P_j^n \max(mF_j + R_{s-m}^{n-1}, R_s^{n-1}) \tag{4}$$

When $n \succ 0, 0 \prec s \prec c$, seat request are made by groups:

$$R_s^n = P_0^n R_S^{n-1} + \sum_{j=1}^{k} P_j^n \max(R_s^{n-1}, ma_{ij}F_j + R_{s-m}^{n-1}) \tag{5}$$

$P_0^n = 1 - \sum_{j=1}^{k} P_j^n$ indicates the probability of the reservation request is zero at the n-th stage. Merge formulas 4 and 5 can be obtained:

$$R_s^n = P_0^n R_S^{n-1} + \sum_{j=1}^{k} P_j^n \max(\beta mF_j + R_{s-m}^{n-1}, R_s^{n-1}, \theta ma_{ij}F_j + R_{s-m}^{n-1}) \tag{6}$$

$$\beta = \begin{cases} 1 & m < 10 \\ 0 & or\ else \end{cases}, \quad \theta = \begin{cases} 1 & m \geq 10 \\ 0 & or\ else \end{cases}$$

Table 1 Demand of air travelers

Class level	Price (discount)	Mean	Standard deviation
1	1,280 (1)	12	3
2	960 (0.75)	26	5
3	640 (0.5)	80	15

Table 2 Demand of individuals

Class level	Price (discount)	Mean	Standard deviation
1	1,280 (1)	12	3
2	960 (0.75)	26	5
3	640 (0.5)	60	11

Table 3 The probability of the reservation request are made at various stages

	Decision stage									
Probability	1–3	4–6	7–9	10–12	13–15	16–18	19–21	22–24	25–27	28–30
P_1^n	0.01	0.02	0.03	0.05	0.03	0.07	0.06	0.10	0.05	0.07
P_2^n	0.02	0.05	0.04	0.06	0.05	0.09	0.09	0.12	0.08	0.11
P_3^n	0.03	0.06	0.07	0.08	0.10	0.09	0.14	0.06	0.05	0.09

Therefore, the expected marginal revenue at n-th stage is:

$$\sigma_m(n, s) = \frac{1}{m}(R_s^n - R_{s-m}^n) \tag{7}$$

If $mF_j + R_{s-m}^{n-1} \geq R_s^{n-1}$ or $ma_jF_j + R_{s-m}^{n-1} \geq R_S^{n-1}$, it means accept the reservation request for the j class level is profitable, no matter for individual or group. The only difference is ticket fare. For individuals, the price is according to their class level, while the group fare enjoy a discount which can be calculated by using the discount formula.

3.3 Example

Suppose an airline operating in a leg, the aircraft capacity is 110. There are three levels of ticket fare, respectively: $F_1 = 1280$, $F_2 = 960$ and $F_3 = 640$.

The deteriorated rate of the ticket value is 0.2 at each class level. Demand of air travelers and individuals are assumed to be normally distributed as shown in Tables 1 and 2. The time when reservation request was put forward is follows the Poisson distribution. Based on the assumption (4) and historical data of airline ticket sales, selling time is divided into 30 stages. Parameters data are obtained by statistical analysis, as shown in Table 3. The probability of the reservation request are made at various stages is present. Generally, group discount rate is fixed at 4.5% of price which is 576 yuan. One of the groups put forward reservation request which size is 20.

Table 4 Results (group size is 20)

Sale stage	Minimum fares	Model revenue	Yield gap
1–12	448	83089.6	327.5
13–24	538	82401.3	439.2
25–30	627	82313.8	351.7

According to figures in Tables 1 and 2, using the EMSR [5] method, we can calculate the optimal expected revenue of the flight which is 81,962.1 yuan. Based on dynamic programming method, using MATLAB software to write a computer program, we obtain the following optimization price and the flight proceeds as well as the results compare with the EMSR method (see Table 4).

On 1–12 sale stage, the ticket price for the group should not less than 448 yuan; on 13–24 sale stage, the lowest price is 538 yuan; and on 25–30 sale stage, is 627 yuan; When change the size of group respectively to 15 and 25, we get the similarly results. From these results, we can see almost all of the model's revenue are better than EMSR's. It means that this method has a high practical value to provide a scientific model to determine the dynamic pricing strategy for the group travelers.

4 Conclusions

From these results, we can see almost all of the model's revenue are better than EMSR's, it possible to effectively increase the flight proceeds.

It means that this method has a high practical value to provide a scientific model to determine the dynamic pricing strategy for the group travelers, it possible to effectively increase the flight proceeds.

References

1. Wu TS et al (2000) How to implement revenue management to group in Airlines Nanjing. Civ Aviat Econ Technol (2):57–58
2. Yang SL (2000) The most profitable methods – management of revenue management. Aviation Industry Press, Beijing
3. Müller-Bungart M (2007) Recent advances in revenue management. Springer, Berlin/Heidelberg, pp 15–17
4. Jerenz A (2008) Modeling the price-based revenue management problem. Gabler
5. Belobaba PP (1987) Air travel demand and airline seat inventory management. Massachusetts Institute of Technology, Cambridge, May 1987
6. Gao Q (2006) Seat inventory control for airline revenue management. Nanjing Aeronautics and Astronautics University, Nanjing

Barrier Identification and Removal Based on Process Analysis

Yichao Liu and Weining Fang

Abstract The need to improve customer satisfaction, operational effectiveness and asset utilization has made effective Operations Management a growing priority for manufacturing organizations. While significant research has been carried out in production process optimization, limited work has been carried out to incorporate business process, culture and manufacture subject matter for improvement. This paper presents the findings of an investigation into the process of production and business within Strong Bond CO., ltd. in Chinese manufacturing industry. Interviews were conducted with experts across manufacturing and multidisciplinary consultancies, the outputs of which were analyzed using general judgment and weight analysis. From this, 21 themes of barriers across three core categories classed as business process, culture and subject matter emerged. The findings ranked by the consequence and remove difficulty show six top barriers. Therefore, some effective countermeasures and proposals for operations management have been made for customer satisfaction and product improvement.

Keywords Operations management • Customer satisfaction • Business process

Supported by RCS2008ZT003

Y. Liu (✉)
State Key Laboratory of Rail Traffic Control and Safety, Beijing Jiaotong University, Beijing 100044, People's Republic of China

School of Mechanical and Electric Control Engineering, Beijing Jiaotong University, Beijing 100044, People's Republic of China
e-mail: annosky@hotmail.com

W. Fang
State Key Laboratory of Rail Traffic Control and Safety, Beijing Jiaotong University, Beijing 100044, People's Republic of China

Z. Zhang et al. (eds.), *LISS 2012: Proceedings of 2nd International Conference on Logistics, Informatics and Service Science*, DOI 10.1007/978-3-642-32054-5_75,

1 Introduction

The design, plan and management of good production logistics are the keys to improve labor production quota and produce quality. Only reasonable production logistics of organization can make the process of business production always in the best state [1]. Strong Bond is a billion-dollar manufacturer of adhesive and applicator products aimed at the worldwide market for adhesives and sealants. It covers all aspects of the adhesives market from large volume industrial and consumer demand to very technical special applications, and it had maintained many proprietary positions in industrial markets since product introduction. But the situation has changed significantly in the last 2 years largely because of the expiration of the basic patents and new developments in adhesive technology. Meanwhile, Strong Bond has had an endless stream of consultants involved in their company. Many of these efforts have produced fruitless results since the change and the exposure it creates is not highly valued at Strong Bond [2].

To improve customer satisfaction, reduce operating costs, improve asset utilization and improve its flexibility, analysis of functions both R&D and Marketing is important [3].

1.1 Key Processes

All of Strong Bond's chemical operations have similar design layouts and comprise two processes named make process and product development process. The key client process mapping is shown in Fig. 1.

The make process (Fig. 2) comprises two main processes: the normal work process and rework process.

The normal work process starts with mixing all chemicals in vessels which is done on the third floor of the production building. Then the reaction process by-products are gravity fed or pumped to a separation operation on the second floor. The gravity auto fed process benefits from the layout of the plant and saves a part of energy. The capacity of this mix and reaction processes varies a lot depending upon the weekly schedule, from this point of view the flexibility of the production is quite good [4]. The third step is to dry the products on the first floor which will be subjected to inspection later on.

After the normal work and subsequent inspection (checking the packages, the quality of product and so on), the failed products are sent to the rework shop. The chemical (mix and reaction) process influences greatly the adhesive and durability response of products. Normally the FPY (First Pass Yield) of all products is around 92–95%. Distillation is then applied to get useful chemicals from the remixing stage, and chemicals are sent to react and dried in following processes, which are the same with "normal work" processes [5].

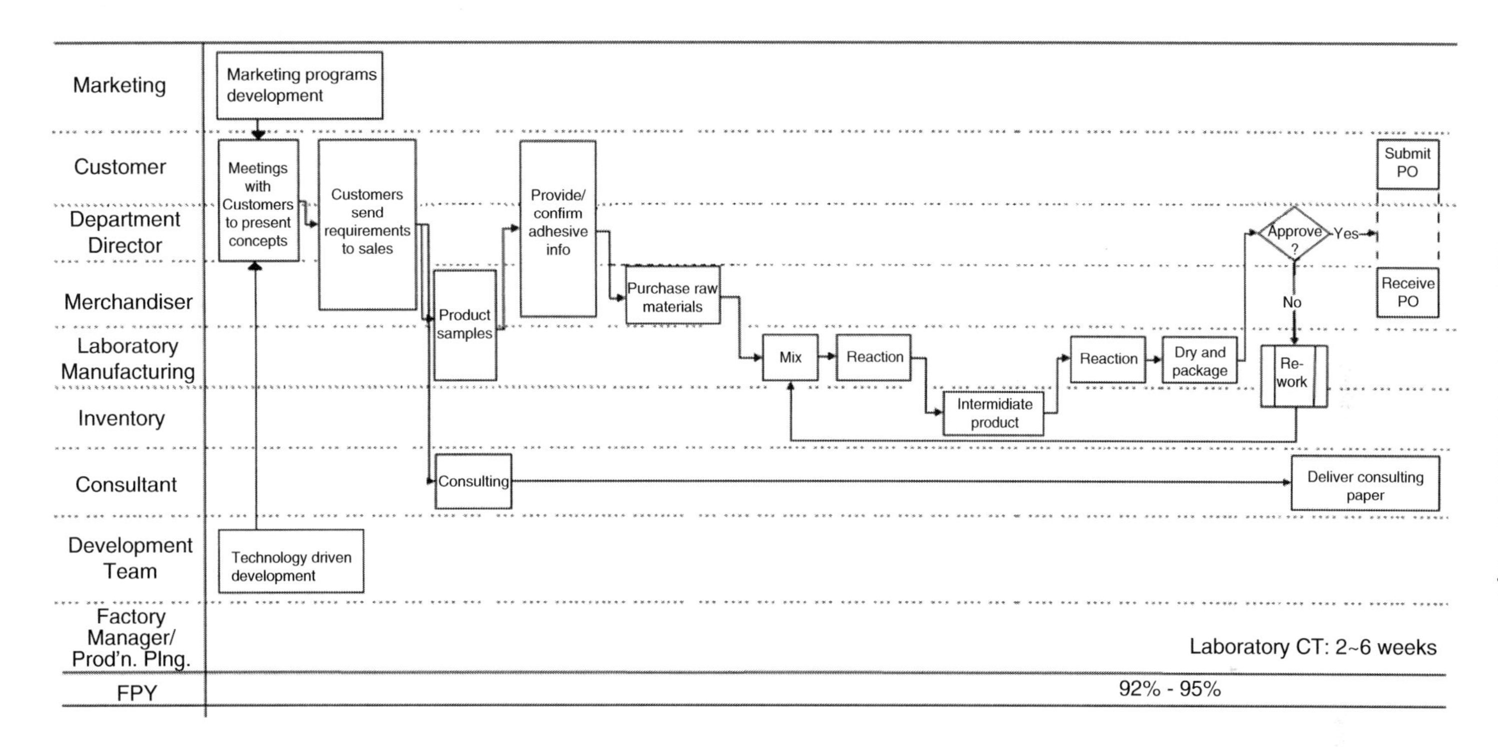

Fig. 1 Key client process mapping

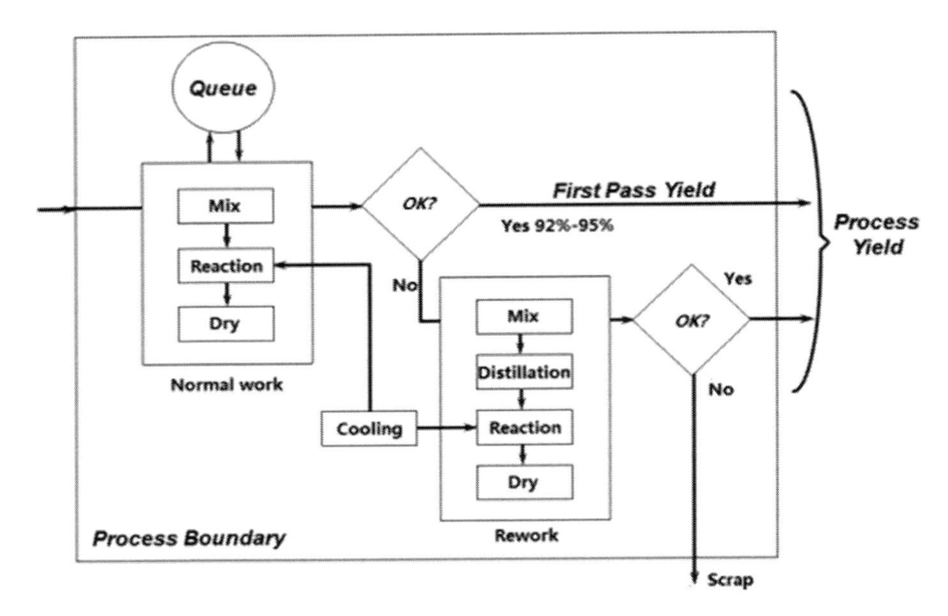

Fig. 2 FPY model of make process

1.2 Product Development Process

Strong Bond has a Product Development Process (PDP) but does not use it very consistently. Some of the major decision points are adhered to, more or less, but it is not the Strong Bond culture to follow a process religiously. It is viewed as a guideline, not a roadmap.

2 Measurement and Implementation

As the barriers of Strong Bond's critical process do not exist separately and taking measures case by case may not work efficiently to solve those problems at corporate-wide [6], a matrix structure of barriers (Fig. 3) can be drawer out.

Thus when many improving projects are in progress to remove barriers, there are three major measures that need to be installed in each of the critical processes which tie to the overall business measures and Strong Bond's objectives: company resources management, culture, and product research and development [7].

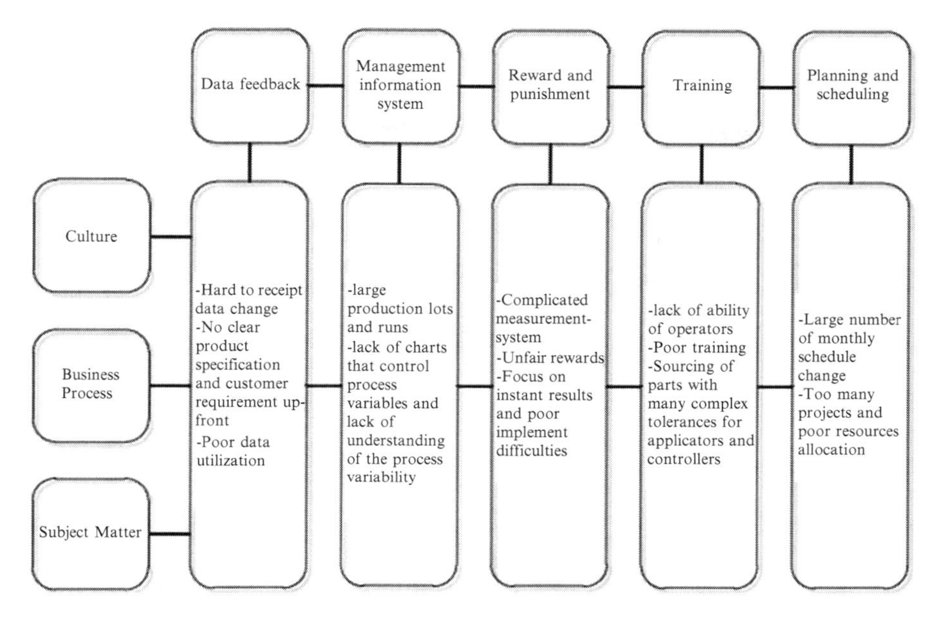

Fig. 3 Matrix structure of barriers

3 Barrier Identification and Removal

3.1 Major Barriers Description

A survey of the impact and remove difficulty of all barriers has been sent to experts across manufacturing and multidisciplinary consultancies, the statistics are concluded in Table 1 (Fig. 4).

The statistical analysis shows that the top six barriers to remove are:

1. Focus on instant results and poor implement difficulties.
2. Functional silos and "nice guy" rules.
3. No clear product specification and customers' requirement.
4. Poor data utilization
5. Superman model and ultra micro-management.
6. Large number of schedule changes

3.2 Strategies to Remove the Barriers

Generally speaking, no matter what the barriers might be, it is important to put together an action-oriented plan to remove them. Here are seven steps that can take to remove these barriers to your investing success.

Table 1 Barrier statistics description

Category	Mark	Barrier description	Who removes	Impact	Difficulty	Rank
Culture	A	Complicated and manipulable measurement system	HR and manager	8.5	6	12
	B	Unfair rewards	HR	8	5.5	9
	C	Reluctance to eliminate personnel	Executive Board	7	7.5	17
	D	Focus on instant results and poor implement difficulties	Manager	8.5	4.5	3
	E	Hard to receipt data change	IT	7	4.5	7
	F	Functional silos and "nice guy" rules	Departments head	7.5	6.5	16
	G	Superman model and ultra micro-management	Manager	6.5	4.5	11
Business process	H	Lack of ownership inventories in different functions	Process	7.5	5.5	13
	I	Paternalistic labor practices	Manager	6.5	5	14
	J	Lack of ability of operators	HR	8	5	6
	K	Poor MIS	IT	7	2	1
	L	Poor training	HR	5.5	3.5	8
	M	Large number of monthly schedule change (planning cycle time and sales power)	Marketing	5.5	6	18
	N	Too many projects and poor resources allocation	Executive Board	8	5.5	9
	O	No clear product specification and customer requirement	Marketing	7	3.5	2
Subject matter	P	Poor data utilization	IT	5.5	3	4
	Q	Large production lots and runs	Manufacturing	5.5	6.5	20
	R	Lack of charts that control process variables and lack of understanding of the process variability	Process and IT	7.5	4.5	5
	S	Poor ability to model the manufacturing process in the pilot stage	R&D	5	5.5	19
	T	Sourcing of parts with many complex tolerances for applications and controllers	Supply	8	6.5	15
	U	Lack of common CAD systems with parts suppliers	IT	5.5	7	21

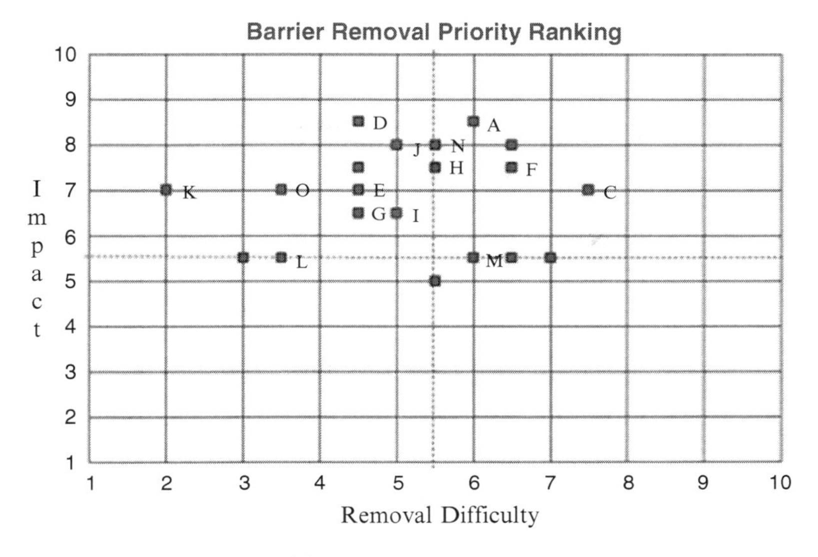

Fig. 4 Barrier removal priority ranking

1. **Learn to monitor the performance**. Measuring the performance creates a track record of what has worked and what has not. This allows the company to identify problems that it repeats.
2. **Once measured the behavior, they can identify what they want to change**. Examine your past trading activity and look for patterns that point to barriers to success. Do you impulsively make decision without doing your homework? Does your rationale for making the business plan prove to be wrong most of the time? The key is to identify the decision-making behavior that hinders your performance [8].
3. **Stay focused on what the Strong Bond needs to change**. Changing one's behavior requires a steadfast focused on what you seek to change. As a manager, he must remain focused on the actions he takes to reinforce the decision-making behavior that he wishes to have. If he feels he is not focused on how to change his behavior, then take a break from his decision until he has regained your focus.
4. **Identify how the company will deal with losses**. Losses are a part of decision. Learning how to deal with them is one of the cornerstones of successful decision behavior. It starts with predefining what your loss looks like through your stop loss and rationale for the trade.
5. **Become an expert at one decision strategy**. There are many ways to assess the market and select the right decision that offer good opportunities. Instead of trying to understand every perspective, it is best to get to know one proven decision strategy [9].
6. **Learn to think in probabilities**. Because the market is in perpetual motion, it places the manager in the position to continually assess the risk-reward of each opportunity. Assessing what is most likely to happen in terms of probabilities will help you make valid decision judgments.

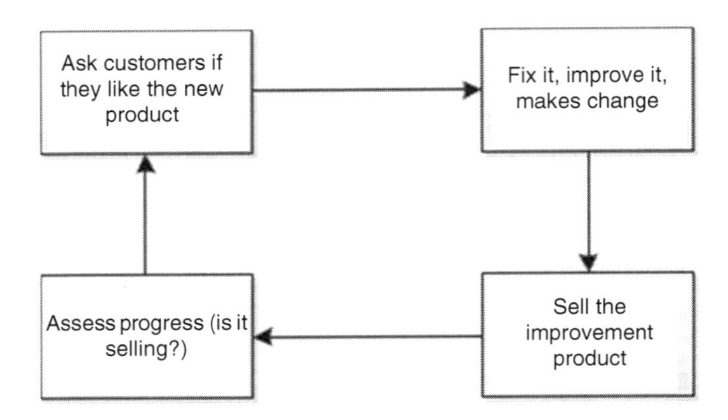

Fig. 5 Customer satisfaction circle

4 Barrier Removal Measurement and Outcome

According to these changes, we can form the cycle below. And in this cycle, customer satisfaction is improved (Fig. 5).

Depending on this working method with the client, the Strong Bond can involve them at different stages in the proposal, implementation, and evaluation stages. They might, after having been left to propose a solution, bring the interested parties together to look at the proposal. Test the proposal themselves with the group by explaining the alternatives they also considered.

By using new production line, cost should be decreased. The changes help integrate the processes in the departments of the company. They can bring together information from the entire company and put it together, also bringing together the external information such that the company has a complete communication chain. The changes also increase efficiency in the company. Since they make the production costs to fall down the company benefits because once the productions costs are down it means the products will be cheaper and the company will have an upper hand than the competitors. The lower the production cost the more customers the company will receive and the more profits.

Employee training is the objective aspect of organizational culture that is most strongly associated with the objective outcomes from sustainable competitive advantage. The statistical results show that employee training within SMEs has a strong association with the positive outcomes that are associated with sustainable competitive advantage. Talent management has a strong association with productivity growth and with the percentage of annual sales derived from new products. Employee participation in empowered work teams shows a strong association with productivity growth and with the percentage of reduction in the total value of inventory throughout the supply chain for the primary product.

References

1. Guojing Xiong, Minjian Zhou (2011) Study about the model of business production logistic and optimizing. Adv Comput Sci Educ Appl 202:519–526
2. Sheriff A, Bouchlaghem D, El-Hamalawi A, Yeomans S (2012) Information management in UK-based architecture and engineering organizations: drivers, constraining factors, and barriers. J Manag Eng 28:170–180
3. Hong D, Suh E, Koo C (2011) Developing strategies for overcoming barriers to knowledge sharing based on conversational knowledge management: a case study of a financial company. Expert Syst Appl 38:417–427
4. Saghiri S (2011) A structural approach to assessing postponement strategies: construct development and validation. Int J Prod Res 49:427–450
5. Ngai EWT, Lai KH, Cheng TCE (2008) Logistics information systems: the Hong Kong experience. Int J Prod Econ 113:223–234
6. Reed R, Defillippi RJ (1990) Causal ambiguity, barriers to imitation and sustainable competitive advantage. Acad Manage Rev 15:88–102
7. Pagell M, Katz JP, Chwen Sheu (2005) The importance of national culture in operations management research. Int J Oper Prod Manag 25:371–394
8. Qingyu Zhang, Vonderembse MA, Jeen-Su Lim (2003) Manufacturing flexibility: defining and analyzing relationships among competence, capability, and customer satisfaction. Oper Manag 21:173–191
9. Wu Chen (2011) Lean manufacturing in a mass customization plant: improvement of Kanban policy and implementation. Massachusetts Institute of Technology, Department of Mechanical Engineering, Cambridge, MA, pp 40–51

Information Technology Investment and Firm Performance in Developing Economies: The Relationship Between Management Practices and Performance

David Phiri and Fang Weiguo

Abstract Over the last two decades, organizations In Namibia and Zambia have been increasingly investing in information technology (IT) largely due to prospects of IT enabled organisational performance improvements. However, despite these assumptions, insufficient validations have been done in the context of these two countries due to limited local research on this topic. The aim of this study is to investigate the performance of local IT investments and the determinants of value creation. A basic premise of this paper is that IT enabled firm performance is influenced by how effective the firm is in using IT resources to support and enhance its core competencies. In accordance with literature the findings indicate that firms in the two countries consider IT investments as vital for superior firm performance. However, IT management practices employed are considerably different from the recommended best practices of IT governance, nevertheless IT investments still performed significantly well partly due to prevalent use of off-the-shelf IT solutions and IT investments leadership and championship by IT and/or line managers. There are also issues with effectiveness of IT deployment and Information system and information quality.

Keywords IT Investments • Firm Performance • IT Impact • IT in Namibia • IT in Zambia • IT firm performance in Developing countries • IT Value creation

1 Introduction

During the last two decades there has been a steady diffusion of information and communication technologies (ICT) in firms from all sectors in Namibia and Zambia. In his research in Namibia in the late 1990s Lubbe Sam [20] notes that a

D. Phiri (✉) • F. Weiguo
School of Economics and Management, Beihang University, Beijing, China
e-mail: mdphiri@gmail.com

Z. Zhang et al. (eds.), *LISS 2012: Proceedings of 2nd International Conference on Logistics, Informatics and Service Science*, DOI 10.1007/978-3-642-32054-5_76,

number of reasons have motivated the trend. He cited among them the increasing intensity of domestic and global competition and ever increasing demand and quest for improved product quality and service delivery. On the part of governments, e-governance is seen as having the potential to improve public service delivery by public institutions towards transparency, accountability, fighting detrimental vices like corruption and responsiveness which is necessary to promote collaborative and joint-up administrations with other stakeholders in the government business [9]. This has led organizations to adopt IT as a means for more efficient and effective ways to achieve their goals. Actually for some organisation IT is an unavoidable necessity for them to continue operating and for most business firms in a possible means for expansion and meaningful participation in international markets.

For developing countries Like Namibia and Zambia the current socio-economic differences from developed economies combine with other complex factors most of which are non-technical to present challenges in IT implementation and exploration [2]. These socioeconomic differences implies that IT implementation and exploration models developed, tested and validated in developed countries cannot be directly replicated to developing countries without further investigation and validation. Perhaps Freeman and Louçã [15] Succinctly puts it better that "successful catch-up has historically been associated not merely with adoption of existing technologies and techniques in established industries but also with innovation, particularly of the organizational kind". Moreover there are remarkable differences in the objectives of firms studied in developed countries and most of those being investigated in the local context.

The organisation of this paper is as follows. The next section elaborates the theoretical framework, and discussion on the research methods followed by presentation of the results and a discussion of the results. Finally the conclusion provides suggestions for considerations by firms and possible further research directions.

2 Theoretical Framework

In recent years, scholars in many fields have sought to rationalise and explain how investments in IT resources and IT capabilities[1] by firms can affect performance or rather create value and potentially serve as sources of sustainable growth and competitive advantage. However, defining what IT value really means has been challenging and a source of contention. There are diverse perspectives on the nature of the benefits from IT to organizations as demonstrated by a meta-analysis of IT payoff variables provided by Kohli and Devaraj [19]. Melville et al. [23] define IT business value (ITBV) as: "the organizational performance impacts of information

[1] The firm's capacity to exploit IT resources to improve operational efficiency and effectiveness as well as to explore using IT resources in order to create novel solutions by pursuing new possibilities

technology at both the intermediate process level and the organizational-wide level comprising both efficiency impacts and competitive impacts". Other researchers have argued that this description is not comprehensive and should be extended to include other dimensions like organizational transformation which should also be seen as a component of the business value resulting from IT and also a driver of further change [16]. It will however take a while and more studies before a comprehensives definition is agreed upon, but for now the definition by Melville et al. [23] with consideration of proposed extensions captures salient characteristics of IT value. We also believe it suffices to define IT value for firms in the two countries being investigated.

2.1 IT Productivity Paradox and It's Implication on This Study

Failure by research to show IT enabled productivity gains has been a recurring theme in literature. IT value researchers have for long and continues to struggle with contradiction between remarkable advances in computer power and its usage and the relatively slow growth of productivity at the level of the whole economy, industry and individual firms leading to coining of the term "IT Productivity paradox" [6, 19]. Early empirical studies found that there was either no relationship, or a slightly negative relationship, between firm performance and IT investments [3, 6]. Some studies found mixed results for the impact IT investment on ITBV [4, 14].

The picture changed however, as subsequent studies incorporated more sophisticated models and extensive analysis of prior studies in testing the relationship between IT and productivity. Generally it was concluded that data, methodological and analytical problems hid "Productivity-revenues" and that Performance output is sometimes difficult to measure, especially in the service sector [22]. This implication is cardinal when investigating IT enabled productivity gains in Zambia and Namibia to ensure no repeat of initial mistakes.

2.2 Research Conceptual Model Overview

Due to lack of research and IT investment and exploration frameworks verified and tested in Zambia and Namibia we have relied on models used in other Developing countries and drawn ideas from those developed and verified in developed countries as well. We identified a number of variables that are critical success antecedents of IT enabled firm performance enhancement. The antecedents variables basically fall in three categories; (1) Leadership: degree of top management support, investment championship & involvement, (2) IS/IT success: actual appropriate IT Usage, System and Information quality, change management and (3) IT/Business strategy alignment: IT/Business function collaboration, training and IT implementation structures and processes.

3 Hypothesis Development

Firstly we acknowledge that there has been numerous scholarly works which have concluded that IT creates positive value [4, 7, 13, 28]. The specifics and the determinants may still be debatable but there is considerable literature emphasizing the innovative capabilities that IT offer as catalysts and enabler of big improvements of existing business processes and work practices, which, in turn under certain conditions can lead to superior firm performance [5, 8, 25]. The surge in IT investments in Zambia and Namibia can also justifiably be regarded as an indication of IT value creation capabilities. This leads to our first hypothesis which.

Hypothesis 1: There is a positive relation between IT investment and firm performance.

In contrast to past studies that have implicitly assumed that IT capabilities have direct effects on firm performance, most recent understanding is that IT can generate value only if deployed so that it leverages pre-existing business and human resources in the firm via co-presence or complementarity. It is assumed that the impact of IT on organizational performance depends not only on IT as such but rather on the alignment or "fit" of IT with other dimensions of the organization such as its strategy, structure, and business processes [21]. Therefore maximizing the value of IT investments requires that these investments link directly to organisational strategic objectives. Hence our second Hypothesis:

Hypothesis 2: Alignment of IT resources with organisational objectives will positively affect performance.

From the foregoing, it is apparent that the delivery of value from IT is dependent on aligning IT with business strategy which requires prudent IT management. Critical success factors for management of IT for success cited in literature include top management support. Basically, researchers have argued that a top management teams that promotes, supports, and guides the IT function is perceived to enhance IT enabled performance impact [1, 26, 31]. For example, Armstrong and Sambamurthy [1] found that "the use of strategic information technologies could lead to strategic advantage subject to management vision and support. When such support is lacking, IT resources will have little effect on performance, even when substantial investments are made." Thus our third hypothesis posits that:

Hypothesis 3: Strong top management commitment to IT will interact with IT resources to positively affect performance.

IT can provide value not only based on how-much is actually deployed but also how successful IS/IT is. According to the DeLone and McLean Model [12] of Information System success, "an Information System is created, containing various features, which can be characterized as exhibiting various degrees of system and information quality. Next, users and managers experience these features by using the system and are either satisfied or dissatisfied with the system or its information products. As a result of this "use" and "user satisfaction," certain "net benefits" will occur, positive net benefits influences and reinforces subsequent "use" and "user

Table 1 Firm performance indicators used in this study

Category of indicator	Indicator variable
Financial/Firm-level indicator	IT impact on improving firm profitability or enhanced service delivery
Efficiency oriented indicators	IT impact on productivity enhancements and cost reduction
Internal oriented capability	IT impact on improving internal processes and controls
External oriented capability	IT impact on improving relations with customer (satisfaction), suppliers and stakeholder confidence

satisfaction"" [12]. This implies that higher levels of appropriate use as a proxy of success of information systems are necessary for IT to impact performance of a firm. Thus our fourth hypothesis posits that:

Hypothesis 4: High level of appropriate use of IT resources and capabilities will positively affect performance.

4 Research Model

4.1 Firm Performance Indicators

Previous researches have used a number of indicators to measure firm performance. Basically these are performance indicators relate to an organisation's capacity to exploit IT resources (IT capabilities) to achieve factors that are important for superior firm performance. In a comprehensive review, Hulland et al. [17] categorises these IT capabilities into internal and external.

1. *Internal capability*: emphasizes on utilizing IT resources to enhance internal control capabilities, strengthening cooperation among departments and supporting operations, including, automation, management decision support and enhancing IT experience [17]. Internal capability also includes ability to facilitate development of new business options or aiding future technology adoption and innovation.
2. *External capability*: concerned with the ability to adapt to the external environment, the ability to work with external partners (such as upstream and downstream suppliers and clients) for cooperation and information sharing, the capacity of facing the market and customer needs promptly. They are mainly concerned with partnership management, market response and organizational agility [17].

Supported by the literature review described above, we have carefully selected four categories of performance variables as shown in Table 1. We believe the selected performance indicators represent key factors necessary for a firm to excel in the context of Namibia and Zambia.

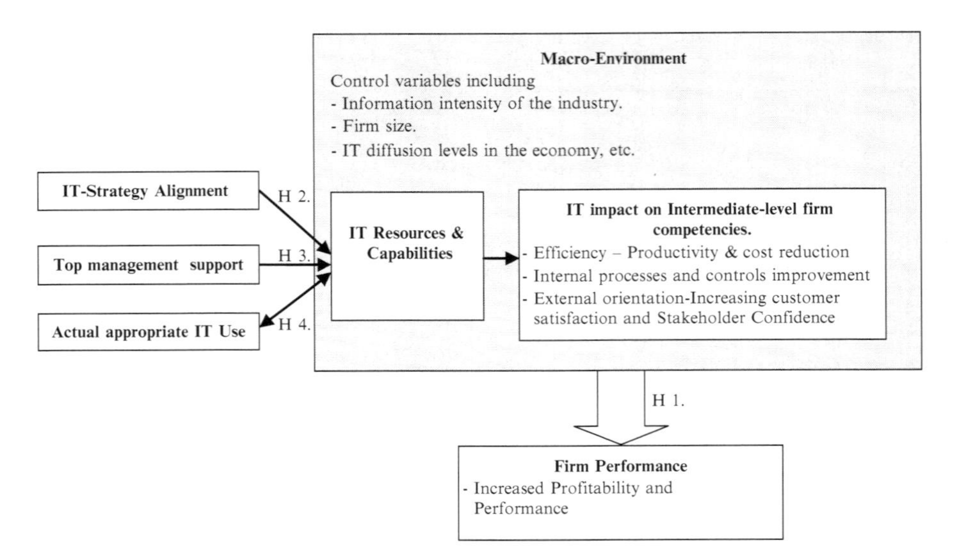

Fig. 1 Research model

4.2 Research Conceptual Model

Guided by detailed review of the relevant literature and we propose an IT value conceptual model that consists of seven (7) compound-variables that reflect antecedents for IT enabled performance and metrics for actual performance outcome at intermediate process level and firm level. We derived a research model shown in Fig. 1. The model includes hypothesized relations between IT resources and IT enabled performance antecedents (Top management support, IT-Strategy alignment and actual appropriate usage as a proxy for IS/IT success) and perceived intermediate and firm level IT enabled benefits. This model is hoped to provide a pertinent framework to investigate the intervening roles our selected IT value determinants.

It should be noted that all inputs and output variables here were regarded to have the same importance, no rating was used distinguish levels of contribution, Furthermore simple average of the variables items was used to form single composite value pertaining to each respective input or output variable. The variables were calculated as follows.

5 Methodology and Data Collection

5.1 Overview of Data Collection and Field Study

To collect data for this study we used a survey instrument that was developed after extensive literature review and subsequently verified and corrected with the help of local IT experts. Essentially most questions that were used required respondents to rate their perception of the impact of IT on different indicators or their rating of how much IT management effort their firm makes. We used 5-point likert scale, with 1 indicating not agreeing and 5 strongly agreeing or 1 indicating no benefits and 5 indicating excellent benefits depending on the questions. However, some open-ended and structured questions relating to details of participating organisations' IT investments (e.g. motivation, type of IT resources, cost, scope and pre and post investment evaluations used) and IT management practices were also used. After the initial data collection, case studies were conducted in four organisations that had participated in phase two. The objective of the case studies were to verify the findings and get deeper understand specific IT value attributes as well as IT governance practices

6 Results and Discussion

The results for the DEA runs and descriptive statistics on our database of 114 records are summarised as shown in Table 2.

6.1 *Limitations*

Before discussing the results, it is important to note the limitations associated with this research. Perhaps the most important limitation is that the results can only be interpreted relative to the inputs and outputs included in this study and restricted to participating firms. Secondly, since we used user perceptions, it is possible some respondents could have misrepresented facts. Nonetheless, we believe we carefully attempted to address each of these limitations as explained in preceding discussions.

6.2 *Analysis of Results and Discussion*

Upon analysis of the results, a number of issues were observed; cardinal among them was that IT governance practices prevailing in Namibia and Zambia differ considerably from the recommended IT governance best practices as well as across different sectors and size of firms. Table 2 presents descriptive statistics for responses to the instrument used during interviews or distributed as a questionnaire. The associations within the research model were tested using Pearson correlations.

It was observed from DEA analysis that Government Departments and Small and Medium (Private) Enterprises (SME) recorded higher proportions of efficient firms compared to Banking/Insurance and Telecommunications/Utilities

Table 2 Results summary

Industry	Number of firms	Number of efficient firms	Percentage of efficient firms (%)	Average % IT-strategy fit (%)	Average %-top Management support (%)	Average %-level of use (%)	Average %-overall perceived impact (%)
Banking/Insurance	21	3	14.3	55.3	65.0	74.0	75.8
Telecommunication and utilities	42	5	11.9	61.1	62.0	69.6	76.6
Government departments	28	11	39.3	32.5	35.1	54.5	65.0
Private SME	23	9	39.1	32.5	9.7	25.3	76.3

as indicated (Table 2). This may seem to contradict other empirical studies that have established Banking industry as one of the most efficient IT users compared to other industries [27]. Nonetheless, since the DEA analysis we performed takes deliberate management efforts as input, it implies therefore that firms that did not do much in terms of management of their resources but recorded some positive impacts were returned as efficient. As seen from summary data in Table 2, most government departments and SMEs recorded significantly high performance impacts but recorded relatively low level of management efforts. For example all firms categorized in the study as SMEs despite making substantial IT investments, performed poorly on measurements on management practices and level actual usage levels.

6.2.1 Hypothesis 1: There is a Positive Relation Between IT Investment and Firm Performance

With regard to the proposed research framework, and consistent with previous studies hypothesis 1 is supported by results. The findings indicate that almost all firms in Namibia and Zambia consider IT as having significant impact on overall firm performance and regard it as strategically essential for organisational survival and growth. As seen from Table 2 in almost all industries IT was perceived as having as high as 76.6% performance impact level on different performance indicators of different firms. Besides direct impact on internal operations, efficiency improvement, external relations and overall organisation performance respondents highlighted a number of other benefits realised from IT investments including:

- Initial investments in IT provided insights into capabilities of IT and most firms indicated that they invested either in further enhancements of initial capabilities or different other systems.
- Most respondents acknowledged that investment in systems supporting internal operations and controls contributed the most benefits.
- Most users acknowledged the strategic need to invest in IT despite lack of immediate payoffs, for the sake of remaining relevant in the new information era.

6.2.2 Hypothesis 2: Good Alignment of IT Resources with Organisational Objectives Will Positively Affect Performance

Apart from establishing whether IT was generating any value the second objectives was to establish the determinants of the value. The results indicate significant (albeit weak) support for IT alignments impact at 0.01 significance level ranging between 0.230 and 0.304 for the all the four output (performance) variables. This finding collaborate findings by other researchers like Chan and Reich [10], Choe [11], and Palmer and Markus [24] who posits IT alignment as moderating factor of IT impact on performance. However, the weak correlation in the local context is somewhat interesting in view of the fact that most researchers have found high correlation between IT alignment and impact. Our observed possible explanation is that IT governance practices in most firms investigated do not meet recommended good practices which we used to operationalise the metrics for IT alignment. As already mentioned we adopted the assertion by Symons [29] and IT Governance Institute [18] that IT alignment results from implementation IT processes (e.g. CobiT framework), structures (e.g. IT Liaison Committees) and facilitation of communication (e.g. Training) as well as Top management participation as part of good IT governance practices. However, most firms did very little towards adopting recommended best practices for achieving IT alignment. Further it was noted that most firms lacked consolidated IT management frameworks and that investments are usually motivated, justified and championed only by IT managers or concerned line managers. In some cases, especially among government departments IT investments were justified outside the firm's management structures. These factors led to low scoring on the IT alignment construct.

Our observation is that the pervasive use of carefully selected Off-the-shelf IT solutions usually chosen after some pre investment selection which in most cases involved vendors of different solution or consultants demonstrating their product is helping firms in developing countries to get away with fairly Aligned solutions even without employing antecedents of alignment. The other reason could be that only a few firms indicated having invested in integrated organisational-wide systems (e.g. ERPs) which usually need prudent management and cross function participation. In this case active involvement of line managers and/or IT managers and in most cases consultants was sufficient to steer IT investment projects.

6.2.3 Hypothesis 3: Strong Top Management Commitment to IT Will Interact with IT Resources to Positively Affect Performance

Top management support did not produce significant correlation (considering the number of records) with any of the four output variables. However, results indicated higher significant correlation with the other input variables e.g. IT alignment and Level of use at $r = 0.643$ and $r = 0.777$ at 1% significance level respectively. This could be an indication that top management support leads to higher usage levels and good alignment. Our findings were that in most firms we investigated, IT investment project ownership and sponsorship was usually the responsibility of either IT managers or line managers. Firm executives especially in government departments did not participate actively in the initiations and subsequent project supervision to ensure successful implementation. Nonetheless, in most cases dedicated spearheading of the projects by IT managers or Line managers was found to be sufficient for the success of the investments. This highlights an important phenomenon, "when IT or Line managers take charge and act as leaders and champions of IT investments, there is chance they will succeed and deliver value even though very senior managers of firms are not themselves directly involved".

6.2.4 Hypothesis 4: High Level of Actual Appropriate Usage of IT Resources and Capabilities Will Positively Affect Performance

The results indicate no support for IT enabled positive impact on firms as a result of High levels of actual appropriate usage of IT resources. By the way actual IT usage was used as a proxy of IS/IT success. Our findings regarding actual use of IT resources were that in most firms investigated there are still a lot of issues regarding system quality, information quality and service quality as well as firms operating parallel system, both manual and automated which leads to lowered actual usage of IT systems. On system quality the problem is twofold; the prevalent vanilla implementation of off-the-shelf solutions implies no or limited contextualising of the solutions which in turn lowers user perception on usability, the second issue is the poor change management and limited trainings given to users which lead to users perceiving Information systems as not good enough. Regarding Information quality, the problem was most prominent among accounting systems were it was found that most systems did not have sufficient reports, whilst some of the reports were not tailor made to the local reporting formats which lead users to supplement what the systems could offer by requesting for ad-hoc reports queried direct from the databases. And finally service quality scored poorly because most organisations had overstretched IT departments in terms of offering user support and in most cases there were no dedicated IT specialist offering support to system users. Considering the limited amount of training and IT skills this led to poor perception of service quality as most users need extra support to work with Information systems.

Nonetheless, although indicators IS/IT success scored low, respondents generally appreciated IT systems impact on several performance indicators. There is however need for more targeted research investigating IS/IT success in the local context. The results may have been inconsistent or showed no correlation due different perceptions of IT systems which could have been influenced by technological readiness, economic and cultural differences with the setups where other studies have been conducted. There is also need to understand how IT is deployed. Taylor and Todd [30] and DeLone and McLean [12] suggest that effective deployment of IT should help ensure high systems, services and information quality leading to high usage levels.

7 Conclusion

In spite of some limitations the findings of this study have significant implications for Namibia and Zambia and developing countries with similar situations in general. This study contributed to addressing the gap that exists between theoretical frameworks, prior empirical research in developed countries, and contemporary practices of IT investments in developing countries. Most importantly findings add to the evidence that IT business value creation is not a phenomenon idiosyncratic to developed countries only but also applicable to developing countries. The findings also established that IT governance, deployment and implementation practices are remarkably different from those recommended in best practices devised in developed countries illuminating areas that need further systematic research. Much as the results indicate almost unanimous recognition of IT performance impact and IT's diverse potential of adding value, there is need for prudent management of IT if the investments are to be optimised and sustained for longer periods. This does not necessarily mean adopting the standards or practices that have worked elsewhere but rather developing models most suitable to local environment which are well informed on the local value determinants.

References

1. Armstrong C, Sambamurthy V (1999) Information technology assimilation in firms: the influence of senior leadership and IT infrastructures. Inf Syst Res 10(4):302–327
2. Barney J (2001) Is the resource-based. View. A useful perspective for strategic management research? Yes. Acad Manage Rev 26(1):41–56
3. Barua A, Lee B (1997) The information productivity paradox revisited: a theoretical and empirical investigation in the manufacturing sector. Ind J Flex Manuf Syst 9(2):145–166
4. Barua A, Kriebel CH, Mukhopadhyay T (1995) Information technologies and business value: an analytic and empirical investigation. Inf Syst Res 6(1):3–23
5. Bresnahan T, Brynjolfsson E, Hitt LM (2002) Information technology, workplace organization and the demand for skilled labor: firm-level evidence. Q J Econ 11(7):339–376
6. Brynjolfsson E (1993) The productivity paradox of information technology. Commun ACM 35 (12):66–77
7. Brynjolfsson E, Hitt LM (1996) Paradox lost? Firm-level evidence on the returns to information systems spending. Manag Sci 42(4):541–558
8. Brynjolfsson E, Hitt LM (1998) Beyond the productivity paradox computers are the catalyst for bigger changes. Commun ACM 41(8):49–55
9. Bwalya KJ (2009) Factors affecting adoption of E-government in Zambia. Electron J Inf Syst Dev Ctries 38(4):1–13
10. Chan EY, Reich HB (2007) IT alignment: what have we learned? J Inf Technol 22(4):297–315
11. Choe J (2003) The effect of environmental uncertainty and strategic applications of IS on a firm's performance. Inf Manag 40(4):257–268
12. DeLone WH, McLean ER (2003) The DeLone and McLean model of information systems success: a ten-year update. J Manag Inf Syst 19(4):9–30
13. Dewan S, Min C (1997) The substitution of information technology for other factors of production: a firm level analysis. Manag Sci 43(12):1660–1675
14. Francalanci C, Galal H (1998) Information technology and worker composition: determinants of productivity in the life insurance industry. MIS Q 22(2):227–241
15. Freeman C, Louçã F (2001) As time goes by: from the industrial revolutions to the information revolution. Oxford University Press, Oxford
16. Gregor S, Martin M, Fernandez W, Stern S, Vitale M (2006) The transformational dimension in the realization of business value from information technology. J Strateg Inf Syst 15(3):249–270
17. Hulland J, Wade MR, Antia K (2007) The impact of capabilities and prior investments on online channel commitment and performance. J Manag Inf Syst 23(4):109–142

18. IT Governance Institute (2003) Board briefing on IT governance. Retrieved 06 Mar 2011, from http://www.itgi.org
19. Kohli R, Devaraj S (2003) Measuring information technology payoff: a meta-analysis of structural variables in firm-level empirical research. Inf Syst Res 14(2):127–145
20. Lubbe S (2000) Information technology investment approaches in Namibia: six case studies. Inf Technol dev 9:3–12
21. Mata FJ, Fuerst WL, Barney JB (1995) Information technology and sustained competitive advantage: a resource-based analysis. MIS Q 19(5):487–505
22. Melone PN (1990) A theoretical assessment of the user-satisfaction construct in information systems research. Manag Sci 36(1):76–91
23. Melville N, Kraemer K, Gurbaxani V (2004) Information technology and organizational performance: an integrative model of ITBV. MIS Q 28(2):283–321
24. Palmer JW, Markus ML (2000) The performance impacts of quick response and strategic alignment in specialty retailing. Inf Syst Res 11(3):241–259
25. Ravichandrasn T, Lertwongsatien C (2005) Effect of information systems resources and capabilities on firm performance: a resource-based perspective. J Manag Inf Syst 21 (4):237–276
26. Ross JW, Beath CM, Goodhue DL (1996) Develop long-term competitiveness through IT assets. Sloan Manage Rev 38(1):31–42
27. Shafer SM, Byrd TA (2000) A framework for measuring the efficiency of organizational investments in information technology using DEA. Omega 28(2):125–151
28. Soh C, Markus ML (1995) How IT creates business values: a process theory synthesis. In: Proceedings of the sixteenth international conference on information systems, Amsterdam, 10–13 Dec 1995, pp 29–41
29. Symons C (2005) IT and business alignment: are we there yet? Forrester, retrieved from http://www.forrester.com/ on 20 Dec 2011
30. Taylor S, Todd PA (1995) Understanding information technology usage: a test of competing models. Inf Syst Res 6(2):144–176
31. Wade M, Hulland J (2004) Review: the resource-based view and information systems research: review, extension, and suggestions for future research. MIS Q 28(1):107–142

A Comparative Study on Predict Effects of Railway Passenger Travel Choice Based on Two Soft Computing Methods

Yan Xi, Li Zhu-Yi, Long Cheng-Xu, Kang Shu, Gao Yue, and Li Jing

Abstract The travelling factors acting on the railway passengers changes greatly with the passengers' choice. With the help of the modern information computing technology, the factors were integrated to realize quantitative analyze according to the travel purpose and travel cost. The detailed comparative study was implemented with the two soft computing method: genetic algorithm, BP neural network. The two methods with different idea, applicable range applicable and the key parameters set were also studied in this model. The analyzed methods were also proved effective and applied for predicting the railway passengers travel choice through the empirical study with soft-computing supporting.

Keywords Railway Passenger • Travel Choice • Genetic Algorithm • BP Neural Network • Comparative

1 Introduction

Railway is an effective way to solve the rapid transportation problem of large number of passengers on a major thoroughfare. In recent years, along with the accelerating process of urbanization in China, there is increasingly demand for transportation between big cities, and China is stepping into the period of great construction and development of railway. The completion of different transportation modes between cities is increasingly fierce especially railway and aviation. In order to better planning and coordination of the overall transportation system, and

Y. Xi (✉) • L. Cheng-Xu • K. Shu • G. Yue • L. Jing
School of Economic and Management, Beijing Jiaotong University, Beijing 100044, China
e-mail: xyan@bjtu.edu.cn

L. Zhu-Yi
Computer Science Engineering Champaign, University of Illinois, Urbana Champaign, Urbana, IL, USA

Z. Zhang et al. (eds.), *LISS 2012: Proceedings of 2nd International Conference on Logistics, Informatics and Service Science*, DOI 10.1007/978-3-642-32054-5_77,

better construction and operation of railway network, so that the railway can make a bigger social and economic benefits, and one of the key issues is derived from the understanding of railway passengers' travel choices.

2 Analysis of Present Research

There are researches for passengers' travel choices, but because of different starting points, apart from the great differences among the above related research methods, models, experimental conclusions and theoretical details, they are also not suitable for the deep-seated reveal of railway passengers' travel choice problem. In existing prediction models, quantitative factors are only for the macro data in a specific section (such as certain administrative areas), ignoring some important details of travel choice that would be considered by residents, such as soft factors like comfort, punctuality, safety etc. Thus the problem information can be considered incompletely, missing some important factors, what's more, not making a scientific and objective analysis for the weight of the factors, and the discussion of the constrains is also very vague in the prediction model. So it is necessary to make comparative research and discussion about the application scope, the relevant parameters and constraints of different soft computing methods. In this paper, Genetic Algorithms and Neural Network are two examples of soft computing methods for a comparative research [1–10].

3 Study on Railway Passenger Travel Choice Prediction Based on Genetic Algorithm

Genetic algorithm (GA) is a random search method. It decreases the effect of original values greatly through crossover and mutation operations, and it can easily find out the global optimal results.

Passengers make travel choices will be influenced by some objective, potential factors. Through analyzing the passengers' characteristics and different factors' effects on passengers, we got a flow chart of passenger travel.

3.1 Model

The problem was described as: there were m travel modes and n batches of passengers (category) waiting to be distributed.

Before the target allocation, the key considerations of each batch of the target and each travel mode's weight on each target has been evaluated and sorted.

J-approved visitors' "travel value" is w_j, i-approved travel mode's weight on j-approved target is p_{ij}, and each travel mode's "trial" benefit value on each target is $u_{ij} = w_j * p_{ij}$. Among them, u_{ij} stands for each batch of the passenger's size of the degree of the effectiveness of the "trial". The purpose is to meet the basic principles of the target allocation and pursuit of the overall effectiveness of the best, which is seeking $\max\left(\sum_{j=1}^{n} u_{ij}\right)$.

3.2 Methods

This paper used binary encoding and the number of individuals was 40. In addition, the max number of generations was 50 and the generation gap was 0.90.

This paper used PN instead of passenger numbers and TV instead of travel value.

Based on the numerical analysis of questionnaires, the standard value of the price dimension is 3.16, the standard value of the time dimension is 2.69 and the standard value of the environmental dimension is 4.47.

Choices for passengers to choose for travel: (1) EMUs, (2) Direct train, (3) Coach, (4) Aircraft, (5) MICE.

3.2.1 According to the Purpose

Passengers for business always focus on convenience and comfort, but have low sensitivity of the cost, so the weight for the price dimension is 0,for the time dimension is 0.6 and for the environment is 0.4, that is, for the purpose of business, "travel value" is 0.6*2.69 + 0.4*4.47 = 3.402. Passengers for tourism often focus on comfort and fare levels, and have high sensitivity of the cost, so the weight for the price dimension is 0.3 and for the environment is 0.7, that is for the purpose of tourism, "travel value" is 0.7*4.47 + 0.3*3.16 = 4.077. Work, school and other commuter passenger traffic often takes fare for the primary consideration, and has certain requirements on punctuality, so the "travel value" for the passengers whose purpose is going to work is 0.6*2.69 + 0.4*3.16 = 2.878, passengers to school's "travel value" is 0.2*2.69 + 0.8*3.16 = 3.066. Passengers for home have low requirements for the comfort, often focus on the price level and have high sensitivity for the costs, so the "travel value" is 1*3.16 = 3.16. Passengers for transfer have high requirements for time, so the "travel value" is 1*2.69 = 2.69. Other passengers' weights are similar, that is, its "travel value" is 0.3*2.69 + 0.3*4.47 + 0.4*3.16 = 3.412 (Tables 1 and 2).

Table 1 Travel value: according to the travel purpose

PN	1	2	3	4	5	6	7
TV(w_j)	3.402	4.077	2.878	3.066	3.16	2.69	3.412

Table 2 Weights: according to the travel purpose

i-j (p_{ij})	1	2	3	4	5
1	0.44	0.01	0.01	0.53	0.01
2	0.19	0.23	0.31	0.22	0.05
3	0.23	0.36	0.18	0.11	0.12
4	0.15	0.42	0.26	0.12	0.05
5	0.15	0.52	0.19	0.13	0.01
6	0.31	0.41	0.26	0.01	0.01
7	0.25	0.25	0.25	0.25	0.25

3.2.2 According to the Cost Mode

Because the travel time is included in the cost of production, passenger traffic at public expense will pay more attention to time and look for convenient, fast and punctuality while they are choosing the travel mode. Therefore they have higher selection bias of civil aviation and high-speed railway. Moreover, they have higher requirements about the frequency of the mode of transportation, departure time and arrival time. And they are less sensitive to the travel cost. So the weight for the time dimension is 0.42 and for the environment dimension is 0.58, that is, the "travel value" is 2.69 + 4.47*0.42*0.58 = 3.7224. However, passenger traffic at their own expense will have lower requirements for the quality of transport and higher sensitivity for the travel costs because they must pay for themselves. So the "travel value" is 0.8 + 0.2*3.16*4.47 = 3.422 (Tables 3 and 4).

3.2.3 According to Income

The previous data showed that middle-income and less income stream of passengers have a preference for the traditional existing rail or road. And for the travel costs are the main considerations, they have a relatively low requirement for the transport quality such as comfort, convenience and punctuality. In addition, they have a relatively high degree of sensitivity for the fare level, so the fluctuations in fares will cause great changes in the passenger traffic distribution. So the "travel value" for the lower-income is 0.9*3.16 + 0.09*2.69 + 0.01*4.47 = 3.1308. And the "travel value" for the low-income passenger is 0.9*3.16 + 0.082*2.69 + 0.02*4.47=3.1486. The "travel value" for the middle-income passenger is 0.5*3.16 + 0.25*2.69 + 0.25*4.47 = 3.37.

Passengers who have high and higher income will take comfort, convenience and punctuality into consideration because they have high abilities to pay, and they always select high quality transportation services, such as high-speed railway, civil aviation. What's more, these passengers are less sensitive to the cost, so a certain

Table 3 Travel value: according to the cost mode

PN	1	2
(w_j)	3.7224	3.422

Table 4 Weights: according to the cost mode

i-j (p_{ij})	1	2	3	4	5
1	0.28	0.39	0.05	0.27	0.01
2	0.12	0.43	0.11	0.17	0.17

Table 5 Travel value: according to income

PN	1	2	3	4	5
TV (w_j)	3.1308	3.1486	3.37	3.5007	3.6327

Table 6 Weights: according to income

i-j (p_{ij})	1	2	3	4	5
1	0.01	0.37	0.11	0.01	0.01
2	0.01	0.57	0.31	0.02	0.01
3	0.01	0.03	0.41	0.05	0.01
4	0.68	0.02	0.07	0.42	0.43
5	0.28	0.01	0.09	0.50	0.54

range of fluctuations in travel mode will have a little influence on them. So the "travel value" for those high-income is $0.21*3.16 + 0.39*2.69 + 0.4*4.47 = 3.5007$, and the "travel value" for the higher-income is $0.15*3.16 + 0.36*2.69 + 0.49*4.47 = 3.6327$ (Tables 5 and 6).

3.3 *Forecast Analysis*

Firstly, because the background is under the spring festival, some predictions may not match exactly with the normal. During the spring festival, different passenger may have different considerations with the usual, such as the price, the time and the environmental dimension. Many people may take "as long as arrive the destination" into main consideration in the pessimistic circumstance, but not for the other factors.

3.3.1 According to the Purpose

Because the passengers for tourism have higher requirements for the time and environment, moreover, most of them are free trips; their sensitivity to the price level is relatively lower. The predicted result showed they would choose plane for their next trips, this is more objective. Passengers for home and transfer have higher requirements for time, so the predicted results are more objective.

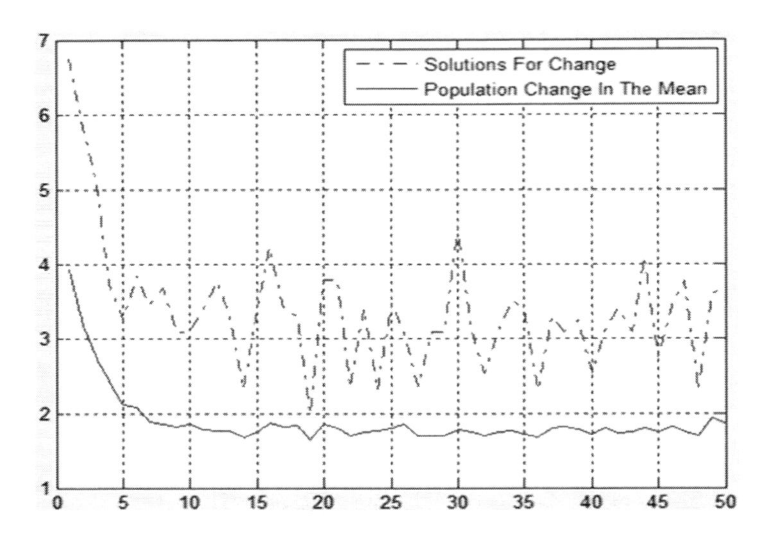

Fig. 1 Predictions: according to the travel purpose

Figure 1 is the application made by Matlab. It is a change tracking map of the total effective value and the mean of the population.

3.3.2 According to the Cost Mode

Passengers who travel at their own expense have low sensitivity for the price, because they do not pay the fees, they will take time or comfort into consideration. So the predictions are more objective. However, the train is less expensive compared with other mode of travel, the passengers who travel at their own expense choosing the direct train is more realistic.

Figure 2 is the application made by Matlab. It is a change tracking map of the total effective value and the mean of the population.

3.3.3 According to Income

In real life, most low-income passengers will choose direct trains as their travel mode for the price is lower, and this is very beneficial to those low-income passengers. However, middle-income passengers will be on a more balanced consideration of all aspects, so the predictions are more realistic. High and higher-income passengers are less sensitivity to the price, what they care about are time, comfort, so the predictions is also close to reality.

Figure 3 is the application made by Matlab. It is a change tracking map of the total effective value and the mean of the population.

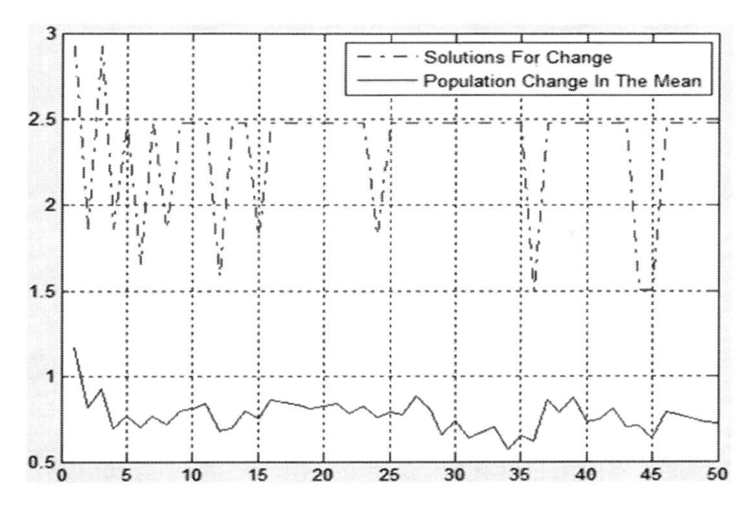

Fig. 2 Predictions: according to the cost mode

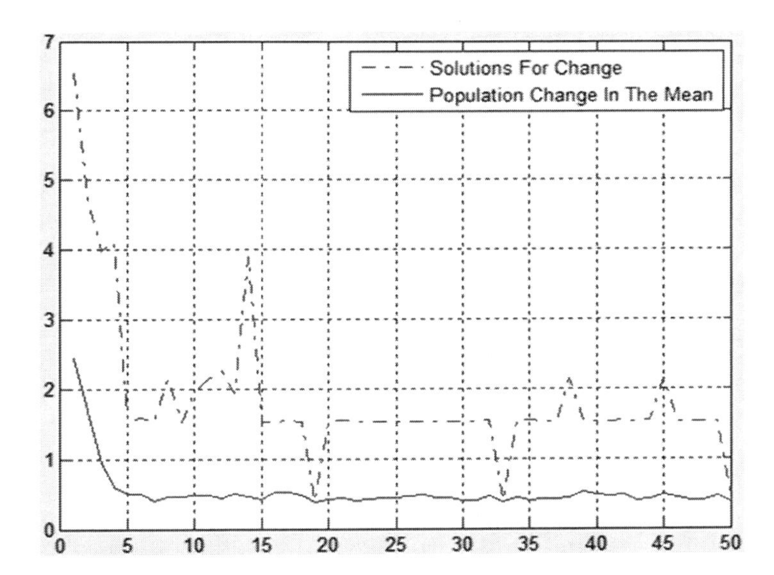

Fig. 3 Predictions: according to income

4 Study on Railway Passenger Travel Choice Prediction Based on BP Neural Network

When using the BP neural network model to forecast the passengers' behaviors, two main steps are built including data processing and the neural network establishment. The concrete algorithm is as follows (Fig. 4):

In order to evaluate the forecast result, the survey data was used to conduct a simulation research based on Matlab 6.6. A BP Neural network consisting of four

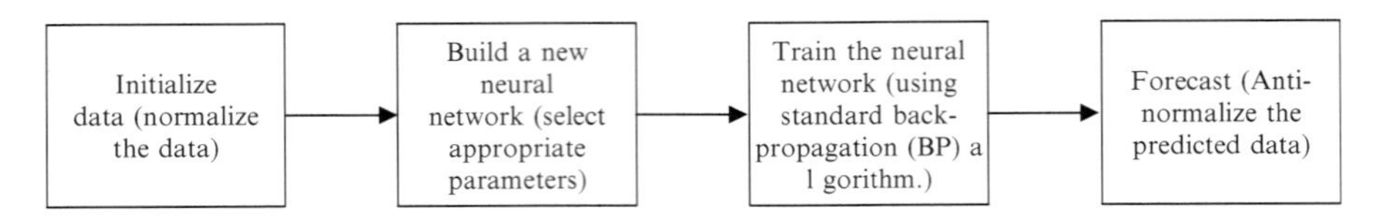

Fig. 4 Algorithm process of the BP neural network forecast model

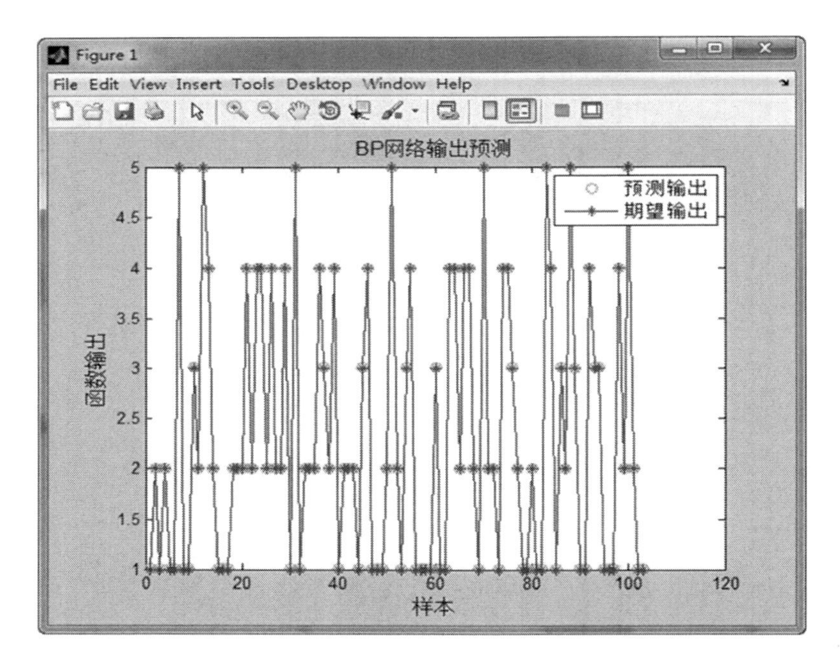

Fig. 5 The BP neural network's forecast result on passengers' travel choices

input nodes, one output node and five nodes in the hidden layer was built and trained to forecast the environmental factors' impacts on High-speed railway passengers' travel choices, and the result is shown in Fig. 5. Figure 6 represents the forecast error in the training and forecasting process with the neural network.

After 1,000 times of learning process, the BP Neural network achieved the best performance with a minimum error of 2.2884e-009, and the network came to a convergence ending the training process. The forecast result concluded in this research is scientific and accurate, demonstrating the validity of using BP neural network in analyzing environmental factors' effects on High-speed railway passengers' travel choices.

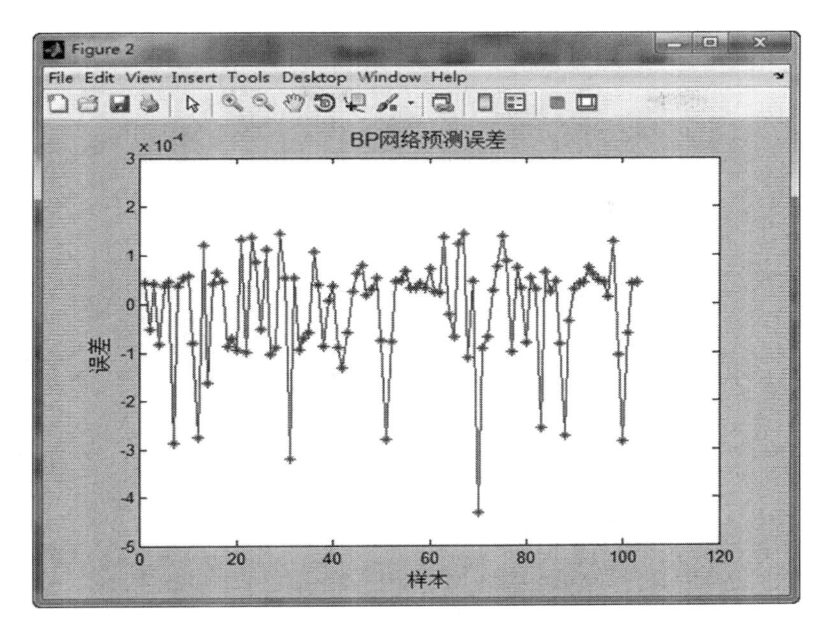

Fig. 6 The forecast error of the BP neural network model

5 Conclusion

We could prove through the empirical study that the soft computing methods applied above are both effective for predicting the railway passenger travel choice. However, the basic idea, the range applicable and the key parameters set in the model are different.

But, in terms of the environmental factors affecting traveling, it is certain that the factors we choose might be incomplete, and we can't eliminate the influences of other factors such as price and time besides the environmental factors on the choices of travel modes. What is more, a better solution of choosing parameters for the established BP neural network may exist. How to establish a more effective neural network model remains a valuable subject. The research is based on the popular topics on environmental factors in recent years, establishing a BP neural network model with a small error and high precision, which could be a beneficial and promising explore in this area. To reiterate, a further experiment and research is expected.

Through the analysis above, the detail from comparative study could be provided for researchers and managers and be applied in the practice according to the actual demand.

Acknowledgment This paper is supported by the National Natural Science Foundation.

References

1. Yao R (2010) The regional economy comes into the era of high-speed railway. Zhejiang Econ 5:43
2. Xingji Jin, Weiwei Jia (2008) Review of researches on artificial neural network. Journal of Jilin Radio and Television University 40(1):65
3. Yingwei Ren, Lingling Song, Shouren Wu (2009) Application of BP neural network in traffic engineering. Traffic Eng 20:136–137
4. Jinqiang Xu, Yuqing Zhang (2006) Urban traffic volume prediction based on BP neural network. Automot Electron 5:23
5. Shiqi Qiu, Qi Wang (2009) Freeway traffic incident detection based on BP neural network. China Meas Test 35(2):48–52
6. He Wei-hui, Wang Jia-Lin (2009) The improvement and application of real-coded multiple-population genetic algorithm. Chin J Geophys [J] 52:2644
7. Ma Yong-jie, Ma Yi-de, Jiang Zhao-yuan, Sun Qi-guo (2009) Fast genetic algorithm and its convergence. Syst Eng Electon [J] 31(3):714
8. Qiang Li-xia, Yan Ying (2006) Analysis on the difference of demand for passenger transport of high-speed railways between home and abroad. Railw Transp Econ [J] 28(9):18–21
9. Wu Qun-qi, Xu Xing (2007) Mechanism research on travelling choices of passengers. J Chang'an Univ (Social Science Edition) [N] 9(2):13–16
10. Chen Zhang-ming, Ji Xiao-feng (2008) Study on features of railway passengers' travel activities. Railw Transp Econ [J] 30(11):23–25

Application of Integer Programming in the Surgical Scheduling

Lihui Dai, Zhenping Li, and Liang Zhao

Abstract This paper addresses the problem of operating room (OR) scheduling by taking into account the factors such as surgeons and his(her) surgical patients with type and time, the admissibility of different surgical types to a certain operating room and the available time of operating rooms. In order to improve operating rooms efficiency and shorten the waiting time of surgical patients, we propose two 0–1 integer programming models, as well as the simulated data, and get the relatively ideal scheduling scheme.

Keywords Operating room scheduling • Integer programming • 0–1 planning • Assignment model

1 Introduction

Increasing costs of healthy services pressure hospital managers into more concerning about the operation efficiency of their organization. In hospital, the operating room department is one of the most significant and expensive resources. According to statistics, more than two-thirds hospitalized patients required surgery. Operating room utilization is directly related to the turnover of the hospitalized patients. Not only the inefficient utilization of OR losses the economic and social benefits of the hospital, but also will increase the economic burden of patients [1].

Supported by Beijing University of Chinese Medicine

L. Dai (✉)
School of Management, Beijing University of Chinese Medicine, Beijing, China
e-mail: wangdlh@gmail.com

Z. Li
School of Information, Beijing Wuzi University, Beijing, China

L. Zhao
Management Office, Peking University Third Hospital, Beijing, China

Z. Zhang et al. (eds.), *LISS 2012: Proceedings of 2nd International Conference on Logistics, Informatics and Service Science*, DOI 10.1007/978-3-642-32054-5_78,

In addition, the hospital OR is always the bottleneck of medical process. Moreover, the surgery which is related to the administrative offices is increasingly intense to the surgery space and the time competition, along with the increasing of specialized and complex degree as well as the new application of technology, the OR operation cost is unceasingly climbing [2]. How to use the appropriate managerial technique method by arrangement of the existing OR and its corresponding resources to increase the efficiency of OR, decrease time and reduces the operation cost? This is one of the important topics in current Chinese hospital management domain.

One of the most direct and effective methods improving the OR working efficiency is to optimize surgical scheduling reasonably. A surgical scheduling is to determine the time and the place of the various departments of the surgery refers to a specific scheduling rules in a scheduling period, and to arrange the appropriate medical staff and medical equipments [3]. On the one hand, the operating room scheduling can solve varied OR management decision-making problems at different levels, on the other hand, it may solve the problem of allocating all the opening hours of all the operating theater between the different surgical intervention departments during a certain period, allocating 1 day opening hours of a certain operating theater among different surgeons and patients in a scheduling cycle as well as arranging all the surgery orders on certain OR opening day [4]. This paper shall discuss the issues according to the second category of OR scheduling.

Brecht Cardoen, Erik Demeulemeester and Jeroen BeliÄen [4] provided an updated overview on operating room planning and scheduling that captures the recent developments in this rapidly evolving area. In some papers, they used an programming method to solved many OR scheduling problems with the reality of hospitals [5–9]. Mostly hospitals obtained the significantly decrease time and reduces the operation cost by their methods.

2 OR Scheduling Problem

The available time for hospital ORs is coordinated by all surgical clinical departments. However, the OR time which hospitals allocate to different clinical departments is relatively fixed in a certain period of time. It usually assign the surgery opening hours to specific clinical departments per week. Since the hospitals will normally reserve certain operating rooms for emergency surgery, in this paper, we will consider the problem of elective surgery arrangements. Namely, each department should arrange the data and the place for its elective surgeries according to distribution of surgery type and time, pre-determined surgeons (team) responsible for every patient needing surgery. Therefore, surgeons (team) and surgical patients should be arranged coordinately rather than only for surgical patients. For example, when the Ors are free, a patient may not be arranged for the surgery because of the confliction.

2.1 Problem Description

A department is known to have M-team of doctors responsible for the K-type surgery, the surgery types which each surgeon team is responsible for, and the status of patients waiting for surgery (the number of days of admission, disease, etc.) are known, the expected time and useable ORs of various kinds of surgeries are known, the daily opening hours for every surgery departments are known, then how to arrange the surgeries of different departments to try to shorten the preoperative length of stay?

2.2 Problem Assumption

In this study, according to the actual situation of a hospital in Beijing, we assume the following:

Operating rooms:

1. The Ops are equipped with independent nurses and anesthetists. Their working hours are consistent with the actual use of operating room time;
2. Certain types of surgery for surgical equipment and personnel may have the special requirements. So, different operating room is equipped with the different surgical equipment and personnel. Assume that the type of surgery which can be carried out in each surgery is pre-qualified.

Doctors:

1. For each doctor, the expected time of implementation of the same type of surgery may be not the same;
2. The patient who is in charged by one doctor may need to implement different types of surgery. So, the surgery which is responsible by the same doctors may not be carried out in a particular operating room;
3. Every doctor arrange surgery in a surgery days, the longest duration does not exceed the maximum value of the operating room daily normal opening hours;
4. Try to arrange the same doctor in a surgery days using an operating room. If they want to improve operating room efficiency and use a different operating room, then the operating rooms should not be exceeded to two;

Patients:

1. Only arrange the surgery for those patients who stayed in the hospital and are satisfied all the surgery conditions;
2. Each patient operation time only has the relationship with the type of surgery and the doctor who is responsible for his surgery.

2.3 The Aim of Problem

The problem aims to shorten the total time of patients with surgery. Namely, it improves the operating room utilization efficiency, as much as possible to arrange the surgery within one business day; taking into account the principle of fairness, that is, as far as possible the first admission of patients preferred arrangements surgery. Therefore, the initial goal of the model sets all the arrangements for the day of surgery in patients with the longest total length of stay. The number of total hospitalization time from that day to arrange surgery and its been hospitalized joint decision, so this goal has the efficiency and fairness of the two principles. However, we consider the various surgical daily opening hours are limited. So under the above objectives, some types of surgery may be delayed because of the long time operating time. For example, we suppose that the T1 surgery needs 5 h, the T2 operation needs 2 h, and the opening hours of the operating room are in total 10 h. Then, the patients who stayed in the hospital 3 days may be arranged for the surgery after those who stayed in the hospital 2 days. Obviously, this is contrary to the principle of fairness to a first-come-first-served basis. To avoid some delayed surgery, we give greater weight for some types of surgery which need much time in the objective function. Three Operating room scheduling 0–1 programming model.

3 The Programming Model for Operating Room Scheduling

3.1 Explanation of Symbols

$\boldsymbol{m}$: the number of surgeons in the department
$\boldsymbol{R}$: the number of Ors in the department
$\boldsymbol{K}$: the number of surgery types
$\boldsymbol{n}_{ik}$: the number of people qualified for the kth surgery responsible by group ith doctor
$\boldsymbol{b}_{ikj}$: the number of days in the hospital of jth patient for the kth surgery
$\boldsymbol{t}_{ikr}$: the time for the ith doctor, for the kth surgery in the rth surgery room, if it is not available $\boldsymbol{t}_{ikr} = 100$
$\boldsymbol{t}_k$: the average (standard) time for kth surgery
$\boldsymbol{M}_r$: the open hour for the rth surgery
$\boldsymbol{M}$: the maximum time of surgery for every group of doctor

$$y_{ikjr} = \begin{cases} 1 & \textit{Arrange } k^{th} \textit{operation, for } j^{th} \textit{patient, using } r^{th} \textit{OR by group } i^{th} \textit{doctor} \\ 0 & \textit{otherwise} \end{cases}$$

$$i = 1, 2, \ldots, m;\ j = 1, 2, \ldots, n_i;\ r = 1, 2, \ldots, R$$

$$x_{ir} = \begin{cases} 1 & i^{th} \textit{ surgeon used } r^{th} \textit{ OR} \\ 0 & \textit{otherwise} \end{cases}$$

3.2 Build the Following Models

$$\max\ C = \sum_{r=1}^{R} \sum_{i=1}^{m} \sum_{k=1}^{K} \sum_{j=1}^{n_{ik}} t_k\, b_{ikj}\, y_{ikjr}$$

$$s.t. \begin{cases} \sum_{i=1}^{m} \sum_{k=1}^{K} \sum_{j=1}^{n_{ik}} t_{ikr} y_{ikjr} \le M_r, \quad r = 1, 2, \ldots, R & (1) \\ \sum_{r=1}^{R} \sum_{k=1}^{K} \sum_{j=1}^{n_{ik}} t_{ikr} y_{ikjr} \le M, \quad i = 1, 2, \ldots, m & (2) \\ \sum_{r=1}^{R} y_{ikjr} \le 1, \quad i = 1, 2, \ldots, m;\ k = 1, 2, \ldots, K;\ j = 1, 2, \ldots, n_{ik} & (3) \\ \sum_{k=1}^{K} \sum_{j=1}^{n_{ik}} y_{ikjr} \le 100 x_{ir}, \quad i = 1, 2, \ldots, m;\ r = 1, 2, \ldots, R & (4) \\ \sum_{r=1}^{R} x_{ir} \le 2, \quad i = 1, 2, \ldots, m & (5) \\ y_{ikjr} = 0, 1; \quad x_{ir} = 0, 1 & \end{cases}$$

Subject to:

1. Time used in the rth surgery room cannot exceed the maximum open time.
2. The surgery time for each group doctor cannot exceed ***M***
3. Each patient has at most one surgery everyday
4. and 5. The surgery rooms are at most two for each group of doctor

All eligible patients into the surgical schedule on the day the principle of maximizing the total benefits in accordance with arrangements between all the open surgeries, the doctors group and the group into the day surgery is scheduled

Table 1 Standard time of each operation type and doctor group (hour)

	Operation type 1	Operation type 2	Operation type 3	Operation type 4
Doctor group 1	1.8	2.4	2.0	–
Doctor group 2	1.7	–	2.1	3.0
Doctor group 3	–	2.5	–	2.8
Doctor group 4	1.7	2.2	–	–
Doctor group 5	1.8	–	2.2	2.9
Standard time	1.75	2.37	2.10	2.90

Table 2 Daily opening hours to the department in each operating room (hour)

	Monday	Tuesday	Wednesday	Thursday	Friday
Operating room A	10	10	0	10	10
Operating room B	10	10	10	0	10
Operating room C	10	0	10	10	10

Table 3 Usage of the various types of surgery on the operating room

	Operating room A	Operating room B	Operating room C
Operation type 1	√	√	√
Operation type 2	√	√	√
Operation type 3	×	√	√
Operation type 4	×	×	√

patients to determine the rth surgery should be arranged, as well as surgery between the date of total operative time. Failed in the day arrangements for the surgery patients, the number of hospitalized days and the next day's surgery is scheduled and unified with the second day of patients with surgical conditions in accordance with the above model.

4 Example

Assume that a department of the hospital, a five doctors group is responsible for a total of four types of surgery, the doctor is responsible for the type of surgery and at different times, are shown in Table 1, the standard time of each type of surgery is the mean value of the operation time for the doctor group; the time of using the operation room of three departments are relatively fixed, but the sections of the opening hours may be different, see Table 2, and the different types of surgery can be used in surgery varies, see Table 3.

Table 4 Surgical arrangements for eligible patients on Monday

Doctor group	Patient number	Operation type	Available operation room	Days of hospitalization	Operating room arrangements	Operation time
1	1	1	A/B/C	3	A	1.8
	2	1	A/B/C	3	A	1.8
	3	2	A/B/C	1	A	2.4
	4	3	B/C	1	–	–
2	1	1	A/B/C	1	B	1.7
	2	1	A/B/C	1	B	1.7
	3	3	B/C	4	B	2.1
	4	4	C	1	–	–
3	1	2	A/B/C	3	C	2.5
	2	2	A/B/C	1	–	–
	3	4	C	3	C	2.8
4	1	1	A/B/C	1	A	1.7
	2	2	A/B/C	3	A	2.2
5	1	1	A/B/C	1	–	–
	2	1	A/B/C	3	C	1.8
	3	3	B/C	1	B	2.2
	4	3	B/C	4	B	2.2
	5	4	C	1	C	2.9

Note: "–" in column 6, 7 means no operation that day

Table 5 Surgical arrangements for eligible patients on Tuesday

Doctor group	Patient number	Operation type	Available operation room	Days of hospitalization	Operating room arrangements	Operation time
1	1	1	A/B/C	1	B	1.8
	2	2	A/B/C	2	A	2.4
	3*	3	B/C	2	B	2
2	1	1	A/B/C	1	A	1.7
	2	3	B/C	1	B	2.1
	3	4	C	2	–	–
	4*	4	C	2	–	–
3	1	2	A/B/C	1	–	–
	2	2	A/B/C	1	–	–
	3*	2	A/B/C	2	A	2.5
4	1	1	A/B/C	1	A	1.7
	2	1	A/B/C	2	A	1.7
5	1*	1	A/B/C	2	B	1.8
	2	3	B/C	2	B	2.2
	3	4	C	1	–	–

Note: "*" in column 2 means arrangements for patients transferred from the previous day

Table 6 Surgical arrangements for eligible patients on Wednesday

Doctor group	Patient number	Operation type	Available operation room	Days of hospitalization	Operating room arrangements	Operation time
1	1	1	A/B/C	2	–	–
	2	2	A/B/C	3	B	2.4
2	1	1	A/B/C	1	–	–
	2	3	B/C	2	–	–
	3*	4	C	3	C	3
	4*	4	C	3	C	3
3	1	2	A/B/C	2	B	2.5
	2*	2	A/B/C	2	B	2.5
	3*	2	A/B/C	2	B	2.5
4	1	1	A/B/C	1	–	–
5	1	4	C	3	C	2.9
	2*	4	C	2	–	–

Table 7 Surgical arrangements for eligible patients on Thursday

Doctor group	Patient number	Operation type	Available operation room	Days of hospitalization	Operating room arrangements	Operation time
1	1	1	A/B/C	2	A	1.8
	2*	1	A/B/C	3	A	1.8
	3	3	B/C	2	C	2
2	1*	1	A/B/C	2	A	1.7
	2*	3	B/C	3	C	2.1
	3	3	B/C	1	–	–
3	1	4	C	2	C	2.8
4	1*	1	A/B/C	2	A	1.7
	2	1	A/B/C	1	–	–
	3	2	A/B/C	3	A	2.2
5	1	1	A/B/C	1	–	–
	2*	4	C	3	C	2.9

Table 8 Surgical arrangements for eligible patients on Friday

Doctor group	Patient number	Operation type	Available operation room	Days of hospitalization	Operating room arrangements	Operation time
1	1	2	A/B/C	2	B	2.4
2	1	1	A/B/C	1	B	1.7
	2*	3	B/C	2	C	2.1
3	1	4	C	3	B	2.8
4	1	1	A/B/C	1	C	1.7
	2*	1	A/B/C	2	C	1.7
	2	2	A/B/C	2	B	2.2
5	1*	1	A/B/C	2	C	1.8
	2	3	B/C	2	C	2.2

Applied algorithm by Lingo software, simulated the data above from Monday to Friday, the arrangement is shown in Tables 4, 5, 6, 7 and 8.

5 Discussion

In this paper, 0–1 programming model which minimizes patients waiting time is used for surgery scheduling; it considers the principle of fairness on first-come-first-served basis, and taking into account the resource utilization of the operating room.

In this paper, we consider the problem, t_{ikr} time for the ith doctor, for the kth surgery in the rth surgery room is a very important parameter in the operating room arrangement. Its value for the surgeon group, type of surgery and the patients' individual circumstances are closely related, and thus this time is often not identified. The estimated duration of activities in practical applications can be estimated by using the "Program Evaluation Review Technique" in [9].

M, opening hours of rth operating rooms, can be adjusted according to various hospitals and departments of patients. If there are more patients waiting for operation, M can be appropriately extended to avoid the phenomenon of surgery backlog.

References

1. Cao X, Dong J et al (2003) Discussion of statistical indicators and methods of operating room utilization efficiency. China Health Stat 20(2):82–83
2. Zhao L, Jin C et al (2008) Management practices to maximize efficiency and effectiveness of operating room. Chin Hosp Manag 28(10):43–45
3. Shu W, Luo L (2008) Study of programming-based surgical scheduling. Technol Mark 2:42–44
4. Cardoen B, Demeulemeester E, Beliën J (2010) Operating room planning and scheduling: a literature review. Eur J Oper Res 201:921–932
5. Jerrold HM, William ES (2011) The surgical scheduling problem: current research and future opportunities. Prod Oper Manag 20(3):392–405
6. Blake JT, Dexter F, Donald J (2002) Operating room managers' use of integer programming for assigning block time to surgical groups: a case study. Anesth Analg 94(1):143–148
7. Blake T, Donald J (2002) Mount Sinai hospital uses integer programming to allocate operating room time. Interfaces 32(2):63–73
8. Dexter F, Epstein RH, Marsh HM (2001) Statistical analysis of weekday operating room anesthesia group staffing at nine independently managed surgical suites. Anesth Analg 92:1493–1498
9. Zhou X (2004) Project management tools of the series ten – program evaluation review technique: estimating project tasks duration. Proj Manag Tech 4:64–65

Ranking the Technical Requirements of the Airport for Maximum Passenger Satisfaction

Sema Kayapınar and Nihal Erginel

Abstract Airport service quality is crucial due to the increasing traveling by plane, nowadays. There are some studies on the passengers satisfaction in literature. But lots of them are related to determine the passengers' needs and expectations and their importance levels. This paper presents the service quality approach for ranking the technical requirements that meet passengers' needs and expectations for the Anadolu University Airport in Turkey. Firstly, the service quality is measured with the questionnaire. The questionnaire is organized to collect the passenger's both expectations and perceptions, and is evaluated by using SERVQUAL model. Then, Quality Function Deployment (QFD) approach is used for setting the relationships between the passenger requirements and the technical requirements, and between technical requirements of the airport. Finally, the technical requirements are ranked by the calculation method for the maximum passenger satisfaction.

Keywords Service quality of the airport • Servqual model • Quality Function Deployment

1 Introduction

In today's raising travel around the world, travel by plane is one of the most preferred traveling types. So, passenger satisfaction is the curricle topic for the airport management. Several authors mentioned the importance of passenger satisfaction like as follows: "Passenger satisfaction is a key performance indicator for the operation of an airport. International airports located at different regions or countries by and large do not compete with one another" [1]. Fodness and Murray proposed a model on service quality in airports by using passenger expectations

S. Kayapınar • N. Erginel (✉)
Department of Industrial Engineering, Anadolu University, Eskisehir 26555, Turkey
e-mail: nerginel@anadolu.edu.tr

Z. Zhang et al. (eds.), *LISS 2012: Proceedings of 2nd International Conference on Logistics, Informatics and Service Science*, DOI 10.1007/978-3-642-32054-5_79,

which were combined by the qualitative research. Quantitative research was used to develop a self-report scale to measure passenger expectations of airport service quality, to test dimensionality and to evaluate scale reliability and validity [2]. Humphreys et al. [3] has reviewedcurrentpractice in performancemeasurement of airports. According this study, the performance measurement can be divided into three main categories which named as business measures, service measures, and environmental measures. Yeh and Kou [4] developed a new fuzzy multi attribute evaluation model with an effective algorithm to obtain an overall service performance index for each of Asia-Pacific's 14 major international airports, based on multiple passenger service attributes. Service attributes were named by "comfort", "processing time", "convenience", "courtesy of staff", "information visibility", and "security".

In this study, the expectations and the perceptions of the passengers at Anadolu University Airport in Turkey are investigated with the questionnaire and the gap between them is analyzed by the SERVQUAL model. After that, passengers satisfactions criteria are classified and handled as a WHAT part of the first house of QFD. Gap values are used considering the importance levels of passengers satisfactions. Then technical requirements (HOW's of the QFD) are determined to meet to the passenger's needs and expectations, and the relationships between them are determined by QFD team. Finally, the technical requirements are weighted and ranked by considering these relations.

2 SERVQUAL and QFD Methods for Customer Satisfaction

2.1 Designing the Questionnaire According to SERVQUAL

Developed questionnaires are arranged according to the dimensions of servquals. The tool for measuring the airport quality derived from six dimensions named as "terminal facilities", "personnel", "accessibility", "responsiveness", "easy access to service", and "assurance". Servqual dimensions [5] are modified as regarding of the airport service quality features. The first dimension Tangible was divided to three parts named as a "terminal facilities", "personnel" and "accessibility". Empathy dimension is also covered in "responsiveness". A new dimension "easy access to service" was also added. Six modified dimensions include 30 service quality attributes developed by the airport experts and the review of the past literature.

The questionnaire is a restructure of the original Servqual model in order to fit the airport industry. 30 items-questionnaires include airport service quality dimensions consistent with the Servqual dimensions. Questions addressing perceptions and expectations were rated by using 5-point Likert scale. The scale is conducted as 1 = unimportant to the 5 = very important. For purpose of data analyses and hypothesis testing, SPSS 16.00 was used.

Table 1 Reliability result

Service quality dimensions	Expectation Cronbach alpha	Perception Cronbach alpha
Total	0.873	0.885
Terminal facilities	0.771	0.726
Personnel	0.910	0.872
Accessibility	0.742	0.814
Responsiveness	0.919	0.911
Easy access to service	0.886	0.784
Assurance	0.851	0.844

The local international airport placed in the middle-part of Turkey, was chosen as an application. This airport passenger can reach only two countries that are Istanbul in Turkey, Belgium and other countries with connecting flights from Istanbul. In terms of destination, the most popular is Belgium, especially in summer days. The survey was carrying out during 1 month in 2011 at May-June. It targeted the passengers who used the airport at least one time and more. The questionnaire was made face to face with 150 passengers waiting in lounge and 135 passengers' questionnaire was applied correctly. The participation was voluntary. The response rate was almost 90%. The questionnaire was examined by passengers at the lounge and their contributions were collected. The questionnaire was initially tested by 30 passengers. After re-designing of the questionnaire, the reliability (Cronbach's alpha were 0.873 for expectations and 0.885 for perceptions) was suitable. Each dimension's reliability was also calculated. According to the reliability result, the content validity of survey was viewed adequately (see Table 1).

2.2 *Analyzing the Questionnaire and Handling the SERVQUAL Scores*

According to the analyses of demographics, the gender distributions were 36.3% female and 63.7% male. The majority of respondents were Turkish (86.7%) and other Nationality (13.3%). The passengers (48.9%) traveled to Istanbul, Belgium (43.7%) and other destinations (connection flights) (7.3%). Descriptive statistical methods were used in this study. The means, standard deviations, the difference were computed for each attributes. The Servqual score (PM-EM) for each customer needs was calculated by the difference between perception means – expectation means. In accordance with the results of the study, the customer attributes with the first two high expectations and perceptions scores were Q30-To feel safe and peaceful at the airport and Q29-Small number of damage and baggage loss. The larger gaps scores were, Q9-Avaibility of fight information display (−2.2889); Q5-Variety and number of shopping stores (−2.2518); Q7-Availability of ATM cash machine and exchange office (−2.2148); Q10-Avability of call/internet service to reply passenger desire and problems. All of the larger gaps are contained within the

Table 2 Servqual score of each dimension

Factors	Perception mean	Expectation mean	Gaps mean	t-test	p-value
Terminal facilities	2.43	3.99	−1.544	31.456	0.000
Personnel	3.74	4.16	−0.4173	5.202	0.000
Accessibility	3.16	4.15	−0.9853	16.082	0.000
Responsiveness	3.48	4.02	−0.5333	5.541	0.000
Easy access to service	3.35	4.22	−0.8666	10.661	0.000
Assurance	3.90	4.31	−0.4099	6.001	0.000

"Terminal Facilities" dimensions. This result shows that the passengers expected a higher level of service with the "Terminal Facilities".

Gaps means (Servqualscore) of each dimension was given by Table 2. It revealed that "Terminal Facilities" (−1.544) has the highest gap mean while the "Assurance" has the minimum gaps mean (−0.4099). According to the results, airport management must be developed firstly in terminal facility's needs. At the end, the overall Servqual scores for service quality is found as −0.7921.

Paired t-test was set to investigate whether there are any significant differences between overall means for the six quality dimensions.

Hypotheses

$$H_{01} : \mu_D = 0$$
$$H_{11} : \mu_D \neq 0$$

The results of test are given in Table 2 and evaluated with a 5% significant level. P-values in Table 2 are smaller than the significant level (p value $= 0.00 < \alpha = 0.5$); H_{01} is rejected, there is a significant difference between expectation and perception means for all dimensions.

2.3 Connection of the Results of the SERVQUAL Model and QFD Approach

The connection of SERVQUAL model and QFD approach is obtained by the House of Quality. House of Quality (HoQ) is the first house of QFD. He customer needs and expectations are handled on the left column ("Why's section") of the HoQ. In the service quality of airport studies, passengers' needs and expectations are set on the Why's section of the HoQ. The importance levels of passengers' needs were also got from questionnaire with 5-scale Likert. The importance and satisfaction level was assigned by the Gaps (Perception mean – Expectation Mean) of each passenger's needs and expectations.

Table 3 Ranking of the technical attributes according to their weights

Ranking order	Technical requirements	Weights
1	The number of the restaurants and cafes	8.26
2	The number of shopping centers	7.50
3	The number of flight information boards	7.45
4	The number of signboards	6.25
5	SMS information service	6.23
6	The number of passenger service vehicles	5.98
7	Capacity of wi-fi	4.68
8	Presence of web-site	4.48
9	The number of emergency phones	4.40
10	The number of technical training	4.39

2.4 Determining the Technical Requirements and Constructing the Relationship Matrix

The technical requirements for the airport service are determined by brainstorming carried out by the members of the QFD team that include the manager and technical persons of the airport. They were assigned in the "Hows" section of House of Quality. Each of passenger's needs is related individually to at least one technical requirement. The QFD team evaluated the relationship between passengers' needs and technical requirements. This matrix between the rows and the columns was defined using a value (9 for strong relationship, 3 for medium relationship and 1 for weak relationship) to each cell of these two attributes. Also, the interaction between technical requirements were determined by QFD teams and set to the roof of HoQ. In this application, positive and negative degrees of relationship were considered for each pair of the technical requirements.

2.5 Ranking the Technical Requirements of the Airport

The importance of technical requirements and relative weights of each of the technical requirements are computed by multiplying the importance levels and the value in relationship matrix and summing these values based on the columns. These weights were converted to the relative weights of each technical requirement by normalization method. These weights are located at the bottom of the relationship matrix.

According to the relative weights, the number of restaurants and cafes, the number of flight information boards, and the number of shopping centers have greater priority more than others in this case (Table 3). Therefore, this airport should consider the airport design projects. This information can help the airport experts to improve the service design projects.

3 Conclusion

Airport population has increased day by day. The airport management endeavor to improve passenger satisfaction. This study has attempted to develop a conceptual service quality approach for the airport services. Proposed "Service Quality Approach" is connected with SERVQUAL and QFD to determine the weights of technical requirements. Also, airport managers can easily reach the improvements which should be made for the maximum passengers satisfaction. Further of this study, cost of improvement can be considered with linear programming models.

References

1. Doganis R (1992) The airport business. Routledge Press, London
2. Foddness D, Murray B (2007) Passengers' expectations of airport service quality. J Serv Mark 21(7):492–506
3. Humphreys I, Francıs G, Fry J (2002) Performance measurement in airports: a critical international comparison. Public Work Manag Policy 6(4):264–275
4. Yeh CH, Kuo Y (2005) Evaluating passenger services of Asia-Pacific international airports. Transp Res Part E 39:35–48
5. Parasuraman A, Zeithaml V, Berry L (1998) SERVQUAL: a multiple-item scale for measuring consumer perceptions of service quality. J Retail 64(1):12–40

The Impacts of Network Competence, Knowledge Sharing on Service Innovation Performance: Moderating Role of Relationship Quality

Zhaoquan Jian and Chen Wang

Abstract This research contributes to existing literature by examining how network competence (NC), knowledge sharing (KS) and relationship quality (RQ) affect service innovation performance (SIP). The sample used in this empirical research is drawn from the Pearl River Delta of China. The results show that: (1) Enterprise's network competence has a distinct positive impact on SIP; (2) Knowledge sharing partially mediates the effect of network competence on SIP. (3) Relationship quality positively moderates the effect of network competence on knowledge sharing, and the effect of knowledge sharing on SIP. (4) Relationship quality does not positively moderate the effect of network competence on SIP. These results enrich current understanding of the relationships among network competence, knowledge sharing, relationship quality and service innovation performance.

Keywords Service Innovation (SI) • Service Innovation Performance (SIP) • Network Competence (NC) • Relationship Quality (RQ) • Knowledge Sharing (KS).

1 Introduction

In the twenty-first century, the era of innovation-based "knowledge economy" has set in; the creation, spread and application of knowledge has become the main driving force to promote the progress of the times; the scale, complexity and interdependence of today's service systems have been driven to an unprecedented level. The rising significance of service and the accelerated rate of change mean that service innovation (SI) is now a major challenge to practitioners in business and

Z. Jian (✉) • C. Wang
School of Business Administration, South China University of Technology,
Guangdong 510640, P.R. China
e-mail: jianzq@163.com; lynn_gofighting@qq.com

Z. Zhang et al. (eds.), *LISS 2012: Proceedings of 2nd International Conference on Logistics, Informatics and Service Science*, DOI 10.1007/978-3-642-32054-5_80,

government as well as to academics in education and research. A better understanding of service systems is required.

Most previous studies focus on manufactory industry, with little attention paid to A service industry. Empirical findings in the innovation literature are limited and inconclusive regarding SI antecedents [1]. Based on existing literatures, this paper aims to explore in depth the interacting influence of network competence, relationship quality, and knowledge sharing on SIP, which contributes to the theory of service science and RBV and provides some managerial implication for company to achieve high SIP.

2 Theory Foundation and Research Hypotheses

2.1 *Influence of Network Competence on Service Innovation Performance*

The initial view of service innovation (SI) is attributed to Schumpeter [2]. Later, the notion came to be regarded as the set of innovations in service processes for an organization's existing service products. Build on the existing literature, we summarize SI as enterprises' intangible activities formed in the process of service, using a variety of innovative ways to meet customer needs and maintain competitive advantage. Topics such as the performance measurement of SI projects remain under-researched. Fizgerald et al. found that SIP is multi-dimensional, not only can reflect the effectiveness of the company's operations, but also reflect a project plan, or the level of the overall development process [3].

The concept of network competencies (NC) and capabilities is derived from the Resource Based View of the firm. We argue that the ability of a firm to develop and manage relations with key suppliers, customers and other organizations and to deal effectively with the interactions among these relations is a core competence of a firm. Previous research had made some achievements on the relationship between NC and SIP. Möller, Kristian and Aino Halinen proposed a network management framework and showed that unique and dynamic networks can improve SIP [4]. Through an empirical research, Ritter and Gemünden found that NC has a strong positive impact on technological collaborations and firm's innovation success [5]. Based on previous studies, we propose:

Hypothesis 1: Network competence is positively related to SIP.

2.2 *Mediating Effects of Knowledge Sharing*

Contingency theory has been used in many contexts, particularly in the field of strategic actions and organizational structure. It examines the effects of related variables (e.g., strategy and business model) on firm performance [6]. We delineate two fundamental strands of contingency theory: the "fit-as-mediation" view and the "fit-as-moderation" view [7]. According to the fit-as-mediation view, when faced with keen competition, organization's predominant response is to aggressively pursue innovation through collaboration. However, we focus in this study on the aspects of a firm's knowledge sharing that account for the effect of network competence on SIP.

Recent research attempts to understand alliance activities from a knowledge-based perspective and posits that the sharing of knowledge becomes central to develop new processes, products, or services in alliance [8, 9]. The argument that inter-firm collaboration enhances innovation practices has gained wide acceptance. Thus, we propose that well-developed mechanism of knowledge sharing can enhance innovation practices and act as a mediator between network competence and service innovation performance. Following contingency theory, we suggest the following hypothesis:

Hypothesis 2: Knowledge sharing mediates the impact of network competence on SIP.

2.3 Moderating Effects of Relationship Quality

Structural contingency theory emphasizes both external and internal fit. Thus, the second strand of contingency theory is the fit-as-moderation view [6]. This view proposes that a firm's performance is attributable to a match between its strategic behaviors and environment conditions. In this view, firm performance is the dependent variable, network competence and knowledge sharing the predictor variables, and relationship quality the contextual variables.

"Relationship quality" was firstly proposed and defined by Crosby, who defined relationship quality as the overall evaluation of the strength of buyer-seller relationship [10]. Based on the previous study, it's an integrated and realistic way to understand the relationship of the intangible resources with SIP [11]. Ritter and Gemünden suggested that with the development of the network, only when managing network relationships, can the company promote information sharing between partners, learn from each other better and have complementary advantages to improve effectiveness and efficiency [5]. The following hypotheses are offered for testing:

Hypothesis 3: Relationship quality positively moderates the influence of knowledge sharing on SIP.

Hypothesis 4: Relationship quality positively moderates the influence of network competence on knowledge sharing.

Hypothesis 5: Relationship quality positively moderates the influence of network competence on SIP.

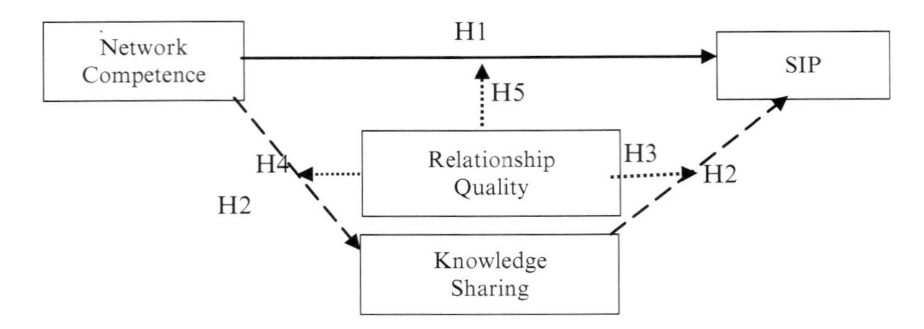

Fig. 1 The research model

3 Method

3.1 Research Framework

Based on previous research findings, interviews and group discussions, the research framework model shown in Fig. 1 is developed.

3.2 Variable Definitions and Measurement

In order to ensure the validity and reliability of measurement tools, we adopt the used scale in published literatures as much as possible, and do some proper modification according to this research's purpose. Before the final questionnaire and investigation, we do some pilot study to evaluate the questionnaire design and accuracy of the word expression, and then make corresponding modifications.

The scale measuring network competence is based on questionnaire of Ritter and Gemünden [5], including task implementation and qualification. The measurement scale of relationship quality is mainly according to Roberts et al. [12], Garbarino and Johnson [13]. Knowledge sharing is in the light of studies of Davenport and Prusak [14], Gupta and Govindarajan [15]. SIP makes reference to the studies of Bock et al. [16], Storey and Kelly [17].

3.3 Research Samples

This study analyzes data at the firm level. Both the sample and the variables used in this analysis come from the Pearl River Delta of China firms' survey. The sample includes six high-tech industries. From July, 2010 to January, 2011, we had sent out 485 questionnaires by mail or door to door interview, with the response rate of 60.2%, and the effective response rate of 50.1%. Structure of the sample firms is sufficiently diverse and heterogeneous.

3.4 Reliability and Validity of the Samples

The scale was developed from prior research and interviews with practitioners. All constructs were measured using a five-point Likert scale to assess the degree to which the respondent agreed or disagreed with each items. Factor loadings, Composite reliability (CR), and Cronbach's alpha are indicative level of measurement reliability. CR value above 0.5 indicates adequate reliability; the least value of CR in the survey exceeds 0.83, which suggests an acceptable level. In this study, the Cronbach's alpha values of each constructs exceed the suggested level of 0.7, showing internal consistency of each construct.

On the validity, the items in the questionnaires of this research are all from the published literatures, and we also did some modification according to some experts and pre-test. We assessed the factorial validity through CFA. A construct with either loadings of indicators above 0.5, or a significant t-value above 2.0, or both, is considered to have convergent validity. We assessed convergent validity using the average variance extracted (AVE), which for the constructs all exceeded 0.50, confirming satisfactory convergent validity. The values of the square root of the AVE for the measures in the diagonal were all greater than the correlations among the measures off the diagonal. Hence, discriminant validity was satisfactory.

4 Results

4.1 Results for the Direct Effects

A bootstrapping technique was used to determine the significance of the structural paths. The path coefficients for the research constructs are expressed in a standardized form. The predictive power of the research model was assessed by examining the explained variance (R2) for the endogenous constructs. For most firms, the positive relationship between network competency and SIP was significant ($b = 0.54$, $t = 10.07$, $p < 0.001$). Thus, Hypotheses 1 was supported. Therefore, the significant hypotheses explained a substantial amount of the variance in the endogenous constructs.

4.2 Results for Mediating Effects

To assess the extent of mediation in the model, we followed Andrews et al. [18], who indicated that four specific criteria must be met. In this study, the independent variable was NC, KS being the proposed mediating variables, and SIP being the dependent variable. As shown in Table 1, Model 1 did not include the mediator of KS. Model 2 results showed that entering the mediator of KS indeed decreased the

Table 1 Regression results

Model	Independent variables	Beta	t	R2	F	ΔR2
1	Constant		0.06	0.30	101.41***	
	NC[1]	0.54	10.07***			
2	Constant		0.07	0.42	87.62***	0.126
	NC	0.40	7.43***			
	KS	0.38	7.23***			
3	Constant		−0.20	0.45	38.32***	0.025
	NC	0.35	6.44***			
	KS	0.36	6.56***			
	RQ	0.11	2.07*			
	RQ*NC	−0.09	−1.52			
	RQ*KS	0.14	2.41*			
4	Constant		0.00	0.15	42.3***	0.149
	NC[2]	0.39	6.50***			
5	Constant		−0.62	0.20	20.05***	0.052
	NC	0.35	5.65***			
	RQ	0.19	3.16**			
	RQ*NC	0.13	2.24*			

Note: In Model 1, 2, 3 knowledge sharing as mediator, relationship quality as moderator, SIP as dependent variable; in Model 4, 5 relationship quality as moderator, knowledge sharing as dependent variable
$^{*}p < 0.05$; $^{**}p < 0.01$; $^{***}p < 0.001$

impact of NC from b = 0.54 to 0.40. In particular, the impact of NC on SIP was diminished, indicating partial mediation. Correspondingly, KS partially mediated the relationship between NC and SIP; thus, Hypothesis 2 was supported.

4.3 *Results for Moderating Effects*

The moderating effects models (see Table 1) tested the extent to which relationship quality moderated the main effect hypothesized in Hypothesis 1 and 2. We mean-centered NC, KS, SIP and RQ. Then we added the interaction terms from Model 2 to Model 3. As shown in Table 1, Model 3 indicated that the interaction term of RQ × KS had a significant positive moderating effect on the association between KS and SIP; Model indicated that the interaction term of RQ × NC had a significant positive moderating effect on the association between NC and KS. Thus, Hypothesis 3 (b = 0.14, p < 0.05) and Hypothesis 4 (b = 0.13, p < 0.05) were supported, which confirms the moderating role of relationship quality. However, Hypothesis 5 was not supported. It comes out a surprise the RQ's moderating effect is negative. When the RQ is higher, NC will not influence SIP so much, it imply that relationship and competence are alternative in China, although the moderating effect doesn't have a statistical significance (b = −0.09, P > 0.10) (see Figs. 2 and 3).

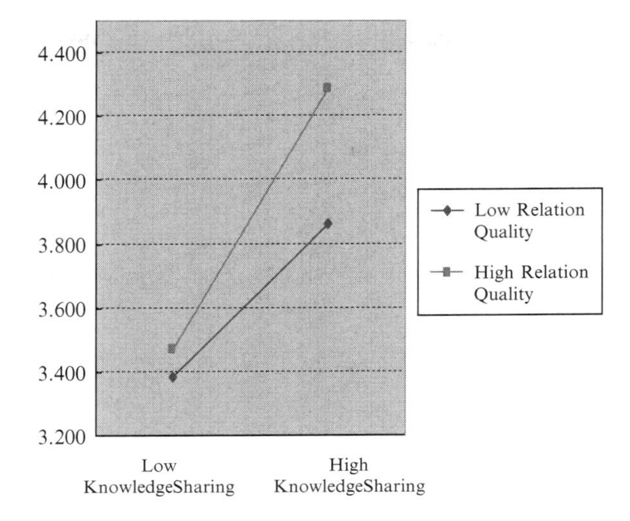

Fig. 2 Moderating effect of RQ on the relation of KS and SIP

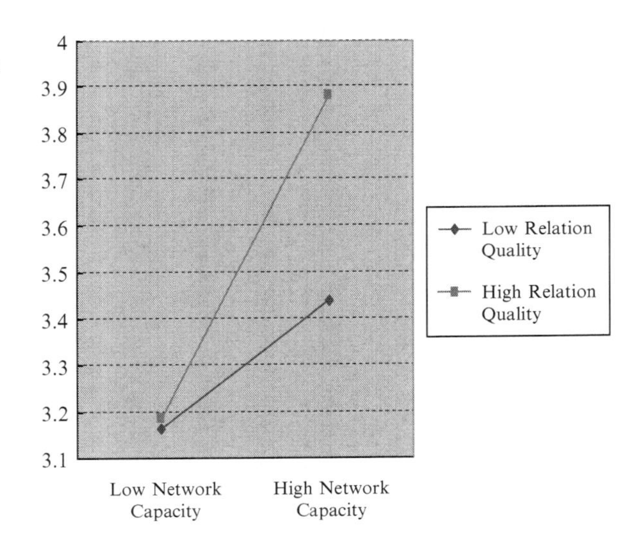

Fig. 3 Moderating effect of RQ on the relation of NC and KS

5 Conclusion

With the literature review and case interview, this paper constructs a theoretical model and studies the relationships among NC, KS, SIP and RQ, selecting 102 high-tech firms in Pearl River Delta as the empirical research sample. Statistical analyses present some interesting findings as follows: (1) Enterprise's NC has a distinct positive impact on SIP, and KS partially mediates the effect of NC on SIP. (2) RQ has a positive moderating effect on the relations between NC and KS, and between KS and SIP. (3) The hypothesis that RQ positively moderates the relation between NC and SIP is rejected. These results enrich current understanding of the relationships among NC, KS, RQ and SIP. The research has some new findings as well as some questions that need further discussion. (1) This study is based on the six high-tech industries. In the future, it is better to make empirical analysis to some other industries and compare the differences. (2) The samples are all from Pearl River Delta of China, thus, the successive study could do a more extensive investigation in other areas, such as the Yangtze River Delta of China, Bohai Economic Rim, etc. New findings may appear in the further empirical research.

Acknowledgment This research is supported by the National Natural Science Foundation of China (No. 70872030 & No. 71090403/71090400).

References

1. Ordanini A, Parasuraman A (2011) Service innovation viewed through a service-dominant logic lens: a conceptual framework and empirical analysis. J Serv Res 14(1):3–23
2. Schumpeter JA (1934) The theory of economic development: an inquiry into profits, capital, credit, interest and the business cycle. Harvard University Press, Cambridge
3. Fizgerald L, Johnston R, Silvestro R, Brignall TJ, Voss C (1991) Performance measurement in service business. CIMA, London
4. Möller KK, Halinen A (1999) Business relationships and networks: managerial challenge of network era. Ind Mark Manag 8:28–49
5. Ritter T, Gemunden HG (2004) The impact of a company's business strategy on its technological competence, network competence and innovation success. J Bus Res 57:548–556
6. Galbraith J (1973) Designing complex organizations, reading. Addison-Wesley, Reading
7. McAdam R (2002) Knowledge management as a catalyst for innovation within organizations: a qualitative study. Knowl Process Manag 7(4):233–241
8. Fleming L, Chen D, Mingo S (2007) Collaborative brokerage, generative creativity, and creative success. Adm Sci Q 52:443–475
9. Hoang H, Rothaermel FT (2005) The effect of general and partner-specific alliance experience on joint R&D project performance. Acad Manage J 48(2):332–345
10. Crosby LA, Cowles D (1990) Relationship quality in service selling: an inter-personal influence perspective. J Mark 54(7):54–82
11. Jaw C, Lo JY, Lin YH (2010) The determinants of new service development: service characteristics, market orientation, and actualizing innovation effort. Technovation 30 (4):265–277
12. Roberts K, Varki S, Brodie R (2003) Measuring the quality of relationships in consumer services: an empirical study. Eur J Mark 37:169–196
13. Garbarino E, Johnson MS (1999) The different roles of satisfaction, trust, and commitment in customer relationships. J Mark 63(2):70–87
14. Davenport TH, Prusak L (1998) Working knowledge: how organizations manage what they know. President and Fellows of Harvard College, Boston
15. Gupta AK, Govindarajan V (2002) Knowledge flows within multinational corporations. Strateg Manag J 21:473–496
16. Bock GW, Zmud RW, Kim YG, Lee JN (2005) Behavioral intention formation in knowledge sharing: examining the roles of extrinsic motivators, Social Psychological Forces, and Organizational Climate. MIS Q 29(1):87–111
17. Storey C, Kelly D (2001) Measuring the performance of new service development activities. Serv Ind J 21(2):71–90
18. Andrews JC, Netemeyer RG, Burton S, Moberg DP, Christainsen A (2004) Understanding adolescent Intentions to smoke: an examination of relationships among social Influences, prior trial behaviors, and anti-tobacco campaign advertising. J Mark 68(3):110–123

Study on Profits Distribution of Internet of Things Industry Value Chain Led by Operators

XiYan Lv, RunTong Zhang, and Haizhou Sun

Abstract The paper summarized Significance and principles of the distribution of benefits of Internet of Things and analyzed the different role of leader and other subsidiary of the participants of value chain, then constructed the distribution of profits model of Internet of Things value chain led by the operators based on the game theory. By comparing the benefits between the operators led value chain and all participants on the equal status value chain, we show that total profits of value chain has an advantage when it is led by operators.

Keywords Internet of Things • Value chain • Profits distribution • Game theory

1 Introduction

The Internet of Things refers to a global network infrastructure, linking physical and virtual objects through the exploitation of data capture and communication capabilities. This infrastructure includes existing and evolving Internet and network developments. It will offer specific object-identification, sensor and connection capability as the basis for the development of independent cooperative services and applications [1, 2].

However, Most of the studies focus on the separate player of the Internet of Things, research for the value chain as a whole is still in its infancy. As global markets grow increasingly efficient, competition no longer takes place between individual businesses, but between entire value chains collaboration [3, 4].

X. Lv (✉) • R. Zhang • H. Sun
School of Economics and Management, Beijing Jiaotong University, Beijing 100044, China
e-mail: lvxiyan@bjtu.edu.cn; rtzhang@bjtu.edu.cn; 123wsshz@163.com

Z. Zhang et al. (eds.), *LISS 2012: Proceedings of 2nd International Conference on Logistics, Informatics and Service Science*, DOI 10.1007/978-3-642-32054-5_81,

2 The Internet of Things Industry Value Chain and Its Leading player

According to Porter's value chain theory, every enterprise is the joint of the value chain. From the perspective of the value, the industrial value chain could be defined as the activities cost chain during the process of economic activities related to the execution of the productive management and operation. Hence, the value chain of the Internet of Things is, without any doubt, the profit chain formed due to series of economic activities [5].

From the business perspective, the composition of the industry value chain includes the sensor/chip manufacturers, network equipment providers, software and application developers, system integrators, network providers and operators. The industry value chain will generally be divided into three parts of the upstream, downstream and midstream.

The leading player of industry value chain refers to the key player in a leading position and a driving role in the development of the industrial chain. According to the current research on the Internet of Things, Operators can be selected to take the leading role in the industrial chain and can promote the industry alliance formed by the upstream and downstream cooperation [6, 7].

3 The Profit Distribution Model: Operators Led vs. All Players on the Equal Status Value Chain

3.1 Assumptions and Variables

1. The total income of Internet of Things industry value chain is R, total profits is π
2. The efforts made by operators ($A1$) and other players of the industrial chain ($A2$) for the gains of Internet of Things, are recorded as a_1, a_2
3. The fixed remuneration to other players paid by operators is T
4. The profit sharing ratio between $A1$ and $A2$ in the total income of the Internet of things industry chain, are recorded as $1 - S$ and S
5. The fixed and variable costs of $A1$ and $A2$ recorded as $C(a_1)$ and $C(a_2)$, fixed cost coefficients are C_1 and C_2, variable cost coefficients are α, β. The further assumption is: variable costs of $A1$ and $A2$ and total income of Internet of Things industry value chain are quadratic function of the level of effort, and plus all the functional coefficient of 0.5 is for the computation of simple convenience.

From the above assumption of total income R and profits of Internet of things industry chain π can be expressed as:

$$R(a_1, a_2) = \frac{1}{2}(a_1 + a_2)^2 + (a_1 + a_2)$$

$$\pi(a_1, a_2) = R(a_1, a_2) - C(a_1) - C(a_2)$$

The profits of *A1* and *A2* are:

$$Q_1 = (1 - s)\cdot R - T - C(a_1)$$

$$Q_2 = T + s\cdot R - C(a_2)$$

The variable costs of *A1* and *A2* are:

$$C(a_1) = C_1 + \frac{1}{2}\alpha {a_1}^2$$

$$C(a_2) = C_2 + \frac{1}{2}\beta {a_2}^2$$

3.2 The Game Model and Equilibrium Solutions

Parties of the Internet of Things industry value chain will choose different strategies in the competition for their own interests, it will form different game behaviors. This section will study the two cases: sequence game of Operators led industry chain and game of all players on the equal status in the industry chain [8].

3.2.1 The Profit Distribution Model: Operators Led Industry Chain

In the game of Operators led industry chain, the leading operators can determine the proportion of the distribution of profits $(1 - S)$ and maximize their own interests by adjusting efforts and actions, other players will decide their own level of effort according to the value determined, and the prerequisite is: the other members of the whole industry chain must be profitable.

According to Stackelberg game backward induction method, the game should start from the second stage.

Firstly, in order to get the profit maximization of other members of the industry chain

$$\frac{\partial Q_2}{\partial a_2} = s[(a_1 + a_2) + 1] - \beta a_2 = 0$$

Then:

$$a_2 = \frac{s(a_1 + 1)}{\beta - s}$$

Secondly, Operators adjust their own efforts and the determination of S to pursue their own maximized interests according to the other members of the level of effort:

$$\begin{cases} \dfrac{\partial Q_{a_1}}{\partial a_1} = \dfrac{\beta^2(1-s)(a_1+1)}{(\beta-s)^2} - \alpha a_1 = 0 \\ \dfrac{\partial Q_{a_1}}{\partial s} = \dfrac{2\beta^2(1-s)(a_1+1)^2 - (\beta-s)(\beta^2 a_1{}^2 + 2\beta^2 a_1 + 2\beta s - s^2)}{2(\beta-s)^3} = 0. \end{cases}$$

From above equation, the optimal effort s of Operators a_1^1 and Profit sharing ratio s^1 can be deduced, then the optimal effort of other members of industrial chain is a_2^1

3.2.2 The Profit Distribution Model: All Players on the Equal Status in the Industry Chain

In the case of all players on the equal status in the industry chain, Operators and other member companies determine a common recognition of the profit sharing ratio by negotiation, then participating members pursue to maximize their own profits with the appropriate strategy.

$$\frac{\partial Q_{a_1}}{\partial a_1} = (1-s)[(a_1 + a_2) + 1] - \alpha a_1 = 0$$

$$\frac{\partial Q_{a_2}}{\partial a_2} = s[(a_1 + a_2) + 1] - \beta a_2 = 0$$

Then Optimal effort of *A1* and *A2 are*:

$$a_1^2 = \frac{(1-s)\beta}{\alpha\beta - (1-s)\beta - s\alpha}$$

$$a_2^2 = \frac{s\alpha}{\alpha\beta - (1-s)\beta - s\alpha}$$

Thus, when the System Integrator and Operator are in static game, the optimal strategy, namely Nash equilibrium solution is:

$$\frac{\partial \pi}{\partial s} = \frac{\partial \pi}{\partial a_1} \cdot \frac{\partial a_1}{\partial s} + \frac{\partial \pi}{\partial a_2} \cdot \frac{\partial a_2}{\partial s} = 0$$

The profit sharing ratio of Operators is:

$$1 - s = \frac{\beta - 1}{(\alpha - 1) + (\beta - 1)}$$

The profit sharing ratio of other members is:

$$s = \frac{\alpha - 1}{(\alpha - 1) + (\beta - 1)}$$

The only one Nash equilibrium is (a_1^2, s^2, a_2^2)

3.2.3 The Data Analysis of Models

In this section, simulation data will be used to compare efforts, the profit sharing ratio and total income of different game model.

Assumption: both parties in the game are risk-neutral; fixed cost C_1 of $A1 = 0.2$, Cost coefficient of $A1$: $\alpha = 4$; fixed cost C_2 of $A2 = 0.1$, Cost coefficient of $A2$: $\beta = 2$; the fixed remuneration to other players paid by operators $T = 0.02$.

Then by the above formula the following game data can be obtained: in the game of Operators led industry chain, profits of operators Q_1 is greater than profits of other members Q_2 ($0.1015 \succ 0.0254$); in the case of all players on the equal status in the industry chain, Q_2 is greater than Q_1 ($0.2253 \prec 0.2857$), the total profits of Internet of Things industry value chain π_1 is greater than π_2.

4 Conclusion

The paper is dedicated to the application of the game theory upon the quantitative model research of the Internet of Things industrial value chain. From the above comparisons, it indicates that when the Operators obtain a dominant position in the industry value chain, it will get more profit and total profit of whole industry will be greater than the case of all players on the equal status.

The conclusions of the study are based on two participants of the industrial chain and not cover all the participants, while the real case was often a number of industrial participants will interact in a complex network. These issues means there is further research requirements on the profit distribution of Internet of Things industry value chain.

Acknowledgements This Research was supported by "the Fundamental Research Funds for the Beijing Jiaotong University" under grant No. 2012JBM049.

References

1. Gershenfeld N, Krikorian R, Cohen D (2004) The internet of things. Sci Am 291(4):76–81
2. Du Jin (2012) Application of "internet of things" in electronic commerce. Int J Dig Content Technol Appl 6(8):222–230
3. Yao Weixin (2006) Atomic models of closed-loop supply chain in e-business environment. Int J Bus Perform Manag 8(1):24–35
4. Debo L, Savaskan C, Wassenhove LN (2003) Chapter 12: Coordination in closed-loop supply chains. Quantitative approaches to reverse logistics, Springer, Berlin
5. Murayama T, Hatakenaka S, Hashimoto M (2006) Simulation-based evaluation of forward and reverse supply chain. Trans Jpn Soc Mech Eng Part C 72(8):2621–2628
6. Huang Haiping, Wang Ruchuan, Qin Xiaolin, Sun Lijuan, Jin Yichao (2011) A novel internet of things' paradigm for smart home using agent-based solution with hybrid intelligence. J Dig Content Technol Appl 5(10):136–151
7. Cox A (1999) Power, value and supply chain management. Supply Chain Manag 4(4):167–175
8. Tayur S, Magazin M, Ganeshan R (1999) Quantitative models for supply chain management. Kluwer Academic Publishers, Boston, pp 229–336

Equilibrium Services of Telecom Operators: An Idea of Service Resources Allocation

Mengru Shen, Feng Luo, and Jianqiu Zeng

Abstract Based on the service resources allocation problems that exists in telecom operators' practical work, the idea of "equilibrium production" is applied from manufacturing to service, and an idea of "equilibrium service" is put forward, which aims at finding out the optimal allocation path and strategy of service resources, proposes equilibrium service model with the value of practical application and makes scientific and reasonable suggestions for telecom operators to enhance their service quality.

Keywords Equilibrium Service • Service Management • Service Resource Allocation

1 Introduction

In recent years, the trend of mobile communication business is becoming markedly homogeneous, while service-level competitions become more and more intense. With the constant level of resources quantity, how to effectively allocate service resources is becoming an important issue for all telecom operators. With the diversification of service channels and forms, "inequilibrium service" problems in telecom operators have become increasingly prominent,and the traditional "balance of production" theories cannot meet the demand of market development and theoretical support any more.

Supported by China National Key Technology R&D Program "The application demonstration of home shopping service based on HD interactive TV platform" (No.2011BAH16B07)

M. Shen (✉) • F. Luo • J. Zeng
School of Economics and Management, Beijing University of Posts and Telecommunication, Beijing, P.R. China
e-mail: buptsmr@gmail.com

Z. Zhang et al. (eds.), *LISS 2012: Proceedings of 2nd International Conference on Logistics, Informatics and Service Science*, DOI 10.1007/978-3-642-32054-5_82,

Therefore, an idea called "equilibrium service" is proposed in this paper based on the "balance of production", which focuses on service resource allocation problems in telecom operators for improving customer satisfaction, effective use of resources and service competitiveness. Based on optimal resources allocation theories, the following contents are focuses on: (1) Proposing "equilibrium service" concept and tracing the source of its theory. (2) Forming the theoretical model of "equilibrium service", summarizing the impact factors. (3) Summarizing the in equilibrium service problems and their forms, and proposing strategies.

2 Service Status of Telecom Operators

Productions and services in telecom industry are easy to be imitated, which drives operators to pursue differentiation and facilitation in services providing. However, the non-optimum allocation of service resources in operators limited the promotion of benefit and the sustainable development.

Lacking of equilibrium service often leads to unbalanced customer satisfaction between different service channels, and this deficiency mainly reflects in service proportion, service quality, service cost, service pattern, service superposition and other aspects. On account of a large number of investigations, expressive forms of inequilibrium service are summed up in following aspects: (1) Services quality and customers' satisfaction between channels are unbalanced. (2) Distribution of services resources between physical channels and electronic channels lacks of dynamic equilibrium. (3) Benefits of physical channels are not balanced between self-operated channels and cooperative-operated channels. (4) Efficiency of services providing is not balanced between channels. (5) Current assessment mechanisms cannot effectively promote dynamic equilibrium in services resources allocation.

3 Concept of Equilibrium Service

3.1 From Equilibrium Production to Equilibrium Service

Service resources allocation is very important for operators. Only in accordance with proper allocation of service resources, business value and ensure sustainable development can be achieved. The concept of equilibrium service is based on the optimal allocation of service resources, meeting customer needs and responding to the competition as preconditions, its idea is inspired from equilibrium production of manufacturing, its ideological nature is equilibrium in time and space, to the best use of every aspect of resources in enterprise and its production capacity, technical capacity, material supply and equipment capacity, and solve the problems of resources allocation optimizing, which is a meaningful creation of General Equilibrium Theory.

Service as a kind of product also faces to resources allocation optimizing problems. But over the past years, due to its characteristic (intangibility, inseparability, heterogeneity and perishability), it was thought that services could not be provided in balanced time and space like tangible products. However, the optimal allocation and lean management thinking and theories can still provide important ideas of reference and methodological basis for the problems focused on in this paper.

Domestic and foreign scholars and institutions have raised some similar ideas or concepts, which provided references for this study. American service management portfolio PZB [1] proposed the thinking of "appropriate services" in their tolerance regional models (zone of tolerance, ZOT),which reflected an idea of "moderate degree in services"; Heskett and other five Harvard Business School professors, their service management group in 1994 proposed the service profit chain model, focusing on how to improve service productivity, and discussed how to effectively control inputs and outputs in service enterprises [2]; Ctrip.com brought six sigma methodologies originating from manufacturing into service providing and proclaimed the slogan "Provide services as producing tangible products"; Zhang Wentao [3] in his study proposed an idea "moderate service", from the perspective of service-based productivity, gave a thinking that enterprises should put a moderate idea into active management, in order to achieve the sustainable development; Moreover, some domestic scholars also put some concept resembling equilibrium services theory, such as equilibrium demand and supply of services [4] and [5], analysis of differential quality in services [6] and formulating the measurement index [7] and so on.

Previous studies were mostly longitudinal analysis in a single field, and these studies rarely take into account the impact of competition on service acting and customer satisfaction. Meanwhile, previous studies on service resource allocation were mostly from a single angle to match the enterprise service capacities to meet customer expectations, not put enterprise service capacities, customer service expectations and external service competition into a unified research framework. This study is based on this new perspective, concerned about these three factors on the combined effect of service allocation. "Equilibrium service" in the dimension of the service contact points considers the three factors of services efficiency, customer satisfaction and service competition, studying on how to achieve the equilibrium between service efficiency and customer satisfaction in telecom operators.

Particularly, the equilibrium mentioned in "equilibrium service" refers to the equilibrium of the service effect (equilibrium in service efficiency and satisfaction at different touch points), not the equilibrium of service resources inputs and means in time and space (the provision of services still comply with the principle of differentiation).

3.2 Implication of Equilibrium Services

In this paper, equilibrium service is defined as in the dynamic competitive environment, premising on the satisfaction of customers, optimizing the service resource allocation between touch points, grasping the principle of effectiveness, efficient

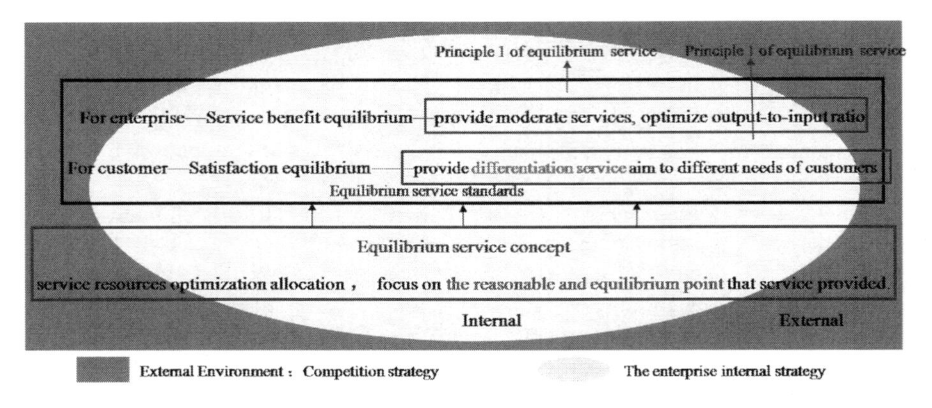

Fig. 1 Illustration of equilibrium services

and moderate, to achieve equilibrium and sustainable development process combined with steadily improvement of customers' satisfactions…As a management idea, equilibrium services focused not only on the results of distributing service resources, but also on the behavioral strategy of the process in service resources distribution, especially on the evolution of service methods in the dynamic challenging environment.

Equilibrium services theory includes three implications: firstly, the company should achieve the equilibrium benefit of services and optimize the cost-benefit ratio; secondly, the company should pursue the equilibrium satisfaction and supply differentiated services to different customers; thirdly, considering the challenging environment in the market, the company should carry on the dynamic equilibrium service strategy to deal with the effect which the strategy from competitors has on the customer perception and telecom operators (Fig. 1).

3.3 Effective Factors of Equilibrium Services

Based on the related literature and theory, three main factors on supplying equilibrium services are proposed in this paper: Capacity, competition and expectation, and their relations are showed in Fig. 2. Operators should provide services appropriately, distribute the services resources effectively, and then satisfy the customers' expectation and deal with the competitions from target market.

3.4 Applications of Equilibrium Services

Equilibrium services model is formed by the following preconditions: firstly, the services area could be confirmed by the capacity-expectation equilibrium; secondly, in the equilibrium area, the touch points of an operator and its competitors

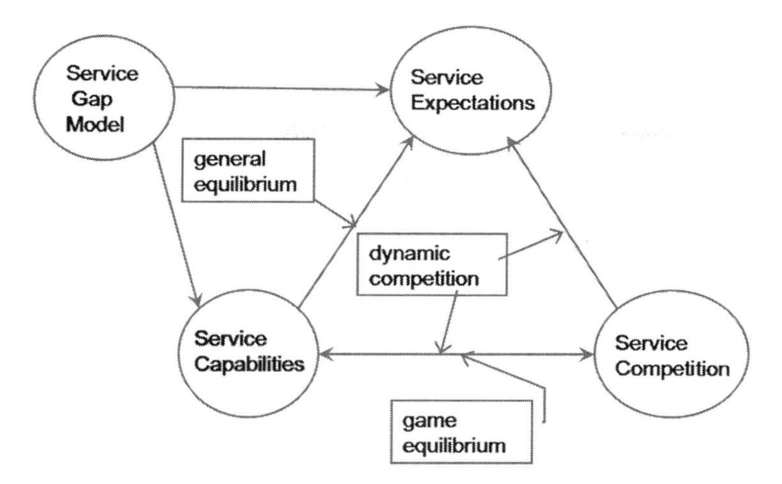

Fig. 2 Three main factors in equilibrium services

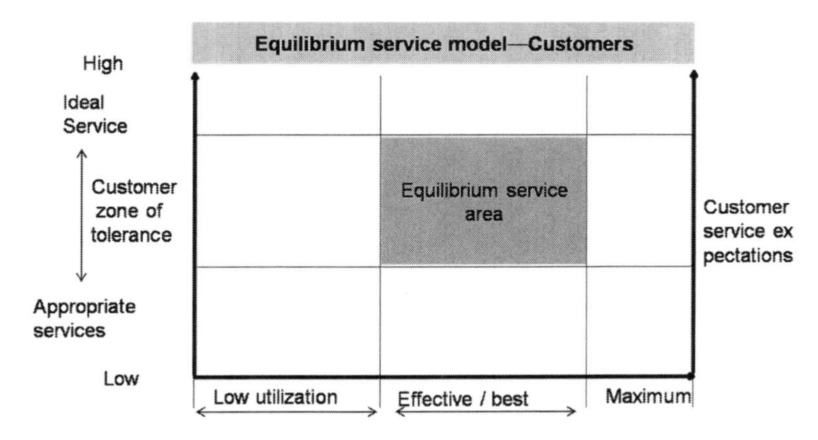

Fig. 3 Model 1 (For customers)

as well as that the customers expected could all be measured; thirdly, the capacities of providing services either in an operator or in its competitors could be stable in a specific period, and the equilibrium services can be achieved through the deployment of customer touch points. The equilibrium services model is set by X, Y axis; the X axis is the service providing capacity the Y axis is customers' expectation for services, just as Figs. 3 and 4 showed. Through evaluation of the resources of the operator and its competitors in the same touch point, we can achieve the benefit equilibrium between the operator, its competitors and customers satisfaction, and also achieve partition management, effectively. As Fig. 5 showed, the scope of

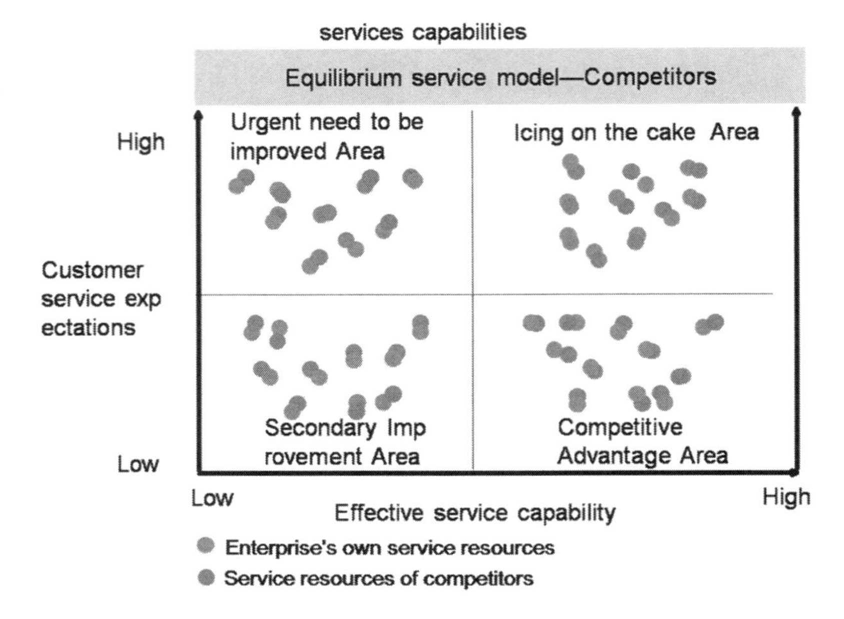

Fig. 4 Model 2 (For competitors)

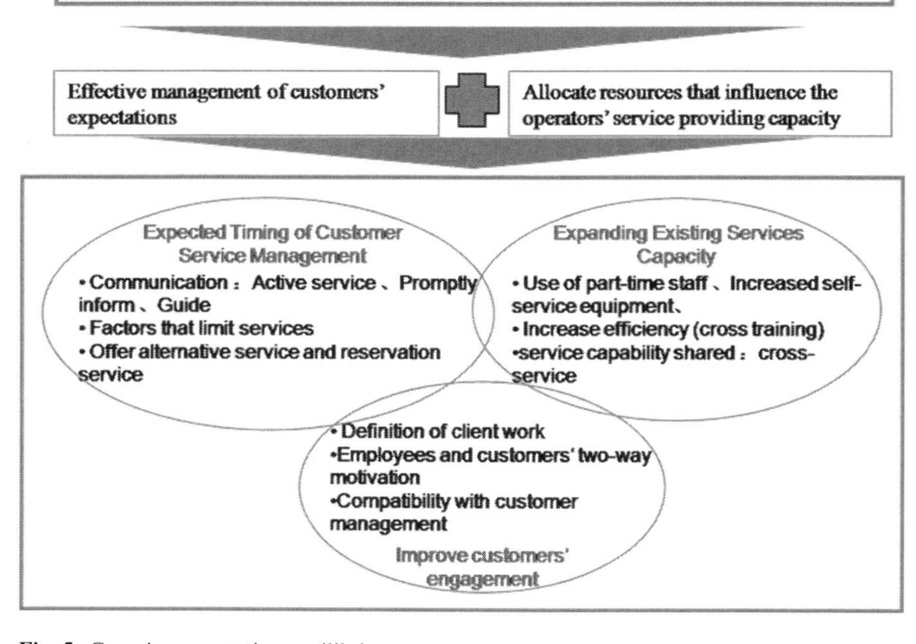

Fig. 5 Capacity-expectation equilibrium

services resources could be divided into four sections: marvelous section, challengeable section, improvable section, and the troublesome section.

Referring to the divided section in equilibrium services model and the customers' expectations, the operator could combine the results with the optimization cost of distributed services resources so as to achieve capacity-expectation equilibrium. Considering the touch point distribution situation of competitors, the operator also could be aware of the differences or the commons between the competitors and itself and assess the rationality of these differences; if the difference is unreasonable, the operator could revise the allocation scheme of touch points, satisfying the customers' expectation and achieve the capacity-competition equilibrium in the dynamic competitive environment.

4 Strategies of Equilibrium Services in Telecom Operators

Two aspects of achieving equilibrium services in telecom operators are proposed in this paper, the capacity-expectation equilibrium and the capacity-competition equilibrium.

There are three points in the capacity-expectation equilibrium: firstly, customers' expectation could be satisfied by resources allocation that influence the operators' service providing capacity; secondly, through effective management of customers' expectation, the capacity-expectation equilibrium could be achieved based on service resources allocation between touch points.; thirdly, through assembling the synergy of service capacity and customers' expectation, the operator can achieve the desired equilibrium effects. Equilibrium services strategy formulating is focused on under the situation that consumers' expectation is beyond the capacity one operator owned. As Fig. 5 shows.

Start from the capacity-competition equilibrium at different service touch points, a management system is proposed here, which means taking effective and feasible service resources allocation measures to respond to other competitors' strategy. This system includes six procedures: Diagnosis and analysis of service processes; Decomposition and definition of customer touch points; Distribution and testing of the customer touch points; Commitment and conversion of the service value; Adjustment and optimizing of the service organization; Service monitoring and evaluation. As Fig. 6 shows. Through suitability analysis, operators could provide services to target customers properly by the feasible channels in the appropriate time.

5 Conclusions

Following conclusions are made in this paper: (1) "Equilibrium service" means optimization of service resource allocation in order to meet customers' demand, and a sustainable developing process in which customer satisfaction is steadily improved.

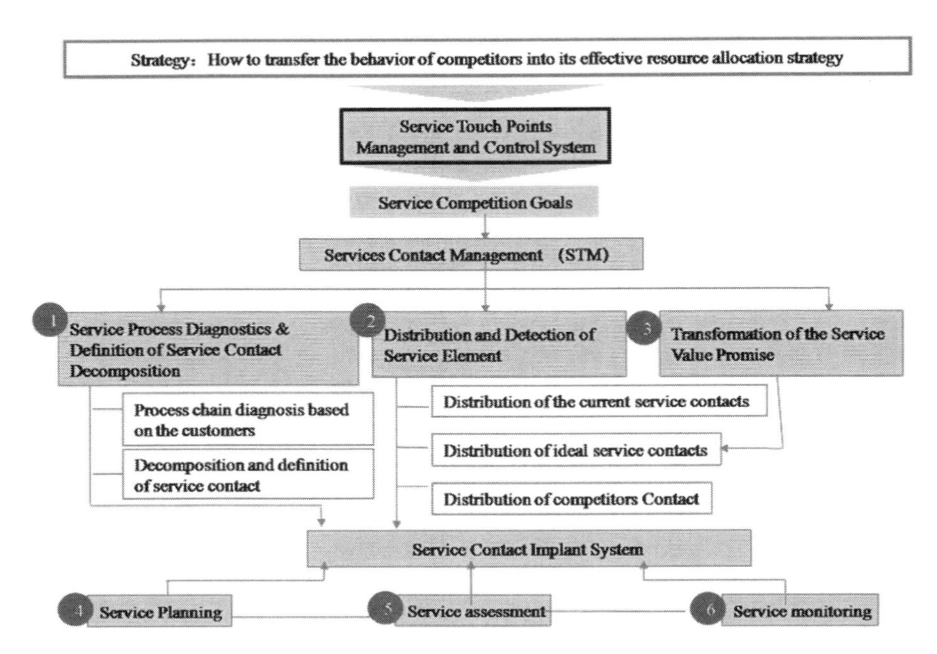

Fig. 6 Capacity-competition equilibrium

(2) "Equilibrium Service" reflects the balance in three areas: Telecom operators achieve a balance in service efficiency; customers achieve a balance in satisfaction degree; and considering the impact from competitors on its own strategy and customer satisfaction, a telecom operator achieves a dynamic equilibrium. (3) There are three affect elements for "Equilibrium service": Service capacity, service competition, and service expectations. Based on this conclusion, "Equilibrium Service" can also be grouped into five factors: Profit capacity, market mining, potential for innovation, collaboration and satisfaction contribution. (4) We can improve the equilibrium level of service from the following two aspects: First, establish the equilibrium between service capacity and service expectations; second, establish the equilibrium between service capacity and service competition.

Acknowledgment The authors would like to thank all the anonymous reviewers for their constructive comments, which undoubtedly served to improve this paper.

References

1. Parasuraman A, Zeithaml VA, Berry LL (1985) A conceptual model of service quality and its implications for future research. J Mark 49:41–50
2. Heskett JL, Jones TO, Loveman GW, Sasser WE, Schlesinger LA (1994) Putting the service-profit chain to work. Harv Bus Rev 72(2):164–174

3. Zhang W (2006) Appropriate services: a theoretical construct based on service productivity. Manag World 3:152–153
4. XiaojunXue (1994) Match of supply and demand. Manag Sci 6:42–46
5. XiaojunXue (1996) Increase productivity relying on customers. Decis-Making Draw 4:28–31
6. Ying Chai (2004) Elastic analysis of the service quality gap. J Tianjin Univ Commer 4:19–22
7. ZheXu, YiqingZhao, Haiqiong Wu (2004) Research on the gap evaluation model of service quality management. China Qual 4:44–48

A Review on Tourist Satisfaction of Tourism Destinations

Yining Chen, Hui Zhang, and Li Qiu

Abstract The present paper is a detailed sort-out and critical review of the foreign literature on tourist satisfaction at destinations in the last decade. Four dimensions that the relevant research surrounded are identified: tourist satisfaction theoretical model; the relation between tourist satisfaction and loyalty, expectation and service quality; the tourists' cultural backgrounds and cultural differences of the tourist satisfaction; and evaluation model of tourist satisfaction at destinations. The limitations of the previous research are briefly discussed, and future research directions are suggested attempting to provide reference and inspiration for the relevant domestic research and the tourism industry.

Keywords Tourism destination • Tourist satisfaction • Influencing factor

1 Introduction

The term "tourist satisfaction" in tourism research derived from "customer satisfaction" in marketing. Since the 1960s in the last century, the scholars have conducted a great deal of research on tourist satisfaction from the perspectives of quality management and repurchase intention, preliminarily accomplishing some theoretical models [7, 21, 27]. Based on the previous research, Pizam et al. [45] pioneered in applying the concept of customer satisfaction in the tourism study, which developed into a hot issue in the tourism study. At the beginning, the "tourist satisfaction" research centered around the product and service, for example, studies on the influencing factors on the satisfaction serving to improve the service quality of the hotels, hostels

Y. Chen (✉) • H. Zhang
School of Economics and Management, Beijing Jiaotong University, Beijing, China
e-mail: ynchen@bjtu.edu.cn

L. Qiu
School of English and International Studies, Beijing Foreign Studies University, Beijing, China

Z. Zhang et al. (eds.), *LISS 2012: Proceedings of 2nd International Conference on Logistics, Informatics and Service Science*, DOI 10.1007/978-3-642-32054-5_83,

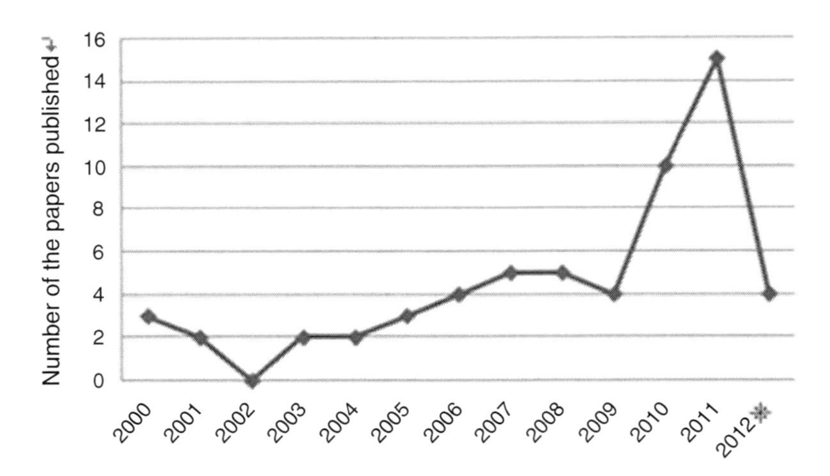

Fig. 1 Annual distribution of research papers published on tourist satisfaction (Note: only literature published in the first season are included in the 2012 category)

and tourist sites [16, 18, 31, 33, 48]. Yet in the recent years, an increasing number of scholars consider the competition among tourism destinations has evolved from one of the tourism resources, products and the tourism industry, to one of comprehensive strength, and one of the important facets to evaluate the tourism destination is precisely the tourist satisfaction [51].

The enhancement of tourist satisfaction not only has positive effects on the tourism service provider and the destination reputation, but also strengthens the tourist loyalty, lowers the price elasticity, lowers the future transaction cost and increases the productive force. Therefore, much attention is laid upon the measurement of tourist satisfaction among countries and regions. The present paper presents a sort-out and critical review of a collection of the foreign relevant literature on tourist satisfaction at destinations, concluding their research perspectives, content, limitations and possible future research direction, attempting to offer reference and inspiration for the relevant domestic research and the tourism industry.

2 An Overall View of the Literature

Tourist satisfaction has become a hot topic for tourism research in the recent years. From foreign language databases like Wiley Blackwell, SCI (Web of Science) ISTP&ISSHP, Elsevier Science and Compendex (Ei village) etc., 68 research papers directly related to tourist satisfaction at destinations have been detected (up to March 23, 2012) with key words of "tourism satisfaction", "destination satisfaction" and "tourism satisfaction" etc. Among them, 13 are from Wiley Blackwell, 11 from SCI (Web of Science) ISTP&ISSHP, 35 from Elsevier Science and 9 from Compendex (Ei village). Sixty one of them were able to be retrieved, and only two were published before 2000 (1999 and 1991 respectively). The rest 59 pieces of papers' annual distribution is shown in Fig. 1.

The research content concentrates on tourist satisfaction theoretical model; the relationship between tourist satisfaction and loyalty, expectation and service quality; the tourists' cultural background and cultural difference concerning the tourist satisfaction; and evaluation model of tourist satisfaction at destinations. In terms of the data analysis method, structural equation model, multiple linear regression, factor analysis, cluster analysis, correlation analysis (contingency table) and one-way ANOVA are among the methods used in the research. In terms of the publisher, most of the research papers were published on mainstream academic journals like "Tourism Management", "Annals of Tourism Research" and "International Journal of Tourism Research".

Table 1 is a selective summary of the theoretical models, analysis methods and research subjects in the representative papers.

3 Research Concentration

3.1 Theoretical Models of Tourist Satisfaction Evaluation

Kozak [29] listed four theoretical models on tourist satisfaction evaluation: expectation-performance model, importance-performance model, expectancy disconfirmation model and performance-only model. Expectation-performance model usually employs the SERVQUAL scale, in which the tourist expectation is measured first and then the perception on the service, referring the gap between the two to the service quality [39, 40]. SERVQUAL is widely applied and holds a certain degree of reliability and validity, but it entails a major flaw that the equaling relationship between the service quality and the gap between the tourist expectation and service performance is not well justified. There are scholars who suggest the tourist evaluation of service does not really depend on the gap between the expectation and the performance, but on the actual perception of the service [9, 61], in other words, regardless of the expectation of the destination before the tour, the tourist satisfaction is decided by the actual experience at the destination. This SERVPERF evaluation method based on performance only has been empirically proved to be better than SERVQUAL in terms of reliability, validity and prediction ability [26, 29].

Oliver and Swan [38] put forward the outcome-input model (EQUITY), which states the tourist satisfaction is determined by the comparison between what the tourists have received and the time, money and energy spent. When the tourists consider the outcome of the tour is more than the input, the tour experience is highly regarded, resulting in high satisfaction, and vice versa. Latour and Peat [30] came up with the NORM theory. Locating the reference points is crucial in using this model, for they determine the norms for judging the service quality. Tourist dissatisfaction comes into play as a result of disconfirmation relative to these norms. The references points can be an ideal trip desired, or other alternative

Table 1 Features of seven representative papers on tourist satisfaction

Paper	Destination	Data analysis	Theoretical model	Survey location	Influencing factors
Yu and Goulden [64]	Mongolia	VI	Brayley model (1989)	Airport	Tourist attraction, facility, service, price etc.
Song et al. [51, 52]	Hong Kong	IV	Expectation performance model, ACSI	Airport hotel, attraction retail shops, etc.	Tourist attraction, accommodation, restaurants, transportation, immigration, retail shops etc.
Kozak [29]	Turkey, Morocco	III	Performance model	By mail	Accommodation, local transportation, sanitation, hospitality and service, facility and activity, price, language convenience, airport service
Truong and Foster [59]	Vietnam	II	HOLSAT model	By tour operators, airlines	Attractions, activities, amenities, accommodation and accessibility
Master and Prideaux [34]	Queensland	VI	Importance-performance model	By mail	Cultural factors: restaurants, language, shopping, accommodation, negotiation etc.
Hui et al. [26]	Singapore	II	Expectancy disconfirmation model, performance model	Airport	Service staff, overall convenience, prices, accommodation and restaurants, tourist attraction, culture, climate, image and commercial product etc.
Truong and King [60]	Vietnam	V	Importance-performance model	Attractions, hotels and bars etc.	Tourist attraction, facilities, diversity, prices, tourist product quality, service and security

Note: I = structural equation model, II = multiple linear regression, III = factor analysis, IV = cluster analysis, V = correlation analysis (contingency table), VI = one-way ANOVA

destinations or places visited in the past. Tourists compare current travel destinations with these reference points and the difference between the present and the past experiences can be a norm used to evaluate tourist satisfaction. Tribe and Snaith [58] suggested the HOLSAT (Holiday Satisfaction) model, the theoretical foundation of which is also the expectation-performance model (disconfirmation approach). Unlike other models [49, 56], HOLSAT is able to measure tourist satisfaction at a destination rather than a specific service provider. Instead of a fixed menu of attributes, it adopts a most appropriate suite of attributes for a particular destination to evaluate the tourist satisfaction. Truong and Foster [59] utilized HOLSAT in a case study on Australian holidaymakers in Vietnam, and it was indicated that the HOLSAT model is a valuable tool that can be used to evaluate the satisfaction of tourists with particular destinations.

3.2 Regional and Cultural Differences of Tourist Satisfaction at Destinations

In recent years, a cross-cultural perspective as how tourists with different cultural backgrounds differ in their service quality evaluation has been a research hot topic as well [6, 13, 29, 44, 62]. Tourists from different countries put emphasis on varying aspects of the tourism service, and it is likely that they have distinct satisfactions toward the same service. Therefore, an understanding of the tourists' cultural backgrounds will help the destinations to design culture-oriented marketing and service. Pizam and Ellis [44] suggested two approaches to study the tourist behavior cross-culturally, either indirectly, say, study based on the local tour guides' and citizens' perception of the tourists' cultural differences, or directly, by studying the behavior of tourists from different countries or regions. Both approaches have been used in the previous research. Turner et al. [62] suggested in their research, a study of tourists from Japan, U.S., Australia and Chinese mainland in Melbourne, that though there appeared to be different emphases on the tourism service for regionally diverse tourists, no causal relationship between the difference and the satisfaction level was found. Yu and Goulden [64] conducted an analysis on international tourists' satisfaction of their travel experience with tourist attractions, prices, service, facilities, destination image perception, revisit and recommendation intentions, and the results were compared to find regional similarities and differences. The findings showed a diversity of the tourists' evaluations in cultural and historical tourist attractions, local tourism staff, facilities, and service quality and nightlife activities depending on their regional backgrounds. Still, there might be contingency in such findings due to the target destinations. Aiming to examine whether there was regional difference of tourists satisfaction at the same destination, Kozak [29] carried out a comparative study on the satisfaction of the British and German tourists in Turkey and Morocco, and the research result suggested the British tourists tended to score higher for almost all the satisfaction influential factors than the German tourists. Truong and King [60] pointed out when it comes to destination marketing, not only the tourists' regional difference but also the language background should be taken into account. Overall, research centering the cultural differences at tourism destination is still at its infancy development stage. On the one hand, the current and previous research mostly distinguish tourists groups by countries, not yet concerning any sub-culture; on the other hand, to what degree do the cultural backgrounds attribute to the different satisfaction levels shown by origin countries is yet to be explored.

3.3 Relations Between Tourist Satisfaction and Expectation, Quality, Value and Loyalty

Quite a few scholars have focused on research of relations between tourist satisfaction and service quality. Soutar [53] believes the service quality has direct influence on the tourist satisfaction, and consequently enhancement of the service quality will increase tourist satisfaction. Many other studies also have indicated that the service quality influences the client satisfaction, as well loyalty and post-purchase behavior [1, 11, 37]. The tourists' service perception is positively related to the satisfaction level, but the effect does not work reversely; well-perceived quality not only enhances the client satisfaction, but also stimulates mouth-to-mouth advertisement, and eases price sensitivity [15, 17]. Other scholars hold the view that quality perception is merely one of the factors influencing satisfaction [41], and since different destinations demand quite distinct travel costs, quality perception and value perception were considered two dimensions that influence tourist satisfaction [53]. Thus, there is a positive correlation between value perception and tourist satisfaction, which means when the tourist consider what they receive is worth the time and money they spend, their satisfaction level is likely to increase [12, 43, 52]. As to the relation between expectation and satisfaction, there are two opposite views. One sees expectation and satisfaction as negatively correlated, that is, a raise in the expectation does little or even no effect on the tourist satisfaction [1, 8, 11]. The other view believes the relation between the two dimensions should be studied in relation to the analysis framework of satisfaction [11, 22, 54]. Lee et al. [32] did a research on Chinese tourists in South Korea, and analyzed the relationships among tourist expectation, motive, quality, satisfaction, complaint and loyalty, concluding a positive effect of tourist motive on quality perception, a negative correlation between satisfaction and complaint, and a non-prominent positive correlation between satisfaction and loyalty when the correlation coefficient is 0.05.

3.4 Identification of Influencing Factors of Tourist Satisfaction at Destinations

The definition of tourist destination is being fiercely discussed among scholars at the moment. Pearce [42] defined destination as an amalgam of products and services in one location that can draw visitors from beyond its spatial confines. Hu and Richie [24] proposed tourist conceptualized the tourism destination as a package of tourism facilities and services, which like any other consumer product, is composed of a number of multi-dimensional attributes. Smith and Olson [50] suggested the tourism service plays an important role in the tourist experience, and inputs from different destinations lead to different outputs. Other scholars believe other factors like the urban and social environments also significantly affect tourist satisfaction, such as local hospitality, language convenience, urban composition and demographic density etc [3, 4, 35]. Other influencing factors on tourist experience and destination perception include economic factors like exchange rate, company market behavior and pricing [14, 35], cultural factors [10, 46], political factors like VISA policy and political stability [19, 20, 47, 57]. Kotler et al. [28] concluded six factors that affect the macro-environment of the destination: demography, economy, nature, technology, politics and culture.

Bowen [5] identified six attributes of the influencing factors of tourist satisfaction: expectation, performance, disconfirmation, attribution, emotion and finally, equity. The effect of the tourists' past experience is likely to be neglected. Some scholars suggested "The Halo Effect" plays a role in tourist satisfaction, that is, their opinion on one single aspect might determine the overall evaluation of the whole tourism product. Untidy bathroom might lead to dissatisfaction toward the entire tour, while excellent tour guiding can result in high overall satisfaction though there is discontent toward other part of the tour, hence it is significant to measure the satisfaction on single aspect of the tour.

Huges [25] interpreted the above theory as an indication of the significance to identify the dimensions that can determine the overall perception, which will facilitate taking out measures to reduce the halo effect. Noam et al. [36] proposed the two-factor theory, stating the satisfaction influencing factors consist of insaxtmental factors and expressive factors. Insaxtmental factors are related to the features and functions of the product, and are basic and indispensable (the absence will lead to tourist dissatisfaction), but they do not have prominent contribution to tourist satisfaction, for example the transportation means to get to the travel destination. Expressive factors are related to the value manifestation and particular features of the product, and bring about contribution to tourist satisfaction, but the absence of expressive factors will not lead to dissatisfaction, for example luxurious transportation facilities and special service. The two-factor theory of tourist satisfaction is a major finding of the tourist satisfaction classification research, which directs the destination to enhance the tourist satisfaction efficiently.

Master and Prideaux [34] found out through study on the Taiwanese tourists Queensland Australia, which the effect of the cultural factors on the inbound tourist satisfaction was not prominent, and the service quality is more decisive to the success of the international tourism destination than multicultural integration [63].

Song et al. [52] evaluated the tourist satisfaction at six inbound tourism-related sectors, and calculated the indexes of overall satisfaction at destination. The six tourism-related sectors include attraction sites, hotels, immigration, restaurants, retail shops and transportation. The evaluation model is a useful attempt to measure tourist satisfaction at destinations, but still, from the perspective of the tourists, satisfaction at each single industry fails to adequately represent their opinion on the entire tour experience, since other factors like the cultural traditions, climate and infrastructure also play important roles in tourist satisfaction.

Truong and Foster [59] believed it was more complex to evaluate tourist satisfaction at destination than at one single tourist service provider, while the expectation-disconfirmation model adopted in the previous research focused on the tourist service provider, neglecting the tourist overall perception. The evaluation of the tourist destination is not supposed to be a simple sum-up of the tourist satisfaction at each service provider, but should involve factors unrelated to single service sector but crucial to overall satisfaction. These factors include visible ones like product, price and urban views, and invisible ones like service quality, local hospitality [2, 23, 35].

4 Conclusion and Future Research Directions

Since the end of 1970s, the research on tourist satisfaction at destinations has gone through from concept establishing, factor analysis to evaluation model stages, gaining fruitful findings on the mechanism of tourist satisfaction, identification and classification of the influencing factors and the satisfaction measuring method. These findings reveal that the tourist satisfaction theory extensively adopts the client satisfaction theory developed in the service and management studies, with the expectation-disconfirmation theory from Pizam et al. [45] being the theoretical foundation. The major research methods include structural equation model, factor analysis, ambiguous assemble and multiple regression. The research concentration evolved from single-dimension study of satisfaction at tourist attractions, hotels and restaurants etc. [16, 18, 31, 33, 48], to more comprehensive analysis, for example, comparative studies of regionally diverse tourists' satisfactions at one or multiple destinations from a cross-cultural perspective [6, 31, 64], and satisfaction evaluation method of aggregating satisfactions toward each tourism-related sector [52]. With regard to the influencing factors, tourist satisfaction at destinations is a multi-dimensional concept, and the evaluation of it is more complex than satisfaction at one single tourist service provider. Issues like the comprehensiveness of tourist satisfaction at destinations, the uniqueness of the effect of interaction between the tourists and destinations have not yet been well explored, and no related systematic theoretical framework has been established.

There are some cutting-edge research topics concerning tourist satisfaction at destinations. Firstly, a city can be taken as a tourist destination to build up satisfaction evaluation system and carry out case study. The tourist demands and consuming habits are undergoing major changes with diversification. The number of tourists who visit a city and its peripheral areas as tourist destination instead of attraction sites are increasing substantially, which highlights the significance of elevating the city tourism competitiveness and attractiveness by improving the city tourist infrastructure and service system, enhancing the city image and increasing the tourist satisfaction at the city as a tourist destination. Secondly, the study on tourist satisfaction at destinations calls for innovation in research method. Chinese Tourism Academy published since 2009 the "Tourist Satisfaction Ranking of 50 Domestic Cities", consisting of tourist questionnaire survey, Internet survey and tourist complaint statistics. The ranking is a good attempt of innovation for evaluating tourist satisfaction, but it is an approach hard to realize in the international context, for the tourist complaint statistics and face-to-face tourist survey data are hard to collect. Due to the difficulty of sample collecting, previous research mostly adopted post hoc analysis, and data was usually collected at the airport and tourist facilities (hotels and attraction sites etc.). The time and spatial limitation of the survey might lower the sample and data quality. Innovative research method will gives impetus for the research of tourist satisfaction at destinations. Another cutting-edge topic can be the influence of destination image on tourist satisfaction. The destination image is not only an important factor to the tourists' travel decision and plans, but also to the tourist satisfaction. Research on the relations and interaction mechanism between destination perception and tourist satisfaction and loyalty, along with comparative analysis of tourist markets and sub-markets, will assist the tourism administrative departments of the destinations to produce effective marketing plans and image promotion strategies.

Acknowledgments This work was supported by the Fundamental Research Funds for the Central Universities (No.2011JBM035) and Beijing philosophy and Social Science Fund(No.11JGC103).

References

1. Anderson EW, Sullivan MW (1993) The antecedents and consequences of customer satisfaction for firms. Mark Sci 12(2):125–143
2. Augustyn MM, Ho SK (1998) Service quality and tourism. J Travel Res 37(1):71–75
3. Bitner MJ (1997) Evaluating service encounters: the effects of physical surroundings and employee responses. J Mark 54(2):69–82
4. Bitner MJ, Hubert AR (1994) Encounter satisfaction versus overall satisfaction versus quality. In: Rust RT, Oliver RL (eds) Service quality: new directions in theory and practice. Sage, Thousand Oaks, pp 72–94

5. Bowen D (2001) Antecedents of consumer satisfaction and dissatisfaction (CS/D) on long-haul inclusive tours-a reality check on theoretical considerations. Tour Manage 22(1):49–61
6. Bowen D, Clarke J (2002) Reflection on tourism satisfaction research: past, present and future. J Vacat Mark 8(4):297–308
7. Cardozo RM (1965) An experimental study of consumer effort, expectation and satisfaction. J Mark Res 2(8):244–249
8. Chan LK, Hui YV, Lo HP, Tse SK, Tso GK, Wu ML (2003) Consumer satisfaction index: new practice and findings. Eur J Mark 37(5/6):872–909
9. Churchill GA, Suprenant C (1982) An investigation into the determinants of customer satisfaction. J Mark Res 19(4):491–504
10. Cohen E (1989) The commercialization of ethnic crafts. J Des History 2(2):161–168
11. Cronin JJ, Taylor SA (1992) Measuring service quality: a reexamination and extension. J Mark 56(3):55–68
12. Cronin JJ, Brady MK, Hult GT (2000) Assessing the effects of quality, value, and customer satisfaction on consumer behavioral intentions in service environments. J Retail 76(2):193–208
13. Crotts JC, Erdmann R (2000) Does national culture influence consumers' evaluation of travel services? a test of Hofstede's model of cross-cultural differences. Manag Serv Qual 10 (6):410–419
14. Dieke PUC (1991) Policies for tourism development in Kenya. Ann Tour Res 18(2):269–294
15. Ekinci Y (2004) A review of theoretical debates on the measurement of service quality: implications for hospitality research. J Hosp Tour Res 26(3):199–216
16. Foster D (2000) Measuring customer satisfaction in the tourism industry. In Proceeding of third international and sixth national research conference on quality measurement, The Center for Management Quality Research at RMIT University, Australia
17. Gonzalez MA, Comesana LR, Brea JF (2007) Assessing tourist behavioral intentions through perceived service quality and customer satisfaction. J Bus Res 60(2):153–160
18. Haber S, Lerner M (1999) Research notes: correlates of tourism satisfaction. Ann Tour Res 26(1):197–200
19. Hall CM (1997) Tourism and politics: policy, power and place. Wiley, Chichester
20. Hall CM (1997) The politics of heritage tourism: place, power and the representation of values in the urban context. In: Murphy PE (ed) Quality management in urban tourism. Wiley, Chichester, pp 91–101
21. Hartman RS (1973) The Hartman Value Profile (HVP): manual of interpretation, research concepts. Southern Illinois Press, Muskegon
22. Hellier PK, Geursen GM, Carr RA, Rickard JA (2003) Customer repurchase intention: a general structural equation model. Eur J Mark 37(11/12):1762–1800
23. Hsu C (2003) Mature motor coach travelers' satisfaction: a preliminary step toward measurement development. J Hosp Tour Res 27(3):291–309
24. Hu Y, Richie JRB (1993) Measuring destination attractiveness: a conceptual approach. J Travel Res 25–34
25. Huges K (1991) Tourist satisfaction: a guided "cultural" tour in North Queensland. Aust Psychol 26(3):166–171
26. Hui TK, Wan D, Ho A (2007) Tourists' satisfaction, recommendation and revisiting Singapore. Tour Manage 28:965–975
27. Hunt HK (1977) Conceptualizing and measurement of consumer satisfaction and dissatisfaction. Marketing Science Institute, Cambridge, MA
28. Kotler P, Bowen J, Makens J (1996, 2003) Marketing for hospitality and tourism. Prentice-Hall, Englewood Cliffs
29. Kozak M (2001) A critical review of approaches to measure satisfaction with tourist destinations. In: Mazanec JA, Crouch GI, Ritchie JRB, Woodside AG (eds) Consumer psychology of tourism, hospitality and leisure, vol 2. CABI Publishing, New York, pp 303–319
30. Latour SA, Peat NC (1979) Conceptual and methodological issues in consumer satisfaction research. Adv Consum Res 6(1):431–437

31. LeBlanc G (1992) Factors affecting customer evaluation of service quality in travel agencies: an investigation of customer perceptions. J Travel Res 30(4):10–16
32. Lee S, Jeon S, Kim D (2011) The impact of tour quality and tourist satisfaction on tourist loyalty: the case of Chinese tourists in Korea. Tour Manage 32:1115–1124
33. Macintosh G (2002) Building trust and satisfaction in travel counselor/client relationships. J Travel Tour Mark 12(4):59–73
34. Master H, Prideaux B (2000) Culture and vacation satisfaction: a study of Taiwanese tourists in South East Queensland. Tour Manage 21(5):445–450
35. Murphy PE, Pritchard M (1997) Destination price-value perceptions: an examination of origin and seasonal influences. J Travel Res 35(3):16–22
36. Noam T, Avivit C, Moti K (2004) Using ratings and response latencies to evaluate the consistency of immediate aesthetic perceptions of web pages. In: Proceedings of the third annual workshop on HCI research in MIS, Washington, DC
37. Oliver RL (1980) A cognitive model of the antecedents and consequences of satisfaction decisions. J Mark Res 17(4):460–469
38. Oliver RL, Swan JE (1989) Consumer perceptions of interpersonal equity and satisfaction in transactions: a field survey approach. J Mark 53(2):21–35
39. Parasuraman A, Zeithaml VA, Berry L (1985) A conceptual model of service quality and its implications for future research. J Mark 49(4):41–50
40. Parasuraman A, Zeithaml VA, Berry L (1988) SERVQUAL: a multiple item scale for measuring consumer perceptions of service quality. J Retail 64(1):12–40
41. Parasuraman A, Zeithaml VA, Berry LL (1994) A concept model of service quality and its implications for future research. J Mark 49(3):41–50
42. Pearce DG (1997) Competitive destination analysis in Southeast Asia. J Travel Res 35(4):16–24
43. Petrick JF, Backman SJ (2002) An examination of the construct of perceived value for the prediction of golf travelers' intentions to revisit. J Travel Res 41(1):38–45
44. Pizam A, Ellis T (1999) Customer satisfaction and its measurement in hospitality enterprises. Int J Contemp Hosp Manage 11(7):1–18
45. Pizam A, Neumann Y, Reichel A (1978) Dimensions of tourism satisfaction with a destination area. Ann Tour Res 5:314–322
46. Prentice R (1993) Motivations of the heritage consumer in the leisure market: an application of the Manning-Haas demand hierarchy. Leisure Sciences 15(4):273–290
47. Richter LK (1989) The politics of tourism in Asia. University of Hawaii Press, Honolulu; Hall CM (1997) Tourism and politics: policy, power and place. Wiley, Chichester
48. Ryan C (1994) Researching tourist satisfaction: issues, concepts and problems. International Thomson Business Press, London
49. Ryan C, Cliff A (1997) Do travel agencies measure up to customer expectations? an empirical investigation of travel agencies service quality as measured by SERVQUAL. J Travel Tour Mark 6(2):1–32
50. Smith RK, Olson LS (1994) Tourist shopping activities and development of travel sophistication. Vis Leis Bus 20(1):23–33
51. Song H, Li G, Van der Veen R, Chen JL (2010) Assessing mainland Chinese tourists' satisfaction with Hong Kong using the tourist satisfaction index. Int J Tour Res 13(1):82–96
52. Song H, Van der Veen R, Li G, Chen JL (2012) The Hong Kong tourist satisfaction index. Ann Tour Res 39(1):459–479
53. Soutar JN (2001) Service quality, customer satisfaction and value: an examination of their relationships. In: Kandampuly J, Mok C, Sparks B (eds) Service quality management in hospitality, tourism and leisure. The Haworth Press, New York
54. Spreng R, Droge C (2001) The impact on satisfaction of managing attribute expectations: should performance claims be understated or overstated? J Retail Consum Serv 8(5):261–274
55. Stevens BF (1992) Price value perception of travelers. J Travel Res 21(2):44–50

56. Suh SH et al (1997) The impact of consumer involvement of the consumers' perception of service quality-focusing on the Korean hotel industry. J Travel Tour Mark 6(2):33–52
57. Teye VB, Leclerc D (1998) Product and service delivery satisfaction among North American cruise passengers. Tour Manage 19(2):153–160
58. Tribe J, Snaith T (1998) From SERVQUAL to HOLSAT: holiday satisfaction in Varadero, Cuba. Tour Manage 19:25–34
59. Truong TH, Foster D (2006) Using HOLSAT to evaluate tourist satisfaction at destinations: the case of Australian holidaymakers in Vietnam. Tour Manage 27:842–855
60. Truong TH, King B (2009) An evaluation of satisfaction levels among Chinese tourists in Vietnam. Int J Tour Res 11:521–535
61. Tse DK, Wilton PC (1988) Models of consumer satisfaction: an extension. J Mark Res 25:204–211
62. Turner LW, Reisinger Y, McQuilken L (2001) How cultural differences cause dimensions of tourism satisfaction. J Travel Tour Mark 11(1):79–101
63. Yetton PW, Craig JF, Johnston KE (1995) Fit, simplicity & risk: multiple paths to strategic IT change, in AGSM, Working Paper, presented at AGSM
64. Yu L, Goulden M (2006) A comparative analysis of international tourists' satisfaction in Mongolia. Tour Manage 27:1331–1342

Practical Research and Establishment for Product Quality Evaluation System Based on AHP Fuzzy Comprehensive Evaluation

Xi Xi and Qiuli Qin

Abstract In order to reduce the subjective prejudice and uncertainty in evaluating product quality, a new model is proposed by comprehensively using AHP method, weighted comprehensive evaluation and fuzzy comprehensive evaluation. AHP is used to analyze the structure of product quality evaluation problem and determine weights for evaluation criteria. After structure judge matrix, sequencing calculation and concordance examination, evaluation methods such as fuzzy synthesis evaluation are used to calculate the integrated quality evaluation result of each product. A practical example of a product has been used to illustrate the theoretical qualitative proposed evaluation model.

Keywords AHP (Analytic Hierarchy Process) • Product quality • Evaluation criteria • Empirical analysis

1 Introduction

To evaluate the quality of basic products is a key problem of quality management in manufacturing enterprises [1]. Product quality can be divided into many quality aspects and each aspect can be subdivided into more specific criteria [2]. It's difficult to get a scientific judgment result only through subjective judgment.

This research is aimed at the characteristics of manufacturing products. It applies AHP evaluation, weighted comprehensive evaluation and fuzzy comprehensive evaluation to the process of product quality comprehensive evaluation to design a product quality evaluation model which converts the subjective judgment into scientific decision-making result.

X. Xi (✉) • Q. Qin
School of Economics and Management, Beijing Jiaotong University,
Beijing, People's Republic of China
e-mail: 11125188@bjtu.edu.cn; qlqin@sohu.com

Z. Zhang et al. (eds.), *LISS 2012: Proceedings of 2nd International Conference on Logistics, Informatics and Service Science*, DOI 10.1007/978-3-642-32054-5_84,

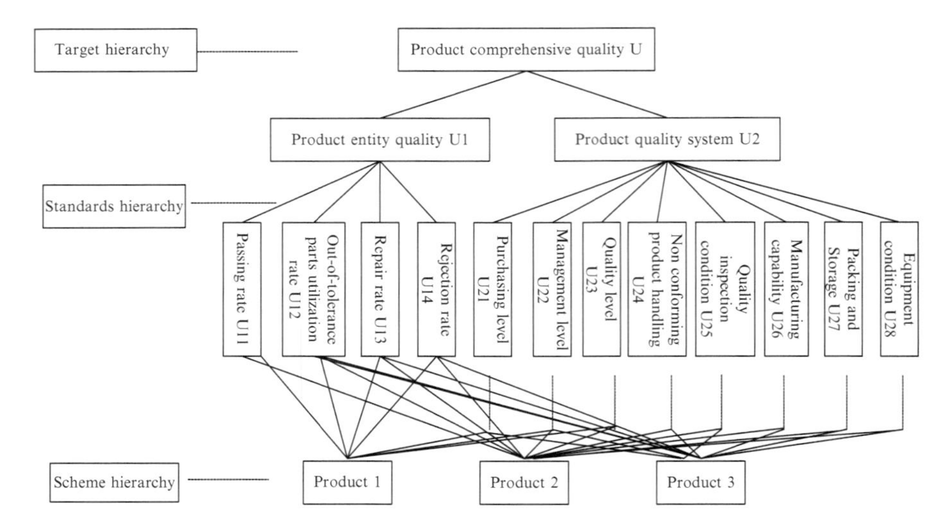

Fig. 1 System structure model

2 Establishment for Product Quality Evaluation System

The product quality evaluation system is based on the study of a large motorcycle production enterprise. The first hierarchy criterion is product comprehensive quality, divided into two second hierarchy criteria as product entity quality and product quality system. The entity quality includes passing rate, out-of-tolerance parts utilization rate, repair rate and rejection rate. The product quality system is divided into purchasing level, management level, quality level, nonconforming product handling, quality inspection condition, manufacturing capability, packing and storage and equipment condition. The third hierarchy criteria can still further divided into the forth hierarchy criteria.

The product quality evaluation system structure model is shown in Fig. 1, the forth hierarchy criteria will be given in Table 1 due to the space limit.

3 The Products Quality Evaluation Model

3.1 Setting Up Quality Evaluation Hierarchy Criteria

According to the comprehensive evaluation hierarchical structure, set up quality evaluation hierarchy criteria. For example, the first evaluation hierarchy criterion is U = {U1, U2}, the second evaluation hierarchy criteria are U1 = {U11, U12, U13, U14} and U2 = {U21, U22, U23, U24, U25, U26, U27, U28}. By analogy, the evaluation criteria at all hierarchies are designed, as shown in Table 1.

Table 1 Quality evaluation hierarchy criteria

First hierarchy		Second hierarchy		Third hierarchy		Forth hierarchy	
Product quality	U	Product entity quality	U1	Passing rate	U11	–	
				Out-of-tolerance parts utilization rate	U12		
				Repair rate	U13		
				Rejection rate	U14		
Product quality	U	Product quality system	U2	Purchasing level	U21	Purchasing documents	U211
						Selection of qualified supplier	U212
						Quality assurance agreement	U213
						Inspection methods agreement	U214
						Raw materials inspection planning and control	U215
						Raw materials quality record	U216
						Methods to resolve the dispute	U217
				Management level	U22	The degree of quality improvement	U221
						Quality organization or institution	U222
						Quality responsibility system	U223
						Quality documents and plan	U224
				Quality level	U23	Manufacturing and assembly quality	U231
						Appearance quality	U232
						Basic performance	U233
						Reliability	U234
				Nonconforming product handling	U24	Integrity of handling procedure steps of nonconforming product	U241
						Handling measure	U242
						Documents record	U243
						Create file	U244

(continued)

Table 1 (continued)

First hierarchy	Second hierarchy	Third hierarchy		Forth hierarchy	
		Quality inspection condition	U25	Inspection system perfection	U251
				Integrity of inspection measure	U252
				Timeliness of inspection data transmission	U253
				Inspection personnel	U254
		Manufacturing capability	U26	Key process	U261
				Process audit	U262
				Process capability	U263
		Packing and Storage	U27	Damp proofing and tightness	U271
				Label, traceability and durability	U272
				Warehouse area and ventilation	U273
				Regular inspection system	U274
				Process to file creation	U275
		Equipment condition	U28	Applicability and accuracy	U281
				Preventive maintenance plan	U282
				Integrality and unity	U283
				Equipment advancement	U284
				Purchasing and update plan	U285

Table 2 Criteria scores

Criterion	Score
U11	0.90
U12	0.68
U13	0.78
U14	0.84

3.2 *Judge Matrix Structuring, Weight Calculation and Concordance Analyzing*

Pairwise comparing each pair of criteria, deciding the degree of importance of each criterion to structure the judge matrix. Calculate the weights of each criteria and analyzing the concordance of judge matrix. Only when judge matrix passed concordance examination can the weights of criteria be scientific [3].

3.3 *The Evaluation Method*

The evaluation methods commonly used are direct evaluation method (total score method), weighted comprehensive evaluation method and fuzzy comprehensive evaluation method. Direct evaluation method is used in the evaluation for the criteria U11–U14 in this system. Weighted comprehensive evaluation method is used in the evaluation of criteria U and U1. And fuzzy comprehensive evaluation method is used in the evaluation for criteria U2, U21–U28 and all the forth hierarchy criteria.

4 The Empirical Analysis

The evaluation method mentioned above is used to evaluate the quality of a product that supplier offered. According to the quality record of the supplier, we can get the scores of the criteria U11–U14, as is shown in Table 2.

Experts are invited to grade the forth hierarchy criteria as excellent, good, ok or bad. Then the scores matrix of forth hierarchy criteria will be determined by the proportion of each grade, as is shown in Table 3.

Enterprise quality department is invited to pairwise each pair of criteria that under the same criterion and in the same hierarchy. If the judge matrix gets through the concordance analyzing, then the weight can be calculated.

The third hierarchy criteria score of product quality system are calculated with fuzzy mathematical evaluation method. The result is shown in Table 4.

The product entity quality criteria score is calculated with weighted comprehensive evaluation method. The result is shown below:

$$b1 = 0.813$$

Table 3 Forth criteria scores matrix

Criterion	Score matrix	Criterion	Score matrix
U211	(0.5,0.3,0.2,0)	U244	(0.3,0.2,0.4,0.1)
U212	(0.4,0.1,0.2,0.3)	U251	(0.5,0.4,0.1,0)
U213	(0.5,0.4,0.1,0)	U252	(0.2,0.4,0.1,0.3)
U214	(0.3,0.4,0.1,0.2)	U253	(0.4,0.2,0.4,0)
U215	(0.7,0.1,0.2,0)	U254	(0.5,0.2,0.2,0.1)
U216	(0.6,0.2,0.1,0.1)	U261	(0.5,0.2,0.3,0)
U217	(0.5,0.4,0.1,0)	U262	(0.7,0.2,0.1,0)
U221	(0.3,0.3,0.2,0.2)	U263	(0.4,0.3,0.2,0.1)
U222	(0.5,0.4,0.1,0)	U271	(0.6,0.2,0.2,0)
U223	(0.2,0.2,0.5,0.1)	U272	(0.6,0.4,0,0)
U224	(0.4,0.5,0.1,0)	U273	(0.4,0.1,0.5,0)
U231	(0.6,0.4,0,0)	U274	(0.5,0.2,0.3,0)
U232	(0.7,0.3,0,0)	U275	(0.5, 0.5,0,0)
U233	(0.6,0.2,0.2,0)	U281	(0.5,0.2,0.3,0)
U234	(0.6,0.2,0.2,0)	U282	(0.6,0.3,0.1,0)
U241	(0.4,0.2,0.3,0.1)	U283	(0.7,0.2,0.1,0)
U242	(0.5,0.4,0.1,0)	U284	(0.8,0.2,0,0)
U243	(0.6,0.2,0.2,0)	U285	(0.5,0.1,0.4,0)

Table 4 Third criteria scores matrix

Criterion	Score matrix	Criterion	Score matrix
U21	(0.494, 0.267, 0.149, 0.090)	U25	(0.408, 0.354, 0.153, 0.085)
U22	(0.334, 0.316, 0.231, 0.119)	U26	(0.544, 0.207, 0.241, 0.007)
U23	(0.628, 0.320, 0.052, 0)	U27	(0.551, 0.258, 0.191, 0)
U24	(0.438, 0.245, 0.253, 0.065)	U28	(0.606, 0.212, 0.182, 0)

The product quality system criteria score is calculated with fuzzy mathematical evaluation method. The result is shown below:

$$b2 = B2 \times Yw = (0.490, 0.277, 0.178, 0.055) \times (0.95, 0.83, 0.68, 0.30) = 0.833$$

Finally, the comprehensive product quality score is calculated with weighted comprehensive evaluation method. The result is shown below:

$$b = A \times P = A \times (b1, b2)^{T} = (0.667, 0.333) \times (0.813, 0.833)^{T} = 0.819$$

The comprehensive quality score of the product is 0.819. Other products comprehensive quality scores can be calculated as the same. Finally, the selection of the products will be made by selecting one which gets the highest score.

5 Conclusion

Product quality Comprehensive evaluation is very complicated. This study is to develop a scientific quality evaluation system which comprehensively applies AHP method, weighted comprehensive evaluation and fuzzy comprehensive evaluation, to convert subjective judgment into scientific decision-making result. The product quality evaluation system is scientific, and the result of it has practical applicability for manufacturing enterprises in supplier selection.

References

1. Shaofang Xie, Xiaochun Lu, Feng Ruan (2009) Establishment of plastic die design quality evaluation system. Mod Manuf Eng 347(8):61–64
2. Shenghai Qiu, Hua Yan, Ningsheng Wang (2005) Research for comprehensive evaluation of enterprise group partner in agile manufacturing. Mod Manuf Eng 296(5):17–21
3. Holder RD (1990) Some comments on the Analytic Hierarchy Process. Oper Res Soc 41(11):1073–1076

Influence of Actors in Alliance Game Based on Social Network Analysis Theory and Its Application

Yuanguang Fu

Abstract Shapley value reflects the influence of the actors in alliance game. Traditional method of calculating the Shapley value takes the actors' absolute power only. In this paper, Social Network Analysis (SNA) method is introduced to concern about the interactive function. The traditional Shapley value is modified by relatively point centrality. An example given at the end of the paper shoes the process of the method.

Keywords Social network analysis (SNA) • Relatively point centrality • Alliance game • Shapley value

1 Introduction

Shapley is one of the founders of the game theory. In 1953 he studies the problem of non-strategy multi-person cooperative game theory [1]. Shapley value is a well-known solution concept in this problem. Suppose a situation where if some economic agents make up a cooperative relationship, i.e., a coalition, then they can get more gains than those if they do not do so. In such situations, one of people's interests is how much share each of them should get by forming the coalition. Shapley value shows a vector whose elements are agents' share derived from some reasonable bases [1–3].

Shapley value has been investigated by a number of researchers. Based on Shapley value, Owen adopted another approach characterizing axiomatically a coalitional value called now Owen value [4, 5]. In this case, the unions play a quotient game among themselves, and each one receives a payoff which, in turn, is

---

Y. Fu (✉)
School of Public Affairs, University of Science and Technology of China, 230026 Hefei, People's Republic of China
e-mail: fyg@ustc.edu.cn

Z. Zhang et al. (eds.), *LISS 2012: Proceedings of 2nd International Conference on Logistics, Informatics and Service Science*, DOI 10.1007/978-3-642-32054-5_85,

shared among its players in an internal game. Both payoffs, in the quotient game for unions and within each union for its players, are given by the Shapley value. In addition to the initial one, many other axiomatic characterizations of the Owen value can be found in the literature [6–12].

Shapley value reflects the players' power in the alliance. It is a solution of the classical cooperative game theory. Classical cooperative game based on two assumptions. One suggests that players participate some certain union completely, i.e., each player either join in a union or not. It does not be other probability for players participating or not-participating in an alliance. The other suggests players in the game be completely independent, without exchange of information, matter and energy. So the players make the decision all by themselves.

Actually, it is impossible for players being completely independent. Exchange of information, material and energy is unavoidable, which constitute a network. Considering the problem in the context of social network, seem more accurate to describe. This paper would introduce the method of Social Network Analysis (SNA), to modify Shapley value.

The organization is as follows. In second section, the framework of the paper is stated and a minimum of preliminaries is provided. In third section, we introduce SNA and modify the traditional Shapley value. A simply case is studied in fourth section. Finally, the fifth section collects some conclusions and describes prospect of the research.

2 Preliminaries

In this paper, we consider cooperative games with the set of players $N = \{1, 2, \ldots, n\}$. A coalition T is a nonempty subset of N, which is identified with a function v from N to $\{0, 1\}$. Function v standing a contribution function, $v(T)$ means the payoff level of coalition T. For any coalition S, if $v(S \cap T) = v(S)$ satisfied, T is called the carrier of the game. Let π be an arrangement of N, define a game (N, π, υ), for any coalition $S = \{i_1, i_2, \ldots, i_s\}$, $U(\pi(i_1), \pi(i_2), \ldots, \pi(i_s)) = v(S)$ satisfied.

Theorem 1 (Optimum or Validity). *Let $\Phi_i(T)$ denotes the payoff sum of all players.*

If T is the carrier of v, then $\sum_{i=1} \Phi_i(v) = \Phi_i(T)$;

Theorem 2 (Symmetry). *If two players substitute each other, the payoff is unalterable, i.e. for any π of N, $\Phi_{\pi_i}(\pi v) = \Phi_i(v)$ satisfied.*

Theorem 3 (Additivity). *The payoff of the sum of two games equals to the sum of the payoff of two games. i.e. For $u, v \in G$, $\Phi_i(u + v) = \Phi_i(u) + \Phi_i(v)$ satisfies.*

It could be proved that Shapley value is unique theoretically when the three theorems are satisfied.

$$\Phi_i(v) = \sum_{i \in S \subseteq N} \frac{(|S|-1)!(n-|S|)!}{n!}[v(S) - v(S\backslash\{i\})] \tag{1}$$

Where $|S|$ denotes the number of the players involved in the game, $S\backslash\{i\}$ denotes the set S without the player i.

3 Social Net and Modified Shapley Value

In a general opinion, it was Brown, the English anthropologist, who firstly put forward the conception of 'social net', which is the set made up of many nodes and segments between the nodes [13]. The node can be a person, an organization, a team and even a country. In that, the SNA theory can be used to study different unit.

Social network analysis method uses absolute centrality and relative centrality to measure the individual status and influence in the network. Absolute centrality A_i refers to the numbers of connecting segments of individual i to others. More times individual connect to others, greater effect it would show and bigger the absolute centrality becomes. The absolute centrality expression is as follows

$$Ai = \frac{o_i + i_i}{2} \tag{2}$$

Where o_i stands for out degree and i_i for in degree.

When we analysis individual position in network and its behavior tendency, absolute center of pure research are not of much significance. In this paper Shapley value is studied based on the relative centrality, which is normalized results of absolute centrality, and the expression is as shown in (3).

$$R_i = \frac{A_i}{\sum_i A_i} \tag{3}$$

In an alliance game, Shapley value and centrality in the network determined individual influence. So Shapley value can be modified with relative centrality.

$$\begin{aligned} \Gamma'_i &= \Phi_i(v) \cdot R_i \\ &= \left\{ \sum_{i \in S \subseteq N} \frac{(|S|-1)!(n-|S|)!}{n!}[v(S) - v(S\backslash\{i\})] \right\} \cdot \frac{A_i}{\sum_i A_i} \end{aligned} \tag{4}$$

After normalizing (4), we can get (5).

$$\Gamma_i = \frac{\Gamma'_i}{\sum_i \Gamma'_i} \tag{5}$$

And we get the result Γ_i that has been modified. Connected with the status in networks of individual i, it would be able to show the real influence of individual.

The following steps show the process of the method.

Step 1: Calculate the Shapley value of individual i by (1);
Step 2: Calculate the absolute centrality and relative centrality of individual I by (2), (3);
Step 3: Calculate the modified Shapley value by (4), (5).

4 Case Studies

Considered a family of three generations includes seven people: grandpa, grandma, grandpa in law, grandma in law, father, mother and daughter. Usually when they have some decisions to be made, they would vote though negotiation. Generally subject can get through if more than four persons agree. However, being spoiled by adults, the daughter's suggestion would get through if anyone of the members agrees. Let's sign the grandpa, grandma, grandpa in law, grandma in law, father, mother and daughter with 1,2,3,4,5,6,7 corresponding. Let's note $N = \{1, 2, 3, 4, 5, 6, 7\}$. And we will get:

$$\begin{aligned}
&v(S) = 1, \forall S \subseteq N, |S| \geq 4,\\
&v(1,7) = 1, v(2,7) = 1, v(3,7) = 1, v(4,7) = 1, v(5,7) = 1, v(6,7) = 1,\\
&v(1,2,7) = 1, v(1,3,7) = 1, v(1,4,7) = 1, v(1,5,7) = 1, v(1,6,7) = 1,\\
&v(2,3,7) = 1,\ v(2,4,7) = 1,\ v(2,5,7) = 1, v(2,6,7) = 1, v(3,4,7) = 1,\\
&v(3,5,7) = 1, v(3,6,7) = 1, v(4,5,7) = 1, v(4,6,7) = 1, v(5,6,7) = 1
\end{aligned}$$

Under other conditions, we get $v(S) = 0$

All family members influence could be quantitative analyzed by Shapley index. For example, for individual i, we should calculate the following subsets that contains 1: {1,2,3,4}, {1,2,3,5}, {1,2,3,6}, {1,2,4,5}, {1,2,4,6}, {1,2,5,6}, {1,3,4,5}, {1,3,4,6}, {1,3,5,6}, {1,4,5,6}, {1,7}.

$$\Phi_1(v) = 10 \times \frac{(4-1)!(7-4)!}{7!} + \frac{(2-1)!(7-2)!}{7!} = \frac{2}{21}$$

And in the same way we can get:

$$\Phi_2(v) = \Phi_3(v) = \Phi_4(v) = \Phi_5(v) = \Phi_6(v) = \frac{2}{21}, \quad \Phi_7(v) = \frac{9}{21}.$$

So Shapley value of the family members is described as $\left(\frac{2}{21}, \frac{2}{21}, \frac{2}{21}, \frac{2}{21}, \frac{2}{21}, \frac{2}{21}, \frac{9}{21}\right)$.

The calculation process shown above does not take the relationship in networks into account. After observing the relations between the family members, the figure of relation networks can be drawn as Fig. 1.

Fig. 1 Relations between the family members

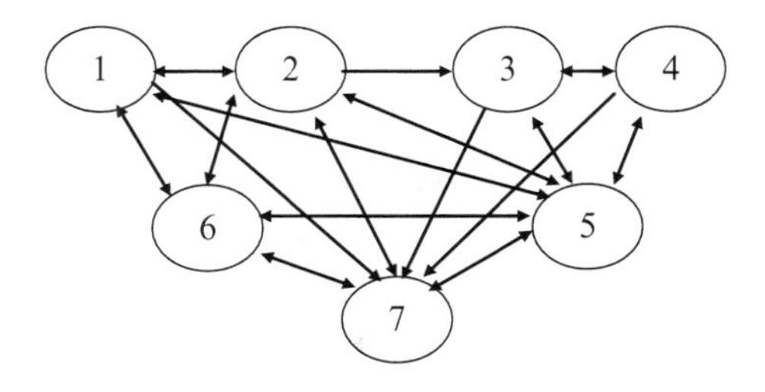

According to (2), (3), we can calculate the centrality of the family members. The result of absolute centrality is shown as following:

$$A_1 = \frac{7}{2}, A_2 = \frac{9}{2}, A_3 = \frac{6}{2}, A_4 = \frac{5}{2}, A_5 = \frac{12}{2}, A_6 = \frac{8}{2}, A_7 = \frac{9}{2},$$

And relative centrality as

$$R_1 = 0.125, R_2 = 0.161, R_3 = 0.107, R_4 = 0.089, R_5 = 0.214, R_6 = 0.143,$$
$$R_7 = 0.161$$

According to Eqs.(4) and (5), modified Shapley value can be calculated and described as:

$$\Gamma_1 = 0.080, \Gamma_2 = 0.103, \Gamma_3 = 0.068, \Gamma_4 = 0.057, \Gamma_5 = 0.137, \Gamma_6 = 0.091,$$
$$\Gamma_7 = 0.462$$

By comparison with the results, we can see centrality has obvious effect on the power of members. Before modified by centrality, the Shapley values of grandpa, grandma, grandpa in law, grandma in law, father and mother equal to each other. On the contrary, the modified Shapley values are quite different. The tense relationship between parents-in-law and daughter-in-law (there is no information communication between node 3 and node 6, node 4 and node 6 as shown in Fig. 1) greatly weakened the grandparents' influence. Father has the capability to deal with both parties (node 5 has a two-way communication to any other node as shown in Fig. 1). So he has a bigger influence than the sum of that of grandma and grandpa. Granddaughter is the core of the family and she has more influence than that before modified.

5 Summaries

Study on the influence of individual in alliance game, should be in the framework of SNA. It is a helpful attempt to modify the Shapley value by relative centrality in this paper. The thought of the method is clear, simply to apply. From the case, we can see that modified results accord with reality well. From the perspective, this method can also be used in the enterprise alliance, integration of production, education and research and other related fields.

It is the first time to combine the game theory and social network analysis method, when the author study individual's network status considered the centrality only. Therefore the method is not perfect. More thorough discuss would be completed in future research.

Acknowledgments Y. F. acknowledges financial support from Fundamental Research Funds for the Central Universities (Grant No. WK216000002) and China postdoctoral Science Foundation on the 50th grant program (Grant NO. 2011M501043).

References

1. L.S. Shapley (1953) A value for n-person games. In: Contributions to the theory of games, ed. 28. Annals of mathematics studies, vol 2. Princeton University Press, Princeton, pp 307–317.
2. Aumann RJ, Shapley LS (1974) Values of Non-atomic games. Princeton University Press, Princeton, pp 76–92
3. Roth AE (1988) The Shapley value: essays in honor of Lloyd S. Shapley. Cambridge University Press, Cambridge, pp 18–26
4. Owen G (1977) Values of games with a priori unions. In: Henn R, Moeschlin O (eds) Mathematical economics and game theory. Springer, Berlin/Heidelberg/New York, pp 76–88
5. Owen G (1995) Game theory, 3rd edn. Academic Press, Orlando, San Diego, pp 127–155
6. Albizuri MJ (2008) Axiomatizations of Owen value without efficiency. Math Soc Sci 55(1):78–89
7. Amer R, Carreras F (1995) Cooperation indices and coalition value. TOP 3(1):117–135
8. Amer R, Carreras F (2001) Power, cooperation indices and coalition structures. In: Holler MJ, Owen G (eds) Power indices and coalition formation. Kluwer, Boston, pp 153–173
9. Hamiache G (1999) A new axiomatization of the Owen value for games with coalition structures. Math Soc Sci 37(3):281–305
10. Hart S, Kurz M (1983) Endogeneous formation of coalitions. Econometrica 51(4):1047–1064
11. Vázquez-Brage M, van den Nouweland A, Garcı´a-Jurado I (1997) Owen's coalitional value and aircraft landing fees. Math Soc Sci 34(3):273–286
12. Winter E (1992) The consistency and potential for values with coalition structure. Games Econ Behav 4(1):132–144
13. Wasserman S, Faust K (1994) Social network analysis: methods and applications. Cambridge University Press, Cambridge, pp 162–187

Customers' Equilibrium Balking Strategies in an M/M/1 Queue with Variable Service Rate

Le Li, Jinting Wang, and Feng Zhang

Abstract We consider a fully observable single-server Markovian queue with variable service rate, where the customers observe the queue length and the state of the server upon arrival. We assume that the arriving customers decide whether to join the system or balk based on a natural reward-cost structure. With considering waiting cost and reward, we study the balking behavior of the customers and derive the corresponding Nash equilibrium strategies for all customers.

Keywords Balking • M/M/1queue • Nash equilibrium strategies

1 Introduction

Recently, the economic analysis of customer behavior on the performance of a queueing system has been studied extensively. Early works on the M/M/1 model with a reward-cost structure include Naor [7] and Edelson and Hildebrand [4]. Hassin and Haviv [6] dealt with Equilibrium threshold strategies in queues with priorities. Until now, there is a increasing number of papers that deal with the economic analysis of the balking behavior of customers in variants of the M/M/1 queue, see e.g. Economou and Kanta [2] (M/M/1 queue with unreliable server), Sun et al. [8] (M/M/1 queue with setup/closedown times), Wang and Zhang [9] (M/M/1 queue with delayed repairs), Guo and Hassin [5] (strategic behavior and social optimization in markovian vacation queues) and Economou and Kanta [3] (the single-server constant retrial queue).

The present paper is to study the equilibrium behavior of the customers regarding balking in the framework of an M/M/1 queueing model. The balking behavior

L. Li • J. Wang (✉) • F. Zhang
Department of Mathematics, Beijing Jiaotong University, 100044 Beijing, People's Republic of China
e-mail: 11121753@bjtu.edu.cn; jtwang@bjtu.edu.cn; zhangfeng@bjtu.edu.cn

Z. Zhang et al. (eds.), *LISS 2012: Proceedings of 2nd International Conference on Logistics, Informatics and Service Science*, DOI 10.1007/978-3-642-32054-5_86,

of customers in an M/M/1 queueing model that model transportation stations is important. However, such systems often evolve in random environment, i.e. there is some external process that influences the service rates. We will investigate the single server Markovian queue with variable service rate in the present study and determine equilibrium balking strategies for the customers under full information.

The paper is organized as follow. In Sect. 2, we introduce the model and the re-ward-cost structure. In Sect. 3, we consider the equilibrium strategies for fully observable queues. In Sect. 4, some conclusions are given.

2 Description of the Model

We consider a transportation station with an infinite waiting room where customers arrive according to a Poisson process with rate λ. We assume the service alternates between two states with state space $I = \{1, 0\}$ that are exponentially distributed at rates ζ and θ respectively. When the service is at state 1, customers are served at a rate of μ. As natural factor or human factor influences the rate of service, the customers are served at a lower rate of μ_0 with service state 0. We assume the service times are both exponentially distributed and $\mu_0 < \mu$.

We represent the state of the station at time t by the pair $(N(t), I(t))$, where $N(t)$ and $I(t)$ denote the number of customers and the state of the server (1:normal working state; 0:working with a slower rate), It is clear that the process $\{N(t), I(t) : t \geq 0\}$ is a continuous-time Markov chain with non-zero transition rates by

$$\begin{array}{lll} q_{(n,i)(n+1,i)} = \lambda & & n = 0, 1, \ldots; i = 0, 1 \\ q_{(n,1)(n-1,1)} = \mu & q_{(n,0)(n-1,0)} = \mu_0 & n = 1, 2, 3, \ldots \\ q_{(n,0)(n,1)} = \theta & q_{(n,1)(n,0)} = \zeta & n = 0, 1, 2, \ldots \end{array}$$

We assume every customer receives reward of R utility units after service. In addition, there exists a waiting cost of C utility units per time unit that a customer remains in the system including the time of waiting in the queue and being served. We also assume that customers are risk neutral and maximize their expected net benefit. Their decisions are irrevocable that retrials of balking customers and reneging of entering customers are not allowed. In the next section we obtain equilibrium customer strategies for joining/balking in the fully observable information when customers know the exact state of the system (n, i) upon arrival.

3 Equilibrium Threshold Strategies

In this section, we show that there exist equilibrium strategies of threshold type in the fully observable case where customers are informed both the number of customers present and the state of the server upon arrival. A pure threshold strategy is specified by a pair $(n_e(0), n_e(1))$ and the balking strategy has the form 'while arriving at time t, observe $(N(t), I(t))$; enter if $N(t) \leq n_e(I(t))$ and balk otherwise'.

Theorem 1. *In the fully observable M/M/1 queue with variable service rate, there exist a pair of thresholds $(n_e(0), n_e(1))$, such that the strategy 'observe $(N(t), I(t))$ upon arrival, enter if $N(t) \leq n_e(I(t))$ and balk otherwise', and $(n_e(0), n_e(1)) = (\lfloor x_0 \rfloor, \lfloor x_1 \rfloor)$, where $x_i$is the unique root of equation*

$$S(n,i) = an + b_i c^{n+1} + d_i = 0, \quad i = 0, 1 \tag{1}$$

where

$$a = -\frac{C}{\mu\mu_0 + \mu_0\zeta + \mu\theta}\left(\mu_0 + \zeta + \theta + \frac{\mu_0\zeta(\mu - \mu_0)}{\mu\theta + \mu_0\zeta}\right) \tag{2}$$

$$b_1 = -\frac{\mu_0\zeta(\mu - \mu_0)C}{(\mu\theta + \mu_0\zeta)^2} \tag{3}$$

$$b_0 = \frac{\mu\theta(\mu - \mu_0)C}{(\mu\theta + \mu_0\zeta)^2} \tag{4}$$

$$c = \frac{\mu\mu_0}{\mu\mu_0 + \mu_0\zeta + \mu\theta} \tag{5}$$

$$d_1 = R - \frac{C}{\mu\mu_0 + \mu_0\zeta + \mu\theta}\left(\mu_0 + \zeta + \theta - \frac{\mu\mu_0^2\zeta(\mu - \mu_0)}{(\mu\theta + \mu_0\zeta)^2}\right) \tag{6}$$

$$d_0 = d_1 - \frac{C(\mu - \mu_0)}{\mu\theta + \mu_0\zeta} \tag{7}$$

Proof: It is obvious that for an arriving customer, his expected net reward if he enters is $S(n,i) = R - CT(n,i)$, where $T(n,i) = E[S|N^- = n, I^- = i]$ denotes his expected mean sojourn time given that he finds the systems at state $(N(t), I(t))$ upon his arrival. Then we have the system:

$$T(n,0) = \frac{1}{\mu_0 + \theta} + \frac{\mu_0}{\mu_0 + \theta}T(n-1,0) + \frac{\theta}{\mu_0 + \theta}T(n,1) \quad n = 1, 2, \ldots \tag{8}$$

$$T(0,0) = \frac{1}{\mu_0 + \theta} + \frac{\theta}{\mu_0 + \theta}T(0,1) \tag{9}$$

$$T(0,1) = \frac{1}{\mu + \zeta} + \frac{\zeta}{\mu + \zeta}T(0,0) \tag{10}$$

$$T(n,1)=\frac{1}{\mu+\zeta}+\frac{\mu}{\mu+\zeta}T(n-1,1)+\frac{\zeta}{\mu+\zeta}T(n,0)\quad n=1,2,\ldots \tag{11}$$

We can use (8), (9), (10) and (11) to obtain for $\forall n\in N$

$$\begin{aligned}T(n,1)=&\left(\frac{\mu_0+\zeta+\theta}{\mu\mu_0+\mu_0\zeta+\mu\theta}+\frac{\mu_0\zeta(\mu-\mu_0)}{(\mu_0\zeta+\mu\theta)(\mu\mu_0+\mu_0\zeta+\mu\theta)}\right)n\\&+\frac{\mu_0\zeta(\mu-\mu_0)}{(\mu_0\zeta+\mu\theta)^2}\left(\frac{\mu\mu_0}{\mu\mu_0+\mu_0\zeta+\mu\theta}\right)^{n+1}\\&+\frac{1}{\mu\mu_0+\mu_0\zeta+\mu\theta}\left(\mu_0+\zeta+\theta-\frac{\mu\mu_0^2\zeta(\mu-\mu_0)}{(\mu\theta+\mu_0\zeta)^2}\right)\end{aligned} \tag{12}$$

$$\begin{aligned}T(n,0)=&\left(\frac{\mu_0+\zeta+\theta}{\mu\mu_0+\mu_0\zeta+\mu\theta}+\frac{\mu_0\zeta(\mu-\mu_0)}{(\mu_0\zeta+\mu\theta)(\mu\mu_0+\mu_0\zeta+\mu\theta)}\right)n\\&-\left(\frac{\mu\theta(\mu-\mu_0)}{(\mu\theta+\mu_0\zeta)^2}\right)\left(\frac{\mu\mu_0}{\mu\mu_0+\mu_0\zeta+\mu\theta}\right)^{n+1}\\&+\frac{1}{\mu\mu_0+\mu_0\zeta+\mu\theta}\left(\mu_0+\zeta+\theta-\frac{\mu\mu_0^2\zeta(\mu-\mu_0)}{(\mu\theta+\mu_0\zeta)^2}\right)+\frac{\mu-\mu_0}{\mu\theta+\mu_0\zeta}\end{aligned} \tag{13}$$

Thus, $S(n,i)$ can be expressed as the statement of Theorem 1.

$$\begin{aligned}&S(n,1)-S(n-1,1)=-C\frac{1}{\mu\mu_0+\mu_0\zeta+\mu\theta}\\&\left(\mu_0+\zeta+\theta+\frac{\mu_0\zeta(\mu-\mu_0)}{\mu\theta+\mu_0\zeta}-\frac{\mu_0\zeta(\mu-\mu_0)}{\mu\theta+\mu_0\zeta}\left(\frac{\mu\mu_0}{\mu\mu_0+\mu_0\zeta+\mu\theta}\right)^n\right)\end{aligned} \tag{14}$$

As $0<\frac{\mu\mu_0}{\mu\mu_0+\mu_0\zeta+\mu\theta}<1$, $S(n,1)-S(n-1,1)$, is a monotone decreasing function in n. When $n=1$,we have $S(1,1)-S(0,1)<0$, so $S(n,1)-S(n-1,1)<0$, $n=1,2,3\ldots$. Similarly we can show that $S(n,0)-S(n-1,0)<0$, so $S(n,i), i=0,1$, is a monotone decreasing function. Let x_i be the unique solution of the equation $S(n,i)=0$. Hence, the threshold values $(n_e(0),n_e(1))=(\lfloor x_0\rfloor,\lfloor x_1\rfloor)$ such that an arriving customer decides to enter if and only if $N(t)\le n_e(I(t))$ when he observes $(N(t),I(t))$ upon arrival.

4 Conclusions

In this paper we considered the problem of analyzing customer strategic behavior, in the fully observable M/M/1 queue with variable service rate, where customers decide whether to join the queue or balk upon arrival. We study the equilibrium threshold strategies of all customers.

Acknowledgements This work is supported by National Natural Science Foundation of China (No.11171019), Program for New Century Excellent Talents in University (NCET-11-0568) and the Fundamental Research Funds for the Central Universities (No. 2011JBZ012).

References

1. Burnetas A, Economou A (2007) Equilibrium customer strategies in a single server Markovian queue with setup times. Queueing Syst 56:213–228
2. Economou A, Kanta S (2008) Equilibrium balking strategies in the observable single-server queue with breakdowns and repairs. Oper Res Lett 36:696–699
3. Economou A, Kanta S (2011) Equilibrium customer strategies and social-profit maximization in the single-server constant retrial queue. Nav Res Logist 58:107–122
4. Edelson NM, Hildebrand K (1975) Congestion tolls for Poisson queueing processes. Econometrica 43:81–92
5. Guo P, Hassin R (2011) Strategic behavior and social optimization in Markovian vacation queues. Oper Res 59:986–997
6. Hassin R, Haviv M (1997) Equilibrium threshold strategies: the case of queues with priorities. Oper Res 45:966–973
7. Naor P (1969) The regulation of queue size by levying tolls. Econometrica 37:15–24
8. Sun W, Guo P, Tian N (2010) Equilibrium threshold strategies in observable queueing systems with setup/closedown times. Cent Eur J Oper Res 18:241–268
9. Wang J, Zhang F (2011) Equilibrium analysis of the observable queues with balking and delayed repairs. Appl Math Comput 218:2716–2729

Optimal Balking Strategies in Single-Server Queues with Erlangian Service and Setup Times

Ping Huang, Jinting Wang, and Li Fu

Abstract In many service systems arising in OR/MS applications, the server needs a setup time when it is turned on and the service time may be non-Markovian as usually assumed. In the present paper, we study the balking behavior of customers in the single-server queue with two-phase Erlangian service time and setup times. Arriving customers decide whether to enter the system or balk, based on a linear reward-cost structure. We identify equilibrium threshold balking strategies under two distinct information assumptions, i.e. fully observable case and almost observable case.

Keywords $M/E_2/1$queue • Setup times • Balking • Equilibrium strategies

1 Introduction

In recent years, there is an emerging tendency in the literature to study queuing systems from an economic viewpoint. Customers are allowed to make their decisions as to how to queue based on a reward-cost structure. Early works on such topic include Naor [5] and Edelson and Hildebrand [2]. The fundamental results with extensive bibliographical references can be found in Hassin and Haviv [4].

Recently, Guo and Zipkin [3] studied the effect of information on a queue with balking and phase-type service times. In this paper, we restrict our attention to a more specific model where the service times belong to Erlang-2 distribution, in which the service time has two stages and hence it is not memoryless. Therefore the queue length is also a signal about the stage of service. Moreover, we study the system under two different levels of information: Fully observable case and almost

P. Huang • J. Wang (✉) • L. Fu
Department of Mathematics, Beijing Jiaotong University, Beijing 100044, China
e-mail: 11121751@bjtu.edu.cn; jtwang@bjtu.edu.cn; lfu@bjtu.edu.cn

Z. Zhang et al. (eds.), *LISS 2012: Proceedings of 2nd International Conference on Logistics, Informatics and Service Science*, DOI 10.1007/978-3-642-32054-5_87,

observable case. In the latter case, customers observe only the queue length upon arrivals. It has not been studied in Guo and Zipkin [3] and thus gives a new insight into the effect of partially observable information on the equilibrium behavior of the customers.

In the literature, the single-server queues with Erlangian service times have been well studied and the formulae for the mean, variance, the distribution of the number of customers and the sojourn time in a steady state are available, see e.g., Adan and van der Wal [1]. In this paper we consider equilibrium balking strategies under two information cases.

The paper is organized as follows. In Sect. 2, we describe the model and the reward-cost structure. Sections 3 and 4 study the equilibrium strategies for fully observable queue and almost observable queue. Section 5 provides conclusions.

2 The Model

We consider a single server queuing system with infinite waiting room in which customers arrive according to a Poisson Process with rate λ. The service times are assumed to obey the general Erlang-2 distribution (two independent exponential phases with mean $1/\mu_1$ and $1/\mu_2$). The server is deactivated as soon as the queue becomes empty. When a new customer arrives at an empty system, a setup process starts for the server to be reactivated. The time required for setup is also exponentially distributed with rate θ. During the setup customers continue to arrive. We assume that inter-arrival times, service times and setup times are mutually independent.

We represent the state at time t by the pair $(N(t), I(t))$, where $N(t)$ denotes the number of customers in the system and $I(t)$ denotes the state of the server (the state 1 and 2 denote the number of phases to be processed for the job in service, while the empty state can be denoted by 0). Every customer's utility consists of a reward R for service minus a waiting cost C. Customers want to maximize their expected net benefit. At the arrival instant, they make their decisions and the decisions are irrevocable.

Consider the observable case where customers are aware of the queue length upon arrival. We study separately the two information cases regarding whether they observe also the server's status (fully observable and almost observable case).

3 Equilibrium Threshold Strategies for Fully Observable Case

Theorem 1. *In the fully observable $M/E_2/1$ queue with setup times, thresholds exist:*

$$(n_e(0), n_e(1), n_e(2)) = \left(\left\lfloor \frac{R\mu_1\mu_2}{C(\mu_1+\mu_2)} - \frac{\mu_1\mu_2}{\theta(\mu_1+\mu_2)} \right\rfloor - 1, \left\lfloor \frac{R\mu_1\mu_2}{C(\mu_1+\mu_2)} \right\rfloor - 1, \left\lfloor \frac{R\mu_1\mu_2}{C(\mu_1+\mu_2)} - \frac{\mu_1}{\mu_1+\mu_2} \right\rfloor \right)$$

Such that the strategy 'While arriving at time t, observe $(N(t), I(t))$*; enter if* $N(t) \leq n_e(I(t))$ *and balk otherwise' is a weakly dominant strategy.*

Proof. A customer's expected utility if he enters the system is $S(n,i) = R - CT(n,i)$, where $T(n,i)$ denotes his expected mean sojourn time, given that he finds the system at state (n,i) upon arrival. We have the system $T(n,0) = \frac{1}{\theta} + \left(\frac{1}{\mu_1} + \frac{1}{\mu_2}\right)(n+1)$ for $n = 0, 1, 2, \ldots$ and $T(n,1) = \left(\frac{1}{\mu_1} + \frac{1}{\mu_2}\right)(n+1), T(n,2) = \frac{1}{\mu_2} + \left(\frac{1}{\mu_1} + \frac{1}{\mu_2}\right)n$, for $n = 1$ $, 2, 3 \ldots$. Solving $S(n,i) \geq 0$ for n, we obtain that the arriving customer prefers to enter if and only if $n \leq n_e(i)$.

4 Equilibrium Threshold Strategies for Almost Observable Case

Propositions 1. *Consider an almost observable* $M/E_2/1$ *queue with setup times for* $\rho_1 \neq \sigma \neq \rho_2$ *and all of them are different from 1. Customers enter the system according to a threshold strategy 'While arriving at time t, only observe* $N(t)$*; enter if* $N(t) \leq n_e$*, and balk otherwise'. The stationary distribution* $(P(n,i) : (n,i) \in \{0, 1, 2 \ldots, n_e + 1\} \times \{0, 1, 2\})$ *is:*

$$P(n,0) = \frac{\mu_2}{\lambda}\sigma^n P(1,2), n = 0, 1, \ldots, n_e, \quad P(n_e+1, 0) = \frac{\mu_2}{\theta}\sigma^{n_e} P(1,2),$$

$$P(n,1) = \left[\frac{\lambda + \mu_2}{\mu_1}\left((\alpha - \beta)\rho_1^n - (\alpha + \beta)\rho_2^n + 2\beta\sigma^n\right) - \frac{\lambda}{\mu_1}\left((\alpha - \beta)\rho_1^{n-1} - (\alpha + \beta)\rho_2^{n-1} + 2\beta\sigma^{n-1}\right) \right] P(1,2), n = 1, 2, \ldots, n_e,$$

$$P(n_e+1, 1) = \left[\frac{\mu_2}{\mu_1}\left((\alpha - \beta)\rho_1^{n_e+1} - (\alpha + \beta)\rho_2^{n_e+1} + 2\beta\sigma^{n_e+1}\right) - \frac{\lambda}{\mu_1}\left((\alpha - \beta)\rho_1^{n_e} - (\alpha + \beta)\rho_2^{n_e} + 2\beta\sigma^{n_e}\right) \right] P(1,2)$$

$$P(n,2) = \left((\alpha - \beta)\rho_1^n - (\alpha + \beta)\rho_2^n + 2\beta\sigma^n\right) P(1,2), n = 1, 2, \ldots n_e + 1,$$

where

$$P(1,2)=\left[\frac{\mu_1+\mu_2}{\mu_1}\left((\alpha-\beta)\frac{\rho_1(1-\rho_1^{n_e+1})}{1-\rho_1}-(\alpha+\beta)\frac{\rho_2(1-\rho_2^{n_e+1})}{1-\rho_2}+2\beta\frac{\sigma(1-\sigma^{n_e+1})}{1-\sigma}\right)\right.$$
$$\left.+\frac{\mu_2}{\lambda}\frac{1-\sigma^{n_e+1}}{1-\sigma}+\frac{\mu_2}{\theta}\sigma^{n_e}\right]^{-1},$$

$$\alpha=\frac{\mu_1\mu_2(\lambda-\mu_1-\mu_2+2\theta)}{2\sigma\sqrt{(\lambda+\mu_1+\mu_2)^2-4\mu_1\mu_2[\mu_1\mu_2-(\lambda+\theta)(\mu_1+\mu_2-\theta)]}},$$
$$\beta=\frac{\mu_1\mu_2}{2\sigma[\mu_1\mu_2-(\lambda+\theta)(\mu_1+\mu_2-\theta)]},$$

$$\sigma=\frac{\lambda}{\lambda+\theta},\rho_{1,2}=\frac{\lambda\left(\lambda+\mu_1+\mu_2\pm\sqrt{(\lambda+\mu_1+\mu_2)^2-4\mu_1\mu_2}\right)}{2\mu_1\mu_2}$$

Proof. According to the argument, we can give the balance equations. By using the standard approach for solving a non-homogeneous linear difference equation with constant coefficients, we can express all stationary probabilities in terms of $P(1,2)$,The remaining probability, $P(1,2)$, can be found from the normalization equation.

Propositions 2. *Following the assumptions and conclusions in Proposition 1, the net benefit of a customer that observes n customers and decides to enter is given by*

$$S(n)=R-C\left(\frac{1}{\mu_1}+\frac{1}{\mu_2}\right)(n+1)$$
$$+\frac{\frac{C\lambda}{\mu_1\mu_2}\left[(\alpha-\beta)\left(\frac{\rho_1}{\sigma}\right)^n-(\alpha+\beta)\left(\frac{\rho_2}{\sigma}\right)^n+2\beta\right]-\frac{C}{\theta}}{1+(\alpha-\beta)\rho_1\left(\frac{\rho_1}{\sigma}\right)^n-(\alpha+\beta)\rho_2\left(\frac{\rho_2}{\sigma}\right)^n+2\beta\left(\rho_1+\rho_2-\frac{\rho_1\rho_2}{\sigma}\right)},$$
$$n=0,1,\ldots,n_e \tag{1}$$

$$S(n)=R-C\left(\frac{1}{\mu_1}+\frac{1}{\mu_2}\right)(n+1),$$
$$+\frac{\frac{C\lambda}{\mu_1\mu_2}\left[(\alpha-\beta)\left(\frac{\rho_1}{\sigma}\right)^n-(\alpha+\beta)\left(\frac{\rho_2}{\sigma}\right)^n+2\beta\right]-\frac{C}{\theta}-\frac{C\lambda}{\theta^2}}{1+\frac{\lambda}{\theta}+(\alpha-\beta)\rho_1(1-\rho_2)\left(\frac{\rho_1}{\sigma}\right)^n-(\alpha+\beta)\rho_2(1-\rho_1)\left(\frac{\rho_2}{\sigma}\right)^n+2\beta\left(\rho_1+\rho_2-\rho_1\rho_2-\frac{\rho_1\rho_2}{\sigma}\right)},$$
$$n=n_e+1 \tag{2}$$

Proof. The expected net reward, if he enters, for a customer that observes n customers is $S(n) = R - CT(n)$, where $T(n)$ denotes his expected mean sojourn time given that he finds n customers in the system just before his arrival. And $T(n) = T(n,0)\Pr(I^- = 0|N^- = n) + T(n,1)\Pr(I^- = 1|N^- = n) + T(n,2)\Pr(I^- = 2|N^- = n)$.

Where $T(n,i)$ is obtained by the proof of Theorem 1 and $\Pr(I^- = i|N^- = n)$ is the probability that an arriving customer finds the server at state i, given that there are n customers, and it can be obtained by the stationary probabilities given in Proposition 1. Now the substitution of $T(n)$ in $S(n) = R - CT(n)$ yields (1) and (2).

Next, we establish the existence of equilibrium threshold policies in the almost observable case. Define the sequences $(f_1(n) : n = 0, 1, 2\ldots)$ and $(f_2(n) : n = 0, 1, 2\ldots)$: $f_1(n)$ and $f_2(n)$ are equivalent to the expression of $S(n)$ in (1) and (2) respectively.

It is easy to get that $f_1(0) = f_2(0) = R - C(1/\theta + 1/\mu_1 + 1/\mu_2) > 0$ and $\lim_{n\to\infty} f_1(n) = \lim_{n\to\infty} f_2(n) = -\infty$. Hence, there exist n_U such that

$$f_1(0), f_1(1), \ldots, f_1(n_U) > 0 \text{ and } f_1(n_U + 1) \le 0 \tag{3}$$

On the other hand, $f_1(n) \ge f_2(n)$, $n = 0, 1, 2, \ldots$. So $f_2(n_U + 1) \le 0$ while $f_2(0) > 0$. Hence there exists $n_L \le n_U$, such that

$$f_2(n_L) > 0 \quad \text{and} \quad f_2(n_{L+1}), \ldots, f_2(n_U), f_2(n_U + 1) \le 0 \tag{4}$$

Theorem 2. *In the almost observable $M/E_2/1$ queue with setup times, all pure threshold strategies 'observe $N(t)$;enter if $N(t) \le n_e$ and balk otherwise', for $n_e \in \{n_L, n_L + 1, \ldots, n_U\}$ are equilibrium strategies.*

Proof. We consider a tagged customer at his arrival instant, then his net benefit if he observes n customers and decides to enter is given by (1 and (2). It is easy to see, based on (1) and (3), that the customer prefers to enter when he finds $n \le n_e$ customers. Similarly, based on (2) and (4), we see that the customer prefers to balk when he finds $n = n_e + 1$ customers. Such a strategy is an equilibrium.

5 Conclusion

To summarize, we analyzed customer behavior in equilibrium in an $M/E_2/1$ queue with setup times where customers decide whether to join upon arrival and identified equilibrium threshold balking strategies under two distinct information assumptions.

Acknowledgements This work is supported by National Natural Science Foundation of China (No. 11171019), Program for New Century Excellent Talents in University (NCET-11-0568) and the Fundamental Research Funds for the Central Universities (No. 2011JBZ012).

References

1. Adan I, van der Wal J (1998) Difference and differential equations in stochastic operations research, online notes. URL:http://www.win.tue.nl/~iadan/
2. Edelson NM, Hildebrand K (1975) Congestion tolls for Poisson queueing processes. Econometrica 43:81–92
3. Guo P, Zipkin P (2008) The effects of information on a queue with balking and phase-type service times. Nav Res Logist 55:406–411
4. Hassin R, Haviv M (2003) To queue or not to queue: equilibrium behavior in queueing systems. Kluwer Academic Publishers, Boston
5. Naor P (1969) The regulation of queue size by levying tolls. Econometrica 37:15–24

The Antecedents and Consequences of Service Climate in G2B e-Government Service Providers: A Case Study of China

Dongyuan Wang, Zirui Men, Hong Ge, and Yuqiang Feng

Abstract This study explores the antecedents and consequences of service climate in G2B e-government service providers. We employed a case study method in this exploratory research with the objectives to identify key influential factors of service climate from both organizational factors and psychological factors and its impact on service quality. Interviews with frontline service employees and their managers revealed ten factors from organization such as training, and confirmed the impact of positive psychological capital. Our findings highlight the positive psychology and some special factors like IT-based management as predictors of service climate and provide a new managerial insight on how to improve service quality. At last, limitations and suggestions for future research directions are discussed.

Keywords Service climate • Service quality • Psychological capital • Job resource • Case study • G2B e-government

1 Introduction

E-government, referring to the use of wired-Internet technology by public-sector organizations to better deliver their services and improve their efficiency, has achieved significant development globally [1]. G2B e-government (G2B) plays an important role in China because of mandatory use to the enterprise, such as

---

This research was funded by the Chinese National Natural Science Foundation (Project No: 71172157).

D. Wang (✉) • Z. Men • H. Ge • Y. Feng
Management Science and Engineering, School of Management,
Harbin Institute of Technology, Harbin, China

Skycloud Co., Ltd, Cixi, China
e-mail: dongyuanhit@gmail.com; menzr@chinaskycloud.com; hge@hit.edu.cn; Fengyq@hit.edu.cn

Z. Zhang et al. (eds.), *LISS 2012: Proceedings of 2nd International Conference on Logistics, Informatics and Service Science*, DOI 10.1007/978-3-642-32054-5_88,

"Golden Tax Revenue" and some other projects headed by the acronym "Golden". It is because of mandatory use to the enterprise, enhancing and optimizing G2B e-government service quality is very important to improve enterprise acceptance and satisfaction. Usually the concrete implement method is that, G2B e-government is entrusted and authorized to a third party, namely e-government service providers.

In service encounters, frontline service employees exchange with the enterprise frequently and directly, whose attitude and behavior will directly affects the customer perceived service quality. Schneider et al. pointed out that service climate refers to employee perceptions of the practices, procedures, and behaviors that get rewarded, supported [2]. Organizations should pay more attention to service quality of employee and provide required resource to build better service climate. Then (1) which factors will influence service climate in the Chinese G2B e-government service providers' context? (2) How employee's psychological factors affect service climate when service employee is responsible for mandatory product and service? To solve these problems, we design a case study protocol and interviewed with frontline service employees, than we build a model about the antecedents and consequences of service climate in e-government service providers.

2 Literature Review

Service climate refers to employee perceptions of the practices, procedures, and behaviors that get rewarded, supported, and expected with regard to customer service and customer service quality. Schneider et al. proved the relationships between service climate and service quality [2, 3]. Some academics have also confirmed that service climate will influence customer satisfaction and company profits [4, 5]. While service quality is the most important and most studied in service science. Parasuraman, Zeithaml and Berry proposed a multiple-item scale for measuring consumer perception of service quality according to their 5GAP model, called SERVQUAL [6].

Schneider pointed out that the organizations need to provide the necessary resources, training and policies and remove barriers in the service process to improve the quality and climate of service. Job resources refer to those physical, psychological, social, or organizational aspects of the job [7]. Different scholars defined corresponding resources in their research such as job autonomy, career development, participation in decision-making and so on [8]. Luthans proposed that positive psychological capital consists of "who you are", beyond economic capital (what you have), human capital (what you know) and social capital (who you know) [9]. The potential of human is unlimited, and its roots lie in a person's psychological capital, such as confidence, hope, optimism, and resilience.

3 Methodology

Case study, an important research method in social sciences, is one of the main methods of empirical research. Following the case study protocols and guidelines [10, 11], a case study protocol was developed that specified the ideal profiles of company and employees, the interview protocol, and the open-ended questions based on the literature and the focal phenomenon of interest, e.g. "How did you perceive that company attach importance to the quality of services, from what, why?".

The research team selected a typical G2B e-government service provider in China, named as H and the only authorized unit to be responsible for Golden Tax Project's technical services. H has more than 600 employees, of which more than 350 frontline service employee. H have won first prize in the comprehensive evaluation of the quality of service in 2008, 2009 and 2011. The data was collected by face-to-face interviews and telephone interviews following the case study protocol. All interview recordings were recorded and organized into the text material for analysis. The employees interviewed are numbered by E1, E2, E3, E4 and E5 in this paper.E1 is a vice-president and responsible for company's service quality. E2, E3 and E4 are service station manager and responsible for an service station. E5 is a frontline service employee. Half of the interviewees are middle-level staff as they had rich grass-roots working experience and a good understanding of company's strategy.

4 Case Analysis and Findings

After in-depth analysis of interview transcripts, we summed up factors influencing service climate from organizational aspect and psychological capital and service climate's impact on service quality. Next our research findings will be displayed and complemented by a large number of the original evidence.

When asked what affected service quality of the employees in the interview process, all of the interviewees mentioned the influence of contingent reward. The senior leadership E1 said "*...As the bonuses are depended on the quality of service, now there are less dissatisfied and invalid receipts....* ". The middle manager E2 said " *...Our Company started with service. If there is a complaint, the responsible employee will be immediately dismissed. Recently our company will soon rank technical experts...*" The grassroots manager E3 mentioned that "*We pay employees for performance, rate star employees of front-line and have competition of reserve cadre...*" From the interview, we can found that contingent rewards have a significant impact on the quality of service.

Through analyzing the interview text, we also found that other job resources, such as support from colleague, IT-based management, professional development, regulatory pressure and supervisory coaching, leading to concern the perception of the

Table 1 Influencing factors of organization psychology from interviews

Influencing factors	E1	E2	E3	E4	E5
Training			√	√	√
Support from Colleague			√	√	
Professional development		√	√	√	
Partition in decision-making				√	
Performance feedback	√				√
IT-based management	√		√	√	
Team climate		√	√	√	√
Contingent reward	√	√	√	√	√
Supervisory coaching		√	√	√	
Regulatory stress		√	√		√
Self-efficiency		√	√		
Hope			√	√	
Resilience		√	√	√	
Optimism			√	√	√

(supervisory coaching), also had influence on the employees perceived he importance of the service quality. All organizational factors are given in Table 1 in details.

Compared with organizational factors, the psychological factors of employees is not easy to express. Through analysis of interview text and observing the behavior of employees in the interview process, positive psychological factors have been verified to some degrees. Resiliency was defines by Luthans as "the capacity to rebound or bounce back from adversity, conflict, failure, or even positive events, progress, and increased responsibility". When asked what the service quality of employees is important, E2 says "*conversation, manners, state of mind, resiliency is important for both frontline service employees and managers.... employees should delve into the subject when encountering problems or technical difficulties...*" *when talking about service encounter, E5 expressed "... (In the service process) sometimes the problem is not attributable to our products, which will produce conflict. At this time we give customers a clear explanation patiently.*"

Optimism is to make a positive attribution for the future and current success. Self-efficiency is one's conviction or confidence about his or her abilities to mobilize the motivation, cognitive resources, and courses of action needed to successfully execute a specific task within a given context. And Hope is a positive motivational state that is based on an interactively derived sense of successful (1) agency (goal-directed energy) and (2) pathways (planning to meet goals). These three positive psychological factors, as three dimensions of psychological capital, were also confirmed in the interviews. All psychological factors are given in Table 1 in details.

Essentially, service climate is the importance of service quality perceived by employees. A number of interviewees mentioned that the company paid a lot of attention on service quality. E2 expressed "*Our employee will be immediately dismissed when an employee receives complaints from customer...*" E3 described "*... the company's development strategy is (quality) services, and the CEO attaches great importance to the quality service ... Employees' life, transport*

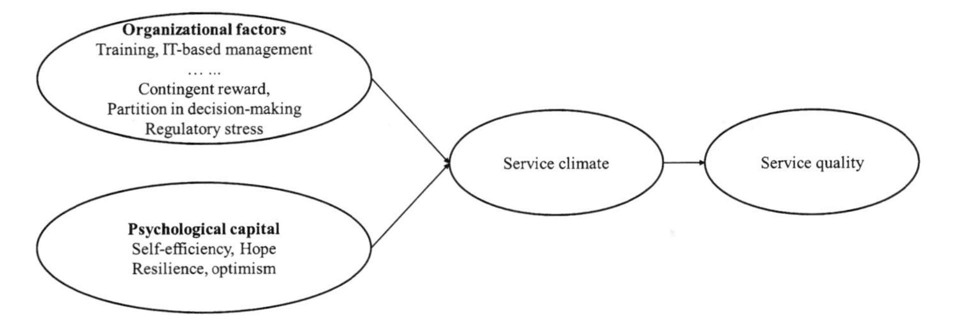

Fig. 1 The antecedents and consequences of service climate model

and treatment had been significantly improved... Company is changing employees' service attitude..." We can conclude that the quality of service and high customer satisfaction come from the company's strong service climate.

We summarized the antecedents of service climate from organizational aspect and positive psychological capital aspect and analyzed the impact of service climate on the quality of service. The antecedents and consequences of service climate can be described using the model as shown in Fig. 1.

5 Conclusion

This study explores the antecedents and consequences of service climate in G2B e-government service providers through case study. We summarized the ten organizational factors and psychological capital factors on the service climate, also confirmed the influence of service climate on service quality. From the perspective of management implication, managers can build better service climate from the aspects of organization and staff psychology to improve service quality. There are some limitations in this study, but showing new directions clearly for further research. The conclusion is drawn from a typical G2B e-government service provider. The limitation provides future research points and ample research space for the future research. According to the proposed theoretical model in this paper, the authors have formulated the questionnaire and will implement a quantitative study of the model.

References

1. Trimi S, Sheng H (2008) Emerging trends in m-government. Commun ACM 51(5):53–58
2. Schneider B, White SS, Paul MC (1998) Linking service climate and customer perceptions of service quality: tests of a causal model. J Appl Psychol 83(2):150

3. Schneider B, Bowen DE (1985) Employee and customer perceptions of service in banks: replication and extension. J Appl Psychol 70(3):423
4. Schneider B et al (2005) Understanding organization-customer links in service settings. Acad Manag J 48:1017–1032
5. Kralj A, Solnet D (2010) Service climate and customer satisfaction in a casino hotel: an exploratory case study. Int J Hosp Manag 29(4):711–719
6. Parasuraman A, Zeithaml VA, Berry LL (1988) SERVQUAL: a multiple item scale for measuring consumer perceptions of service quality. J Retail 64(1):12–37
7. Bakker AB et al (2003) Job demands and job resources as predictors of absence duration and frequency. J Vocat Behav 62(2):341–356
8. Xanthopoulou D et al (2009) Reciprocal relationships between job resources, personal resources, and work engagement. J Vocat Behav 74(3):235–244
9. Luthans F, Luthans KW, Luthans BC (2004) Positive psychological capital: beyond human and social capital. Bus Horiz 47(1):45–50
10. Yin RK (2009) Case study research: design and methods. vol 5. Sage publications INC, Thousand Oaks
11. Liu L et al (2011) From transactional user to VIP: how organizational and cognitive factors affect ERP assimilation at individual level. Eur J Inf Syst 20(2):186–200

Equilibrium Analysis of the Markovian Queues with Repairs and Vacations

Ceng Li, Jinting Wang, and Feng Zhang

Abstract A single server Markovian queue with repairs and vacations is studied in this paper. Under a linear reward-cost structure, we investigate the behavior of customers with various levels of information regarding the system state. Equilibrium strategies for the customers under different levels of information are derived and the stationary behaviors of the system under these strategies are investigated.

Keywords Equilibrium strategies • Breakdowns • Vacations

1 Introduction

Studies about the economic analysis of queueing systems can go back at least to the pioneering work of Naor [5] who analyzed customers optimal strategies in an observable M/M/1 queue with a simple reward-cost structure. After the work of Naor, Edelson and Hildebrand [3] investigated this model. Then surveys on this subject were showed in [1, 2, 4] respectively. The paper is organized as follows. In Sect. 2, we describe the dynamics of the model and the reward-cost structure. In Sect. 3, we derive the equilibrium strategies for the customers under different levels of information. Finally, in Sect. 4, some conclusions are given.

C. Li • J. Wang (✉) • F. Zhang
Department of Mathematics, Beijing Jiaotong University, Beijing 100044, China
e-mail: 10121819@bjtu.edu.cn; jtwang@bjtu.edu.cn; zhangfeng@bjtu.edu.cn

Z. Zhang et al. (eds.), *LISS 2012: Proceedings of 2nd International Conference on Logistics, Informatics and Service Science*, DOI 10.1007/978-3-642-32054-5_89,

2 Model Description

We consider a single-server queue in which customers arrive according to a Poisson process with rate λ. We assume that the service times are exponentially distributed with rate u. The server alternates between on and off periods that are also exponentially distributed at rates ξ and θ respectively.

And the server is deactivated and begins a vacation as soon as the queue becomes empty. When a new customer arrives at an empty system, a setup process starts for the server to be activated. The time required to setup is also exponentially distributed with rate β. We assume that inter-arrival times, service times and setup times are mutually independent.

We represent the state at time t by the pair $(N(t); I(t))$, where $N(t)$ denotes the number of customers in the system and $I(t)$ denotes the state of the server. Hereinafter, we define a breakdown period at state 0 and a work period at state1 as well as a vacation period at state 2. We assume that every customer receives reward of R units for completing service. Moreover, there exists a waiting cost of C units per time unit.

In the next sections we obtain customer optimal strategies for joining/balking. We distinguish two cases with respect to the level of information available to customers at their arrival instants:

(1) Fully observable case: Customers are informed about the queue length as well as the sever state $(N(t), I(t))$;
(2) Fully unobservable case: Customers are not informed about the queue length $N(t)$ nor the sever state $I(t)$.

3 Equilibrium Strategies

In this section we show that there exist equilibrium strategies in the two information cases that we defined above.

3.1 Fully Observable Case

We begin with the fully observable case in which there exist equilibrium strategies of threshold type.

Theorem 1. *In the fully observable $M/M/1$ queue with repairs and vacations, there exist a triple of thresholds*

$$(n_e(0), n_e(1), n_e(2)) = \left(\left\lfloor \left(\frac{R}{C} - \frac{1}{\theta}\right) \frac{\theta\mu}{\theta + \xi} \right\rfloor - 1, \left\lfloor \frac{R}{C} \frac{\theta\mu}{\theta + \varepsilon} \right\rfloor - 1, \right.$$
$$\left. \left\lfloor \left(\frac{R}{C} - \frac{1}{\beta}\right) \frac{\theta\mu}{\theta + \varepsilon} \right\rfloor - |1 \right). \quad (1)$$

Such that the strategy 'While arriving at time t, observe (N(t), I(t)); enter if $N(t) \leq n_e(I(t))$ *and balk otherwise' is a weakly dominant strategy.*

Proof. Consider an arriving customer, his expected net reward if he enter is $S(n; i) = R - CT(n; i)$, where $T(n; i)$ denotes his expected mean sojourn time given that he finds the system at state $(N(t); I(t))$ upon his arrival. We have

$$T(n, 0) = \frac{1}{\theta} + T(n, 1), \quad T(n, 2) = \frac{1}{\beta} + T(n, 1), \qquad n = 1, 2, \ldots, \quad (2)$$

$$T(n, 1) = \frac{1}{\mu + \varepsilon} + \frac{\varepsilon}{\mu + \varepsilon} T(n, 0) + \frac{\mu}{\mu + \varepsilon} T(n - 1, 1), \qquad n = 2, 3, \ldots. \quad (3)$$

We denote T_0 by the expected mean service time of a customer, then we get

$$T_0 = \frac{\mu}{\mu + \varepsilon} \cdot \frac{1}{\mu} + \frac{\varepsilon}{\mu + \varepsilon} \left(\frac{1}{\theta} + T_0 \right), T(1, 1)$$
$$= \frac{\mu}{\mu + \varepsilon} \left(\frac{1}{\mu} + T_0 \right) + \frac{\varepsilon}{\mu + \varepsilon} T(1, 0), T(0, 2) = \frac{1}{\beta} + T_0.$$

Then the following equations are obtained easily

$$T(n, 0) = \frac{1}{\theta} + (n + 1) \frac{\theta + \xi}{\theta\mu}, T(n, 1) = (n + 1) \frac{\theta + \xi}{\theta\mu}, \qquad n = 1, 2, \ldots, \quad (4)$$

$$T(n, 2) = \frac{1}{\beta} + (n + 1) \frac{\theta + \xi}{\theta\mu}, \qquad n = 0, 1, \ldots. \quad (5)$$

By solving $S(n, i) \geq 0$ for n, we obtain that customers decide to enter if and only if $n \leq n_e(I(t))$ where n_e (0), n_e (1) and n_e (2) are given by Theorem 1.

3.2 Fully Unobservable Case

In this section we proceed the fully unobservable case. With a mixed strategy, an arriving customer joins with a certain probability q, and the effective arrival rate, or joining rate is λq.

Proposition 1. *Consider the fully unobservable model of the M/M/1 queue with repairs and vacations, the expected mean sojourn time of a customer who decides to enter is given by*

$$E[W] = \frac{(\xi\beta - \xi\theta - \theta^2)\lambda q + \theta(\xi\beta + \theta\beta + \theta\mu)}{\theta\beta[\theta\mu - \lambda q(\xi + \theta)]}. \tag{6}$$

Proof. Let $p(n, i)$ be the stationary distribution of the corresponding system and the balance equations are presented below.

$$(\lambda q + \theta)p(1,0) = \xi p(1,1), \tag{7}$$

$$(\lambda q + \theta)p(n,0) = \lambda q p(n-1,0) + \xi p(n,1), \quad n = 2,3,\ldots, \tag{8}$$

$$(\mu + \xi + \lambda q)p(1,1) = \theta p(1,0) + \mu p(2,1) + \beta p(1,2), \tag{9}$$

$$\begin{aligned}(\mu + \varepsilon + \lambda q)p(n,1) = \theta p(n,0) + \mu p(n+1,1) + \beta p(n,2) \\ + \lambda q p(n-1,1), n = 2,3,\ldots,\end{aligned} \tag{10}$$

$$\lambda q p(0,2) = \mu p(1,1) \tag{11}$$

$$(\beta + \lambda q)p(n,2) = \mu p(n-1,2), \quad n = 1,2,\ldots. \tag{12}$$

Define the partial stationary probability generating function of the system as $G_i(z) = \sum_{n=0}^{\infty} p(n,i)z^n, (|z| \leq 1, i = 0,1,2)$ and we can get $G_0(z)$, $G_1(z)$ and $G_2(z)$ by solving the above balance equations. Then the mean number of the customers in the system is as follows

$$\begin{aligned}E[N] &= \sum_{n=0}^{\infty} n(p(n,0) + p(n,1) + p(n,2)) = G'_0(z)|_{z=1} + G'_1(z)|_{z=1} + G'_2(z)|_{z=1} \\ &= \frac{(\xi\beta - \xi\theta - \theta^2)\lambda^2 q^2 + \theta(\xi\beta + \theta\beta + \theta\mu)\lambda q}{\theta\beta[\theta\mu - \lambda q(\xi + \theta)]}.\end{aligned} \tag{13}$$

Hence, the expected mean sojourn time of a customer who decides to enter upon his arrival can be obtained by using Little's law $E[W] = E[N]/\lambda q$.

Theorem 2. *In the fully unobservable model of the M/M/1 queue with repairs and vacations, a unique Nash equilibrium mixed strategy 'enter with probability q_e' exists, where q_e is given by*

$$q_e = \begin{cases} q_e^*, & if \quad C\left(\frac{1}{\beta}+\frac{\theta+\xi}{\theta\mu}\right)\frac{\lambda(\xi\beta-\xi\theta-\theta^2)+\theta(\xi\beta+\theta\beta+\theta\mu)}{\theta\beta[\theta\mu-\lambda(\xi+\theta)]} \\ 1, & if \quad R \geq C\frac{\lambda(\xi\beta-\xi\theta-\theta^2)+\theta(\xi\beta+\theta\beta+\theta\mu)}{\theta\beta[\theta\mu-\lambda(\xi+\theta)]} \end{cases},$$

$$where \ q_e^* = \frac{R\theta^2\mu\beta - C\theta(\xi\beta+\theta\beta+\theta\mu)}{\lambda\left[C(\xi\beta-\xi\theta-\theta^2)+R\theta\beta(\xi+\theta)\right]}. \tag{14}$$

Proof. We consider a tagged customer at his arrival instant. If he decide to enter the system, the expected net benefit he gets is

$$S(q) = R - C\frac{(\xi\beta-\xi\theta-\theta^2)\lambda q+\theta(\xi\beta+\theta\beta+\theta\mu)}{\theta\beta[\theta\mu-\lambda q(\xi+\theta)]}, \tag{15}$$

where $q \in [0,1]$. We have

$$S(0) = R - C\left(\frac{1}{\beta}+\frac{\theta+\xi}{\theta\mu}\right), \tag{16}$$

$$S(1) = R - C\frac{\lambda(\xi\beta-\xi\theta-\theta^2)+\theta(\xi\beta+\theta\beta+\theta\mu)}{\theta\beta[\theta\mu-\lambda(\xi+\theta)]}. \tag{17}$$

The expected mean sojourn time is strictly increasing for $q \in (0,1)$, so when $R \in \left(C\left(\frac{1}{\beta}+\frac{\theta+\xi}{\theta\mu}\right), C\frac{\lambda(\xi\beta-\xi\theta-\theta^2)+\theta(\xi\beta+\theta\beta+\theta\mu)}{\theta\beta[\theta\mu-\lambda(\xi+\theta)]}\right)$, (15) has a unique root in (0,1) which gives the first branch of (14). When $R \in \left[C\frac{\lambda(\xi\beta-\xi\theta-\theta^2)+\theta(\xi\beta+\theta\beta+\theta\mu)}{\theta\beta[\theta\mu-\lambda(\xi+\theta)]}, \infty\right)$, $S(q)$ is positive for every q. In other words, the unique equilibrium point is $q_e = 1$ in this case, which gives the second branch of (26).

4 Conclusion

In this paper we studied the equilibrium strategies of the M/M/1 queues with repairs and vacations. The equilibrium balking strategies were investigated for the two kinds of queues. To the author's knowledge, this is the first time that repairs and vacations are introduced into the queue system at the same time.

Acknowledgements This work is supported by National Natural Science Foundation of China (No. 11171019), Program for New Century Excellent Talents in University (NCET-11-0568) and the Fundamental Research Funds for the Central Universities (No. 2011JBZ012).

References

1. Burnetas A, Economou A (2007) Equilibrium customer strategies in a single server Markovian queue with setup times. Queueing Syst 56:213–228
2. Economou A, Kanta S (2008) Equilibrium balking strategies in the observable singleserver queue with breakdowns and repairs. Oper Res Lett 36:696–699
3. Edelson NM, Hildebrand K (1975) Congestion tolls for Poisson queueing processes. Econometrica 43:81–92
4. Guo P, Hassin R (2011) Strategic behavior and social optimization in Markovian vacation queues. Oper Res 59:986–997
5. Naor P (1969) The regulation of queue size by levying tolls. Econometrica 37:15–24

Fresh Produce Supply Chain Management Decisions with Circulation Loss and Options Contracts

Chong Wang and Xu Chen

Abstract Considering the circulation loss of the fresh produce, we investigate management decisions with options contracts in a two-stage supply chain in which a fresh produce supplier sells to a retailer with Stackelberg model. We derive the retailer's optimal option ordering policy and the supplier's optimal pricing policy. Taking the integrated supply chain as the base model, we get that options contracts cannot coordinate the fresh produce supply chain when the retailer only orders options.

Keywords Fresh produce • Supply chain • Options contracts • Management decisions

1 Introduction

The competition is intensified in today's market. Fresh produce's short life cycles result in little or no salvage value at the end of the selling season as well as amazing circulation loss both in quality and quantity, which makes supply chain coordination more critical. Therefore, it is urgent and important to apply supply chain's management theories and methods managing fresh produce's high risk.

C. Wang
School of Management and Economics, University of Electronic Science & Technology of China, Chengdu 610054, China

School of Economics and Management, Sichuan Agricultural University, Chengdu 611130, China
e-mail: wc000500@163.com

X. Chen (✉)
School of Management and Economics, University of Electronic Science & Technology of China, Chengdu 610054, China
e-mail: xchenxchen@263.net

Z. Zhang et al. (eds.), *LISS 2012: Proceedings of 2nd International Conference on Logistics, Informatics and Service Science*, DOI 10.1007/978-3-642-32054-5_90,

Agricultural products supply chain management has generated significant interest lately. As a kind of perishable goods, fresh produce has received a great deal of attention especially. Lowe and Preckel [1] points out that the supply chain in the food and agribusiness sector is characterized by long supply lead times combined with significant supply and demand uncertainties, and relatively thin margins. They review some of the literature on applications of decision technology tools for a selected set of agribusiness problems and conclude by outlining what they see as some of the significant new problems facing the industry. Ahumada and Villalobos [2] present an operational model that generates short term planning decisions for the fresh produce industry. See Banker et al. [3] and Shen et al. [4] for more discussion.

The research of supply chain risk management involves the application of contracts and options. Lariviere and Porteus [5] consider a simple supply-chain contract in which a manufacturer sells to a retailer facing a newsvendor problem and the lone contract parameter is a wholesale price. Cachon [6] discusses revenues-sharing contracts. See Fu et al. [7] and Chen and Shen [8] for more discussion.

The remainder of this paper is organized as follows. Problem description and assumptions are presented in Sect. 2. In Sect. 3, we take the integrated supply chain as the base model. In Sect. 4, we focus on the retailer and supplier's optimal ordering and pricing policies respectively. Fresh produce supply chain coordination is considered in Sect. 5. We conclude our findings in Sect. 6 and highlight possible future work.

2 Problem Description and Assumption

We consider a two-stage supply chain in which the supplier is a Stackelberg leader selling a fresh produce to the retailer who is the follower. Before the beginning of the selling season, the retailer purchases options from the supplier at per unit price of w. Each option gives the retailer the right (not the obligation) to buy one unit of product at the exercise price of αw after demand has been observed. α is an exogenous parameter and $\alpha > 0$. Then, the supplier decides the order and exercise prices according to the retailer's order quantity Q and the stochastic market demand. The supplier delivers the products at per unit cost of c and the retailer sells the product at per unit price of p which is known. Considering the features of fresh produce, we invite β $(0 < \beta < 1)$ to imply the circulation loss in quantity.

Let random variable D be the demand which is non negative with a mean of u. Let Q_I be the supply quantity in an integrated supply chain. The probability density function is $f(\cdot)$, cumulative distribution function is $F(\cdot)$. Let F be differentiable, increasing and $F(0) = 0$. Let $[x]^+ = \max(0, x)$, $\bar{F}(\cdot) = 1 - F(\cdot)$.

We assume that the unsold fresh produce has no salvage value at the end of the selling season, and each of the shortages that do not meet demand incurs a unit shortage cost of g. To avoid trivialities, suppose $p > w + \alpha w > c$.

3 The Integrated Supply Chain

To begin, consider the integrated supply chain as the base model. This model is a useful benchmark. In the integrated supply chain, there is no intermediate links such as wholesale. The integrated supply chain's profit function, denoted $\pi_I(Q_I)$, is

$$\pi_I(Q_I) = p\min[D, Q_I(1-\beta)] - cQ_I - g[D - Q_I(1-\beta)]^+ \tag{1}$$

The integrated supply chain's expected profit, denoted $E[\pi_I(Q_I)]$, is

$$E[\pi_I(Q_I)] = [(p+g)(1-\beta) - c]Q_I - (p+g)\int_0^{Q_I(1-\beta)} F(x)dx - gu \tag{2}$$

Proposition 1. In the integrated supply chain, there exists a unique optimal supply quantity $Q_I{}^*$ given by

$$Q_I^* = \frac{1}{1-\beta}F^{-1}\left[1 - \frac{c}{(p+g)(1-\beta)}\right]$$

Proof. Because

$$dE[\pi_I(Q_I)]/dQ_I = [(p+g)(1-\beta) - c] - (p+g)(1-\beta)F[Q_I(1-\beta)]$$

$$d^2E[\pi_I(Q_I)]/dQ_I{}^2 = -(p+g)(1-\beta)^2 f[Q_I(1-\beta)] < 0$$

So, $E[\pi_I(Q_I)]$ is concave in Q_I, i.e., in the integrated supply chain, there exists a unique optimal $Q_I{}^*$ by solving Eq. (4) =0. The proof is complete.

4 The Decentralized Supply Chain

4.1 *The Retailer's Optimal Ordering Policy with Options Contracts*

Now we consider the retailer's optimal ordering policy with options contracts. The retailer now orders only options. The retailer's profit function, denoted $\pi_R(Q)$, is

$$\begin{aligned}\pi_R(Q) =& p\min[D, Q(1-\beta)] - \alpha w\min[D, Q(1-\beta)] \\ &- g[D - Q(1-\beta)]^+ - wQ\end{aligned} \tag{3}$$

The retailer's expected profit, denoted $E[\pi_R(Q)]$, is

$$E[\pi_R(Q)] = (p + g - \alpha w)\int_0^{Q(1-\beta)} \bar{F}(x)dx - wQ - gu \tag{4}$$

Proposition 2. With options contracts in a fresh produce supply chain, there exists a unique optimal option order quantity Q^* for the retailer given by

$$Q^* = \frac{1}{1-\beta}F^{-1}\left[1 - \frac{w}{(p+g-\alpha w)(1-\beta)}\right]$$

Proof. The proof of Proposition 2 is similar to the proof of Proposition 1.

4.2 The Supplier's Optimal Pricing Policy with Options Contracts

After the fresh produce retailer announces Q^*, the supplier can decide the option order price w and the option exercise price αw. The supplier's profit function, denoted $\pi_S(w)$, is

$$\pi_S(w) = wQ^* - cQ^* + \alpha w \min[D, Q^*(1-\beta)] \tag{5}$$

The supplier's expected profit, denoted $E[\pi_S(w)]$, is

$$E[\pi_s(w)] = (w - c)Q^* + \alpha w\int_0^{Q*(1-\beta)} \bar{F}(x)dx \tag{6}$$

Proposition 3. With options contracts in a fresh produce supply chain, when $w \geqslant c$, *there exists a unique optimal option order price w^* for the supplier given by*

$$w^* = \frac{c(1+\alpha(1-\beta)\bar{F}(Q^*(1-\beta)))}{(1+\alpha(1-\beta)\bar{F}(Q^*(1-\beta)))^2 - \left(Q^* + \alpha\int_0^{Q^*(1-\beta)} \bar{F}(x)dx\right)(1-\beta)r(Q^*(1-\beta))}$$

And the optimal option exercise price is αw^*, meanwhile $0<\alpha/w^* - 1$.

Proof. The proof of Proposition 3 is similar to the proof of Proposition 1.

Proposition 3 implies that, in order to avoid the market risk caused by demand uncertainty, the premise of the fresh producer willing to provide options contracts is that the option's order price be no low than the product's supply cost.

5 Supply Chain Coordination with Options Contracts

Proposition 4. Options contracts cannot coordinate the fresh produce supply chain when the retailer only orders options.

Proof. With Propositions 1 and 2, let $Q^* = Q_I^*$. Thus, when the contracts parameters satisfy $w = c - \alpha cw/(p + g) < c$, options contracts can coordinate the fresh produce supply chain. But this condition is contrary to the condition which the supplier is willing to provide options contracts i.e., $w \geq c$ in Proposition 3. So options contracts cannot coordinate the fresh produce supply chain when the retailer only orders options. The proof is complete.

6 Conclusion and Suggestions for Further Research

In this paper, consider the characteristics of fresh produce and use the Stackelberg model, we investigate the role of options contracts and management decisions for the fresh produce supply chain. This research can be extended to a case that the retailer orders both products and options to see whether options contracts can coordinate the fresh produce supply chain.

Acknowledgments The authors thank the editor and the referees for careful reading the paper. This research is partially supported by Youth Foundation for Humanities and Social Sciences of Ministry of Education of China (No. 11YJC630022), the Fundamental Research Funds for the Central Universities (No. ZYGX2009X020), and Sichuan Province Key Technology R&D Program (No. 2012FZ0003).

References

1. Lowe TJ, Preckel PV (2004) Decision technologies for agribusiness problems: a brief review of selected literature and a call for research. Manuf Serv Oper Manag 6:201–208
2. Ahumada O, Villalobos JR (2011) Operational model for planning the harvest and distribution of perishable agricultural products. Int J Prod Econ 133(2):677–687
3. Banker R, Mitra S (2011) The effects of digital trading platforms on commodity prices in agricultural supply chains. MIS Q 35(3):599–611
4. Shen DJ, Lai KK, Leung SC et al (2011) Modelling and analysis of inventory replenishment for perishable agricultural products with buyer-seller collaboration. Int J Syst Sci 42(7):1207–1217
5. Lariviere MA, Porteus EL (2001) Selling to the newsvendor: an analysis of price-only contracts. Manuf Serv Oper Manag 3(4):293–305
6. Cachon GP (2003) Handbooks in operations research and management science: supply chain management. North-Holland, Amsterdam
7. Fu Q, Lee CY, Teo CP (2010) Procurement management using option contracts: random spot price and the portfolio effect. IIE Trans 42(11):793–811
8. Chen X, Shen ZJ (2012) An analysis on supply chain with options contracts and service requirement. IIE Trans 44(10):805–819

Analyzing Competing Behaviors for Graduate Scholarship in China: An Evolutionary Game Theory Approach

Jiang Wu, Hui Zhang, and Tiaobo He

Abstract This paper aims to analyze the gaming behaviors of graduates in scholarship competition in China by using an evolutionary game theory approach. It is found that graduates in the majors with a relatively small number of candidates and symmetric information tend to connive to equally share the fund of scholarship. However, graduates in the majors with a large number of students and asymmetric information attempt to compete for different level of financial support by differentiating their performance.

Keywords The graduate scholarships • Competition • Evolutionary game

1 Introduction

In 2006, some Chinese universities, including Harbin Institute of Technology, Huazhong University of Science and Technology and Xi'an Jiao Tong University began to reform mechanisms, in such areas as graduate tuition, scholarships and financial aid mechanism [1]. In 2007, Peking University, Tsinghua University, Fudan University, and the other 15 top universities under Ministry of Education of the P.R.C. no longer distinguished between the free and self-financed graduates, but a unified latter. By 2009, all universities under direct jurisdiction of the Ministry of Education had done [2].

Graduate education is one kind of quasi-public goods with high marginal cost, so it should adopt the principle of "the beneficiary pays expenses" [3]. Because those reforms of the partial or full tuition policy will bring negative impact on research enthusiasm of graduates, a diversified funding system for the graduates should be

J. Wu (✉) • H. Zhang • T. He
School of Statistics, Southwestern University of Finance & Economics,
610074 Chengdu, P.R. China
e-mail: wujiang@swufe.edu.cn; 210120100009@swufe.edu.cn; 210120100003@swufe.edu.cn

Z. Zhang et al. (eds.), *LISS 2012: Proceedings of 2nd International Conference on Logistics, Informatics and Service Science*, DOI 10.1007/978-3-642-32054-5_91,

established to solve graduate education costs to ease these problems [4]. It was found that in Three Gorges University in southwest China, there is some phenomenon such as the single scholarship source, different review scale, and the lack of fairness and justification. Therefore, some advice is given to diversify the sources for graduate scholarship, and to improve the scholarship evaluation program [5].

Although the reforms of graduate scholarship allocation system has some effect on motivating the graduate students to study hard, Liu shows that a majority of graduates still hold a negative attitude for reforms of graduate scholarship system [6]. The reason is that all students are assumed as "economic man", but in fact not all graduates are so. By using evolutionary game theory this paper analyzes competing behaviors for graduate scholarship under the conditions of asymmetry information and symmetry information respectively.

2 An Evolutionary Game Model of Graduate Scholarship Competition Under Asymmetric Information

2.1 Proposition

1. The total number of graduate is M (M is big enough), the interaction between major groups is random with limited rationality.
2. Learning aptitude of each graduate in some major is the same.
3. The individual scholarship competition strategy mainly has two kinds: One is the competitive strategy (s^1) to study diligently to obtain the scholarship, and the other non- competitive strategy (s^2) with a negative indifferent manner.
4. Whether the individual chooses the competitive strategy or non-competitive strategy, both sides gain equals.

Assuming the scholarship amount is V. The ratio of group A, who choose competitive strategy, is p. The ratio of group B, who choose non-competitive strategy, is q ($p + q = 1$). During scholarship evaluation, we consider the graduate groups with the same major make the non-cooperative repeated game, whose payoff matrix is shown in Table 1:

2.2 Evolutionarily Stable Strategy Equilibrium Analysis

Based on the evolutionary game theory [7], the individual fitness by choosing the competitive strategy s^1 is:

$$f\left(s^1; s\right) = p\frac{V}{2} + qV \tag{1}$$

Table 1 Payoff matrix of graduate scholarship competition under asymmetric information

Group B Group A	Competitive strategy (s^1)	Non-competitive strategy (s^2)
Competitive strategy (s^1)	$\frac{V}{2}, \frac{V}{2}$	$V, 0$
Non-competitive strategy (s^2)	$0, V$	$\frac{V}{2}, \frac{V}{2}$

The individual fitness by choosing the non-competitive strategy s^2 is:

$$f\left(s^2; s\right) = q\frac{V}{2} \tag{2}$$

Then the overall average fitness is given by

$$f(p; s) = pf\left(s^1; s\right) + qf\left(s^2; s\right) \tag{3}$$

Group replicator dynamic equation by adopting the competitive strategy s^1 is:

$$\frac{dp}{dt} = p\left[f\left(s^1; s\right) - f(p; s)\right] = p(1-p)\frac{V}{2} \tag{4}$$

Group replicator dynamic equation by adopting the non-competitive strategy s^2 is:

$$\frac{dq}{dt} = q\left[f\left(s^2; s\right) - f(q; s)\right] = -q(1-q)\frac{V}{2} \tag{5}$$

The Jacobian of this system matrix is given by

$$J = \begin{pmatrix} \frac{V}{2}(1-2p) & 0 \\ 0 & -\frac{V}{2}(1-2q) \end{pmatrix} \tag{6}$$

The system composed of Eqs. (4) and (5) only has a partial equilibrium point $(1, 0)$, so the corresponding evolutionary stable strategy is: When that some major has a large number of students, finally all of them will choose competitive strategy in order to get more scholarships.

3 An Evolutionary Game Model of Graduate Scholarship Competition Under Symmetric Information

3.1 Proposition

1. The total number of graduates in some major is M (M is small), the interaction between that major groups is random and graduates are of limited rationality.

Table 2 Payoff matrix of graduate scholarship competition under collusive agreements

Group B / Group A	Competitive strategy (s^1)	Non-competitive strategy (s^2)
Competitive strategy (s^1)	$\frac{V}{2}-u_2, \frac{V}{2}-u_2$	$V-u_2, u_1$
Non-competitive strategy (s^2)	$u_1, V-u_2$	$\frac{V}{2}+u_1, \frac{V}{2}+u_1$

2. After they reach that collusive agreements, individuals who comply with those agreements can get more time to do other things, whose utility is expressed as u_1, on the other hand, individuals who do not follow the collusive agreement will be subject to the exclusion of other people among groups, which result in a certain psychological losses, whose utility is expressed as u_2.
3. This item is the same as (2), (3) and (4) of Proposition 2.1.

The ratio of group A, who choose competitive strategy, is p. The ratio of group B, who choose non-competitive strategy, is q. Then the payoff matrix is shown in Table 2:

3.2 Evolutionarily Stable Strategy Equilibrium Analysis

The individual fitness by choosing the competitive strategy s^1 is:

$$f(s^1;s) = p\left(\frac{V}{2} - u_2\right) + q(V - u_2) \tag{7}$$

The individual fitness by choosing the non-competitive strategy s^2 is:

$$f(s^2;s) = pu_1 + q\left(\frac{V}{2} + u_1\right) \tag{8}$$

Then the overall average fitness is given by

$$f(p;s) = pf(s^1;s) + qf(s^2;s) \tag{9}$$

Group replicator dynamic equation by choosing the competitive strategy s^1 is:

$$\frac{dp}{dt} = p\left[f(s^1;s) - f(p;s)\right] = p(1-p)\left(\frac{V}{2} - u_1 - u_2\right) \tag{10}$$

Group replicator dynamic equation by choosing the non-competitive strategy s^2 is:

$$\frac{dq}{dt} = q\left[f(s^2;s) - f(q;s)\right] = q(1-q)\left(u_1 + u_2 - \frac{V}{2}\right) \tag{11}$$

Four system equilibrium points from Eqs. (10) and (11) are composed of (0,0), (0,1), (1,0), (1,1), the system Jacobian matrix is given by

$$J = \begin{pmatrix} (1-2p)\left(\dfrac{V}{2} - u_1 - u_2\right) & 0 \\ 0 & (1-2q)\left(u_1 + u_2 - \dfrac{V}{2}\right) \end{pmatrix} \tag{12}$$

The system composed of Eqs. (10) and (11) has two partial equilibrium points (1,0) and (0,1), so the evolutionary stable strategy is:

If $u_1 + u_2 < \frac{V}{2}$, the collusion gain will be not big or individual mental loss of competitive strategy will get very small, finally all of them will choose competitive strategy.

If $u_1 + u_2 > \frac{V}{2}$, non-cooperative individual mental losses by choosing the competitive strategy will get large enough, and the individual who initially choose competitive strategy will eventually choose the non-competitive strategy.

4 Conclusion

This paper finds that the size of the group and individual mutual information symmetry will directly influence individual behaviors for strategic choice. That is, graduate individual will adopt the different strategy based on the different number of graduate groups, symmetry information and asymmetry information.

Acknowledgments This Paper was supported by the National key Discipline Program of Statistics, 211 Project for the Southwestern University of Finance and Economics (Phase III) and Graduate Education Research Foundation of the Southwestern University of Finance and Economics (2011).

References

1. Zhong XC (2005) The tuition and equity of graduate education. Huazhong Normal University, Wuhan, pp 31–32
2. Liu ZZ (2009) A butterfly effect caused by the reformation of graduate education. China Electr Power Educ 11:53–55
3. Xu WH, Zhang QJ (2002) The rationality of the tuition for graduate education. China High Educ Res 4:33–34
4. Li L (2005) A discussion about the tuition of graduate education in the perspective of science. J Univ Sci Technol Beijing 6:59–62
5. Yang YQ (2010) A discussion about the tuition of graduate education for local colleges. J Changchun Univ Sci Technol 1:92–93
6. Liu C (2010) Advantages and disadvantages of the master scholarship system. J Changchun Univ 6:111–114
7. Friedman D (1991) Evolutionary games in economics. Econometrica 59:637–666

A Risk Control Method Based on Two-Factor Theory

Lu Ming and Fan Yunxiao

Abstract Risk management as an effective means to prevent accidents has been widely applied in coal mines in China. Risk control is the final stay point of risk management. As the key part to reduce risk it plays an important role in improving safety production level of coal mine. Because of lacking theory supports, it is difficult for the current methods to control risks systematically and specifically. Based on two-factor theories and the organizational structure, this paper will present a risk control method through analysis of physical condition failures and human errors.

Keywords Two-factor theory • Risk control • Physical condition failure • Human error

1 Introduction

Our coal mine safety has a grim situation due to its complicated technological process, severe and changeable operating environment and the low awareness level of the workers [1]. Risk control as the final stay point of risk management plays an important role in eliminating hazards and reducing risks.

With the prevalence of risk management, a variety of hazard identification cards have been widely applied. But a deep-seated mechanism study of hazard causations was lacked [2]. Therefore, the existing risk control was empty, vague and separated from practice. To a certain extent, the identified hazards are unable to be exactly eliminated. Round and round, the risk management is top-heavy, so accident prevention is like a blank check, no practical effectiveness is gained.

From flooding to explosive agents and the risk of asphyxia, miners are exposed to some of the most hostile working conditions of any occupation. Due to the

L. Ming (✉) • F. Yunxiao
School of Engineering & Technology, China University of Geosciences (Beijing), Beijing, P.R. China
e-mail: luming134263@163.com; fanyxiao@cugb.edu.cn

Z. Zhang et al. (eds.), *LISS 2012: Proceedings of 2nd International Conference on Logistics, Informatics and Service Science*, DOI 10.1007/978-3-642-32054-5_92,

complicated production process and the harsh production conditions in coal mines, the problem of safety production in coal mine is more significant, complicated and difficult to solve compared with any other industry. With the growing complexity of systems, working conditions have become more difficult to understand and predict how they interact together. Therefore, more attention should be paid to the control of the physical condition failures.

Nevertheless, the majority of accidents cannot be solely attributed to adverse working conditions. For instance, a study by the US Bureau of Mines found that nearly 85% of all mining accidents identified human error as a causal factor. Clearly, if safety is to be improved, it is also vital to control human errors.

Based on two-factor theories, this paper will develop a scientific and reasonable risk control method targeting on physical condition failure and human error.

2 Analysis of Two-Factor Theory

The unsafe acts of workers and the unsafe conditions in workplace which one plays a dominant role in accidents is a controversial issue among scholars. Therefore, different accident-causing theories were proposed respectively targeting on human errors and physical condition failures.

2.1 Theories Based on Human Error

From the early accident proneness theory proposed by M. Greenwood and H.H. Woods in 1919 to the later domino theory proposed by H.W. Heinrich in 1980, human error was regarded as a main causation of accidents. Even though some the accidents were caused by unsafe conditions, these unsafe conditions were taken for the results of human errors [3].

Focused on human error, some human error models were respectively proposed by Goller in 1969, Surry in 1969 and Wigglesworth in 1972. These models not only highlight the effects of human error, but also explain why workers make mistakes. Based on the models above, a human error model targeting on gold mine was established by Lawrence in 1974. All of these models defined human error as inappropriate and wrong responses to stimulations.

Reason explained why workers make mistakes through another way. Errors are seen to emerge from psychological factors in individuals such as aberrant mental processes, including forgetfulness, inattention, poor motivation, carelessness, negligence, and recklessness [4]. This kind of models include Norman's schema activation error model, Reason's Generic Error Modeling System and Rasmussen's model of human malfunction. These models typically attempt to identify the nature and frequency of the errors made by workers within complex systems, the ultimate aim being to propose operator-focused strategies and countermeasures designed to reduce variability in human behavior.

All the human error based theories are focused specifically upon human behavior, rather than the inadequate physical conditions.

2.2 Theories Based on Physical Condition Failure

With the growing complexity of systems, it is aware that in addition to unsafe acts, there must be certain unsafe conditions when accident happened. R. Skiba thought that workers and mechanical equipments are two important factors in accidents. For some industries, unsafe conditions of mechanical equipments play a more important role. Accidents can be greatly reduced through improving the safety and reliability of system.

Gibson and Haddon et al. proposed a theory of causation regarding the unplanned release of energy as a cause. The theory clarified the physical nature of accidents and pointed out that to prevent injuries is to prevent unplanned release of energy and to prevent human exposure to energy.

Based on Haddon's theory, Michael Zabetakis proposed a new domino theory of accident causation [3]. This theory treated unplanned release of energy or hazardous substance as a direct cause of accidents. Then unsafe acts and unsafe conditions were regarded as the direct causes of the release of energy. Mechanical equipments are the carriers of energy. So the probability of accidents can be reduced through improving equipments and technologies.

In the 1970s, the labor bureau of Japan came up with the "Orbit Intersecting Theory" after a investigation [5]. Then a catastrophe model for accident-causing was proposed by Xinming Qian in 1995 [6]. These two theories both argued that accidents were attributed to unsafe acts of humans and the unsafe physical conditions.

Whether the human factor or the physical condition factor is in a safe state, by improving the condition of either one factor, the safety production level and labor productivity can be raised by a big margin [6].

Whether the theories arguing human error as a cause or the theories arguing physical condition failure as a cause, leaving aside who playing a dominant role in accident, we can be sure that both of the two factors have certain influences on accidents. Neither of they should be ignored. We should start from both sides, rather than focusing on a certain factor solely.

3 Analysis and Control of Physical Conditions Failure in Coal Mine

The probability of physical condition failures in coal mine is far higher than other industries due to its complicated technological process, adverse and changeable operating environment and various mechanical equipments.

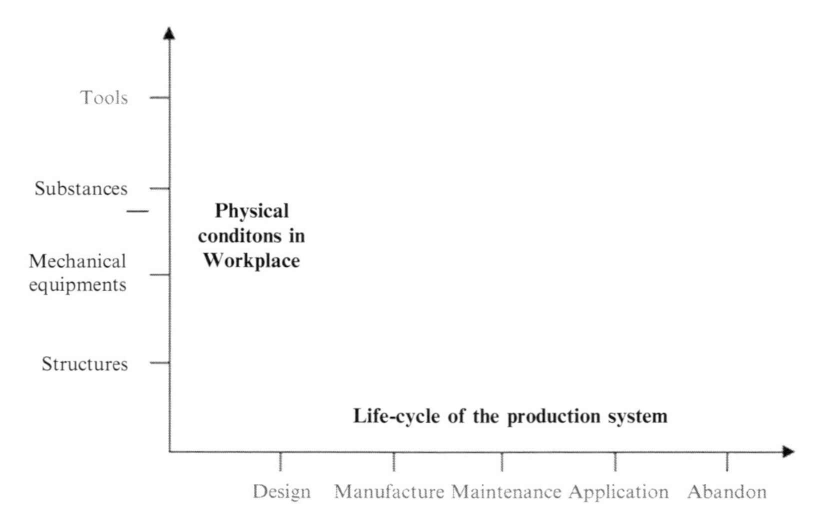

Fig. 1 Two-dimensional diagram of the life-cycle of the physical conditions

3.1 Analysis of Physical Conditions Failures

In man–machine system, the mechanical equipments, substance, production objects and other factors of production are collectively referred to as physical conditions [7]. According to their different functions in the man–machine operating system, these factors are divided into four types: structures, mechanical equipments, substances and tools. The event chain of these physical conditions is design → manufacture → maintenance → application → abandon [5].

Combining physical conditions and life-cycle together we get a two-dimensional diagram, as is shown in Fig. 1. Making sure that each intersection in Fig. 1 is safe, the physical conditions failures can be avoided.

3.2 Control Methods of Physical Conditions Failures

In general, enterprises are consisted of decision layer, management layer and operating layer. The employees of different layers have different ways to control the failures due to their different functions and authorities. Table 1 shows a hierarchy risk control model of physical condition failures based on the above organizational structure.

Based on the analysis above, a three-dimensional diagram of physical conditions risk control was developed shown in Fig. 2. Not only the control objectives can be clear, the causation of the failures can also be found through this three-dimensional control method. Then the risk control will not be that empty, vague and separated from practice.

Table 1 A hierarchy risk control model of physical condition failures

Layers	Control methods	Control objectives	Detailed ways
Decision layer	Improvement	Essential safety	Decision makers should try to eliminate the hazards in workplace, such as replacing the toxic and hazardous material, designing and purchasing the equipments of high reliability
Management layer	Management	Reducing risk	If some hazards can not be eliminated, then managers need to reduce the risk through prevention, reduction and isolation measures
Operating layer	Supervision	Controlling residual risk	Some residual risks may still exist in workplace after the above controlling. Then operators are required to monitor, supervise and communicate these risks constantly

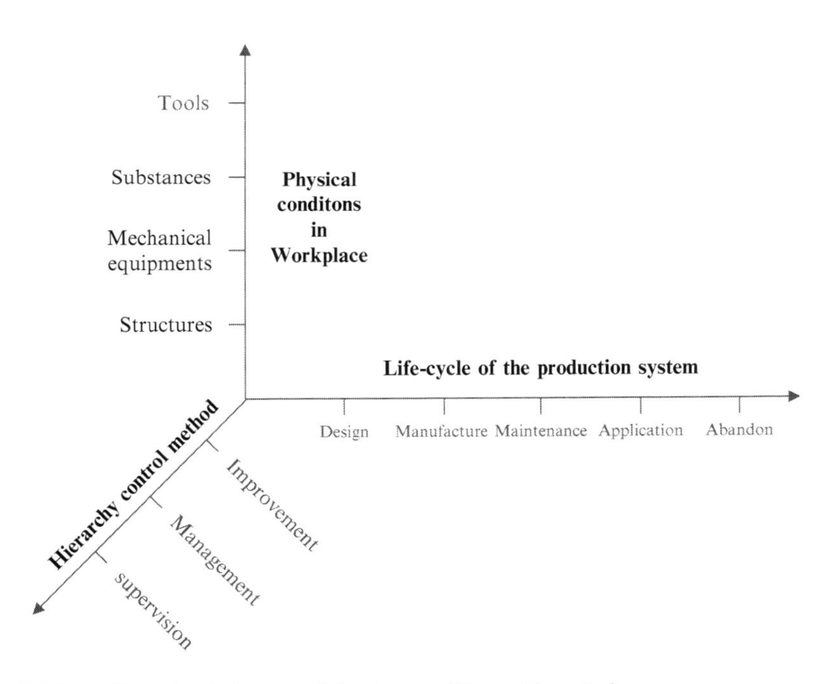

Fig. 2 Three-dimensional diagram of physical conditions risk control

4 Analysis and Control of Human Error in Coal Mine

Physical factors are the prerequisites to keeping safe. And human factors are internal factors [7]. Controlling human errors effectively is a significant way to prevent accidents.

4.1 Analysis of Human Errors

The human factors in the coal mines of China more commonly refer to the so-called "three violations". A survey was conducted among the front-line workers in a coal mine. Through the analysis of the questionnaires, it is found that 11% of the workers took violating actions because they do not know that the acts are violations of the relevant rules and regulations. So this kind of violations is named as "ignorant fearlessness". These workers are mainly new employees or transferees from other positions. Another 64% of the workers knew their behaviors were illegal, but in the past these behaviors had never lead to any accidents, so they took the violating actions fearlessly. So this kind of violations is named as "knowingly and willful". These workers are mostly older employees who have some experience. The rest 25% of the workers took the violating behaviors due to misjudgment and negligence. So this kind of violations is named as "misstep".

Unsafe actions are driven by the above three consciousnesses. Through a forum with workers, it is found that the negative consciousnesses generated due to capabilities, physical and emotional stimulations, such as lower technical level, fatigue, bad health condition, agitated mood and low spirit etc.

The mentioned fatigue, bad health condition and other factors are negative stimulations to the workers according to the Surry mode [3], the Ferrell theory [3] and the Petersen accident-incident causation model [3]. Thus negative stimulations make the workers generate consciousnesses of leaving things to chances and the consciousnesses of trying facile ways and so on. Therefore, unsafe actions were taken.

In order to eliminate human errors, firstly, negative stimulations should be eliminated from root. Secondly, the way that these stimulations evolve into unsafe actions should be cut off, that is to transform the negative consciousness of workers.

4.2 Control Methods of Human Errors

During the daily safety inspections, the underlying factors are rarely corrected. A hierarchy control method which is similar to the controlling of physical condition failures is developed, shown in Table 2. In this way, risk control will be done retroactively when something goes awry, rather than scratching the surface of a problem.

5 Conclusion

Risk control plays an important role in the process of eliminating and reducing risk. To eliminate risks effectively, the ideas and objectives of risk control should be exactly clear. Under the guidance of two-factor theories, the following results were got targeting risk control in workplace.

Table 2 A hierarchy risk control model of human errors

Layers	Control methods	Control objectives	Detailed ways
Decision layer	Organization management	Avoiding negative stimulations	Constructing safety culture Arranging work rationally Clarifying safety responsibility Granting safety rights
Management layer	Behavior management	Suppressing negative consciousnesses	Training and education Motivation mechanism Enacting reasonable safety rules and codes
Operating layer	Mandatory correction	Guarding against unsafe actions	Adequate supervision

1. Physical condition factors in workplace and the life-cycle of production system were analyzed. Combining the two-dimensional diagram of life-cycle of the physical conditions with hierarchy risk control of failures, a three-dimensional diagram of physical conditions risk control was developed.
2. Based on the Surry model, the Ferrell theory and the Petersen accident-incident causation model, the causation of human error in coal mine was analyzed. According to this causation, a hierarchy control method of human errors was developed.

References

1. Zhang'ai, Yanlixia (2011) Shallow talk on accident causation theory of coal mine in China. Manag Technol Medium-sized Small Enterp 1:250
2. Cao Yuankun, Wang Guangjun (2011) Business risk management development course and new understandings of the research trends. Contemp Financ Econ 1:89–90
3. Heinrich HW (1980) Industrial accident prevention. McGRAW-HILL Book Company Inc, New York/London, pp 21–26
4. Reason J (2000) Human error: models and management. BMJ 320:768–770
5. Sui Pengcheng, Chen Baozhi, Sui Xu (2005) Safety principles. Chemical Industry Press, Beijing, pp 55–60
6. Qian Xinming, Chen Baozhi (1995) The catastrophe model for accident-causing. Chin Saf Sci J 2:2–4
7. Li Xuandong, Fu Gui, Lu Bai (2005) Two-factor theory and prevention of safety accidents. J Liaoning Tech Univ 5(2):771–774

Dynamic Quantitative Analysis on Chinese Urbanization and Growth of Service Sector

Congjun Cheng

Abstract This paper is based on the index number of value-added of tertiary industry and town population/total population ratio during the years of 1978–2010 in China. It have made sophisticated researches on the relationship between development of domestic service industry and urbanization by applying ADF test, cointegration test, error correction model, Granger causality test, finally drawing the conclusions as follow: (1) urbanization is the important power of service industry. (2) The level of urbanization has Granger causality relationship with the development of service industry recently, but the effect is one direction, which means the promotion of urbanization level improves the level of service industry.

Keywords Urbanization • Cointegration test • Error correction model • Granger causality test

1 Introduction

"The urbanization of China" and "The high-tech of America" have been called two key factors that affect all mankind in the twenty first century by Stieglitz, and he said that China's urbanization will take the lead in regional economic growth, and will produce most important economic interests [1]. The modern service industry, the tertiary industry, it plays a very important role to promote the economic structure optimization, improve the people's living quality, and increase chances of employment. According to Petty-Clark's law, with a corresponding increase of per capita national income, the labors will transfer the primary industry to the secondary industry and other non-agricultural sectors, and then with the further

C. Cheng (✉)
College of History, Hebei University, Hebei, P.R. China
e-mail: chengcongjun@hbu.edu.cn

Z. Zhang et al. (eds.), *LISS 2012: Proceedings of 2nd International Conference on Logistics, Informatics and Service Science*, DOI 10.1007/978-3-642-32054-5_93,

development of the economy, the labors will transfer the primary industry and the secondary industry to the tertiary industry. So the relationship between the service industries' development and urbanization has attracted more and more scholars' attention.

2 Data Selection

For the purpose of testing the dynamic relationships between urbanization and the growth of the service sector, the paper selects China's data between the year of 1978 and 2010 for quantitative analysis. Besides, in order to eliminate the influence of population, China measures the level of the tertiary industry by the added value of per capita in it. In this thesis, the level of Chinese service sector is analyzed by the index number of value-added of the tertiary industry. SERV refers to the level of Chinese service sector. What's more, the rate of urbanization is town population/total population ratio, which is the common indicator to measure the level of urbanization in the world at present and it is showed by URBA. In order to eliminate the heteroscedasticity of the data, the paper uses LN analysis on SERV and URBA, and then the data is expressed by LNURBA and LNSERV. The data is shown in Table 1:

3 Model Development

Firstly, the paper makes Cointegraion Test between the two variables, which refer to the level of urbanization and the index number of value-added of tertiary industry. At the same time, it starts to do empirical research using time series analysis methods. Among the process of the Cointegraion Test between the two variables, the first step is unit root test and determine the stationary in the two sides. Next, we should conclude that whether the two variables have Cointegraion relationship, then analyze the long-term equilibrium relation between them. In accordance with the error-correction model, this paper tests the magnitude of the deviation between the long-term equilibrium relation and short-term fluctuation in the two variables. With Granger causality test, one can find that whether exist the causal relationship between the two variables, which can be short-term or long-term.

3.1 ADF Test

At first, we should test whether the time series of the two variables is stationary, and whether the fallacy test exists. The purpose is to avoid the problem of "spurious regression". With the soft of Eviews6.0, this paper applying the AIC criterion to determine the optimal lag order [2]. The test style of the differential sequence will

Table 1 The level of urbanization and the index number of value-added of tertiary industry in China between the year 1978 and 2010 (1978 = 100)

Year	SERV	URBA	LNSERV	LNURBA
1978	100	0.1792	4.6052	−1.7193
1979	107.9	0.1896	4.6812	−1.6628
1980	114.3	0.1939	4.7388	−1.6404
1981	126.2	0.2015	4.8379	−1.6020
1982	142.6	0.2113	4.9600	−1.5545
1983	164.3	0.2162	5.1017	−1.5316
1984	196.0	0.2301	5.2781	−1.4692
1985	231.7	0.2371	5.4454	−1.4393
1986	259.6	0.2452	5.5591	−1.4057
1987	296.8	0.2531	5.6931	−1.3740
1988	335.9	0.25814	5.8168	−1.3543
1989	353.9	0.2621	5.8690	−1.3390
1990	362.1	0.2641	5.8919	−1.3314
1991	394.3	0.2694	5.9771	−1.3116
1992	443.3	0.2746	6.0942	−1.2924
1993	497.4	0.2799	6.2094	−1.2733
1994	552.5	0.2851	6.3145	−1.2549
1995	606.9	0.2904	6.4084	−1.2365
1996	664.1	0.3048	6.4984	−1.1881
1997	735.3	0.3191	6.6003	−1.1423
1998	796.8	0.3335	6.6806	−1.0981
1999	871.2	0.3478	6.7699	−1.0561
2000	956.1	0.3622	6.8629	−1.0156
2001	1054.2	0.3766	6.9605	−0.9766
2002	1164.2	0.3909	7.0598	−0.9393
2003	1274.9	0.4053	7.1506	−0.9031
2004	1403.1	0.4176	7.2464	−0.8732
2005	1574.7	0.4299	7.3618	−0.8442
2006	1797.3	0.439	7.4940	−0.8233
2007	2084.6	0.4494	7.6423	−0.7998
2008	2301.4	0.4568	7.7413	−0.7835
2009	2521.5	0.4659	7.8326	−0.7638
2010	2762.4	0.4995	7.9239	−0.6941

Source: *China Statistical Yearbook 2011* [6]

be confirmed by the corresponding principles, and the ADF test is used to determine whether each sequence has a unit root. Results are shown in Table 2:

Obviously, from Table 2, we can conclude that the ADF test value of the value-added of tertiary industry is −3.7456, and it is stationary at 5 and 10%. The ADF test value of urbanization is −1.613985, and it is greater than the critical value which at the level of significance, so it is the non-stationary unit root. After that, through the integration test on the two variables' first order difference, we will discover a fact that the ADF test value in the two valuables' first order difference sequence are −4.0282 and −3.5723, and they are both stationary at the level of 5 and 10%, so it shows a fact that the two variables' first order difference are both stationary.

Table 2 ADF test results for each variable time series

Variable	Type of test (C, T, K)	Test value	Critical value under the level of significance 1%	5%	10%	Test results
LNSERV	(C, T, 3)	−3.7456	−4.3098	−3.5742	−3.2217	Stable
ΔLNSERV	(C, T, 2)	−4.0282	−4.2967	−3.5683	−3.2183	Stable
LNURBA	(C, T, 1)	−1.6139	−4.2845	−3.5628	−3.2152	Unstable
ΔLNURBA	(C, T, 0)	−3.5723	−4.2845	−3.5628	−3.2152	Stable

Note: Δ in the table refers to the first order difference; among the test form (C,T,K), C refers to the constant term; T refers to the time trend; K refers to the lag order; and they are all test equation of unit root. 0 refers to the test equation which excludes the constant term or the time trend

3.2 Cointegration Test

To further analyze the long-term equilibrium relation between urbanization and the development of tertiary industry, the paper makes cointegration test for the two variables. Through the above analysis, it finds that the sequences of the two variables LNURBA and LNSERV meet the premise of the cointegration test, so we can consider that whether cointegration relationship exists in them [3]. Now, with Engle-Granger test (two-step testing method),we can make cointegration test between LNSERV and LNURBA.

First step, using OLS method estimate the long-term equilibrium equation

$$LNSERV_t = 10.30177 + 3.341910LNURBA_t \tag{1}$$

Calculating the non-equilibrium error of the ordinary least-squares estimation to get the sequence:

$$\varepsilon_t = -3.341910LNURBA_t + LNSERV_t - 10.30177 \tag{2}$$

Second step, we should test whether the residuals ϵ_t of the above models are stable sequences, and make unit root test to the estimated residuals ϵ_t of the above-mentioned regressive equations. The absolute value of the ADF test statistic −2.000870 is greater than the absolute value of the critical value when the level of significance at 5, 10%. Therefore, the estimated residuals sequence ε_t is stationary sequence. And it indicates that cointegration relationship exists in LNURBA and LNSERV. It is shown in Table 3:

From the cointegration model, we conclude that the level of urbanization and the growth of the service sector have a positive correlation, and once the urbanization changes 1% each time; it will make the proportion of the service industries' production value increasing to 3.3419%. From the Table 3, we conclude that the absolute value of residual ϵ_t' ADF test is −2.0452, and it is greater than the absolute value of the critical value when the level of significance at 10, 5%, so the cointegration regression equation has a practical significance.

Table 3 The stable test of residuals εt

Variable	Type of test (C, T, K)	ADF test value	Level of significance (%)	Critical value	AIC	DW	Test result
			1	−2.6416			
ε_t	(0, 0, 1)	−2.0452	5	−1.9520	−3.3068	1.8714	Stable
			10	−1.6104			

Note: among the test form (C,T,K), C refers to the constant term; T refers to the time trend; K refers to the lag order; and they are all test equation of unit root. 0 refers to the test equation which excludes the constant term or the time trend

3.3 *Error Correction Model*

The error correction model is the econometric model which has a particular form. According to Engel-Granger theorem, if the cointegration relationship exists in a set of variables, their short-term non-equilibrium relationship will always has the representation of an error correction model, which means that the cointegration regression always can be converted to the error correction model. Besides, according to Granger theorem, results of the two variables' unit root test and the cointegration test, we can express the error correction model of the short-term dynamic equilibrium relationship between the urbanization and the index number of value-added of tertiary industry [3]. It is:

$$\Delta LNSERV_t = \Delta LNURBA - 0.120468\Delta LNSERV_{t-1} + 0.489958\Delta LNURBA_t \\ - 0.035226ecm_{t-1}$$

On the basis of the error correction model, we find that the two variables' short-term dynamic equilibrium relationship refers to that once the urbanization level changes 1 unit each time, the proportion of the service industries' output value will change 0.257279 units in the same direction. In addition, the value is smaller than regression coefficients in the long-term cointegration regression equation, so it indicates that long-term impact of the development of urbanization on the proportion of the service industries' output value is more significant than shot-term impact.

3.4 *Granger Causality Test*

We should make the Grange causality test between the level of urbanization and the development of the service sector by OLS, and the maximum lag order makes 5. The basic approaches are: first, to estimate the degree of explanation, which refers to how the current LNSERV is explained by its value of lag phase, second, to test whether the lag order that introduce the sequence LNUBRA has improved significantly the interpreted degree of LNSERV [3]. Results in Table 4:

Therefore, in lag phase 1, the level of urbanization is the Granger cause of the service sector's development, but the latter is not the Granger cause of the former.

Table 4 Granger causality test between various variables

Granger causality	Lag length	F-value	P-value	Results
LNSERV isn't the Granger reason of LNURBA	1	0.01717	0.8966	To accept
LNURBA isn't the Granger reason of LNSERV	1	5.19375	0.0302	To refuse
LNSERV isn't the Granger reason of LNURBA	2	0.18469	0.8324	To accept
LNURBA isn't the Granger reason of LNSERV	2	1.76908	0.1904	To accept
LNSERV isn't the Granger reason of LNURBA	3	0.84537	0.4832	To accept
LNURBA isn't the Granger reason of LNSERV	3	1.39792	0.2687	To accept
LNSERV isn't the Granger reason of LNURBA	4	1.58781	0.2163	To accept
LNURBA isn't the Granger reason of LNSERV	4	0.88449	0.4910	To accept
LNSERV isn't the Granger reason of LNURBA	5	1.47102	0.2506	To accept
LNURBA isn't the Granger reason of LNSERV	5	1.47750	0.2486	To accept

In lag phase 2, lag phase 3, lag phase 4 and lag phase 5, the two sides have non-Granger causality [4]. From the above, we can conclude that the level of urbanization and the development of service sector have Granger causality recently, but the effect maybe one-way, which is the improvement of urbanization level promotes the development of service sector. In the long-term perspective, the mutual influences of urbanization and service sector have non-Granger causality.

4 Conclusions and Suggestions

According to Chinese data in the year of 1987–2009, we make ADF unit root test, cointegration test, error correction model test, vector auto regression analysis and Granger causality test between urbanization level and the index number of value-added of tertiary industry of China, so the following conclusions can be drown:

First of all, from the cointegration equation, we can find that the equation fitting very well and the coefficients are more significant; from the cointegration model, the level of urbanization and the development of service sector have the positive correlation; finally, we conclude that the level of urbanization promotes the development of service sector, in other words, urbanization is the power of promoting the growth of service sector.

Secondly, from the causality analysis, the level of urbanization and the development of service sector have the Granger causality recently, but the effect is one-way, which the level of urbanization promotes the development of service sector; in the long-term perspective, the mutual influence of urbanization and service sector have non-Granger causality.

Finally, urbanization will lead to changes in the structure of population, and cause the movement of population. So urbanization provides the supply of labor forces to the growth of service sector, and changes in employment structure ultimately promote the development of service sector.

According to the research study and the conditions of the service sectors' development, we can get the following suggestions.

Firstly, we should optimize the industrial structure and promote the development of the service industry [5]. Contorted structure of industry is the germ that causes inconformity between urbanization and the development of service industry. Bidirectional guides are taken to quicken the urbanization courses and promote the development of service industry. From economy development practices, urbanization is the only way which we must pass. Therefore, we need to constantly adjust the industrial structure, strength the development of the tertiary industry, and change the dominant status of the heavy industry. Thus, it helps the rural surplus labor forces changing to non-agriculture forces or realizes the transfer of urbanization, promoting the urbanization process of China.

Secondly, we should eliminate the barrier of urbanization and fully achieve the benign and synergistic relationship between the urbanization and the tertiary industry. In cities, the number of the migrant labor is gradually increased and they make great contribution to the economic development. In rural areas, the government should reinforce the job training to improve the quality of peasants and explode the discrimination of employment and policy which exists in the labor market [5]. With the flowing of the rural surplus labor, the government should set about reform from every aspect such as employment, education, medical care, and social insurance, protecting the legal right of the migrant labors, thus we will achieve the scientific city management. Therefore, we should improve the household, employment and social insurance system and eliminate the barrier between urban and rural areas. In this way, it can stimulate the development of service industry and in return of driving further the speed of the urbanization process.

Thirdly, we should improve the environment for urbanization and the coordination of the tertiary industry [5]. Reasonable and appropriate size of city system provide a stage for the development of the tertiary industry, and gradually forms urban agglomerations or metropolitan area on the basis of major cities, thus it boosts the size and structural adjustment of the modern service industry. Similarly, we should also enrich the development level of tertiary industry and develop the newly raised service industry intensively so as to change the elements and the economic structure of cities, and gain new vigor and power for the development of the urbanization. From the macroscopic view, if the government puts more input on urbanization and the tertiary industry, it will promote mutual development between them.

Fourth, we should adjust and improve the mechanism of the job market, and guide the labor entering the service industry, thus employment structure will get the adjustment. Service industry is an elastic industry which has a high employment rate and a stronger capacity of absorbing the labor; moreover, raising the level of economic development will continue to play a driving effect towards overall employment. The structural change of the urban employment service will also dramatically increase the level of the urban employment service and the quality of employment. Moreover, the increase of the consumer demand will also put forward higher demand for the service industry which refers to improve the quality of living standard as well as the other related industry; it will be the long-term power of service industry development.

References

1. Jiang Xiaojuan (2011) The growth of service industry: the true meaning, the multiple effects and future development. Econ Res 4:4–14
2. Li Chunlin, Song Tanhua (2011) Correlative analysis between service industry and urbanization in Shandong province. Econ Res Guide 1:143–144
3. Wang Yan (2008) Analysis of applied time series. China Renmin University Press, Beijing, p 12
4. Gao Laibin (2010) Quantitative analysis and suggestions between service industry and urbanization in Jilin province. Contemp Econ 13:86–87
5. Yang Shenggang, Yang Jianmo (2010) The relationship between urbanization and tertiary industry – an empirical analysis based on Hunan province. Commer Res 3:143–147
6. National Bureau of Statistics of China (2011) China Statistical Yearbook 2011. China Statistical Press, Beijing

Research on Probability Distribution of Impulse Noise Power Correlation Coefficient in Multi-carrier Communications

Zhu Yong

Abstract Impulse noise has become one of the major interference in multi-carrier communications, in-depth study should be expanded to find new ways of noise suppression for the purpose of system performance improvement. According to impulse noise which has the weibull distribution, the relationship between probability distribution of impulse noise power and the random location of impulse noise is analyzed. The simulation results and analysis show that the reliability of the theoretical derivation, also the position of probability distribution peak indicates impulse noise occurrence frequency in communication channel. The method can be used as means to estimate the severity of impulse noise in the channel.

Keywords Multi-carrier communication • Impulsive noise • Weibull distribution • Correlation coefficient • Distribution probability

1 Introduction

Multi-carrier technology based communication system has been widely used in broadband access networks, wireless communications and power line communication. Noise reduction, channel equalization and resources allocation are research focuses of the multi-carrier communication [1–3]. With the continued improvement of the communication rate and the increase of anthropogenic noise sources, the impact of impulse noise on the performance of multi-carrier communication system can not be ignored. So the study of impulse noise becomes a hotspot of communication theory.

Z. Yong (✉)
Institute of Communication Technology, Ningbo University, Ningbo 315211, China

Institute of Ningbo Technology, Zhejiang University, Ningbo 315101, China
e-mail: sinba@nit.zju.edu.cn

Z. Zhang et al. (eds.), *LISS 2012: Proceedings of 2nd International Conference on Logistics, Informatics and Service Science*, DOI 10.1007/978-3-642-32054-5_94,

Impulse noise is generally with the characteristics of sudden, short-term and strong interference, and constituted by the irregular pulses or noise spikes which have short duration and large magnitude. The occurrence of impulse noise is due to a variety of reasons, including natural phenomena (such as thunder and lightning) and the electromagnetic radiation while the switches and relays of various electronic devices changing their states. Impulse noise study focused on two aspects of the analysis of its statistical properties [4, 5] and how to effectively suppress it [6, 7].

Due to the significant feature of the impulse noise in time domain, the preliminary analysis of the statistical properties of impulsive noise focuses on this target. Some methods for impulse noise detecting and suppression are derived under these features. One approach is to use its character of short-term and large magnitude by setting thresholds to detect the presence of impulse noise and using the time-domain nonlinear preconditioning to limit the impact of the large energy impulse noise. This method is simple and easy to implement, but small impulse noise can not be detected, so the detection sensitivity is very low. Interleaving and error correction coding is another effective method of inhibition of impulse noise. In the case of unknown intensity and occurrence frequency of impulse noise, the interleaver length and error correction coding redundancy must be designed in accordance with the worst case. This paper studies the weibull distribution impulse noise power correlation coefficient, proposes a new method to analyze the probability distribution characteristics for the purpose of detecting the occurrence frequency of impulse noise in multi-carrier signal frame. Combined with interleaving and error correction coding, this method can dynamically adjust the interleaving length and error correction coding redundancy to match the pulse noise characteristics based on test results, and increase the performance of communication system [8, 9].

2 Distribution Characteristics of Impulse Noise Power Correlation Coefficient

The mathematical model of noise interfered signal can be expressed as:

$$r(k) = s(k) + n_{AWGN}(k) + n_{IN}(k)$$

Where $r(k)$ and $s(k)$ are the transmitted and received signals, $n_{AWGN}(k)$ and $n_{IN}(k)$ are additive white Gaussian noise and impulse noise. The simplified model of multi-carrier system is applied as Fig. 1.

Taking the linear property of the DFT into account, frequency domain expression of the noise part can be obtained as:

$$N_k = \sum_{l=0}^{L-1} n_l W_L^{lk}, \quad k = 0, 1, \ldots, L-1$$

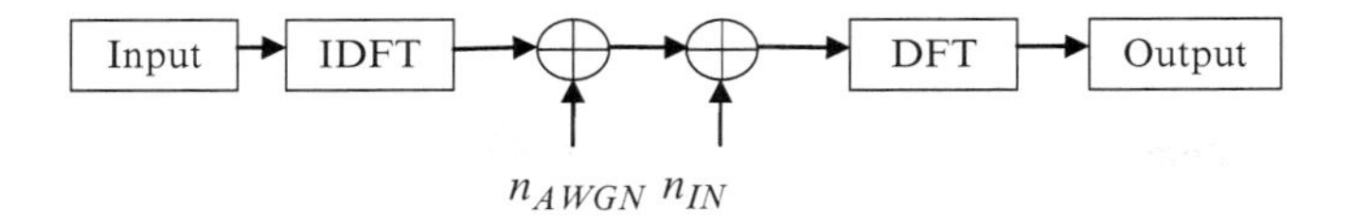

Fig. 1 Simplified model of the multi-carrier communication

Only the case of impulse noise is considered, it is assumed that a frame of the multi-carrier modulation signal contains L symbols, where M symbols interfered by impulse noise. The probability density function (pdf) of Weibull distributed impulse noise n is $f(n) = \frac{1}{2}ba^{-b}|n|^{b-1}e^{-(|n|/a)^b}$, with the mean of zero and the variance of $a^2\Gamma(1+2b^{-1})$, where Γ is gamma function.

According to the Paseval theorem, $P^{(f)} = \frac{1}{L}\sum_{k=0}^{L-1}|N_k|^2$ can be obtained. Because of the conjugate characteristic of Fourier transform, the real and imaginary part of $P^{(f)}$ can be expressed by time domain signals respectively as [10]:

$$\begin{cases} P_r^{(f)} = \frac{1}{2}\left[\sum_s \alpha_s n_s^2 + \sum_{k,l} n_k n_l\right] \\ P_i^{(f)} = \frac{1}{2}\left[\sum_s (2-\alpha_s) n_s^2 - \sum_{k,l} n_k n_l\right] \end{cases}, \quad \alpha_s = \begin{cases} 2, s = 0, \frac{L}{2} \\ 1, s \neq 0, \frac{L}{2} \end{cases} \tag{1}$$

The value of k and l have to match the conditions of $k + l = L$ and $k, l \neq L/2$. Defining the quantity of the impulse noise that correspond to this condition as K, so K must be even. Hence their means are respectively:

$$\begin{cases} E(P_r^{(f)}) = \frac{1}{2}\sum_s \alpha_s a^2\Gamma(1+2b^{-1}) \\ E(P_i^{(f)}) = \frac{1}{2}\sum_s (2-\alpha_s) a^2\Gamma(1+2b^{-1}) \end{cases} \tag{2}$$

The variances of the real and imaginary part are:

$$\sigma^2_{P_r^{(f)}} = \frac{1}{4}\left(\sum_s \alpha_s^2 a^4\left[\Gamma\left(\frac{4}{b}+1\right) - \left(\Gamma\left(\frac{2}{b}+1\right)\right)^2\right] + 2Ka^4\left(\Gamma\left(\frac{2}{b}+1\right)\right)^2\right) \tag{3}$$

$$\sigma^2_{P_i^{(f)}} = \frac{1}{4}\left(\sum_s (2-\alpha_s)^2 a^4\left[\Gamma\left(\frac{4}{b}+1\right) - \left(\Gamma\left(\frac{2}{b}+1\right)\right)^2\right] + 2Ka^4\left(\Gamma\left(\frac{2}{b}+1\right)\right)^2\right) \tag{4}$$

So define the correlation coefficient as follows:

$$\rho^{(f)} = \frac{E(P_r^{(f)}P_i^{(f)}) - E(P_r^{(f)})E(P_i^{(f)})}{\sigma_{P_r^{(f)}}\sigma_{P_i^{(f)}}} \tag{5}$$

Depending on the presence (or not) of noise impulses in the positions of $s = 0$ and $s = L/2$, three cases are analyzed here and Eq. 5 can be rewritten as:

$$\rho_i^{(f)} = \frac{(M-i)a^4\left(\Gamma(1+\frac{4}{b}) - \left[\Gamma(1+\frac{2}{b})\right]^2\right) - 2Ka^4\left[\Gamma(1+\frac{2}{b})\right]^2}{\sigma_{P_r^{(f)}}\sigma_{P_i^{(f)}}}, \quad i = 0, 1, 2 \tag{6}$$

where i represents the number of noises that occur on $s = 0$ and $s = L/2$.

With the definition of correlation coefficient and the follow-up deducing, the value of $\rho_i^{(f)}$ is decided by impulse noise number and their locations, and is discrete and limited.

3 Probability Distribution of Correlation Coefficient

Considering the circumstance that the number of impulse noise M in a frame is fixed, the problem can be count as M random impulse noise inject to the L symbol. Under the analysis of the possible correlation coefficient of three cases, different K would result in different $\rho^{(f)}$. So the analysis of the distribution of probability is actually converted to researching on the randomness of the impulse noise location. The occurrence probability of different correlation coefficients is

$$\Pr(\rho_{i,K}^{(f)}) = (C_{L/2-1}^{K/2} \sum_{l=0}^{M-K-i} C_{\frac{L}{2}-\frac{K}{2}-1}^{l} C_{\frac{L}{2}-\frac{K}{2}-1-i}^{M-K-i-l})/C_L^M, \quad i = 0, 1, 2 \tag{7}$$

where $\rho_{i,K}^{(f)}$ is the correlation coefficient that i impulses occur on the locations of $s = 0$ and $s = L/2$, while K is known.

Then considering the number of impulse noise to occur randomly, the value of the probability depends on the environment where multi-carrier system is applied. Usually the random process of the pulse generation can be seen as Bernoulli model, so the probability while a frame is distorted by M pulses can be written as

$$\Pr(M) = C_L^M p^M (1-p)^{L-M} \tag{8}$$

where p is the occurrence probability of a single impulse noise. So the probability of different correlation coefficient is

$$\Pr(\rho) = \sum_{M=0}^{L} \Pr(M) \Pr(\rho|M) \tag{9}$$

4 Simulation and Analysis

Before the simulation, some parameters such as the length of a frame (L) and the number of impulse noise (M) that occur in a frame are decided. The locations of impulse noise are considered to be uniform distributed. In order to comprehensively analysis of the correlation coefficient distribution under different conditions, and to obtain the relationship between the probability distribution to the channel characteristics, it is divided into three kinds of circumstances. Firstly, the theoretical analysis and the simulation results are compared to verify the reliability of theoretical derivation. Afterwards, the cases with different values of L and M are simulated to show the significance of the probability distribution of impulse noise power correlation coefficient.

1. The Comparison of The Theory and Simulation
   As shown in Fig. 2, the theory and simulation are compared under three groups of parameters, the curves are well matched with the same value of L and M expect for slightly difference while L = 16 and M = 5. The reason of this difference can be attributed to two aspects. One is that the difference between the neighbor correlation coefficients is very small, and prone to error in judgment. The other is due to the limit of simulation conditions, the quantity of the random sample is slightly insufficient. It causes the acquisited correlation coefficient getting a little deviation. Taking the deviation reasons into account, the similar correlation coefficients are emerged together. The results show that two sets of curves are very consistent, so it illustrates the theoretical analysis is correct and effective. Considering the reliability of the theoretical model and huge amount of computation while L is large during the simulation, such as there should be nearly 2×10^{16} cycles with L = 64, M = 20, so the following analysis will be based on theory mode.
2. Same *L* and Different *M*
   Different conditions with same L and different M are shown in Fig. 3. The results show with the increase of impulse noises, in the other words, the channel being more severely interfered, the peak position of the probability of different correlation coefficients is getting closer to -1. By determining the peak position of correlation coefficient probability distribution, it can be used to estimate the severity that impulse noise interfering the communication channel.

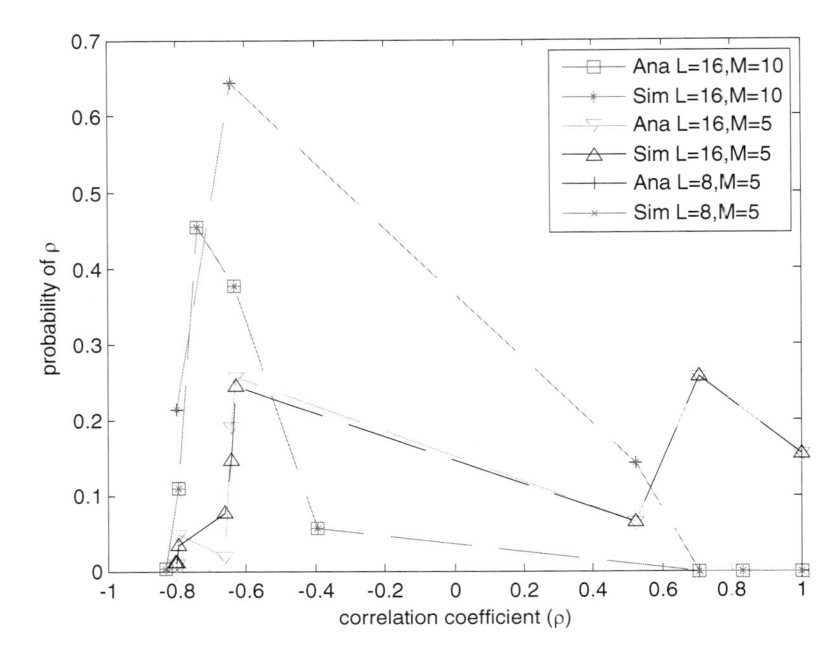

Fig. 2 Comparison of theory and simulation

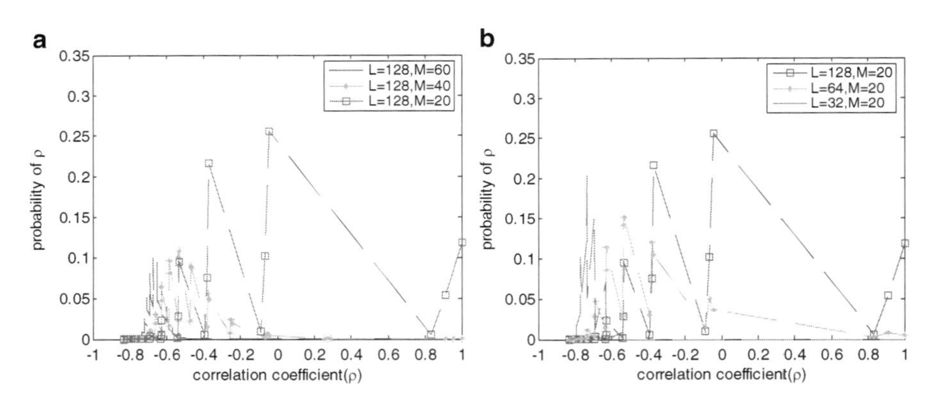

Fig. 3 The probability distribution of the correlation coefficient with (**a**) same L and different M (**b**) different L and same M

3. Different L, same M
   The results of the analysis under the same L and the different M are shown in Fig. 3b. Compared with Fig. 3a, b, both are very much alike. Actually, the condition that L = 64, M = 20 can be regarded as the situation of 128 symbols with 40 inner impulse noises, so it results in a very similar curve. As for the difference of occurrence probabilities, it can be owed to that different L and M lead to different number of the correlation coefficient. The situation of L = 128, M = 40 has more different numerical correlation coefficient than that of L = 64, M = 20.

5 Conclusion

For the purpose of performance improvement, the analysis and the mitigation of impulse noise in multi-carrier communication are getting more and more urgent. Weibull distributed impulse noise is a credible mathematical model that is verified by measured signals, so it can be applied in theorem analysis. According to the randomness of impulse noise, the correlation coefficient of the impulse noise power and the distribution probability are achieved through strict theoretical derivation. And the simulation results prove the reliability of the theory derived by this paper. Under the analysis of the probability distribution of correlation coefficient, a relationship between the severities of impulse noise to the location of the peak of probability is found. This relationship can serve as a kind of new method to estimate the characteristic of the impulse noise, and have a certain potential application.

References

1. Shaoping Chen, Guangfa Dai, Tianren Yen (2004) Zero-forcing equalization for OFDM systems over doubly-selective fading channels using frequency domain redundancy. IEEE Trans Consum Electron 50:1004–1008
2. Mitola J et al (1999) Cognitive radio: making software radios more personal. IEEE Wirel Commun 6:13–18
3. Li You-Ming, Wang Rang-Ding, Shen Wei, Fang Li-Ming (2008) Fast crosstalk precoder based on tridiagonal matrix splitting. J Commun 29:10–15
4. Henkel W, Kessler T (1994) Statistical description and modelling of impulsive noise on the German telephone network. IEEE Electron Lett 30:935–936
5. Khangosstar J, Li Zhang, Mehboob A (2011) An experimental analysis in time and frequency domain of impulse noise over power lines. In: 2011 I.E. international symposium on power line communications and its applications (ISPLC), Italy, pp 218–224
6. Tatsufumi Shirai et al (2002) A study on reduction of the affection of impulse noise in OFDM transmission. In: Proceedings of the ISPLC 2002, Athens, pp 208–212
7. Pighi R, Franceschini M, Ferrari G, Raheli R (2009) Fundamental performance limits of communications systems impaired by impulse noise. IEEE Trans Commun 57:171–182
8. Zimmermann M, Dostert K (2002) Analysis and modeling of impulsive noise in broad-band powerline communications. IEEE Trans Electromagn Compat 44:249–258
9. Rozic N, Radic J (2010) Distribution of the noise squared norm in OFDM systems interfered by class A impulsive noise. IEEE Commun Lett 14:318–320
10. Rozic N, Chiaraluce F, Radic J (2011) Analysis of the correlation coefficient between component noise squared norms for OFDM systems. IEEE Signal Process Lett 18:311–314

A Modeling of the Description of Urban Residents' Traveling Decision Based on Simple Genetic Algorithm

Chenxu Long, Jing Li, and Heping Dong

Abstract Genetic algorithm is a new optimizing searching method based on biology evolutionary theory. Just as evolution deals in populations of individuals, genetic algorithms mimic nature by evolving huge churning populations of code, all processing and mutating at once. One of the most important points of this paper is how to describe the urban residents making trip modes decisions based on simple Genetic Algorithm. What's more, to establish a proper fitness function is another difficulty of this paper.

Keywords Model • Urban resident • Traveling decision • SGA

1 Introduction

This paper put forward a modeling of the description of urban residents' traveling decision based on Simple Genetic Algorithm (SGA), and checked the application result with the use of Mat-lab and toolbox from the University of Sheffield. There are many researches about how urban residents behave when choosing trip modes [1–5], however, there are few studies discussing the application of the urban residents making trip modes decisions based on GA, and this is where the innovation of this paper locates.

One of the most important points of this paper is how to describe the urban residents making trip modes decisions based on simple Genetic Algorithm. What's more, to establish a proper fitness function is another difficulty of this paper.

C. Long (✉) • J. Li • H. Dong
School of Economics and Management, Information Management and System, Beijing Jiao tong University, Haidian District, Beijing 100044, P. R. China
e-mail: longlonglonglonger@163.com; jingli@bjtu.edu.cn

Z. Zhang et al. (eds.), *LISS 2012: Proceedings of 2nd International Conference on Logistics, Informatics and Service Science*, DOI 10.1007/978-3-642-32054-5_95,

2 A Example of One Urban Resident

This section gives an example of real life to describe how urban residents choose the travel mode based on genetic algorithm.

Scenario Description: A teacher from the Beijing Institute of Architectural Building (hereinafter referred to as Building) goes to the Golden Resources Shopping Mall (hereinafter referred to as the gold source) to have a class on Sunday afternoon. The class ends at 16:30 and the next begins at 18:00. Due to the returning of objects, and making ensure all of the students leave the building safely, the time to go after school is about 16:50. And this time is about the evening, if the teacher doesn't have dinner, the evening class is likely to be subject to impact, so it is best to set aside time for dinner, however, have dinner before the evening class is not necessary. Besides, the teacher must sign before 17:45. As the foregoing analysis, the teacher must make sure that the travel time is limited within 1 hour.

3 A Description of Urban Residents' Decisions Based on SGA

Firstly, convert the above scenarios into the model based on the genetic algorithm. The teacher's decision-making include: Way to travel: bus or taxi? Location: bus stop or other? Dinner: eat or not?

The purpose is to find a combination of these three decisions to produce the highest utility. According to the above-described methods, Table 1 shows the four of them.

In this instance, the initial group of four individual fitness value is given in Table 2 and the fitness is defined as follows.

Based on experience and personal preferences of the teacher, we define the travel satisfaction utility function of this teacher as the function of the travel time and the cost.

$$f\left(x_{ij}, y_{ij}\right) = \sum_{i=1}^{3} {x_{ij}}^2 + \sum_{i=1}^{3} y_{ij} + \varphi \tag{1}$$

$f\left(x_{ij}, y_{ij}\right)$ is the sum of the travel time and the cost corresponds to the decision-making program of this teacher.

Table 3 gives the satisfaction utility every decision gets corresponds to the travel time and cost (The value is from 1 to 10).

Among them, the travel time of choosing a bus is five times that of a taxi.

Table 4 shows an example of a gamble to choose, the first line are four numbers randomly selected from 0 to 700.

Table 5 gives a possible outcome of this new group generated by the reproduction operator with the chose of gamble. And this new group is called the mating pool.

Table 1 Representation of the travel decision-making

Program number	Travel mode	Location	Dinner	Binary representation
1	Bus	Bus stop	Eat	011
2	Bus	Bus stop	Not eat	010
3	Bus	Else	Eat	001
4	Taxi	Else	Not eat	100

Table 2 The fitness for the decision-making programs in the initial population

	Generation 0	
k	String (x_k)	The fitness $f(x_k)$
1	011	162
2	010	132
3	001	169
4	100	237
Sum		700
Minimum		132
Mean		175
Maximum		237

Table 3 The reference of the satisfaction utility

Decision variables	Travel time (x)	Cost (y)
Taxi(1)	10	9
Bus(0)	1	10
Bus stop(1)	3	2
Else(0)	4	2
Eat(1)	9	9
Not eat(0)	10	10

Table 4 The chose according to gamble

Random number	123	655	642	474
The selected string	011	100	100	001

Table 5 The mating pool after copy

	Generation 0			The mating pool	
k	String (x_k)	Fitness $f(x_k)$	$\frac{f(x_k)}{\sum f(x_k)}$	String	$f(x_k)$
1	011	162	0.23	011	162
2	010	132	0.19	100	237
3	001	169	0.24	100	237
4	100	237	0.34	001	169
Sum		700			805
Minimum		132			162
Mean		175			201
Maximum		237			237

Table 6 The results of the copy and crossover operators

	Generation 0			The mating pool		Generation 1		
K	String (x_k)	Fitness f(x_k)	$\frac{f(x_k)}{\sum f(x_k)}$	x_k	f(x_k)	Crossover point	x_k	f(x_k)
1	011	162	0.23	011	162	2	010	132
2	010	132	0.19	100	237	2	101	267
3	001	169	0.24	100	237	–	100	237
4	100	237	0.34	001	169	–	001	169
Sum		700			805			805
Minimum		132			162			132
Mean		175			201			201
Maximum		237			237			267

Table 6 shows the generation of a likely outcome of the application of copy operator and crossover operator from generation 0. The crossover probability is set to 50%.
Compared with generation 0 and generation 1, we can find:
The average fitness changed from 175 to 201;
The fitness value of the best individual changed from 237 to 267.

4 Conclusion

In this instance, the final optimal decision-making program is the first generation of 101: taxi, bus stop and having supper. If we happen to know that 267 is the most highly satisfied with the utility, in this instance, we can stop in the first generation implementation of the genetic algorithm. When the genetic algorithm stops execution, put the current generation of the best individual to specify the results of the genetic algorithm. Of course, the genetic algorithm is generally not performed as in this case to generation to stop, but to dozens of generations, hundreds of generations or more generations.

Acknowledgments This paper is supported by the National Natural Science Foundation (No. 71103014), the basic research and operating expenses (No. 2011JBM032) and the key project of logistics management and technology lab.

References

1. Xu Yongneng, Li Xu-Hong, Zhu Yandong, Shi Shuming (2005) Satisfaction criteria of urban residents travel mode choice model. Comput Commun 4:54–57
2. Liu Huanyu, Song Rui, Xu Wangtu, Han Bi-lin (2009) Optimization of urban public transportation network layer model and genetic algorithms. Urban Public Transp 9:32–36

3. Ming Shijun, Yang Deming (2010) Trip gradient and travel structural relationship. WestChina Univ Technol (Natural Science) 9:75–78
4. Xiong Yuanbo, HuYongJu, Wang Hao (2010) City residents peak time travel mode option. Transp Sci Technol Econ 5:59–61
5. Ma ChangXi, Wen Juanjuan, Li Chuanghong, Li Xiaoli (2009) Big city residents' travel mode decision-making method. J Transp Eng Inf 2:33–39

The Design of Public Information Service System Based on Public Demand

Chunfang Guo and Chongying Sun

Abstract This article analyses the current situation and existing problems of China's public information service system, and conclude that regardless of the public's needs is the problem needs to be solved. Then this paper puts forward to use the Maslow's hierarchy of need to divide the public demand into different levels. At last, according to the personalized demand of the public, it builds a new public information system.

Keywords Public information service • Maslow's hierarchy of need • Application

1 Introduction

Since the beginning of the twenty-first century, with the gradual improvement of China's information level, the network has been integrated into our lives, and with the use of newspapers, television and the World Wide Web, the public has more approaches to get public information. But at the same time the people place higher demand for the public information service. The quality of public information service has a direct impact on public satisfaction and trust on government. Hence, if we want to establish a harmonious, democratic society, the establishment of an efficient public information service system is the key point.

C. Guo (✉) • C. Sun
School of Economics and Management, Beijing Jiaotong University,
Beijing, P. R. China
e-mail: chfguo@bjtu.edu.cn

Z. Zhang et al. (eds.), *LISS 2012: Proceedings of 2nd International Conference on Logistics, Informatics and Service Science*, DOI 10.1007/978-3-642-32054-5_96,

2 The Present Situation of the Public Information Service

With the rapid development of Internet and IT, the country's public information service is also booming, China has set up the Information Management Office, the State Information Center and the Ministry of Industry and Information Technology and other institutions. The establishment of these departments has achieved great success; the most prominent of this is the e-government work. This measure not only makes it more convenient for the public to access to public information, but also helps the government has a better understanding of the needs of the public. As a result, the e-government has greatly promoted China's democratization process and improved the efficiency of government's overall image. But what have to note is that the country's public information service capabilities cannot meet the demand of the people. There are still many problems in public information service which have seriously hampered the development of China's process of democracy.

2.1 Inadequate Information Disclosure

There are still some difficulties in the publicity of China's information, the people cannot get some information that reflects their own interests through the normal channels. Since May 1, 2008, China has promulgated and implemented the People's Republic of China on Open Government Information Regulations, but it met many obstacles in the concrete execution. And we have wasted a great number of information resources and reduced the degree of public trust on the government's work. In addition, although the e-government construction in China has achieved a certain scale, a large part of government departments can't update government websites in time and the public cannot obtain the valuable information from the site, thus the e-government is becoming a poor apology.

2.2 The Government Is Lack of Demand-Oriented Service Concept

Although the government has provided a variety of ways for the public to obtain the public information, most of the public information cannot be achieved from the two-way communication between the public and the government. The government passed the information out, but did not take into account whether it can be accepted by the audience. China's public information service has been gradually emerging "Matthew effect", due to the level of economic development, educational level, age gap, and other reasons, highly educated life have more chances to enjoy the services of public information, while vulnerable groups which have more demand for the public information are hard to access to public information, so that the strong is stronger and the weak is weaker, strengthen the unequal distribution of society [1].

3 Public Information Demand Analysis

3.1 Maslow's Hierarchy of Need

Maslow points out that customers' needs can be divided into five levels, as is shown in Fig. 1, the physical needs are the basic needs of a human, just like food, water and sleep; the safety and security needs include stability, predictability, freedom from fear, harm and injury; the social acceptance needs mean a person feels he is a part of social groups, he want to feel acceptance, approval, and affection from others; the esteem needs mean that people want to receive the recognition from others; and the self-actualization needs describe people try their best to achieve their goals, to be the people they want. According to Maslow's hierarchy of need, people can hardly pursuit a higher level demand if the lower level demand has not been achieved, but because of the difference of personalities, one person may have two or three levels of demand at the same time. And the line between these demand levels is not clear [2, 3].

3.2 The Application of Maslow's Hierarchy of Need in Public Information

According to Maslow's hierarchy of need, we can divide the public information needs into five levels, the physical level includes the analysis of water quality and air quality, the food price and the housing transactions; the safety and security level covers defense security information, the forecast of major natural disasters, and food and drug safety information; the social acceptance level means the government provides the information of social security and social welfare and advocates the public concerns more about the vulnerable groups; the esteem level means the public want to communicate with the government and they can acquire the information that meet their needs; the self-actualization level means the government need to provide more chance for the public to achieve their goals, encourage and support more entrepreneurship and employment.

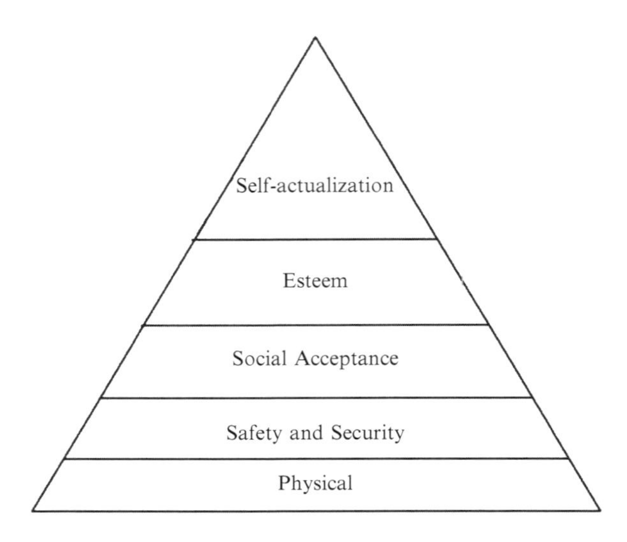

Fig. 1 Maslow's hierarchy of need

4 Design of the Public Information System

In order to meet the personalized needs of the public, we divide the public information service system into three levels which are shown in Table 1, the first level is public information, in this level, the public can find the basic information, such as weather forecast, news and information, and so on. Then we divide the public information into five parts, including social security, public health and medical service, public education and culture service, environment protection service and government information service. These parts constitute the second level, and they cover more items which constitute the third level. The second level is made up by the news, policy and legislation in its own field and the third level contains the personalized information service and the feedback from public.

Through the social security part, we can get information about the policy and legislation of social security. If we want to understand the more detailed information, we need to enter to the third level. The social security is divided into three items: the social assistance, employment information and vulnerable groups' aid. We can take the employment information as an example. If we want to find a job, we can get into the employment information item, then the system will provide us the employment advertisements, some valuable job links and resume format are also provided here.

The public health and medical service contains epidemic situation, disease surveillance and monitoring of hygiene three items. These items can help the public know more about the healthy information about their own physical condition, and give the public advice about how to keep healthy.

The public education and culture service can realize the information sharing and it includes library, distance education and recreational activities. Through this part, people can realize online learning, and it can help people make a study plan.

Table 1 Three levels of the public information service system

The first level	The second level	The third level
Public information	Social security	Social assistance
		Employment information
		Vulnerable groups aid
	Public health and medical service	Epidemic situation
		Disease surveillance
		Monitoring of hygiene
	Public education and culture service	Library
		Distance education
		Recreational activities
	Environment protection service	Environment quality information
		Ecological environment information
		Environmental protection method
	Government information service	Government information disclosure
		Online communication

The environment protection service can give the public a chance to concern the environment quality, and it can also give people advices about the makeup index, dressing index, matching index of the day, through the day's weather condition, such as temperature, humidity and wind.

The government information service can provide the release of the government's data, such as the public finance, tax revenue and government expenditure, in order to meet the need of the society, the system provides online communication item, the public can give advices and opinions to the government [4, 5].

5 Conclusion

With the development of economic internet and computer technology, the public requires higher quality of public information service, and the overall assessment of the public for public information service system is also directly affect the image of the government. While Chinese public information service is still have many shortcomings, and it is not facilitate enough for the public to access the public information. So if the government wants to build a more effective information service system, the prime thing understands the public's needs; however, this article only offers a few commonplace remarks by way of introduction so that others may come up with valuable opinions.

Acknowledgments This research was supported by "the Fundamental Research Funds for the Central Universities" under Grant 2011JBM042.

References

1. Graham Brown (2002) Sydney 2000: an invitation to the world. Olympic Games, pp 16–20
2. Abraham C, Sheeran P (2003) Acting on intentions: the role of anticipated regret. Br J Soc Psychol 42:495–511
3. Hickey AM, Davis AM (2004) A unified model of requirements elicitation. J Manag Info Syst 20(4):65–84
4. Hadi ZA, McBride N (2000) The commercialization of public sector information within UK government departments. Int J Pub Sect Manage 13(7):552–570
5. Adomavicus G, Tuzhilin A (2005) Personalization technologies: a process-oriented perspective. Commun ACM 48(10):83–90

Customer Evaluation Model Based on the Catering Industry's Supply Chain Ecosystem

Rui Chai, Juanqiong Gou, and Guguan Shen

Abstract With the transferring from the state of small scale and decentralization to the state of branding and chaining, Chinese catering industry is more competitive, which intensifies the evolution of industry ecosystem and business development model rapidly. Catering supply chain management services is gradually specialized, and faced with many problems in corporate positioning, business model options and external operations. This paper analyzes the characteristics and evolution path of various types of catering enterprises. The catering industry supply chain ecosystem is designed based on a lot of investigation, which classified related catering enterprises and their relationships. The evaluation models of the catering industry supply chain service providers and customers are established. A case study is presented to illustrate the model application in the design of business model of a supply chain provider.

Keywords Customer evaluation • Catering industry • Supply chain • Ecosystem

1 Introduction

With the rapid development of Chinese economy, the development of Chinese catering enterprises face many brand-new questions brought by the fast transition, especially those about the management of supply chain. Although some professional supply chain service providers are showing gradually, but challenge from the domestic and foreign faced by the development mode is still alive [3]. So in this traditional labour-intensive development mode, how to attain the value of supply chain service, build an enterprise's core competitiveness and long term development strategy rapidly is the main dilemma faced by those enterprises.

R. Chai (✉) • J. Gou • G. Shen
School of Economics and Management, Beijing Jiaotong University, Beijing, China
e-mail: 11125172@mail.bjtu.edu.cn

Z. Zhang et al. (eds.), *LISS 2012: Proceedings of 2nd International Conference on Logistics, Informatics and Service Science*, DOI 10.1007/978-3-642-32054-5_97,

Literatures about the research of the value and supply chain are quite a lot, but precedents about research puts enterprises into business ecosystem are few. China on the business ecosystem theory started relatively late, following that Fan [7] analyzed the effect of the commercial ecosystem competition on enterprises, national industrial planning, technical standards and polices with the start point of the competition style evolution. On the basis of research results of domestic and foreign scholars on the business ecosystem, natural ecosystem, Du and Wang [8] proposed a research framework which compares the two systems from 6 aspects and 27 study such as the external environment, and summarized two kind of system's similarities and differences though analysis, synthesis method.

It can be seen that the previous research on the business ecosystem is very full, but all proceed from a macro perspective, this article will be resolved restaurant industry supply chain ecosystem for catering enterprises at the micro level for market positioning strategy and customer options for catering enterprises.

2 Characteristics and Cost Structure of Catering Industry Supply Chain

In the process of catering industry supply chain, the flow direction from the supplier to the customer is physical products, in the flow process of the whole industrial chain; a primary solid product is attached on a series of economic value. In this huge industry chain, logistics and information flow are two important arteries of the industry chain, and the enterprise takes advantage of variety of resources, combine the two arteries, and support a smooth and harmonious industry chain ecosystem.

Since the specificity of the catering industry products, catering industry supply chain will also show a corresponding feature. From the view of logistics point, the catering industry supply chain is characterized by: (A), quantity, variety; (B), logistics requirement due to seasonal animal and plant products; (C), low logistics profit and low price of produce; (D), logistics difficulties, warehousing, different packaging standards; (E), inevitable loss.

From the perspective of supply chain management, it constant the follow characteristics: (A), food and beverage service supply chain involves many kinds of goods; (B), catering enterprises are all sensitive to the price volatility in commodity, and market price volatility is inevitable; (C), food and beverage supply chain is full of enterprises; (D), the quality of employees at the basic level is low; (E), the cost is high to ensure the quality and avoid two population.

3 Catering Industry's Supply Chain Ecosystem

James F Moore thought, the so-called business ecosystem means: economic union based on the interaction of organizations and individuals. Composed by customers, suppliers, manufacturers and other relevant members of groups complement each other to produce goods and services, which also include the suppliers of funds, the industry association in charge of standards bodies, trade unions, government and semi-governmental organizations and other interested parties [4].

3.1 The Definition of Catering Industry's Supply Chain Ecosystem

Collection of related businesses or organizations to adapt to each other and formed a collaborative relationship, in order to adapt to the changing environment, creating and maintaining sustainable competitive advantage, the role of system both cooperation and competition in pursuit of co-evolution. The article refers to the catering industry supply chain ecosystem, in essence, a value circle by the formation of those factors: the end customers of catering industry, the central kitchen of the catering companies, distribution centers, all organizations involved in the supply and individual, seed companies, including all relevant factors, as well as the interaction between factors interactions [1].

3.2 Participants of the Supply Chain Ecosystem

Like species in nature ecosystem, many participants in the catering industry supply chain ecosystem can be Divide in categorize, hierarchy and function [2]. These participants are divided into four categories: Lowers Participants, Intermediates, Tops Participants and Consumers.

3.2.1 Lowers Participants

Like seed companies, breeding companies, R & D company of the new varieties, plant breeding, rural cooperatives and individual farmers.

These participants are the ones who create valuable enterprises in this small ecosystem. Such enterprises professional lines are high, but lower the threshold of entering the market. Their catering market sensitivity is the lowest in the supply chain ecosystem, the most terminal by the bullwhip effect.

3.2.2 Intermediate

Like production plants, logistics companies, warehousing companies, distributors, IT services companies.

Capacity as partners in these enterprises, it is easy formation of the symbiotic and competitive relationships. It's easy to form a symbiotic and competitive relationship between partners, symbiotic relationship of mutual influence, mutual benefit; the relationship of the competition, companies continue to improve their own competitiveness. If the symbiotic balance with the forces of competition, the evolution is harmonious, supply chain ecosystem will be evolutionary, if the evolution of such individual factors discord, it may result in the destruction or reorganization of the supply chain ecosystem.

3.2.3 Tops Participants

Like catering enterprises, shops, supermarkets, investment companies, the Government Association.

The top participants including the supply chain-end enterprise customers, government agencies and industry associations, play a role in guiding the development of the supply chain. Corporate customers are divided into many forms, the relationship between the client and the client is a parallel competition. This competition between catering companies on the one hand to improve their own competitiveness, on the other hand to save the cost of supply chain management, will lead the market called for supply chain service providers. Government agencies and industry associations for any business on the supply chain, there is no competition. They just play a guiding role and policy support for the development of supply chain.

3.2.4 Consumers

Just have the general public. They play a role like the equivalent of the ultimate decomposers and consumer in the ecosystem of the supply chain, the entire supply chain activities have value and meaning is because of the most common consumer individual.

There are differences between Supply chain ecosystem and nature ecosystem; all things are equal in nature ecosystem. But in the supply chain ecosystem, there have a leading enterprise of these participants for the value of integration.

On the other hand the catering industry supply chain ecosystem participants can be broadly divided into three categories stand on the angle of the leading companies: customers, partners and competitors.

One part of Competitors is the partners of leading business, in order to win customers, open up the market, they will also extend a number of other non-core business in addition to their main business, only the professional of services is different, but they often to obtain some of the favor of customers because of its cheap; the other part is the supply chain services already in the market, each supply chain service providers are likely to become the leading enterprises in their supply system, only share of resources and mode of operation is vary, even some customers

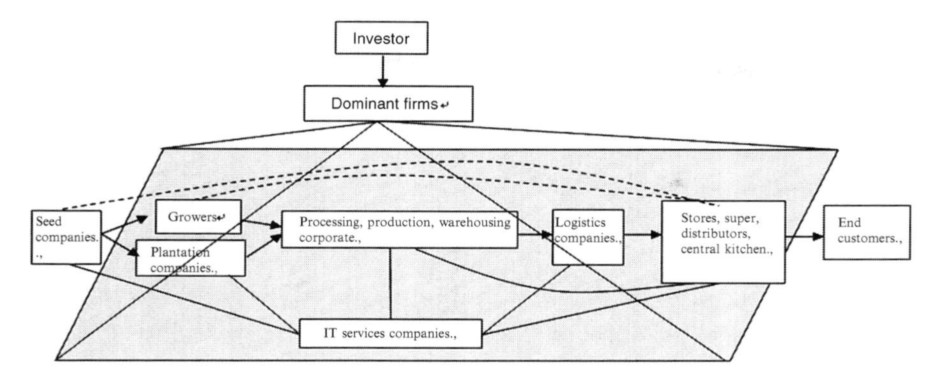

Fig. 1 Catering industry ecosystem model

may become competitors, some of the larger chain catering companies build their own self-built central kitchen from the early stage of development, it has enough ability to provide supply chain services for other catering companies, because of longer operating time, grasped a certain amount of market resources, and supply chain operations have spare capacity.

3.3 Catering Industry's Supply Chain Ecosystem

Balance of nature because of the harmony ecosystem. A system's balance is because of the balance of entropy inside and outside of the system. For enterprises where the supply chain ecosystem, the same need for harmony in order to achieve the multi-win-win situation. The ecosystem of the supply chain must have a leading enterprise, whose task is to jump out of a complex network of contacts, use of its available resources to each node in the supply chain re-integration of the resources, thereby reducing the circulation and reduce loss of circulation, reduce distribution costs (Fig. 1) [5].

4 The Customer Evaluation Model in the Enterprise

Different Supply chain services for different types of customers, allocation of resources in the supply chain there will be differences. Oriented enterprises can throughout the supply chain process modular, different service modules of different services, but can guarantee the integrity of some of the basic business processes. Then restructure of each service module according to the different needs of different customers.

The most salient features of Supply chain services are selecting the appropriate service module and service processes for demand-side services through the form of outsourcing. Remove the elements involved in the traditional supply chain, supply chain service providers have the competitive elements that require special attention: more emphasis on information sharing for supply chain services, efficient and durable operation of the whole supply chain is dependent on capacity management, demand management, relationship management, service delivery management and capital management, integration and coordination of a number of functions, in addition to technical information systems and support of network platforms.

In addition, establish the mode of operations for the enterprise development. The first mode is depends on demands to produce promotional, to promote the progressive realization of the program management model between supply chain partners. Through sophisticated information platform, the program management of each partners, infinitely close to balance production and sales situation. The second model is the independent accounting, project-oriented mode, customer-oriented, project as a carrier independent accounting management model. The third model is a financial-based internal control mode, construction of the internal control of financial-based, production cost accounting, budgets and funds management model.

Finally, planting farmers to customers, divided into four sections: plate of planting and breeding, processing plate, logistics plate, distribution plate. These four sections divided the supply chain business into four parts, link up the four sections of the business through the integration of the enterprise, re-engineering resources, optimize, and achieve partner win-win situation (Fig. 2).

5 Case Study: Design of Supply Chain Service Business Model Based on the Customer Evaluation Model

Supply chain ecosystem, each enterprise population has its own characteristics. Below the top companies populations for example: corporate clients, segment customers.

Supply chain service providers face numerous of market clients, but the need of each client has its own particularity, it has to reflect the difference of needs between different catering enterprises. In its early stages, first is market segmentation, and choosing the corporate who is a long-term cooperation and conducive to the balanced of ecosystem. While understand the real need of the enterprises.

Subdivide the customer needs, it's not difficult to see that each client has its most concern point, these sort roughly divided into four aspects: price-sensitive, quality concerned, supply chain collaboration, supply outsourced. Such as price-sensitive customers are joining enterprises have their own supply chain system, and in the steady state of development. Most of the type of quality concern enterprises is a chain of direct enterprise requirements for high quality, generally processed products. So draw the following five categories of indicators according to customer concerns, customer classification (Fig. 3).

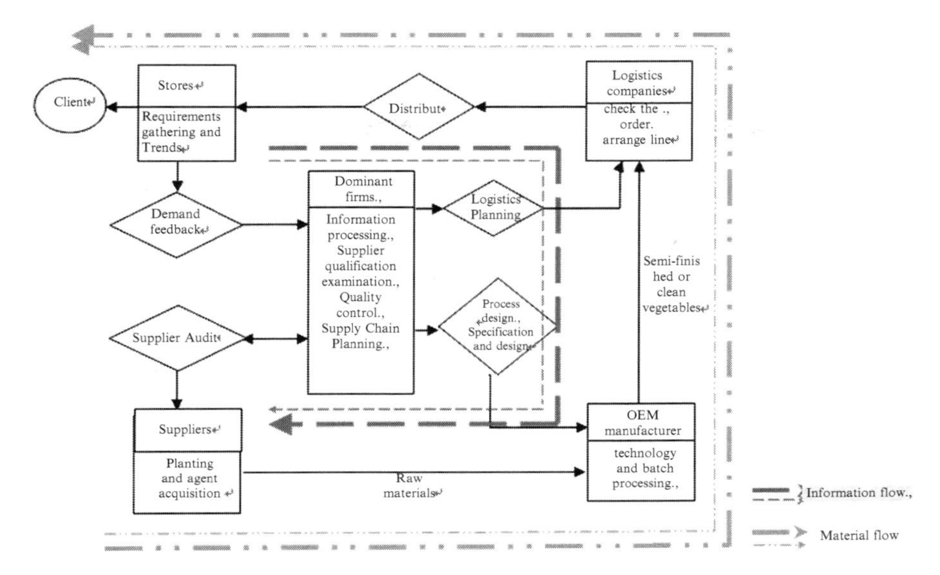

Fig. 2 The ecological model structure diagram of the integrated supply chain

5.1 Evaluation Model Indicators Determining the Weights [6]

A. **The establishment of a hierarchical model**
   Use the theory of hierarchical analysis to segment index of customer evaluation.
B. **The judgment matrix**
   The affiliation of factors between the upper and lower level have been identified after the establishment of a hierarchical model.
   On this basis, we need to make a judgment often relative importance of each factor in each level.
   Judgment matrix is usually written as formula (1) below

$$A = \begin{bmatrix} a_{11} & \cdots & a_{1n} \\ \vdots & \ddots & \vdots \\ a_{n1} & \cdots & a_{nn} \end{bmatrix} \tag{1}$$

C. **Calculate the weight of each index**
   In order to extract useful information from the judgment matrix group to achieve understanding of the law of things, to provide a scientific basis for decision-making, we need to calculate the weight vector of weight vector of each judgment matrix and the judgment matrix synthesis.

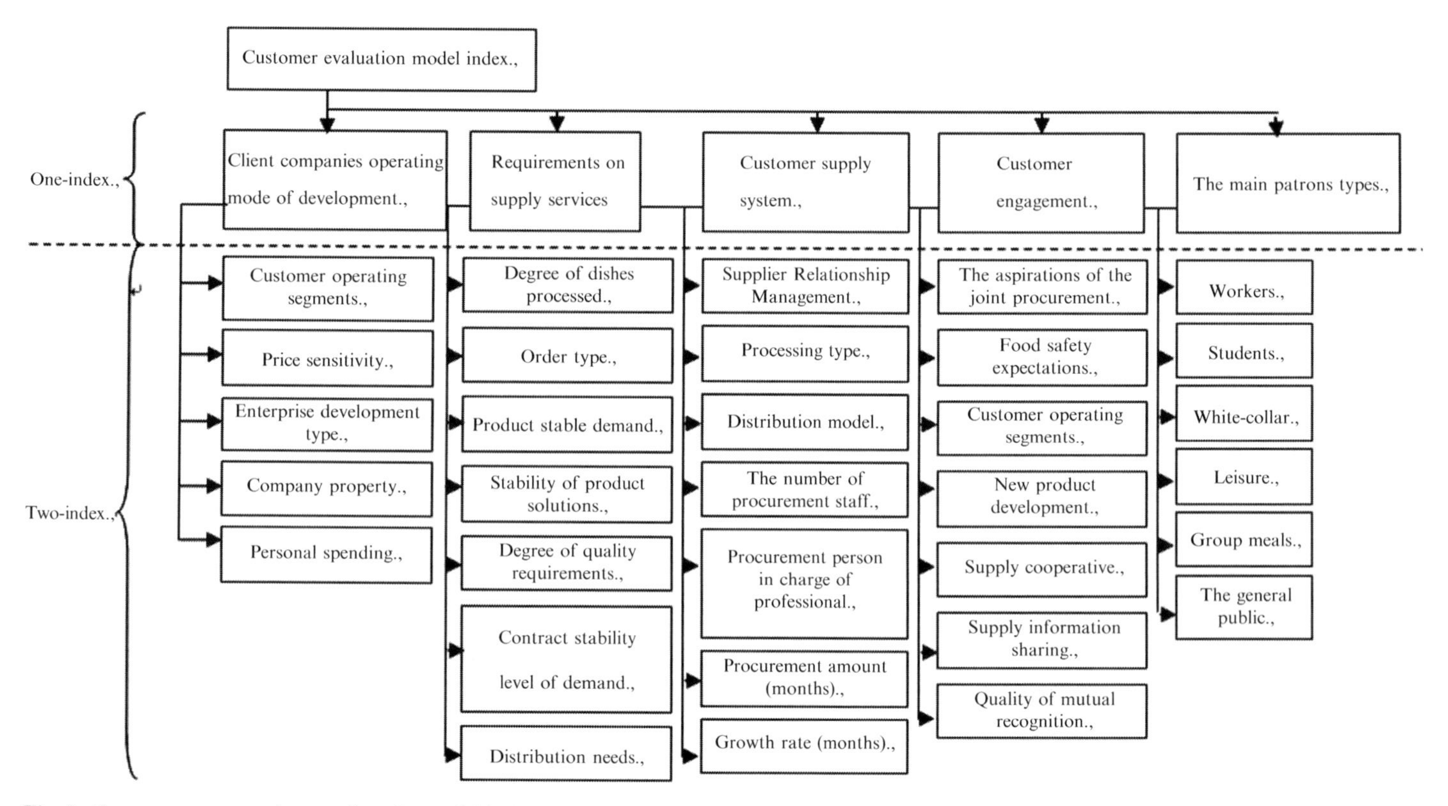

Fig. 3 Customer segmentation to select the model index level diagram

5.2 *Fuzzy Evaluation of Customer*

Due to the particularity of the evaluation of restaurant industry, each customer-facing indicators are different, evaluation to participate in a score of subjective awareness, so the fuzzy comprehensive evaluation method to evaluate quantified.

A. **The establishment of fuzzy comprehensive evaluation factor index set and evaluation criteria set**
   If make the model of choosing Customers of the supply chain in this article as an example. The factors set can be chosen as: U = {Enterprise's own business development models, Requirements for suppliers to supply services, Customer supply system, Customer engagement}. Evaluation set to be chosen as V = {Trade, Concerned about services, Supply chain collaboration, Supply outsourced}.
B. **Determine the fuzzy subset of the weights of evaluation factors Indicators.**
   Under normal circumstances, the impact of various factors on the evaluation of the object is inconsistent, different factors according to where the whole evaluation system in the relative importance of assigning a value. Weight distribution of factors has been calculated by the analytic hierarchy process. Weight distribution of factor set is a fuzzy set on U, generally denoted as: $\tilde{A} = (a_1, a_2, \ldots, a_m)$. Which a_i the i factors U_i weight, and satisfy the normalization condition: $\sum_{i=1}^{n} a_i = 1$.
C. **Solving the judgment result set**
   Fuzzy comprehensive evaluation is built on the basis of the judge set, and therefore judged that the result set should be a fuzzy set in V, denoted by $\tilde{B}$. The evaluation result set by fuzzy comprehensive evaluation model obtained: $\tilde{B} = \tilde{A} \cdot \tilde{R}$
   Subdivide according to customer demand, the Trading enterprises roughly divided into four types: Trade-oriented concerned about the services, supply chain collaboration and supply outsourcing enterprises.
   By the fourth part of the formula: $b_0 = \max(b_1, b_2, b_3, b_4)$ can draw the following conclusions:
   If $b_0 = b_1$, the customer is price-sensitive:
   If $b_0 = b_2$, the client is concerned about the quality:
   And so on.

6 Conclusions

This article used the analytic hierarchy process indicators to give the right weight, but for different enterprises, the weight of the same indicators will be different. In the implementation process should pay attention to the score reasonable. The same type of customers can use the same index weight. In practical applications, the enterprise should combine its own development; have a reasonable choice of the customer base. Their own development should be established on the basis of the harmony and win-win business circle. In today's information explosion, the establishment of a transparent information-sharing platform, logistics platform will create the market, the only way to create customer.

References

1. Iansiti M, Levien R (2004) The keystone advantage: what the new dynamics of business ecosystems mean for strategy, innovation, and sustainability. Harvard Business School Press, Boston, Massachusetts, America, pp 45–189
2. Adizes I (1979) Organizational passages: diagnosing and treating life cycle problem of organization. Organ Dyn 8(Summer):2–25
3. Kemal A. Delie, Umeshwar Dayal (2002) The rise of the intelligent enterprise. ACM Ubiquity Issue 45, 31 Dec 2002–6 Jan 2003
4. Mauboussin MJ (2005) The ecosystem edge. Legg Mason Cap Manag 7:1–11
5. Barkema HG, Baum JAC, Mannix EA (2002) Management challenges in a new time. Acad Manage J 45:916–930
6. Bryson N (1996) Group decision—making and analytic hierarchy process: exploring the consensus—relevant information content. Comput Oper Res 23(1):27–35
7. Baoqun Fan (2005) Competitive mode based business ecosystem and its inspiration [J]. Bus Econ Adm 17(11):3–7
8. Guozhu Du, Botao Wang (2007) A comparative study on business ecosystem and biological ecosystem [J]. J Beijing Univ Posts Telecommun (Social Sciences Edition) 9(5):34–38

A New Signal Coordination Control Model of Joint Area at the Expressway Conventional Network

Xingqiang Zhang, Pan Chen, and Yangyang Xun

Abstract In order to improve urban expressway capacity, a new signal coordination control model of joint area is proposed based on single-point control schemes which include on-ramp, off-ramp, upstream and downstream auxiliary road intersections. The additional lane length of expressway weaving section is used to control the on-ramp and off-ramp. The signal control schemes of upstream and downstream intersections are optimized using the adjustable ratio of auxiliary road on-ramp and off-ramp. VISSIM simulation results show that the proposed coordination control model is feasible.

Keywords Expressway • Conventional road network • Single-point signal control • Signal coordination control

1 Introduction

With increasing traffic demand in recent years, urban traffic congestion problem has become worse. Traffic expressway system can undertake main urban traffic trips because of its rapidness and large capacity. So it's necessary to study the signal coordination control of a joint area for alleviating traffic congestion and improving current road capacity. Zong Tian et al. [1] proposed an integrated control algorithm, based on an adaptive-controlling intersection and a dynamic-control ramp [1]; Zheng Jianhu et al. [2], taking the traffic density and vehicle queue length as inputs, put forward the on-ramp fuzzy control method [2]; Ci Yusheng et al. [3] applied neural network and fuzzy logic algorithm to study the on-ramp control algorithm of urban expressway [3]; Yuan Changliang and Li Honghai [4] used a small-step adjustment algorithm to calculate the signal cycle and green ratio of the

X. Zhang (✉) • P. Chen • Y. Xun
School of Traffic and Transportation, Beijing Jiaotong University, Beijing, China 100044,
e-mail: zhangxq@bjtu.edu.cn; chenpan0704@gmail.com; 467667323@qq.com

Z. Zhang et al. (eds.), *LISS 2012: Proceedings of 2nd International Conference on Logistics, Informatics and Service Science*, DOI 10.1007/978-3-642-32054-5_98,

downstream intersection and the signal timing of the upstream auxiliary road [4]; Chen Xuewen et al. [5], proposed an optimization control method for on-ramp based on the speed and time occupancy [5]. M. van den Berg L etc, based on signal control, proposed a model control framework which took the total travel time of all vehicles in the road network as control objective [6]. Therefore, the paper studied the signal coordination control method in order to improve the urban expressway capacity.

2 Signal Coordination Control

There are four signal control points in the joint area: auxiliary road on-ramp, auxiliary road off-ramp, upstream and downstream intersection of auxiliary road (Figs. 1 and 2).

The WEBSTER method is used to optimize the signal timing of the intersections, and the traffic volume adjustable ratio is used as the signal control index of the on-ramp and off-ramp.

$$r(k) = Q_r(k-1) + K_r(\mathrm{O}_{rc} - \mathrm{O}_{rd}(k-1)) \tag{1}$$

Where $r(k)$ is the traffic volume adjustable ratio of on-ramp (off-ramp) in the k time interval; $Q_r(k-1)$ is the $(k-1)$ traffic volume of on-ramp (off-ramp), (pcu/h); K_r is the adjustment coefficient; O_{rc} is the critical time occupancy of on-ramp (off-ramp), (pcu/h); $O_{rd}(k-1)$ is the $(k-1)$ critical time occupancy of on-ramp (off-ramp), (pcu/h).

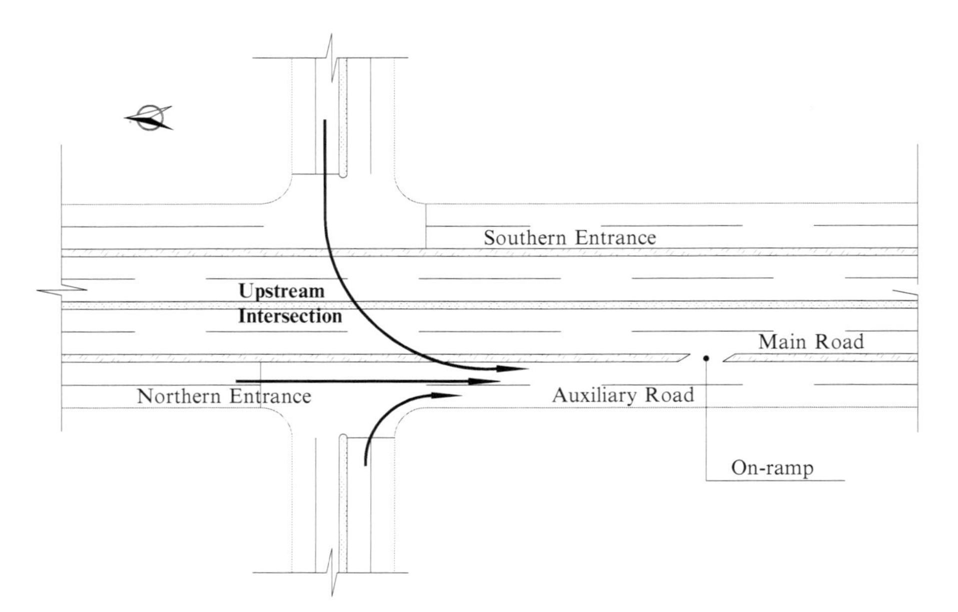

Fig. 1 Upstream intersection schematic diagram

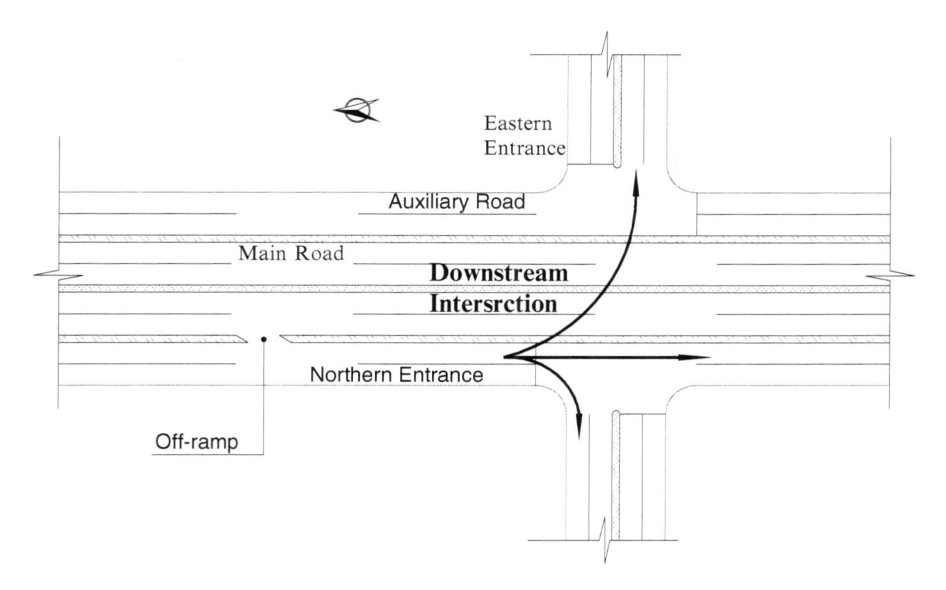

Fig. 2 Downstream intersection schematic diagram

1. Coordination Control of Entrance and Exit
   A new coordination control method is proposed based on the vehicle number that can be contained in the additional lane of weaving area.

$$n_1 = \frac{C_1 r_1}{3,600} \quad n_3 = \frac{q_d - q_3}{3,600} g_{e2} \tag{2}$$

   Where: n_1 is the queuing vehicle number of on-ramp in one cycle (pcu); n_3 is the queuing vehicle number of off-ramp in weaving section (pcu); C_1 is the signal control cycle of on-ramp (s); r_1 is the traffic volume adjustable ratio of on-ramp (pcu/h); q_d is the vehicle arriving rate of off-ramp (pcu/h); q_3 is the traffic volume of off-ramp (pcu/h); g_{e2} is the off-ramp green time (s).

   Suppose that the additional lane vehicle number of main road weaving section is N, the vehicle number of on-ramp in every signal cycle time is n_1, the queuing vehicle number of off-ramp on the main road weaving section is n_3. Then,

   (a) $n_1 + n_3 > N$
   The off-ramp traffic flow should be firstly satisfied, the traffic volume adjustable ratio of auxiliary road is a value calculated by single-point control model.

   (b) $n_1 + n_3 < N$
   This case shows that on-ramp vehicles could be added, the traffic volume adjustable ratio of auxiliary road is a value calculated by single-point control model.

2. Coordination Control of On-ramp and Upstream Intersection

   In order to reduce the on-ramp queue length of auxiliary road, the upstream intersection green time of straight-going and left-turning should be reduced, then

$$S_s g'_s \alpha_s / C_3 + S_l g'_l \alpha_l / C_3 = r_1 \tag{3}$$

   Where S_s is the straight-going saturation volume (pcu/h); S_l is the left-turning saturation volume (pcu/h); g'_s is the adjusted straight-going green time, (s); g'_l is the adjusted left-turning green time (s); α_s is the ratio of straight-going vehicles through on-ramp; α_l is the ratio of left-turning vehicles through on-ramp; C_3 is the intersection signal period (s); r_1 is the traffic volume adjustable ratio of on-ramp (pcu/h).

   The compression rate of the straight-going and left-turning green time is:

$$\frac{g'_s}{g'_l} = \frac{g_s S_s \alpha_s}{g_l S_l \alpha_l} \tag{4}$$

   Where g_s is the original straight-going green time (s); g_l is the original left-turning green time (s).

3. Coordination Control of Off-ramp and Downstream Intersection

   Based on the relationship between the off-ramp traffic volume and the auxiliary road traffic volume, the left-turning and straight-going green time of the downstream intersection could be calculated.

$$q_2 \alpha_{fl} g_{e2} / C_2 + q_3 \beta_l = S_l g_l / C_4 \tag{5}$$

$$q_2 \alpha_{fs} g_{e2} / C_2 + q_3 \beta_s = S_s g_s / C_4 \tag{6}$$

   Where q_2 is the maximum traffic volume of auxiliary road (pcu/h); q_3 is the off-ramp traffic volume (pcu/h); α_{fl} is the auxiliary road left-turning ratio; α_{fs} is the auxiliary road straight-going ratio; β_l is the off-ramp left-turning ratio; β is the off-ramp straight-going ratio; g_{e2} is the off-ramp green time (s); g_l is the downstream intersection left-turning green time (s); g_s is the straight-going green time in downstream intersection (s); C_2 is the off-ramp signal cycle (s); C_4 is the signal cycle in downstream intersection (s); S_l is the left-turning saturation volume (pcu/h); S_s is the straight-going saturation flow (pcu/h).

4. Intersection Control Fusion

   In the coordination control model, the intersection signal control scheme may use two kinds of different regional division at the same time. So it's necessary to fuse the signal control scheme. Two factors are taken into account: (1) no twice-release during one cycle; (2) signal is adjusted to minimize the interference of other phase.

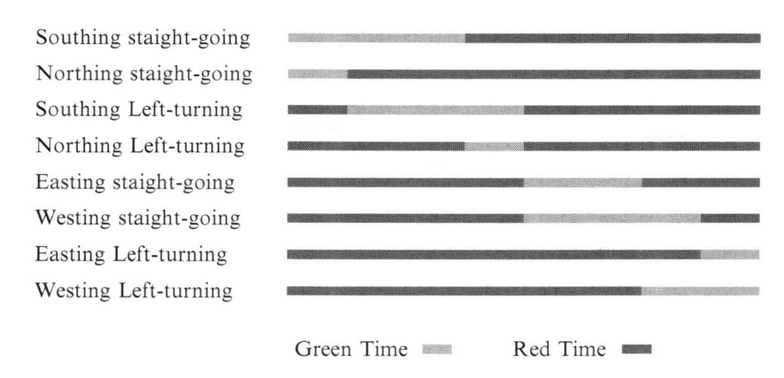

Fig. 3 Four signal phases fusion adjustment scheme

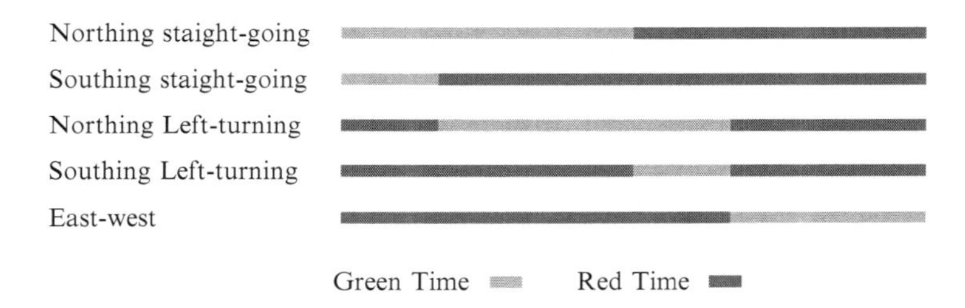

Fig. 4 Three signal phase fusion adjustment scheme

Based on this, the signal adjustment scheme at the upstream intersection is "Stop straight-going phase of northern entrance ahead of time + Lag release left-turning phase of eastern exit" (Fig. 3); the signal adjustment scheme at the downstream intersection using "Stop straight-going phase of southern entrance ahead of time + Lag release left-turning phase of southern entrance" (Fig. 4).

5. Coordination Control of Auxiliary Road
The signals of the off-ramp, upstream and downstream intersection are optimized by the main road coordination control theory.

The off-ramp and on-ramp signal cycle is equal to half of the intersection and it sets twice release, while the green time and red time are determined by the green ratio. The intersection, which has the longest signal cycle, is regarded as the key intersection. The critical intersection signal cycle is the system unified signal cycle length. For the non-key intersection, all excessive green time is added to the main-line phase to improve the capacity of auxiliary road. To ensure the signal coordination control of upstream and downstream intersection, the green phase difference should be determined by average speed and distance.

3 Model Validation

The regional road network, from Wenhui Bridge to Mingguang Bridge, is selected as the model validation case. The signal control parameters, which are determined by the above-mentioned signal coordination control method, can be obtained by traffic survey. Phase sequence is shown in Figs. 3 and 4.

VISSIM is used as the simulation platform, and the simulation interface is shown in Fig. 5. The simulation results show that the traffic volume on the western side of main road increases by 18.9%, and average speed increases by 40.7%. Therefore, the proposed coordination control model is feasible (Table 1).

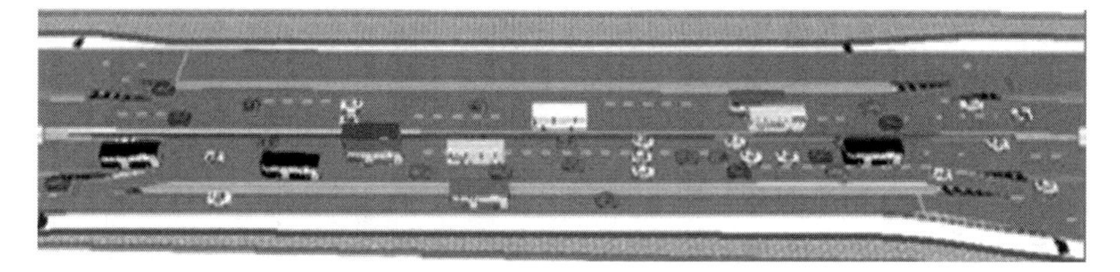

Fig. 5 Operation effect of simulation for weaving section

Table 1 Comparative analysis of expressway western main road

Evaluation index	Traffic volume (*pcu*/5 *min*)			Operation speed (*km*/*h*)		
Time interval	Non-control	Coordination control	Increase %	Non-control	Coordination control	Increase %
8:35–8:40	356	383	7.7	26.26	40.50	54.2
8:40–8:45	339	350	3.5	36.28	56.80	56.5
8:45–8:50	340	361	6.4	22.17	28.95	30.6
8:50–8:55	353	363	2.9	32.64	45.20	38.5
8:55–9:00	373	383	2.9	28.92	45.88	58.6
9:00–9:05	354	443	25.3	29.71	45.92	54.6
9:05–9:10	329	468	42.2	36.31	45.68	25.8
9:10–9:15	364	478	31.4	42.76	57.50	34.5
9:15–9:20	327	438	33.9	42.07	57.05	35.6
9:20–9:25	374	438	17.3	44.21	57.40	29.8
9:25–9:30	374	505	35.1	44.51	57.60	29.4
Average			18.9			40.7

4 Conclusion

In the signal coordination control model of joint area, the additional lane length of weaving section is used to control both the on-ramp and off-ramp traffic volume, and the traffic occupancy is used to determine the adjustable ratio of auxiliary road, and Webster is used to adjust the signal cycle schemes of the upstream and downstream intersection. Moreover, a new intersection signal control fusion method is proposed to optimize the intersection signal control.

From our VISSIM simulation results, the traffic volume on the western side of main road increases by 18.9%, and the average speed increases by 40.7%. Therefore, it is reliable to use the proposed coordination control model.

References

1. Zong Tian (2007) Modeling and implementation of an integrated ramp metering-diamond interchange control system. J Transp Syst Eng Inf Technol 7(1):61–72
2. Zheng Jianhu, Chen Hong, Dong Decun (2006) Study on design and simulation of on-ramp fuzzy controller for urban expressway. J Highw Transp Res Dev 23(12):133–136
3. Ci Yusheng, Wu Lina, Pei Yulong, Ling Xianzhang (2006) On-ramp neuro-fuzzy metering for urban freeway. J Transp Syst Eng Inf Technol 10(3):136–141
4. Yuan Changliang, Li Honghai (2010) Coordinated control strategy of expressway-exit-auxiliary road signal and urban road intersection signal. Road Traffic Saf 10(3):38–44
5. Chen Xuewen, Wang Dianhai, Jin Sheng (2008) On-ramp metering control strategy of urban expressway. J Jilin Univ (Engineering and Technology Edition) 38:43–48
6. Van den Berg M et al (2004) Model predictive control for mixed urban and freeway networks. The 83th annual meeting of transportation research board meeting, Washington, DC

Printed by Publishers' Graphics LLC